Mother Nature

Also by Sarah Blaffer Hrdy:

The Woman That Never Evolved

The Langurs of Abu: Female and Male Strategies of Reproduction

The Black-man of Zinacantan: A Central American Legend

Co-editor with Glenn Hausfater:

Infanticide: Comparative and Evolutionary Perspectives

Mother Nature

A History of

Mothers, Infants, and

Natural Selection

Sarah Blaffer Hrdy

Pantheon Books   New York

Pantheon Books and colophon are registered trademarks of Random House, Inc.

Library of Congress Cataloging-in-Publication Data

Hrdy, Sarah Blaffer, 1946–
Mother nature : a history of mothers, infants, and natural
selection / Sarah Blaffer Hrdy.
p.   cm.
Includes bibliographical references (p.   ) and index.
ISBN 0-679-44265-0
1. Mother and child.   2. Motherhood—Psychological aspects.
3. Natural selection.   4. Parental behavior in animals.   5. Working
mothers.   I. Title
HQ759.H784   1999
306.874'3—dc21   99-13092

Random House Web Address: www.randomhouse.com

Book design by Fearn Cutler de Vicq

Printed in the United States of America

2 4 6 8 9 7 5 3

For Dan,
the wisest choice this female ever made

Mother Nature—who by the bye is an old lady with some bad habits . . .

—George Eliot, 1848

Contents

Preface   xi

PART ONE
Look to the Animals   1

1   Motherhood as a Minefield   3
2   A New View of Mothers   27
3   Underlying Mysteries of Development   55
4   Unimaginable Variation   79
5   The Variable Environments of Evolutionary Relevance   96

PART TWO
Mothers and Allomothers   119

6   The Milky Way   121
7   From Here to Maternity   146
8   Family Planning Primate-Style   175
9   Three Men and a Baby   205
10   The Optimal Number of Fathers   235
11   Who Cared?   266
12   Unnatural Mothers   288
13   Daughters or Sons? It All Depends   318
14   Old Tradeoffs, New Contexts   351

PART THREE
An Infant's-Eye View   381

15   Born to Attach   383
16   Meeting the Eyes of Love   394
17   "Secure from What?" or "Secure from Whom?"   408
18   Empowering the Embryo   419

19   Why Be Adorable?   441
20   How to Be "An Infant Worth Rearing"   452
21   A Matter of Fat   475
22   Of Human Bondage   485
23   Alternate Paths of Development   511
24   Devising Better Lullabies   532

      Notes   543
      Acknowledgments   599
      Bibliography   603
      Index   691

Preface

I have spent my entire adult life engaged in a quest to understand not just *who* I am but *how* creatures like me came to be. That humans evolved at all is a fluke. My own existence, like that of any other person's, is more than a fluke—it is a miracle. Out of the seven million or so egg cells my mother was born with, it was mine that ripened to be fertilized by my father. Against the usual odds, that fetus survived the vagaries of gestation to be born. And what about this creature, this person I would become? What does it mean to be born a mammal, with an emotional legacy that makes me capable of caring for others, breeding with the ovaries of a primate, possessing the mind of a human being? What does it mean for a woman to have descended from ancestors who spent the Pleistocene (the time span between 1.6 million and ten thousand years ago) trying to gather enough food to stay fed and also obtain enough help from others so that her offspring would survive and prosper? What does it mean to be all these things embodied in one ambitious woman? To be a semicontinuously sexually receptive, hairless biped, filled with conflicting aspirations and struggling to maintain her balance in a rapidly changing world?

For better or for worse, I see the world through a different lens than most people. My depth of field is millions of years longer, and the subjects in my viewfinder have the curious habit of spontaneously taking on the attributes of other species: chimps, platypuses, australopithecines. This habit of thinking about mothers in broad evolutionary and comparative—as well as cross-cultural and historical—perspectives distinguishes my examination of motherhood from those of the psychoanalysts, psychologists, novelists, poets, and social historians whose work I build on.

I am trained in anthropology, primatology, and evolutionary theory. I entered graduate school in 1970 at a time when Harvard was still a very male-centered institution, especially in the sciences. Nor was it then fashionable, as it has since become, to focus on the active role of mothers in evolutionary processes. During the years of my education and early career, the genderscape of the natural sciences was transformed by a broader inclusion of women. With their involvement came new emphases and new topics of study. Just a century before, the only women writing about evolution were novelists or commentators on the outer edges of the scientific community, with no impact on mainstream theory. Today that situation has changed.

Even more unlikely than being born, the accident of when and to which family I was born meant that I ended up among the fraction of human females ever permitted the kind of education and opportunities to do research traditionally reserved for men. I had uncommon opportunities to travel, to observe all types of primates in the habitats in which they evolved, and to enter previously all-male scientific domains. Because I wanted to be a mother as well as a scientist, this meant making compromises. But I was never forced to make the really big compromise, to choose between my aspirations and the rewards of marriage, pregnancy, giving birth, and the satisfaction of watching my children become remarkable people in their own right. I owe this luxury to the availability of an unprecedented degree of reproductive choice, especially in the realm of birth control. I could delay the birth of my first child until after my doctoral thesis was published. After the birth of my second child, I could still continue to do research, although it meant giving up fieldwork and turning to topics I could study closer to home. In the case of my last child, I could postpone his birth until after I was forty, confident that genetic screening and, if necessary, an early abortion would protect my family from the liability of aging ovaries.

My melding of vocation with family has been supported by a steadfast partner, resilient children, and the generous hearts of alloparents—all the individuals other than me and my husband who helped us rear our children. As I would learn, mothers have worked for as long as our species has existed, and they have depended on others to help them rear their children.

Like many humans today, I was reared quite differently from the primates I observed in the field. Unlike the mothers in every other ape species, my own mother had no interest in carrying me everywhere she went. Indeed, in the particular tribe of elite Texans into which I was born, the custom of mothers nurturing their own infants had been lost generations before. My mother delegated the care of her infants to others, just as both my grandmothers had done.

My mother's idea of good management was that if a child became too attached to a nanny, it was time to hire a new one, lest maternal control be diminished. This meant that I was reared by a succession of governesses. No one ever doubted that my mother loved her five children. She was under the impression that infants might be born of inferior or superior "stock" (like all southerners, she was fascinated by family lines), but like most college-educated women of her time, she also believed that babies were blank slates, conditioned to act as they were trained to.

I was born in 1946. It was not until several decades later that the writings of the British medical doctor and psychoanalyst John Bowlby began to change the way educated people thought about the needs of children. Bowlby demonstrated that babies are genetically programmed to seek and form an attachment to a trusted figure. Secure attachment to one or more trusted caretakers is an essential aspect of emotional development in humans, just as it is in all other primates.

Along with several recent revisions to "attachment theory," Bowlby's ideas will stand among the greatest contributions made by evolutionary-minded psychologists to human well-being. But Bowlby's insights also produce a series of new—often seemingly irreconcilable—dilemmas for mothers who want to rear emotionally healthy, self-confident children, but who also want lives or careers of their own.

Both pre- and post-Bowlby, a woman's maternal emotions (those concerned with producing and nurturing offspring and keeping them alive) continued to be viewed as separate from and even antithetical to the rest of her—her sexuality and, particularly, her ambition. Yet the compartmentalized way we have learned to think about these emotions has nothing to do with how they evolved.

Such thoughts were far from my conscious mind when, years ago, I found myself in the throes of cyclically experienced sexual feelings. At that time, such restless yearnings were still a largely unacknowledged primate legacy. Indeed, in some circles such feelings were assumed not even to exist in women. Hence I had no framework for interpreting them. I found myself attracted to some men more than others for reasons that I did not understand. Eventually I fell in love (almost from the moment I first set eyes on him in a course, ironically enough, on "fossil man") with a fellow anthropologist. Even though nothing could have been further from my conscious thoughts, he would prove to be a wonderful and very committed father. Yet, at that point, I had no idea how inextricably linked sexual and maternal emotions have been in the course of primate evolution.

When we finally decided that it was a "good enough" time to have children, all I can remember was wanting a daughter and how wonderful it was to be pregnant. Even "doing nothing," I felt creative. Odd as this may sound to some, birth, too, was a euphoric experience, the pain far more nearly fascinating than unbearable. For me, contractions during labor were an opportunity to find out what it feels like to be totally in the grips of all-encompassing biological forces over which my conscious mind had no say. Thinking back to my first glimpse of the slimy creature who emerged head first, *luscious* is the word that comes to mind to describe the daughter of my dreams. I was stunned by the sensual responses she evoked in me. I made up love songs celebrating each beautiful feature, her soft skin and silky hair.

After the first weeks of living with a baby girl who mostly slept or quietly nursed through seminars, it became increasingly apparent that in the world I lived in, caring for a baby was incompatible with concentrated work. A new baby's terrifying vulnerability, the magnitude of the responsibility, and the insatiable demands that kept me on-call twenty-four hours a day, came as a shock. Yet, as a primatologist in the post-Bowlby era, what could I do but turn my life over to her?

I was overwhelmed by contradictory impulses, trapped if I did, damned if I didn't. Not the least of the emotions that bubbled up was a whirring resentment—the kind I identify with the most primitive portions of my brain—toward my daughter's father, by then a medical doctor and infectious-disease researcher who could go off and spend long hours in the lab while I tried to eke out of the daily interruptions enough time to write. I wanted him to invest more, so that I could be free. Yet if I delegated care to others, wouldn't this mean reverting to the ways of my mother's generation, before we understood the attachment needs of infants?

I so desperately wanted to succeed in my chosen profession; yet I didn't want to deprive my daughter of the emotional security I had become convinced she needed. Personal ambition seemed to be on a collision course with my baby's needs. At that time I had no idea how interrelated maternal and professional aspirations actually are.

From recent surveys, I now know that the kind of ambivalence I felt is experienced to varying degrees by most mothers in the United States. Working mothers feel more ambivalence than mothers who stay home with their children. Mothers who work by choice feel more conflicted than those who must work to support their family.[1] Still, the bottom line is: the majority of mothers with preschool-age infants *are* now working outside the home. With ongoing changes in the welfare system, that proportion is growing even higher.

Always, in the back of my own mind, lurked nagging questions. Did the ambivalence I felt mean that I was a bad mother? Would my success come at the emotional expense of my children? Even years later, when my oldest daughter called home her first week in college to say she wasn't sure she belonged there—probably not an unusual response—reflexively I recalled the miserable, distorted face of a diaperless toddler standing outside a bungalow in Rajasthan, India. She was a little girl suffering from a virulent diaper rash in a far-off land, where women with dots on their foreheads or veils across their faces must have seemed very strange. Yet the familiar figure of her mother would not return till evening. Even more painfully vivid was my six-week return to India when she stayed home with her father and a house-keeper. I still recall every intonation of her voice on the other end of the phone when I landed in New York and called home. "Ma-ma" came the heart-breaking cry, with no trace of resentment.

But I've never been content to agonize when I could analyze instead. I set out to use every perspective at my command, and every source of information I could locate, to marshal such evidence as I could bring to bear on the question of what it means to be a mother, and what human infants need from their mothers, and *why.* Even what I failed to learn in time to help me rear my own three children, I could pass on to others.

I was driven to understand my past. For we are not ready-made out of somebody's rib. We are composites of many different legacies, put together from leftovers in an evolutionary process that has been going on for billions of years. Even the endorphins that made my labor pains tolerable came from molecules that humans still share with earthworms.

But I am jumping ahead of my story. At the outset, it would never have occurred to me to articulate the nature of my quest to understand maternal ambivalence, female sexuality, and infant needs in these terms. It was not until college that I learned about evolution. To my amazement, people actually studied the behavior of other animals to learn more about human nature.

In 1968 I was taking an undergraduate course in college given by Irven DeVore, one of the pioneers in the then-fledgling field of primatology. One day he mentioned a report by Japanese primatologists working in India that described bizarre behavior among adult male langurs, a species of monkey I had never heard of. According to the report, these males would grab infants from their mothers and bite them to death. Mothers would fight to keep their infants safe but, in the end, fail. On rare occasions—as I later confirmed—

Fig. 1 *The Formation of Eve,* by Gustave Doré, ca. 1866. *(From* The Doré Bible Illustrations, *Dover Publications, NewYork, 1974)*

they appeared not even to try to defend them. The bereaved mothers seemingly did not even bear a grudge. After the male killed her infant, the langur mother mated with him, which struck me as unaccountably odd.

I had just learned that creatures evolved so as to enhance their reproductive success. But now I was confronted with a description of males killing infants with more than a hint of maternal collusion—behavior that decreased rather than increased infant survival. I was more than intrigued. After graduating from college, I went to India to learn *why* such strange behavior occurred.

Studying infanticide in other primates turned out to be only the beginning of my quest to understand female nature and motherhood in particular. This quest lured me to do research in seven countries over thirty years, drawing on extremely unlikely sources of information—last wills and testaments, documents from foundling homes, folktales, even the pages of phone books—in my effort to learn about parental attitudes in my own species. Along the way, I have come to understand just how flexible parental emotions in humans can be. Whatever maternal *instincts* are, they are not automatic in the sense that most people use that term. Most important, I have learned that even though the world has undergone immense changes since our ancestors lived by foraging, many of the basic outlines of the dilemmas mothers confront remain remarkably constant.

———

For a billion years, ever since egg- and sperm-producing organisms first arose, the genetic futures of males (the "sperm-producing" organisms) have been affected by what females (the "ovum-producing" ones) do, and vice versa. But the selflessness of mothers giving of themselves to offspring has always seemed too vital to the well-being of too many for anyone—scientists included—to be able to examine their behavior dispassionately. And none did. Old biases from many sources burrowed in and nestled at the heart of evolutionary theory, the most coherent and all-encompassing theory that scientists have ever had to explain the living world.

It was no accident that first moralists and then Victorian evolutionists looked to nature to justify assigning to female animals the same qualities that patriarchal cultures have almost always ascribed to "good" mothers (nurturing and passive). Women were assumed to be "naturally" what patriarchal cultures would socialize them to be: modest, compliant, noncompetitive, and sexually reserved. This, I suspect, is the main reason why *sexuality* has always been studied separately from *maternity,* as if sex has nothing to do with maternity or keeping infants alive.

Inquiring women have sensed that there were underlying agendas to the way female nature was depicted, and they responded in different ways. Many identified a male bias within science—particularly biology. For Virginia Woolf, the biases were unforgivable. She rejected science outright. "Science, it would seem, is not sexless; she is a man, a father, and infected too," Woolf warned back in 1938. Her diagnosis was accepted and passed on from woman to woman. It is still taught today in university courses. Such charges reinforce the alienation many women, especially feminists, feel toward evolutionary theory and fields like sociobiology.

Right from the outset of evolutionary thinking, however, a tiny group of women were as Darwinian as they were feminist. George Eliot, a woman who took a man's name because women writers at that time were not taken seriously, was one of them. Eliot, whose real name was Mary Ann Evans, recognized that her own experiences, frustrations, and desires did not fit within the narrow stereotypes scientists then prescribed for her sex. "I need not crush myself . . . within a mould of theory called Nature!" she wrote. Eliot's primary interest was always human nature as it could be revealed through rational study. Thus she was already reading an advance copy of *On the Origin of Species* on November 24, 1859, the day Darwin's book was published. For her, "Science has no sex . . . the mere knowing and reasoning faculties, if they act correctly, must go through the same process and arrive at the same result." I fall in Eliot's camp, aware of the many sources of bias, but nevertheless impressed by the strength of science as a way of knowing.

Unlike superstition or religious faith, a good scientist's underlying assumptions are subject to continuous challenge. Sooner or later in science, wrong assumptions get revised. Nevertheless, some take longer to get corrected than others, as was the case with overly narrow stereotypes about females.

Long ago a wise friend, evolutionary biologist George Williams, warned me that natural selection is an impersonal "process for the maximization of short-sighted selfishness," something far worse than moral indifference. Darwin was of the same mind: "maternal love or maternal hatred, though the latter fortunately is most rare, is all the same to the inexorable principle of natural selection." Natural selection is primarily about differential reproduction, which simply means that some individuals leave more offspring than others. Once we understand that natural selection has neither morals nor values, a concept like "Mother Nature" ceases to be shorthand for romanticized Natural Laws that are more nearly wishful thinking than objective observation of creatures in the world around us.

In the course of revising ideas about mothers, I have had to discard much received wisdom. Compared to what I was taught in graduate school, the mothers who gradually came into focus for me almost seemed a new life-form—alien and utterly different from culturally produced expectations. Mothers were multifaceted creatures, strategists juggling multiple agendas. As a consequence, their level of commitment to each offspring born was highly contingent on circumstances. Realizing this, I have been forced to conclusions about our ancestresses that, given the values I grew up with and still live by, I am poorly equipped to comprehend.

Among the specific questions I address in this book are:

1. What do we mean by "maternal instincts"? And have women "lost" them?
2. If women instinctively love their babies, why have so many women across cultures and through history directly or indirectly contributed to their deaths? Why do so many mothers around the world discriminate among their own infants—for example, feeding a son but starving a daughter?

3. Unlike other apes, humans have been selected to produce offspring
    that are helpless and dependent for so long a time that no woman living
    as our foraging ancestors did could hope to rear a family by herself. Yet
    paternal assistance was then, as now, far from certain. How could there
    have been selection on mothers to produce babies so far beyond their
    means to rear?

4. Given that fathers share the same proportion of their genes with babies
    as mothers do, why didn't fathers evolve to be more attentive to infant
    needs? Are there (as Charles Darwin also wondered) "latent instincts"
    for nurturing in males? And if so, when are they expressed?

5. So far as babies are concerned, fathers range from caring to indifferent.
    Why then do virtually all men take such an intense interest in the
    reproductive affairs of women?

6. And, finally, what is the bottom line on infant needs? Just why did these
    little creatures evolve to be so plump, engaging, and utterly adorable?

History and personal experience of course explain a great deal about how
a mother feels about her baby. But to answer my questions I must travel further back in time, long before a court guaranteed a woman's right to privacy
over what goes on in her womb, before contraception, before formal laws of
any kind, before regulations about infanticide, even before cradles or walls—
back to a time when it was essential for someone to be in continuous contact
with a baby just to keep a wild animal from eating it. Many of the emotions
informing women's reproductive decisions today were shaped in this distant
past, by processes that by current standards can only be viewed as inexorable.

PART ONE

Look to the Animals

Look to the animals for your example.

—Jean-Emmanuel Gilibert, 1770

DAME·NATURE'S·PATENT·FOOD

AS USED BY CAIN AND ABEL

FREE TRIAL   GOOD RESULTS

(Reproduced from *Maternity and Child Welfare*   2 [4] 1918)

Motherhood as a Minefield

*Woman seems to differ from man in her greater tenderness and less selfishness.
Woman owing to her maternal instincts, displays these qualities toward
her infants in an eminent degree; therefore it is likely that she would
often extend them toward her fellow-creatures.*

—Charles Darwin, 1871

B eing a mother has never been simple. Today, modern medicine, safe
water, stored food, pasteurized milk, cradles, and houses with walls
make it easier than ever before to keep a baby alive. Rubber-nippled
baby bottles and daycare centers especially designed and licensed for the care
of the very young provide working mothers, even those with weeks-old
babies, with alternatives to the only two viable options previously available:
keep your baby close or find a wet nurse. The availability of breast pumps and
freezers means that more women can both breast-feed and spend hours separated from their babies.

Above all, there is birth control, which permits a woman to consciously
override her ovaries and choose when, or if, she will bear children. Ultrasound and amniocentesis enable women to spend decades in a career and still
look forward to bearing a healthy infant. Far from simplifying motherhood,
these novel choices have exposed tensions just beneath the cheery surface of
our traditional assumptions about what mothers should be.

Today, mothers in developed countries, and with them fathers and children, enter uncharted terrain. Without anyone raising their hands to volunteer, we have become guinea pigs in a vast social experiment that reveals what
women who can control reproduction really want to do. Children, too, are
finding out what it means to be born to a complex and multifaceted creature
who has an unprecedented range of options. It is an experiment-in-progress,
with two outcomes already apparent. First, the decisions that mothers make
do not always conform to our conventional expectations about innately ten-
der, selfless creatures. Second, whatever today's mother decides is likely to be

controversial in some quarters. Bluntly put, motherhood has become a mine-field, and we are walking through it without so much as a map to guide us.

Politics of Motherhood

The politician who naïvely assumes that motherhood, like apple pie, is still a safe topic quickly learns otherwise. The topic was safe only so long as people took the centuries-old view of self-sacrificing motherhood for granted. This view rested on mankind's assumption that women were designed by nature to be mothers and that they instinctively want to rear every baby they bear. Self-sacrificing motherhood was what women were for, and women in many societies have believed this was their destiny. Overlooked was the huge stake that everyone has in motherhood.

Our sense of self, pride, vulnerability, propriety, and job security, our life-long preconceptions and anxieties, our peace of mind, not to mention our toehold on posterity—all of these depend on what our own mothers, wives, lovers, daughters, and female colleagues do or are expected by others to do. This is why a politician can lose votes for encouraging mothers to stay home, *as well as* for suggesting they return to work; for pointing out that breastfeed-ing is beneficial to infants (which it is), *as well as* for neglecting to mention that it is. One week, newspaper headlines ask, "Is day care ruining our kids?"[1] or decry "A dangerous experiment in child-rearing." Another week, headlines in the same paper will declare, "Infant bonding is a bogus notion" or call for businesses to provide more daycare.[2] At the same time, birth control is still against the law in many countries; and on the sidewalks outside family plan-ning clinics in the United States, near civil war prevails. A visitor to Earth from another planet might well ask how the same creatures that invented sophisticated technology to explore the solar system could display such primitive behavior when it comes to the female reproductive system?

No topic of mother politics is so divisive as abortion, and none elicits more irrational debate. In Washington, D.C., in May 1997, a bill was intro-duced to outlaw a rare type of abortion—the procedure known as dilation and extraction, christened "partial-birth" by opponents. This is an over-whelmingly unpopular, traumatic surgical procedure that no group in the United States advocates, no woman in the world wants, and no doctor is eager to perform. Yet this bill marked the fifty-second time that this particu-lar Congress had debated an abortion-related issue.[3] Disagreement centered on whether this distressing procedure could still be performed even if physicians deemed it necessary to save the woman's life, to guard her health, or to preserve her ability to have viable children in the future. Those who sought the across-the-board ban were not interested in exploring ways to further reduce the need for this rarely performed procedure (one tenth of 1 percent of the 1.5 million abortions performed annually in the United States)[4] by funding more sex education, birth control, and better prenatal care, or by making it easier to get an abortion early on. Banning late-stage abortion was simply their first step to banning all abortions.[5]

The abortion issue is notorious for generating so much "heat" and so little "light." On this particular occasion, one of the senators debating the issue (Rick Santorum, Republican from Pennsylvania) became "so emotional" that the blood vessels leading to his stomach constricted, while those leading to his heart and brain dilated. Responding to signals from the most ancient portions of his brain, his pounding heart caused the face of this deeply threatened mammal to flush "crimson" in preparation for a fight. His voice rose to such a pitch that colleagues had to intervene.[6]

Chances were vanishingly small that any kind of late-stage abortion would ever be applied to anyone he knew. Yet against such odds, the senator had just had a brush with one. He and his wife were informed that the fetus she carried suffered a fatal defect. Even if born alive, the baby, they were told, would not be viable. Infections ensued, and with his wife's life in jeopardy, physicians asked the senator to consider an abortion. The senator, as he reported in a press conference afterward, never even came close to accepting that option. As he saw it, his wife "was in danger of septic shock . . . but she was not in imminent danger."[7]

The abortion debate is ultimately about what it means to be a mother; and the senator, like many humans before him, had his own unusually clear notion of what mothers were for. The couple already had three young children, but this fourth birth was given clear priority over his wife's well-being as well as that of her other children. Fortunately, the mother survived. But, as doctors predicted, the new baby died shortly after birth.

As the debate unfolded, the rush of blood and pounding heart beneath the senator's coat and tie spoke volumes about motivations far deeper, far older, than members of Congress ordinarily consider. Like all humans, and indeed as is typical of the entire Primate order, the senator exhibited an intense, even obsessive, interest in the reproductive condition of other group members. Like other high-status male primates before him, he was intent on controlling when, where, and how females belonging to his group reproduced. One former member of the House of Representatives, however, sensed that there was more at stake than just the issues under debate. "It's very interesting the issues they select," observed Patricia Schroeder of Colorado. "They don't want to intervene in the bodily functions of men."[8]

Schroeder's quip goes to the heart of the matter. Passionate debates about abortion derive from motivations to control female reproduction that are far older than any particular system of government, older than patriarchy, older even than recorded history. Male fascination with the reproductive affairs of female group members predates our species.

Young women of my daughters' generation take for granted a historically unique situation. They regard birth control, precautions against sexually transmitted diseases, women's education and athletic teams, as well as open-ended professional opportunities for women, as innovations here to stay. They view the antiabortion movement in the United States, along with the emergence of powerful political lobbies seeking to substitute "abstinence only" for practical knowledge about human sexuality and reproduction, as too irrational to take seriously. Reports from far-off places like Taliban-controlled Afghanistan, where Islamic fundamentalists seek to deny women personal autonomy (forcing them to stay sequestered in their homes, keep their faces and bodies veiled, and marry as instructed) seem exotic and remote.

It is hard for my daughters and their generation to believe that such forces could ever intrude upon their own lives. Even when the sequestering of women is shown to have measurable costs to the health and well-being of wives and children (as has recently been documented for Afghanistan),[9] they are saddened, but not apprehensive for themselves. They see no connection between innate male desires to control women in earlier times and the attitudes toward women and family that inspire sermons to all-male audiences of "Promise Keepers," or that motivate elected officials to debate endlessly over who has the right to choose whether and when a woman gives birth. Few Westerners take seriously the possibility that old tensions between maternal and paternal interests could explode one day in their own country and transform a world they take for granted. I am not nearly so confident. If age-old pressures are allowed to erode hard-won laws and protections, it is far from certain that the unique experiment we have embarked upon can persist.

Mothers of Us All

With six billion people on the planet, it is easy to forget that we have not always been so numerous. Every person on earth descends from a population living in Africa roughly 100,000 years ago that probably did not number more than ten thousand breeding adults. The Pleistocene epoch, between 1.6 million and 10,000 years ago, was also a time when it was very risky to be born. We are just beginning to understand the full range of hazards and their implications for the attributes babies possess.

Almost all women who reached adulthood became pregnant and bore young. Yet the majority died without a single surviving offspring because so many of those born never grew up. Consider the life history of a hunter-gatherer woman named Nisa. This plucky woman belonged to the !Kung San, a nomadic foraging people who continued to traverse the Kalahari Desert long into the twentieth century, confronting challenges similar to those faced by Pleistocene hunters and gatherers for thousands of years before the invention of agriculture and the domestication of animals for food.

When interviewed for her biography in the early 1970s, Nisa had suffered two miscarriages and borne three daughters and a son—close to the average family size (3.5 children) for a !Kung woman. Two of Nisa's children survived into adolescence, but both died before adulthood. Thirty-six percent of !Kung women would, like Nisa, die without a single surviving offspring.[10] Since all the !Kung women interviewed were postmenopausal at the time, a few more might lose their last child before they themselves died. If we take into account how many women died before they even had a chance to breed, the true proportion of those dying childless is probably higher. I estimate that one-half or more of all !Kung women born died childless.

The death of a child is the most awful occurrence parents can imagine. Fortunately for most of those reading this book, childhood death is a rarity. Unlike Nisa, they live in privileged regions of the globe, at least for the time being, and enjoy an unprecedented standard of living. Nine hundred and ninety-four of every thousand babies born in the United States survive infancy. Yet even though the odds of keeping infants alive have improved astronomically, the chances that a woman in a postindustrial society will die without descendants have not changed that much. In the Sacramento Valley of California, where I live, 40 percent of all grown women who died between 1890 and 1984 left no surviving offspring.[11] But the reasons so many women in the Kalahari and California populations died childless are quite different. In twentieth-century California, many of the women never married. Others could not, or consciously decided not to, have children, or else decided to give birth to only a few. Almost all infants born survived, so that the average number of children per woman (1.6) was close to the number actually born. The circumstances surrounding motherhood have never been more different. Yet, as I will show in this book, from contemporary countries in which women live in a state of ecological release, no longer constrained by having to forage enough food each day to stay alive and with a broad range of reproductive options, to other parts of the world where they are less fortunate, women are constantly making tradeoffs between subsistence and reproduction that are similar in outline.

Quality vs. Quantity

Depending on the marriage customs in the culture she belongs to, a woman may or may not get to choose the father of her children or the time in her life when she first becomes pregnant. Depending on prevailing values, she may or may not treat her sons and daughters, her firstborn versus her last-born, exactly the same way. Yet, by and large, she will decide how much of herself, her time, energy, and love, she will invest in each child. The father has similar choices, although he may make quite different decisions. In the past, such decisions had a direct effect on both a mother's own survival and that of her children. Throughout most of human existence, to be an infant without a mother, or a child without older kin, was to suffer a life-threatening disadvantage. To have become the ancestress of any one of us today, a mother would have had to succeed in rearing at least one offspring to breeding age. In turn, that offspring would have had to produce surviving heirs. This is why quantity has rarely been the top priority for a mother. The well-being of her children and their quality of life, usually inseparable from her own, were primary. When fidelity to his mate means that a man's reproductive success is identical to his only wife's, he is more likely to share her preference for quality over quantity. Otherwise, and especially if he does not intend to invest in his offspring, a man may simply seek to breed with as many women, and hence sire as many offspring, as he can. It is from such ancestors that we inherit our maternal emotions and decision-making equipment. Underlying tensions between males striving for quantity and females for quality (a simplification I will clarify later) are as old as humanity. Yet this tension has risen to the surface and become more conspicuous than ever in the ecological circumstances of a modern world in which women have unprecedented choices.

For example, when young women are given a choice between having children and improving their lot in life, most opt for the latter. At first glance, such a finding seems completely antithetical to predictions from evolutionary theory. In the crass coinage of my Darwinian worldview, success is measured in terms of "fitness," genetic representation in succeeding generations. Access to more resources should translate into people having more children, not fewer. Certainly there is massive documentation that throughout recorded history, quite a few men opted for more.

Kings, emperors, and despots—who had the power to do so—filled their seraglios with fertile women. Feudal lords insisted on droit de seigneur with virgins marrying within their domains, while some American presidents have used their office (literally, the oval one) to enjoy assignations.[12] However, the emphasis on quantity that holds true for male potentates (and surely we don't call them that for nothing) does not hold true for mothers.

Around the world, there is a tendency for people who are better off to have a lower birthrate. This tendency is evident among peasant women in India as well as women in industrialized societies. Witness the declining birthrates in contemporary Japan, or the below-replacement fertility that has long characterized modern France and Italy and is increasingly true in established populations in the United States. Wherever women have both control over their reproductive opportunities and a chance to better themselves, women opt for well-being and economic security over having more children.[13] For many, leaving children every day while they work is a matter of survival, the only way mothers can support their families, or the only way they can secure a decent future for offspring. (A big difference between modern industrial societies and people who live by foraging is that children who must not only be fed but clothed and educated become more costly with age, not less.)

But survival does not explain all the choices. Third world peasants just making a skimpy living on small plots of land will trade the clean air and safer environment of the countryside for squalid urban shantytowns with their glimmer of economic opportunity, accepting the deaths of some children from respiratory or gastrointestinal infections in exchange for some prospect of a "better" life. Far more privileged women also may opt for self-realization over reproduction, forgoing motherhood to become artists, pilots, or scientists.

At first their choices appear counter to evolutionary expectations, until we recall that mothers evolved not to produce as many children as they could but to trade off quantity for quality, or to achieve a secure status, and in that way increase the chance that at least a few offspring will survive and prosper. This is why a closer look at what late-twentieth-century women are doing reveals behavior that is not so much "unnatural" as behavior that is in conflict with conventional expectations—all the myths and superstitions about what women are *supposed* to want.

So how did people in the Western world come to conceptualize female nature and "motherhood" the way we do?

Maternity and Charity

Biologically the word *maternity* refers to conceiving and giving birth, just as paternity refers to the individual who sires an offspring. But in the West, the concept of maternity carries with it a long tradition of self-sacrifice. "Her charity was the cause of her maternitie . . ." reads the *Oxford English Dictionary*'s illustrative sentence for a word in use from at least the seventeenth century. How confidently, then, could eighteenth-century moralists, steeped in the Enlightenment's celebration of God, Reason, Nature, and Man, admonish women (in a famous passage from 1770) to "Look to the animals for your example. . . ."

> Even though [animal] mothers have their stomachs torn open. . . .
> Even though their offspring have been the cause of all their woes, their
> first care makes them forget all they have suffered. . . . They forget
> themselves, little concerned with their own happiness. . . . Woman
> like all animals is under the sway of this instinct.[14]

No matter what their physical condition, the author—French physician Jean-Emmanuel Gilibert—was convinced that women should follow nature's eternal and unchanging precepts by nursing each child they bore. He was not trying to frighten his patients, only remind them of their instinctive and God-given maternal duty to nurture offspring. Gilibert and others like him looked to the animals not to make unbiased empirical observations but to use nature to confirm their own and their society's preconceptions about how humans

Fig. 1.1 *Allégorie de la charité,* by Pierre Daret, 1636. *(Courtesy of the Bibliothèque Nationale, Paris)*

should behave. These men, who were more evangelists than scientists, imposed their moral code on nature rather than allowing creatures in the natural world to speak for themselves.

Gilibert's passionate insistence on suckling derived from his antipathy to the then-widespread practice of wet-nursing, hiring a lactating, often much poorer, woman to breast-feed a mother's baby for her. A vast number of babies during this period were sent away right after birth to a distant rural wet nurse, consigning them to severe hardships and indifferent care. As a result, infant mortality soared. Gilibert was convinced that this "vice" of wet-nursing was responsible for France's declining population (a fascinating distortion of the real situation, about which more later).

In urban areas such as Paris the majority of infants born (95 percent, according to a 1780 police report [discussed in chapter 14]) were sent away to wet nurses. Yet Gilibert knew that humans are mammals, and that women, like all female mammals, have breasts *in order to suckle their young.* To Gilibert, this could only mean that women were intended by God to nurture their young; to shun this duty was unnatural.

Gilibert's grasp of our place in nature derived from the new classification by the great Swedish taxonomist Carolus Linnaeus. Among other things, Linnaeus's *Systema Naturae* of 1735 called attention to the special relationship that links humans to prosimians, monkeys, and apes. They were all lumped together in one order, Primates—Latin for "first ranking."

Underlying Agendas

Something just as primal, however, also linked Gilibert and Linnaeus. The two medical men were united in their assessment of what females were for and in their opposition to wet-nursing. Gilibert had translated one of Linnaeus's anti-wet-nursing pamphlets from Latin (the language of science at the time) into French so that it might reach a wider audience. The title *Nutrix Noverca* can be crudely translated as the cruel or unnatural "step-nurse."

It was to highlight the importance of lactation that Linnaeus identified an entire class of animals—Mammalia—by the odd milk-secreting glands that develop in only half the members of the class. The Latin term *mammae* derives from the plaintive cry "mama" spontaneously uttered by young children in widely divergent linguistic groups. The urgent message, "Suckle me," is universal. By calling mammals Mammals, instead of "sucklers" (as in the German term for mammals, *Säugetiere*), Linnaeus was making his point about both a natural law and the unnaturalness of any woman who deviated from it by failing to nurse.[15]

Our views of "motherhood" (including such scientific-sounding phrases as "*the* maternal instinct") derive from these old ideas and even older tensions between males and females. The fact that most of us equate maternity with charity and self-sacrifice, rather than with the innumerable things a mother does to make sure some of her offspring grow up alive and well, tells us a great deal about how conflicting interests between fathers and mothers were played out in our recent history. Sad to say, these old conceptions about maternity infiltrated modern evolutionary thinking.

Darwinism, Social Darwinism, and the "Supreme Function" of Mothers

According to Genesis, God created first heaven, then earth, then each variety of plant, every species of nonhuman animal, and, on the sixth day, man, and from one of his ribs, or perhaps his thigh, woman. In 1859, Charles Darwin proposed a revolutionary alternative to the biblical account. He titled his alternative genesis *On the Origin of Species.*

Darwin proposed that humans, along with every other kind of animal, evolved through a gradual, mindless, and unintentional process dubbed natural selection. Morally indifferent, natural selection culls and biases life chances with the unintended result that evolution (defined today as the change in gene frequencies over time) takes place. This mindless and "worse than morally indifferent" process geared to the maximization of short-sighted selfishness is what we mean by natural selection. She is the old lady with bad habits, the "Mother Nature" of my title.

Every environment, said Darwin, confronts organisms with challenges to their survival, whether the problem is cold or heat, tropical damp or drought, famine, predators, or limited space. For mothers, these problems become obstacles to keeping their infants alive. Individuals that are best adapted to their current environment survive and reproduce, passing on the attributes they possess to future generations. Losers in the struggle to survive die before they have a chance to breed, or they produce few offspring. Eventually, their line dies out.

The unfortunate and much misused expression "survival of the fittest" to paraphrase this phenomenon was introduced not by Darwin but by his prolific and widely read contemporary, the social philosopher Herbert Spencer. To Spencer, survival of the fittest meant "survival of the best and most deserving."

Indeed, Spencer's popularity was due to the simple take-home message delivered to his privileged audience in Victorian England and America: the advantages you enjoy are well deserved. For him, evolution meant *progress.* The flaw in Spencer's reasoning was to mistakenly assume that environments stay the same, unchanging backgrounds against which "superior," optimally adapted individuals rise to the top and stay there in perpetuity. What Spencer left out were the fluctuating contingencies of an ever-changing world.

Only colored by that oversight could Spencer's *social Darwinism* provide a blanket endorsement of the status quo. By contrast, *Darwinism*—real Darwinian thought, correctly interpreted—ascribes no special place to anyone. No adaptation continues to be selected for outside the circumstances that happen to favor it.

When Darwin adopted Spencer's phrase "survival of the fittest," he meant the survival of those best suited to their current circumstances, not the survival of the best in any absolute sense. To Darwin, fitness meant the ability to reproduce offspring that would, themselves, mate and reproduce. But no matter. Spencer and his followers were gratified that so celebrated a naturalist and experimentalist as Darwin would cite his views, accept his catchy phrase, and endorse heartfelt convictions about essential differences between males and females that derived from Spencer's theory of a physiological division of labor by sex.

The supreme function of women, Spencer believed, was childbearing, and toward that great eugenic end women should be beautiful so as to keep the species physically up to snuff. Because mammalian females are the ones that ovulate, gestate, bear young, and lactate (this much is irrefutable), Spencer assumed that the diversion of so much energy into reproduction had inevitably to lead to "an earlier arrest of individual evolution in women than in men"—a far more dubious extension.[16] Not only were men and women different, but Spencer's females were mired in maternity.

For Spencer, this physiological division of labor by sex meant that men produce, women merely reproduce. Costs of reproduction constrained mental development in women and imposed narrow bounds on how much any one female could vary from another in terms of intellect. Since variation between individuals is essential for natural selection to take place (which is true), Spencer reasoned (wrongly) that there was too little variation among females for proper selection to occur, precluding the evolution in women of higher "intellectual and emotional" faculties, which are the "latest products of human evolution."

Spencer was aware that a woman might occasionally possess a capacity for abstract reasoning. The only such female he personally knew, however, was Mary Ann Evans (the novelist George Eliot), whom he regarded as "the most admirable woman, mentally, I ever met." But Spencer regarded her gifts as a freak of nature, attributable to that trace of "masculinity" that characterized her powerful intellect.[17]

The assumption that education would be wasted on women was, of course, a self-fulfilling prophecy. Denied higher education and opportunities to enter fields like science, how could women *not* fail to excel in them? Eliot herself was one of a minuscule number of women in Europe at that time educated (in her case, largely self-educated) in languages, literature, philosophy,

Fig. 1.2 By the 1860s, Herbert Spencer (1820–1903) fell completely under the sway of his own theory of the physiological division of labor by sex. He abandoned his early support for women's education, an interest he and Eliot shared when they first met in London in 1851. If the function of women was to breed, if women's mental faculties were less evolved, he decided, educating women beyond a certain point would be a waste.

(1858 portrait from Coventry City Library)

and natural science. By regarding her as a masculinized exception, Spencer could reconcile his recognition of this woman's talents with his internalized evolutionary scale, on which women hovered in a fecund, biologically pre-destined limbo somewhere between Victorian gentlemen, on the one hand, and children and savages, on the other.[18]

Women as Breeding Machines

Spencer's validation of the status quo had far broader popular and political appeal than Darwin's more nihilistic perspective ever could. This is one reason why social Darwinism would become so influential. The second, related, reason was that Spencer's theory of the physiological division of labor by sex provided a scientific-sounding rationale for assuming male intellectual and social superiority. Spencer's "scientific" theories were an urgently needed antidote to the rising tide of feminist sentiment—especially in the United States—at a time when women were making real headway in their efforts to obtain the rights to vote and to own property in their own name.

Even before Freud declared that sex is destiny, Spencer and other evolutionists were constructing a complex theoretical edifice based on that assumption. They took for granted that being female forestalled women from evolving "the power of abstract reasoning and that most abstract of emotions, the sentiment of justice." Predestined to be mothers, women were born to be

Fig. 1.3 Progressive and liberal, the French artist Honoré Daumier was nevertheless ambivalent about women who aspired to be more than mothers. The caption to this lithograph read: "The mother is in the heat of writing. The child is in the bath water!" While Daumier read books like *La Physiologie du bas bleu* (Physiology of the Bluestocking), and produced cruel caricatures of freakish and non-nurturing women for his series entitled *Les Bas-bleu,* back in England an early evolutionist, Herbert Spencer, was struck by the fact that "upper-class girls" reproduced less than "girls belonging to the poorer classes," even though the latter were less well fed. Spencer decided that the "deficiency of reproductive power among [the advantaged] may be reasonably attributed to the overtaxing of their brains" and that "the flat-chested girls who survive their high-pressure education" would be "incompetent" to breast-feed.[19] *(From* Liberated Women: The Lithographs of Honoré Daumier, *Alpine Fine Arts Collection, London)*

passive and noncompetitive, intuitive rather than logical. Misinterpretations of the evidence regarding women's intelligence were cleared up early in the twentieth century. More basic difficulties having to do with this overly narrow definition of female nature were incorporated into Darwinism proper and linger to the present day.[20]

Equating a complex organism with a single defining "essence," such as giving birth, is known as essentialism. In 1949, the French writer Simone de Beauvoir sarcastically articulated the essentialist view in *The Second Sex:* "Woman? Very simple, say the fanciers of simple formulas: she is a womb, an ovary; she is a female—this word is sufficient to define her."

Earlier generations of feminists had also responded to Spencer and Darwin. For the most part, however, their voices went unheard. Eliot was one of the few exceptions, although she is far better remembered for her novels than for her critiques of early evolutionary thought. Yet even in her fiction Eliot took every opportunity to slip in rejoinders to Spencer's essentialist views and to demonstrate how multifaceted female nature actually is.

In her first major novel, *Adam Bede* (read by Darwin as he relaxed after the exertions of preparing the *Origin* for publication), Eliot put Spencer's views concerning the diversion of somatic energy into reproduction in the mouth of a pedantic and blatantly misogynist old schoolmaster, Mr. Bartle: "That's the way with these women—they've got no head-pieces to nourish, and so their food all runs either to fat or to brats. . . ."[21]

No doubt there was an edge to Eliot's rebuttals of Spencer. She had once dreamed of sharing her life with him. Opinionated as he was, Spencer was one of the few people in her circle seeking a rational alternative to religious dogma. It was Spencer who introduced her to evolutionary thinking, and he shared her passionate commitment to a scientifically based understanding of human nature.

They met in London in 1851 and soon after, she confessed her love and sent him an extraordinarily direct proposal which still survives.[22] Whatever his real reasons, Spencer's stated reasons for turning her down were eugenic. Eliot lacked, he said, the physical beauty he considered essential for mothers. As he put it: "Nature's . . . supreme end, is the welfare of posterity, and as far as posterity is concerned, a cultivated intelligence based upon a bad physique is of little worth, seeing that its descendants will die out in a generation or two."[23] Eliot, whose nose was long, her jaw pronounced, was too masculine-

Fig. 1.4 Portrait of George Eliot, etching by Paul Adolphe Rajon, 1884, from a photograph taken in 1858. *(From the S. P. Avery Collection; Miriam and Ira D. Wallach Division of Art, Prints and Photographs; New York Public Library; Astor, Lenox, and Tilden Foundations)*

looking to be regarded as pretty, and so far as Spencer's criteria for motherhood were concerned, her robust health and obvious intelligence were not relevant, only her looks.[24]

"The One Animal in All Creation About Which Man Knows the Least"

Spencer was not the only early evolutionist to wear blinders where women were concerned. Guided by a theory of unusual scope and power, Charles Darwin exhibited an uncanny knack for winnowing out kernels of accurate observation from the hodgepodge of anecdotes being sent him by a vast array of hobbyists, pigeon breeders, and sea captains from around the world. Yet he could not shake the biases of a man who had, after all, grown up in a patriarchal world where the most important thing a woman ever did was choose, or be chosen by, a man of means. It did not occur to his Victorian imagination—as it would immediately have occurred to a !Kung forager—just how resourceful and strategic a woman would have to be to keep children alive and survive herself.

Compared with his observations on barnacles, orchids, coral reefs, and even the expression of emotion in his own children, Darwin's observation of women and other female primates, in particular, were at best cursory. Thus in a passage few evolutionary biologists like to recall, and few feminists can bring themselves to forget, did the ever-careful Darwin deliver himself of

Fig. 1.5 Eliot used her novel *Middlemarch* to critique Spencerian notions of eugenic mate choice. Dr. Tertius Lydgate, the rational and positivist man of science, seen here in the recent BBC production, selects his bride, Rosamond Vincy, in accordance with a "strictly scientific view of woman." With her "perfect blonde loveliness" and "lovely little face," as childlike "as if she was five years old," Rosamond proceeds to ruin the besotted Lydgate's life. One can read Eliot's fictional case study as an admonition against social Darwinist illusions about universal ideal types.[25] *(Photo by David Edwards; © BBCWorldwide 1998)*

the opinion that: "whether requiring deep thought, reason, or imagination, or merely the use of the senses and hands, [man will attain] a higher eminence . . . than can woman."[26] Like Spencer, Darwin convinced himself that because females were especially equipped to nurture, males excelled at everything else. No wonder women turned away from biology.

For a handful of nineteenth-century women intellectuals, however, evolutionary theory was just too important to ignore. Instead of turning away, they stepped forward to tap Darwin and Spencer on the shoulder to express their support for this revolutionary view of human nature, and also to politely remind them that they had left out half the species.

In 1875, four years after Darwin's *The Descent of Man and Selection in Relation to Sex* appeared, there came a polite, almost diffident, rejoinder from the American feminist Antoinette Brown Blackwell. "When, therefore, Mr. Spencer argues that women are inferior to men because their development must be earlier arrested by reproduction," she wrote in *The Sexes Throughout Nature*, "and Mr. Darwin claims that males have evolved muscle and brains much superior to females, and entailed their pre-eminent qualities chiefly on their male descendants, these conclusions need not be accepted without question, even by their own school of evolutionists."[27]

Unquestionably, the most brilliantly subversive of these nineteenth-century distaff Darwinians was Clémence Royer, Darwin's petite, blue-eyed French translator. Self-educated like Eliot, Royer was the first woman in France to be elected to a scientific society. Darwin initially admired her as the "oddest and cleverest woman in France" but by the third edition of the *Origin* had lost patience with what he regarded as Royer's presumptuous manner. It particularly irritated Darwin that she criticized his (erroneous) ideas about "pangenesis," Darwin's notion of how maternal and paternal attributes were blended in their offspring. Darwin instructed his publishers to find another translator (a man, who did not do nearly so good a job), essentially firing her. Ultimately, what most unnerved Royer's fellow evolutionists would have been her outspoken views on the "weakening of maternal instinct" in the human species and tactics women use to subvert patriarchal control of their lives.[28]

In France at this time the decline in birthrate, or "demographic transition," that occurred in industrialized countries from the nineteenth century onward was well under way. Frenchmen were both puzzled and deeply concerned. There were plenty of married women of breeding age, many with more than sufficient resources for a family, some even wealthy, yet the censuses continued to register a declining population. Plenty of food, yet little in the way of "brats."

Not in the least puzzled, Royer scoffed at her male colleagues' lack of imagination: "Woman . . . is the one animal in all creation about which man knows the least. . . . a foreign species." When a male scientist describes women, she cautioned, he either extrapolates from his own experience or, worse, engages in an exercise in wishful thinking. Women were simply disguising from men their conscious desire to have few children. Large numbers of women, she believed, were deliberately curtailing conception—an idea that did not at all fit current evolutionary stereotypes about mothers.

Within the French scientific establishment of that time, Royer was doubly subversive—Darwinian in Lamarck's homeland and a maverick female with iconoclastic ideas about motherhood. No other evolutionist in the world, much less a woman, was writing about women who learn to be "mistresses so they do not have to be mothers," or wrote so enthusiastically about new techniques emanating from America for aborting unwanted pregnancies, taking advantage of physicians who have learned to "skillfully kill off the fruit without injuring the tree."[29]

Royer's own book on the origin of man (*Origine de l'homme et des societés*) appeared in 1870. But her most interesting ideas were set down in a later manuscript explaining why maternal instincts were weakened in the human species. Entitled "Sur la natalité" (On birth), it was already in proof for an 1875 edition of the bulletin of the Societé d'Anthropologie de Paris when the journal's editors suppressed its publication. In that suppressed manuscript Royer wrote:

> Up until now, science, like law, has been exclusively made by men and has considered woman too often an absolutely passive being, without instincts, passions, or her own interests; a purely plastic material that without resistance can take whatever form one wishes to give it; a living creature without personal conscience, without will, without inner resources to react against her instincts, her hereditary passions, or finally against the education that she receives and against the discipline to which she submits following law, customs, and public opinion.
>
> Woman, however, is not made like this.[30]

Royer assumed females were active strategists with agendas of their own. A hundred years later (in 1981), unaware of Royer's existence, I would publish a book, *The Woman That Never Evolved,* that made similar points. By then, the intellectual climate had changed. Much more empirical evidence about females was available, so a stronger case could be made. Evolutionary biology

Fig. 1.6 From an 1881 caricature of Clémence Royer, 1830–1902. A colleague at the Societé d'Anthropologie de Paris referred to her as "almost a man of genius," echoing Herbert Spencer's supreme compliment that Eliot was a "large intellect, even masculine."[31] *(Courtesy of Houghton Library, Harvard University)*

did eventually respond to these criticisms, yet in their lifetimes, the effect that these early Darwinian feminists—Eliot, Blackwell, Royer, and a few others—had on mainstream evolutionary theory can be summed up with one phrase: the road not taken. The toll was a costly one.

More than a century would elapse before Darwinians began to incorporate the full range of selection pressures on females into evolutionary analyses and in doing so recognize the extent to which males and females had coevolved, each sex responding to stratagems and attributes of the other. It took far longer than it should have to correct old biases, for evolutionists to recognize just how much one mother could vary from another, and to take note of the importance of maternal effects and context-specific development.

An unfortunate by-product of the delay in correcting long-standing biases in evolutionary theory was that by the last quarter of the twentieth century, when evolutionary paradigms were widened to include both sexes, many women, especially feminists, had already long since abandoned evolutionary approaches as hopelessly biased. Biology itself came to be viewed by women as a field sown with mines, best avoided altogether.[32]

The "Invisible Violence of the Institution of Motherhood"

Today Spencer's ideas are generally out of favor. Yet the "naturalistic fallacy" (mistaking what *sometimes is* for what *should be*) never really disappeared.[33] Spencerian ideas about universal and enduring species-typical standards of physical perfection are alive and well. A 1996 *Newsweek* cover story on the "Biology of Beauty" sported a photo of a perfectly beautiful Barbie-faced Caucasian woman with the caption, "Would you want your children to carry this person's genes?"[34] Within, a partially clad male model with a measuring tape compares the statistics of his female counterpart against a species-specific ideal, while scientists pronounce that "beauty is a signal of fertility and genetic quality" which, as far as I know, is a proposition never tested among humans.[35] Spencer's ideas on the importance of not just youth and good health but "personal beauty" persist. The notion that there is some unchanging status quo is still a given in some circles, almost always with Spencer's initial patriarchal assumptions intact. This status quo is a world where men control the resources women need to breed so that women everywhere—just as in the Victorian England George Eliot described in her novels—choose mates on the basis of wealth, while men choose breeding mates according to how nearly they approximate species-typical standards of beauty, or as Eliot put it: "wives must be what men will choose: men's taste is woman's test"; thus Lord Grandcourt, the Eliot character who most epitomizes patriarchal control of females, "would not have liked a wife . . . whose nails were not of the right shape; nor one the lobe of whose ear was at all too large. . . ." The consequences of such a breeding system where women choose men for inherited resources they did not earn, and men choose women for looks alone (as Darwin but not Spencer recognized) is not only destructive to women but terrible for the viability of offspring. (Lord Grandcourt's beautiful wife never conceives, and he dies young, without legitimate issue.)[36]

Evolutionary psychologists studying mate preferences today throw the

occasional sop to women lest they mind being told they should look like Bar-
bie dolls or else despair of becoming successful breeders. David Buss's recent
book, *The Evolution of Desire,* assured young women that:

> All women today are unique, distinctive winners of a five million year
> Pleistocene beauty contest of sexual selection. Every female ancestor
> of the readers of these words was attractive enough to obtain enough
> male investment to raise at least one child to reproductive age. . . .[37]

Yet, as I will make clear, any Pleistocene woman who relied on looks alone to
pull offspring through was not likely to be a mother very long or leave
descendants.

———

But it wasn't just social Darwinist stereotypes about what females should be,
or instinctively do, or what they are innately incapable of doing, that bothered feminists. Ideas about *what infants instinctively needed mothers to do* generated even more alarm. With her usual magisterial sweep, Simone de Beauvoir
summed up such fears. Biological stereotypes will lead to the "enslavement of
the female to the species and the limitations of her various powers."[38]

Attachment theory—the proposal that human infants have an innate need
for a primary attachment figure in the first years of life, a role that mothers
are uniquely qualified to fill, and that human babies deprived of such attachments suffer irreparable damage—rubbed precisely the spot where evolutionary acid burns deepest into feminist sensibilities. Women seemed to be
offered the choice of putting their lives on hold for years or else becoming
irresponsible mothers. The way many feminists saw it, an infant "attached"
meant a mother enchained.

One obvious way for these feminists to avoid this painful and irreconcilable dilemma was to deny that biology is relevant to human affairs or even
deny that infants have innate needs for highly personalized care. Another tactic was to insist that the human brain and our capacity for culture make us so
different from other animals that humans can learn to be anything they
choose. In fact, humans *can* learn a lot, but not anything we choose, and especially not in such ancient emotional domains as those involved with "love."

Nevertheless, the idea took hold that maternal love was a socially constructed sentiment without any biological basis, a "gift given."[39]

———

About the same time John Bowlby was pioneering an evolutionary perspective on infant development, a darker literature was emerging. The diagnosis of "battered child syndrome" first appeared in the 1960s.[40] From the 1980s onward, however, there was increasing awareness that infant abuse, neglect, abandonment, and infanticide were far more widespread than even those of us who studied such phenomena had realized. I already knew that abandonment and infanticide—both in humans and other animals—stretched far back in evolutionary time. I just had not realized the magnitude of what was going on. Even after I grasped the larger picture, I had trouble admitting to myself what the numbers gleaned from so many independent sources were showing, or what they meant.

Infanticide is not an appealing topic, especially not to women. Nevertheless, many feminists saw in some of these grim statistics a sort of intellectual silver lining. Historical, ethnographic, and demographic case studies documented the existence of many mothers who did not instinctively care for their young. Surely the prevalence of so many non-nurturing mothers undermined "essentialist" arguments about mothers genetically preprogrammed to nurture babies.

If women do not naturally nurture their young, then the parent with XX chromosomes is no more innately equipped for child-rearing than fathers are, ran their argument. Hence, why should the breadwinning mom be expected to stay home from work to tend a sick child, but not the dad? If mothers are no better equipped to be parents than fathers are, mothers need no longer shoulder so much blame when things go amiss. Gone, then, would be the dreaded "judgements and condemnations, the fear of her own power, the guilt, the guilt, the guilt," as Adrienne Rich so graphically identified the real "G-spot" for mothers.[41] And as for babies? One all-too-current feminist response is that we will worry about that another day, when data are "more conclusive."

Whatever moral one derives, clearly one of the West's most cherished ideals and, even in scientific circles, widely accepted pieces of conventional wisdom—the view that mothers instinctively nurture their offspring—has been receiving bad press of late. A rash of poetry and psychoanalytic commentary has also emerged, registering discontent with what Adrienne Rich calls "the invisible violence of the institution of motherhood." What Rich is referring to was the impossible ideal by which mothers not just willingly but "naturally" punch in for twenty-four-hour lifelong shifts of unconditional love.

Now that the generally accepted view that mothers instinctively love their offspring has been toppled, and now that it can be demonstrated (on many fronts) that maternal succor in the human species is anything but automatic or universal, how can we maintain that there is a biological basis to a mother's attachment to her infant? The answer all depends on what we actually mean by "biological basis."

Much lip service has been paid to "Biology," "Instinct," and "Natural Laws" without a great deal of attention paid to how maternal behavior unfolds in the real, everyday environments in which mothers actually live, or in those very different ancient environments in which women evolved. My focus in this book is both this recent historical past and especially the distant past where natural selection worked to shape the human life-forms that mothers and infants have become today. I am interested in whatever windows we have for peering into and reconstructing these worlds. They help me with my quest to more fully understand maternal emotions, infants' needs, and the implications of natural selection for mothers and infants alike. Understanding how creatures like us originated—creatures that are at once mammalian, primate, hominoid (or apelike), and human—helps us to understand the deep as well as recent history of the compromises that being a mother, or a father, inevitably entails. Without such a perspective, we cannot hope to do anything more than sweep the surface of terrain where landmines are buried deep below. Even when it isn't possible to completely defuse a mine—for these dilemmas are tough, sometimes truly irreconcilable—there are advantages to at least knowing where one is buried.

A New View of Mothers

Anything is more endurable than to change our established
formulae about women.
—George Eliot, 1855

The enormous human talent for self-deception should caution us that
anything we conclude so facilely about our species may serve evolutionary
ends we do not recognize.
—Patricia Adair Gowaty, 1998

W hen I entered graduate school in 1970, nineteenth-century
essentialist views still prevailed. Mothers were viewed as one-
dimensional automatons whose function was to pump out and
nurture babies. These stereotypes were especially pervasive in the field of primatology, where the creatures being studied were so similar to ourselves and
where practitioners were unlikely to be exposed to critiques like Eliot's, who
in 1855 expressly warned about "the folly of absolute definitions of woman's
nature."[1]

According to a widely cited 1963 essay on "The Female Primate" by one of
the first women professors of biological anthropology, which was published
in an anthology ironically titled *The Potential of Woman:*

> Her primary focus, a role which occupies more than 70 percent of her
> life, is motherhood. . . . A female raises one infant after another for
> her entire adult life. . . . Dominance interaction is usually minimal in
> the life of a female. She is invariably subordinate to all the adult males
> in the group and seldom if ever contests [their] superior status.[2]

In other words this female potential consisted of her capacity to conceive,
gestate, and suckle babies, period.

No creature epitomized "the female primate" better than old Flo, Jane
Goodall's wonderfully appealing, droopy-lipped chimpanzee mother who
"starred" in a half-dozen National Geographic specials. Flo seemed the model

of maternal patience and devotion, evoking the "Magic Mama" in Marge Piercy's poem of that name, the Mama:

> who sweats honey, an aphid
> enrolled to sweeten the lives of others.
> The woman who puts down her work like knitting the moment
> you speak. . . .[3]

Many behavioral biologists assumed that "the normal [female] always is a mother" and that those "females which had the most difficulty in becoming pregnant are generally those who have very severe antisocial and social problems."[4] Any reluctance or failure to care for offspring, any diversion of her energy to other pursuits, especially a mother's demonstration of competitive intent or aggressive behavior on her part, was viewed as pathological.[5]

Mothers Out of Context

Through the 1960s, comparative psychologists isolated mother rats, hamsters, cats, and other animals in discrete cages with only their offspring for company. Subjects were buffered from the complexities of larger social networks and the need or opportunity to forage (what one might call "breadwinning"). These mother-infant units were eerily reminiscent of model suburban housewives of the same era. Inevitably, separation of the mother and her young from other individuals limited the range of behaviors observed, while the research protocols virtually preordained which behaviors could be reported.

Mother-infant pairs were kept separate from their ecological or social and "political" surroundings.[6] Because food was freely available, little attention was paid to how mothers might differ in foraging abilities or resource defense. Behaviors other than nurturing, such as striving for status, seeking out or avoiding particular males, relations with kin and other group members, were viewed as irrelevant to a female's role as a mother. The virtual realities called "check-sheets" that these early researchers drew up to describe "maternal behavior" recorded how often a mother approached, hovered over, licked, carried, or suckled her infant. Categories describing interactions with other animals were rarely included. In any event, there were usually no animals except babies in the cage with her; hence no attention was paid to the possible significance of other group members who might assist—or interfere—with infant-rearing.

"Maternal behavior" as defined in this narrow way was a convenient operational category to aid in quantitative descriptions of what mothers do to care for their infants. Unfortunately, narrow definitions became synonymous with mothers themselves. It was all too easy to continue to "look to the animals" not to describe how females in nature actually behave but to confirm preconceptions about what it means to be a mother mammal. But new ways of observing animals in their natural habitats, and new ways of thinking about the role individuals play in the evolutionary process, would provide field-workers their first hint that motherhood was going to turn out to be more complicated than could be deduced from such limited studies.

Instead of interchangeable members belonging to a homogeneous class called "mothers," this new life-form would include a wide range of highly variable individuals who dealt with a wide array of situations and challenges. "Real-life" mothers were just as much strategic planners and decision-makers, opportunists and deal-makers, manipulators and allies as they were nurturers. The compromises mothers made and the tactics they employed were everywhere contingent on circumstances rather than being automatic, and might or might not result in nurturing behavior. These were the trends that would eventually transform the way we thought about female primates such as old Flo.

The key to this paradigm shift began almost imperceptibly with the realization by a handful of field biologists that even though Darwinians talk about the origin of *species,* Darwinian natural selection rarely if ever acts at the level of the species. Mothers did not evolve to benefit the species but to translate such reproductive effort as they could muster into progeny who would themselves survive and reproduce. What evolves is not behaviors that benefit groups but behaviors that contribute to the differential reproductive success of individuals—even at a cost to others in the group.

From Group to Individual Selection

David Lack, arguably the first "reproductive ecologist," was among the first evolutionists to analyze the breeding behavior of mothers as individuals. Scientific advisor to the British Trust for Ornithology in the years before World War II, Lack mobilized amateur bird-watchers all over England to collect

Fig. 2.1 Jane Goodall's droopy-lipped, ever-patient and responsive Flo, who nursed her last-born son a half-decade, until the day she died, is justifiably the most famous and admired mother ape in the world. Flo is shown here with her pouting daughter Fifi eyeing her baby brother Flint, along with Flo's then adolescent son Figan, who grooms his mother. Flo's aptitude as a mother is matched, if not surpassed, by her talent as an entrepreneurial dynast of the first order. *(Photo by Hugo van Lawick, courtesy of Jane Goodall Institute)*

data on individually banded swallows, robins, and other birds, an endeavor still going on today. Nests were watched, eggs counted and weighed. Lack paid special attention to how many eggs actually hatched. Of those hatched, how many chicks fledged?

Unprecedented quantities of data on the breeding success of *individually known mothers* living under a broad range of naturally occurring conditions poured in. With characteristic understatement, Lack noted that "individual differences exist and are by no means negligible."[7]

Lack saw no evidence that mothers adjusted fertility so as to benefit their group or species. Rather, mothers—with greater or lesser success— managed their reproductive effort so as to make the best of their own partic- ular circumstances. The bird who laid the most eggs in a given season, or who attempted to rear every egg she produced, did not necessarily fledge the most chicks. Over her entire lifespan, the female who bred all-out did not neces- sarily rear the most surviving offspring. Lack's findings marked the beginning of thinking about reproduction in terms of tradeoffs. New models for under-standing maternal behavior assumed that mothers traded off reproduction in the present against the possibility of doing better in the future, and then tested these assumptions against data from the real world.

The Fundamental Tradeoff in the Life of Mothers

Lack was particularly struck by the way mother birds staggered egg-laying. Among gulls, eagles, herons, boobies, and other birds, mothers lay eggs a day or so apart. Instead of waiting until her clutch is complete, a mother begins to incubate the eggs right away, with the result that the first-laid egg hatches several days before the last one. From that point on, the die is cast, the stage is set. The first chick laid will be the first to hatch, larger and more mature. It will already have received food from the parents by the time later chicks hatch. The older sib then easily intimidates younger siblings to a point of "no contest." Should food become scarce, the weaker soon succumbs to starva-tion.

By optimistically aiming high, then allowing the strongest in her brood to pare it down as needed, mother birds brought brood size in line with food supply by both what they did (lay eggs at intervals but begin brooding right away) and what they did not do (intervene or compensate). It was a flexible system, well suited to fluctuating food supplies. If food was scarce, the brood would be reduced prior to the point of maximum demands on the mother to bring in food for growing chicks. If food was abundant, however, the entire brood survived—including the eggs hatched "on spec."[8]

The cruel efficiency of brood manipulation implied considerably more discretion on the part of females than Herbert Spencer had imagined. Fur-thermore, females were far from interchangeable. Female reproductive suc-cess varied enormously over the course of their lives, and even within single seasons. Such variation meant that females were a sex wide open to selection. And this meant something else. Although a mother's interests would often be identical with those of her brood, this would not always be the case. The same mother who bravely drove away a predator from her nest would not inter-vene to protect the last-hatched chick from a less ferocious but more lethal enemy, its own older sib. This was a highly discriminating mother, whose commitment to her young was contingent on circumstances.

Could natural selection actually favor such mothers? It was several decades before a young ornithologist named Caldwell Hahn decided to test

Lack's hypothesis. Like a post-Enlightenment deus ex machina, she swooped down upon a colony of laughing gulls in a New Jersey salt marsh and switched eggs around so as to rewrite past evolution by producing experimental nests filled with eggs laid on the same day, which would therefore hatch at the same time. Mother gulls with artificially equalized broods were twice as likely to lose their entire reproductive effort as were "control" mothers, whose eggs were also manipulated but rearranged so as to hatch asynchronously—as they would have if left alone. (Eggs had to be moved around in both experimental and control groups so that experimenters could be sure that human intervention by itself did not produce greater mortality.)

David Lack had identified the fundamental tradeoff in the life of mothers: whether to produce many offspring, investing little in each, or produce a few and invest a great deal. This idea of "fitness tradeoffs" laid the groundwork for studying the myriad ways that a mother adjusts maternal investment in line with ecological conditions. Far from self-sacrificing, the mothers in this Lackian paradigm were flexible, manipulative opportunists.

Lack's insights, subsequently refined and expanded by American biologists George Williams and Robert Trivers, laid the groundwork for evolutionary biologists to begin to analyze the evolution of social behavior from the perspective of each participant. As it happened, the langur monkeys whose bizarre infanticidal behavior had so riveted my attention when I learned of it, would provide one of the first tests for the idea that the interests of mothers and fathers did not necessarily coincide and that males and females behaved as they did to promote individual reproductive success rather than the continued survival of the species.

Mothers Coping with Males

In the summer of 1971 I traveled to India, where the Japanese biologist Yukimaru Sugiyama had first witnessed infanticidal behavior in langur monkeys. By this time there was a growing consensus that infanticidal behavior was brought about by human interference and compression of the monkeys' ranges into unnaturally crowded habitats.[9] For the next nine years I would alternate between studying the behavior of these monkeys in a range of habitats—village to forest—and at different population densities, while completing my graduate work and holding various teaching positions.

Early on in my study it became clear that assaults on infants were not random acts of violence by stressed animals. Infants were attacked only by strange adult males, never by males likely to be their fathers. These attacks occurred when males from outside the breeding system took over one of the breeding troops and drove out the resident male. Then, in a relentless and goal-directed manner, the newcomer stalked mothers with unweaned infants and attacked them. Once their infants were eliminated, the mothers became sexually receptive and solicited the new male. But why, I wondered, were mothers "rewarding" such behavior by breeding with the same male that killed their infants? (Note that langur males never copulate unless first solicited; rape is unknown.)

By the end of my first field season, I was forced to set aside my original hypothesis. Rather than pathological, this infanticidal behavior appeared to be surprisingly adaptive behavior on the part of males. By eliminating the off-spring of their predecessors, males induced the mother to ovulate again sooner than she otherwise would have. Thus the killer had compressed repro-ductive access to her into the brief period he was likely to be present in her troop (on average twenty-seven months). From the male's point of view, his behavior was genetically advantageous. But why would any mother go along with it?

The main reason was that even though her species as a whole, and her sex in particular, suffered from retaining the genes of infanticidal males in the population, mothers could not afford to sexually boycott them. By the time she lost an infant she had already invested in, she could not afford the further delay of waiting for a nicer male to show up, one with attributes more benefi-cial to the survival of her species. To postpone ovulating again for that long would put her at a disadvantage in competition with other mothers who went ahead and bred with the infanticidal male. Furthermore, the sons of such a mother would be at a disadvantage in competition with the sons of infantici-dal males who, instead of waiting around for a chance to breed, took matters into their own hands by eliminating impediments to their breeding.

In the case of langur monkeys, a forty-pound male equipped with dagger-like canine teeth has the advantage of size and weaponry over a female, who weighs just over half as much. Even if she evades him for a time, or even if her female relatives intervene on her behalf, as many do, the infanticidal male can try and try again, day after day, until he finally succeeds. The odds are stacked against the mother.

I knew that in some species (hyenas, for example) females evolved to be the same size or larger than males and as a result could protect their infants

Fig. 2.2 A langur male seizes an infant sired by his predecessor. *(Drawing by Sarah Landry)*

better. But animals can evolve larger body size only incrementally, bit by bit over generations. Being just a little bit bigger helps a female hyena, who, in addition to fending off cannibalistic group mates, also does better scrambling at carcasses to stay fed. However, being a little bit bigger is of little use to a leaf-eating monkey that does not compete with the monkey next to her for food. Because growing bigger would require delaying maturity, it might even place a bigger female at a reproductive disadvantage relative to a faster-maturing smaller one. Being bigger might also make her more vulnerable to starvation in famine times. There would be all these drawbacks to growing big without the payoff: being enough bigger so that the mother could prevent a forty-pound male from killing her infant.

Infanticide among langurs provides a vivid example of behavior that clearly did not evolve to benefit the species. The killer gains at the expense of his infant victim, the rival male who fathered it, and the mother, who loses all that she has invested in the offspring up to that point. Among langur monkeys, where this phenomenon was first studied, recurrent male takeovers accompanied by infanticide can lead to a decline in group size over time, potentially even the extinction of a particularly vulnerable group.[10]

Although detrimental to the good of the group, infanticide by unrelated males and collusion by mothers turns out to be far more widespread than anyone thought possible. Reported now for some thirty-five different species of primates belonging to sixteen different primate genera, infanticide is often a significant source of infant mortality. In the most extreme cases, among mountain gorillas studied at Dian Fossey's old study site in the Virunga volcanoes of Zaire, 14 percent of all infants born, and among red howler monkeys studied in Venezuela 12 percent of all infants born, are killed by marauding males.[11] Among langur monkeys studied near Jodhpur, a team of researchers under the direction of Indian primatologist S. M. Mohnot and German primatologist Volker Sommer would go on to estimate that over a twenty-five-year period, *33 percent of all infants born* were killed by invading males.[12] Similarly high infant mortality rates from infanticide are also being reported among one-male troops of chacma baboons living in regions of Botswana. Elsewhere savanna baboons live in multimale troops and infanticide is rare, but at Moremi single males monopolize breeding only for a brief period. These short tenures of male access to fertile females intensify male-male competition for access to fertile females, and increase the selection pressure on males to compress a mother's reproductive career into the brief period he has access to her.[13]

———

Even in normally gentle and herbivorous primates like langurs and gorillas, other members of the same species can be a threat to infant survival every bit as hazardous as lurking predators. This was a new kind of selective pressure on mothers, one not previously dreamed of. I was beginning to think that the threat of infanticide might explain something else peculiar about females.

In the early 1970s, it was still widely assumed by Darwinians that females were sexually passive and "coy." Female langurs were anything but. When bands of roving males approached the troop, females would solicit them or actually leave their troop to go in search of them. On occasion, a female mated with invaders even though she was already pregnant and not ovulating (something else nonhuman primates were not supposed to do). Hence, I speculated that mothers were mating with outside males who *might* take over her troop one day. By casting wide the web of possible paternity, mothers could increase the prospects of future survival of offspring, since males almost never attack infants carried by females that, in the biblical sense of the word, they have "known." Males use past relations with the mother as a cue to attack or tolerate her infant.[14]

Still, how could something so generally deleterious as infanticide evolve? The answer is sexual selection, a process by which same-sex individuals compete with one another for matings. The loser left few or no offspring.

Darwinian Sexual Selection

Darwin argued that any trait that enhances survival and thereby increases the chances that an individual breeds and leaves offspring will be favored by natural selection. But he was aware of some peculiar variations on this theme— traits that evolved even though they did not enhance the survival of the possessor. Such traits seemed to occur most often in males. For example, Darwin was fascinated by the peacock's elaborate tail and the stag's antlers, traits that had no apparent survival value and seemed largely ornamental. Indeed, some of these decorative appendages might actually hinder survival by using up a scarce resource like calcium for making fancy antlers instead of making stronger bones. The flashy and cumbersome tail of a peacock would appear to make the male who possessed it both more conspicuous to its predators and too clumsy to escape them.

A hindrance in terms of survival, Darwin suspected that these traits served a different function. They helped members of one sex compete with one another for mates. Darwin termed this special subset of the evolutionary process *sexual selection,* which he believed had special relevance to human evolution. Thus he titled his major work on the subject *The Descent of Man and Selection in Relation to Sex* (1871).

Sexually selected traits might help individuals outbreed others of the same sex by outfighting them, dominating them and excluding them from the scene (the canine teeth of a langur male probably evolved this way), by making the possessor more attractive to the opposite sex (like the peacock's tail), or by canceling a rival's genetic contribution and replacing it with his own (as did the tendency to kill infants sired by rivals). Sexual selection typically involved two components: male-male competition for access to females, and female choice.

The logic of sexual selection was not just counterintuitive. The idea that females (supposedly passive, after all) were choosing between mates on the basis of some secret logic or seemingly whimsical aesthetic preferences seemed downright incredible. Darwin's most original theory soon fell into disrepute and was largely forgotten.[15] More than a century elapsed before scientists in the 1970s revisited the topic. But when they did, the groundwork would be laid for explaining why male-male competition for mates was so widespread, and why female choice was so important.

Parental Investment and Female Choice

Although Darwin pointed out the importance of sexual selection, he was vague on just why members of one sex (typically males) competed so ferociously for access to the other. The explanation would be laid out in a brilliantly original paper on "Parental Investment and Sexual Selection" by Robert Trivers in 1972.

"Parental investment," according to Trivers, is anything that a parent does to promote the survival of an offspring that also detracts from the parent's ability to invest in other offspring. For most animals, especially mammals, parental investment by mothers and fathers was far from equal. Typically the male contribution consisted of ejaculation, whereas the female's consisted of ovulation, gestation, and lactation—a sequence of costly biological processes that could tie up a mother reproductively for a long time, while the male, if successful, could go right on breeding. This inequality in "relative parental investment of the sexes in their young is the key variable controlling the operation of sexual selection," Trivers wrote in 1972.[16]

Where one sex invests considerably more than the other, "members of the latter will compete among themselves to mate with members of the former. Where investment is equal, sexual selection should operate similarly on the two sexes. . . ." But if not, "the individual investing more (usually the female) is vulnerable to desertion." Since in most species fathers invested less than mothers, competition between males for access to the limiting resource—potential mothers—was intense. This led to no-holds-barred evolution of any trait that helped males in this competition, even if it ultimately hurt the female (as in the case of infanticide). Male efforts to exclude other males, or herd and sequester females, were all outcomes of sexual selection. Often attempts by males to control their mates came at the expense of the viability of mothers and offspring.

The one area, however, where sexual selection worked to increase rather than decrease viability of offspring was female mate choice. As researchers rediscovered this topic, an old objection to Darwin's theory lost some of its power. For female choices were not just aesthetic. Burgeoning research on this topic now indicates that females do sometimes select mates on the basis of cues to the male's genetic merits (what Trivers called "good genes") that

Fig. 2.3 Displaying peacock *(Sarah Blaffer Hrdy/Anthro-Photo)*

can be passed on to offspring—especially in species such as peacocks, where mates provide little else. Furthermore, the more that males in a given population differ genetically from one another, the greater the selection pressure on females to avoid being monopolized by one—not necessarily the best—male.[17]

Powers of discrimination in the peahen, for example, go beyond whimsical aesthetics. True, females comparison-shop and then mate with the peacock whose blue and green train of feathers fans out to reveal the largest eyespots. The "eyes have it," but with a quite practical kicker: recent research by sociobiologist Marion Petrie demonstrates that seemingly arbitrary preferences for fancy tails can have tangible payoffs. Petrie followed the fates of chicks sired by different males as part of a carefully controlled experiment undertaken at England's Whipsnade Park. Offspring of the most ornamented males grew faster and had better survival rates.[18]

Since there is no evidence that male peafowl do anything to help offspring, the most plausible interpretation is that males able to produce the most ornamented tails also provide the most viable genes.

———

The bizarre idea that there may be genetic method to feminine aesthetics receives support from biologists who measure to the millimeter how symmetrical animals are. Potentially symmetrical traits include the outermost eyes on the fanned-out tails of peacocks, side feathers of a swallow's tail, two wings on either side of a male scorpionfly, or, in the case of humans, left versus right ear lobes, cheekbones, jawbones, and elbows.[19]

Scarcely perceptible lopsidedness in bilateral traits like tail feathers is assumed to be a stress-induced deviation from a perfectly symmetrical ideal body plan, brought about by environmental insults—pathogens, food shortage, or inability to metabolize critical nutrients—thus interfering with the organism's development. Indeed, the methodology for measuring such "fluctuating asymmetries" (small, random deviations from perfect bilateral agreement in what should be perfectly symmetrical traits) was initially developed by wildlife biologists worried about the toll on animals taken by changes in their environment, pollution, diseases, or new parasites.

In creatures as diverse as earwigs and scorpionflies, finches, swallows, fish, and at least one mammal—humans—degree of fluctuating asymmetries can be inversely correlated with large body size, freedom from parasites, and some measures of performance such as competing for a food item or a mate. In short, the brightest, most ornamented or symmetrical males are likely to be those best suited to prosper in the environment where they grew up or developed breeding plumage. For females who don't have the option of running lab tests on potential fathers, such up-to-the-minute indices of physical condition provide the next best thing.[20]

Following this logic, entomologist Randy Thornhill hypothesized (and confirmed) that scorpionflies with the most symmetrical wings would have the highest mating success, and that females could detect which males these were just by their scent. Symmetrical males may emit different pheromones, or perhaps they are more effective in commandeering the dead insects and other little prey items that scorpionfly males proffer to potential mates as nuptial gifts, and this is what catches females' attention. Whatever the reason, symmetry, performance, and female preferences were all correlated.

Thornhill teamed up with psychologist Steven Gangestad to find out whether the same principle of "fearful symmetry" that humans find so riveting in art and metaphor has anything to do with how humans choose sexual partners. They applied their calipers to the cheeks, eyes, ankles, and elbows of undergraduates at the University of New Mexico. Men with low scores on fluctuating asymmetries (who, of course, had *no* way of knowing this fact about the various fairly quirky facial and body traits being measured) tended

to be the same subjects who self-reported that
they had found opportunities to engage in sex
earlier, more often, and with significantly more
different partners than did men with higher
scores—but who, as far as most people consciously register, were not per-
ceptibly lopsided. No one knows yet exactly what cues are being used, how
important they actually are in overall human mate choices, or exactly what
any of these burgeoning data sets on human fluctuating asymmetries mean.
However, the evidence suggests that either women somehow pick up tiny
cues of male condition, or self-confidence, or else men in good condition
have an inflated self-image and habitually exaggerate their sexual prowess for
the benefit of prying professors.[21]

Let's say that females really are choosing more symmetrical partners. Are
they after better genes in potential fathers, or after protectors and providers
who are bigger and in better condition? The trouble is, we can't yet tease
genetic effects apart from developmental happenstance. For example,
recently discovered correlations between degree of fluctuating asymmetries
and performance on IQ tests can be explained either by better genes for the
relevant abilities inherited from one or both parents, or by favorable conditions early in life—a healthier environment for physical development inside
the mother's womb.[22]

But Is It Really "Choice"?

Fieldworkers like Marion Petrie and Randy Thornhill rescued Darwin's controversial concept. Female choice would play an important role in burnishing
the emerging image of calculating female strategists, as researchers studying
animals in natural habitats took into account the full range of activities necessary for a mother to reproduce successfully and be "in control."[23] But on
closer examination, some of the apparent "choice" in female choices was
more nearly like compromise, or worse. Often females were responding to
offers they could not refuse. For example, when a male langur kills an infant,

he essentially nullifies any previous choice the mother might have exercised in her selection of the father of the killed offspring. By distorting her options, the infanticidal male constrains the mother to choose him as her next mate. This is why some sociobiologists, such as Patricia Adair Gowaty, are intent on learning not only which male a female chooses to mate with (a topic that was controversial enough in Darwin's time) but which male a female would choose *if her choice were not constrained by other factors.*

Gowaty and a consortium of coworkers have embarked on an ambitious experimental study to discover the consequences of male-imposed constraints on female choice. When, for example, a male hamadryas baboon herds females about, or, for that matter, men lock up wives in seraglios, are there biological costs? Does removal of such constraints (Gowaty terms it "free female choice") result in improved viability for a female's progeny, and hence greater viability for the species?[24]

Early results from a related project—this one under the direction of William Rice at the University of California, Santa Cruz—suggests that the costs borne by females of having mate choices constrained may be considerable. A body of slowly accumulating evidence is showing that not only does female choice have profound evolutionary implications, so does its curtailment.

In a series of staggeringly ambitious experiments, Rice set out to allow males to evolve while forcing females to stand still in evolutionary time. This feat was possible because Rice is an experimental geneticist who works with drosophila, the fast-breeding little fruit flies that materialize in kitchens when bananas are left out too long. Rice established an elaborate experimental design involving thousands of flies over forty-one generations. In each generation, males from the "evolving" population were provided females from the original stock which, through various technically complex manipulations, Rice held more or less genetically constant through time.

Although male fruit flies don't out-compete other males by killing their offspring as langurs do, they use tactics just as dastardly: toxic molecules in male seminal fluid that destroy the sperm of males that subsequently mate with the same female. Unfortunately, after many ejaculations, toxins from the poisons accumulate in the female's reproductive tract, decreasing her fecundity and shortening her lifespan. "The more they mate," Rice observes of the females, "the faster they die." In time, a race of "super males" emerged, especially adapted to this lethal mode of sperm competition, but at a cost. Because females were taken from the breeding stock and held artificially constant, they had no opportunity to evolve defenses to this chemical warfare being practiced by their mates. If, for example, females had only been able to choose among mates, the genes of females who avoided toxic males might have been selected for, or mothers might have evolved otherwise equipped to counter toxic sperm. Instead, females prevented from coevolving with these "sexually antagonistic" mates died at higher rates. Few experiments could have demonstrated more clearly the extent to which the two sexes "evolve substantially in response to one another," and not just in response to the physical environment or to predators.[25]

By the end of the twentieth century, sociobiologists had revealed that females were anything but passive or sexually coy, and certainly not (except in Bill Rice's experiments) less evolved. Females were the genetic custodians of the species, and through their mate choices—when permitted—directed the course of evolutionary change. Parental investment and female choice not only called attention to the burdens that motherhood imposes but also focused attention on the long-standing importance of female autonomy in reproductive decisions, often synonymous with offspring viability. But if female decisions were important in the realm of sexual selection, and critical for selecting progenitors and countering male-male competition, female "decisions" were even more important in the realm of natural selection generally.

Over the course of her life, a female bound for fitness is required to make a series of physiological and developmental "decisions" about how big to grow, when to mature, how soon to reproduce, and how much time to allow between offspring. One of the biggest challenges for understanding selection pressures on mothers that confronted sociobiologists in the early years was getting the balance right between considering traits that are sexually selected (for example, through female choice or male choice) and equally important, if not more important, female traits that are naturally selected because they increased the survival of the mother or her offspring.

The Rare Self-Sacrificing Mother

For each mother, life is a series of turning points and decisions mostly about how best to allocate resources over the course of her lifetime, be it long or short. Should she put all her effort into growing big, producing a large number of offspring all at once, breed in a single fecund burst like the salmon who forage in the ocean, swim upstream, spawn, and die? (This is known as semelparity.) Or should she bear fewer offspring, with births spaced at long intervals over a long life (iteroparity), like chimpanzees or humans do?

A survey of the natural world through the lens of this life-historical approach reveals just how special a creature the self-sacrificing mother envisioned by men like the French physician Gilibert must be. Such mothers exist, but they do not evolve as species-typical universals of the female sex except under the most stringent circumstances. Typically, self-sacrificing mothers are found in highly inbred groups, or when mothers are nearing the end of their reproductive careers. The breed-and-then-die strategy typical of semelparous creatures (who reproduce only once in their lifetime) provides the best examples.

If semelparity is hard to visualize, think of Charlotte, the altruistic spider mother in E. B. White's beloved children's book *Charlotte's Web*. She toils and spins to lay her single pouch of eggs, and when her life's work is done, she dies. This is classic semelparous reproduction.

A human mother who feels put-upon by the onslaught of child-related demands when she arrives home after a long day at work may be heartened to know how much better off she is than one of these more "selfless" mothers. Semelparous mothers are often literally "eaten alive" by their young.

The prize for "extreme maternal care" goes to one of the various matriphagous (yes, it means mother-eating) spiders. After laying her eggs, an Australian social spider (*Diaea ergandros*) continues to store nutrients in a new batch of eggs—odd, oversized eggs, far too large to pass through her oviducts, and lacking genetic instructions. Since she breeds only once, what are they for?

These eggs are for eating, not for laying. But to be eaten by whom? As the spiderlings mature and begin to mill about, the mother becomes strangely subdued. She starts to turn mushy—but in a liquefying rather than a sentimental way. As her tissue melts, her ravenous young literally suck her up, starting with her legs and eventually devouring the protein-rich eggs dissolving within her.

Few things seem quite so antisocial as cannibalism. Yet dining on mother may be the key ingredient to the evolution of these spiders' unusually gregarious lifestyle. By having the bad manners to eat their mother, sated spider-lings are rendered less likely to eat one another. Furthermore, even among selfless mothers, not all are equal. There is room for all sorts of selection on each mother's attributes. The more efficient a mother is at capturing prey, the bigger she gets; the bigger she is, the bigger the banquet she provides and the less inclined her progeny are to eat each other and the more her little canni-bals reap the benefits of a social existence.[26]

How to Succeed in Breeding over a Long Life

Few mammals breed in one semelparous bang. Instead, most are iteroparous, breeding sequentially over a lifetime. Such mothers may produce young singly or in litters. For creatures like ourselves, shaped by this iteroparous potential to breed more than once, it rarely makes sense for a female to put all her eggs in one basket. An iteroparous mother who overshoots the opti-mum clutch-size for her circumstances, whose ovaries are bigger than her larder, may lose her entire brood to starvation or end up so weak she does not breed next season when conditions improve. Worse, she may succumb trying.

Learning just how mothers allocate their time and energy between mak-ing a living, resting, and reproducing, or caring for infants became the goal of primatologist Jeanne Altmann. She and her husband, Stuart, had set out in 1963 to study the ecology and social behavior of baboons on the dusty plains of Amboseli, in Kenya. Their landmark study, which continues to the present day, would also provide the first opportunity to investigate the tradeoffs that primate mothers make. Her research on the ecology of mothers would emphasize the extent to which every baboon mother is a "dual-career mother" spending most of each day "making a living": feeding, walking, avoid-ing predation, while also caring for her infant.[27]

Acutely sensitive to the problem of observer bias, Jeanne Altmann devel-oped techniques for choosing subjects at random. She made sure each indi-vidual was watched for the same number of minutes. Then she used statistical tests to analyze the results. It was the best available antidote to the all-too-human habit of seeing only what we expect to see. In time, such methods for studying free-ranging animals became standard practice for animal behavior-ists, and spread into human behavioral ecology. Instead of relying on asking mothers what they remembered doing, or thought they should do, accurate accounts of what mothers were actually doing became available.[28]

Fig. 2.5 Jeanne Altmann's experiences studying wild baboons at Amboseli sensitized her to the problems of observer bias. Her 1974 critique of observation methods helped revise the terms on which humans "look to the animals." *(Photo courtesy of Jeanne Altmann)*

"In all aspects of the present study one fact recurs: baboon mothers like most primate mothers, including humans, are dual-career mothers in a complex ecological and social setting," Altmann wrote. "They do not take care of their infants while isolated in small houses or cages, as the rest of baboon life goes on. They are an integral part of that life and must continue to function within it. The baboon world affects them, and they it, through their lifetime."[29] With 70 percent of their day going toward making a living, and perhaps another 10 to 15 percent for resting, these baboon mothers were pushing the envelope of their own survival. If they were to breed any faster, they would risk maternal depletion and death.

Altmann's field research shifted the focus away from what had become an overemphasis on male-male competition and mate choice back to natural selection. In balancing her tradeoffs so as to stay alive and breed, almost every aspect of a mother's life was shaped by natural selection.

At the same time, Altmann was managing her own balancing act, raising two children and doing science under harsh field conditions. She also had another concern. The study of mother-infant relations was at that time still widely viewed as the "home economics" of animal behavior, an area of little theoretical significance. Altmann feared that her hard work would not be taken seriously.[30] Instead, her 1980 monograph *Baboon Mothers and Infants* became a classic in the study of life-history tradeoffs.

The Art of Iteroparity (Breeding More Than Once)

A female's lifetime reproductive success (or fitness) depends on luck, of course—everything does—and, as in the case of Altmann's baboons, on the physical constraints of the environment. In an evolutionary sense, how one mother fares relative to another depends on how well she handles the series of tradeoffs she encounters in the course of her entire life. One tradeoff is between growing larger and maintaining herself (somatic effort) versus reproductive effort. The second main tradeoff involves how she allocates bodily resources available for reproduction among offspring. Again, there is a quantity-versus-quality tradeoff. In some species, such as rabbits and galagos, mothers invest little in each infant and breed fast; others, like a chimpanzee or a human, breed slowly over a long life. Others pursue mixed strategies, alternating according to conditions.

Golden hamsters, quintessentially flexible breeders adapted to the irregular rainfall and erratic food supplies of their arid Middle Eastern environment, illustrate the art of iteroparity—or, how to breed successfully more than once over a lifetime. In addition to building a nest, licking her pups clean, protecting and suckling them—all pleasantly maternal-seeming pursuits—a mother hamster may also recoup some of her investment in these pups by eating a few, a time-honored maternal tactic for adjusting litter size in line with prevailing conditions. Among mice (but not hamsters), mothers cull right after birth apparently in an effort to configure litter quality (favoring the heaviest pups) or litter size, occasionally abandoning whole litters if the number of pups falls below a certain threshold, something much larger mammals (lions and bears) also do.[31]

Once embarked upon a reproductive trajectory, mothers face new challenges. How to reconcile conflicting demands of different offspring? Treat

Fig. 2.6 All mothers balance tradeoffs between subsistence and reproduction. *(Photographer unknown)*

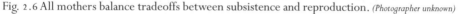

each offspring as equivalent, or value some over others? Should a mother gamble on an offspring now or reserve herself for some future offspring who might be born under more promising conditions, or might perhaps be born a sex that is more advantageous for her to rear? Given that her body is deteriorating over time, when should she throw in the towel, quit producing, and care for her daughter's offspring instead?

The Big Mother Hypothesis

When was it worthwhile to delay reproduction and keep growing? Zoologist Katherine Ralls hypothesized that a big mother could be a better mother. The standard answer to why males are bigger than females is sexual selection: males are selected to be bigger and stronger than rival males. Females remain smaller, the ecologically optimal size for their environment—a sort of default body size. During the 1970s, Ralls challenged the fixation with sexual selection. Playing devil's advocate, she listed all the mammals in which *females grow larger than males,* an eclectic array of moon rats, musk shrews, chinchillas, jackrabbits, cottontails, klipspringers, duikers, water chevrotains,

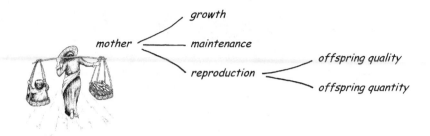

Fig. 2.7 Mother's main "life history tradeoffs"

dik-diks, marmosets, bats, bats, and more bats, and so forth, and then showed
how poorly sexual selection theory accounted for many of the cases.

Increasing fecundity with body mass turns out to be one of the main reasons so many invertebrates, such as spiders, have females larger than their
mates. Among species where mothers produce more eggs or young as they
grow older and bigger (the best examples are fish), females seem to live on
and on until something external kills them. Both fishermen and male fish
seek out these "big mothers"—the former for their cachet, the latter for their
greater fecundity.[32] Depending on the species, big mothers produce bigger
babies, deliver larger quantities of rich milk more quickly (as whales do),
outcompete smaller females so as to monopolize resources available in their
group, or, as spotted hyena females must do, not only defend their place in
the chow line but also defend their infants from carnivorous and extremely
cannibalistic group mates. The fact that bigger mothers make better mothers
probably explains why the blue whale female grows to 196 tons, qualifying
her as the largest mammal in the world.[33]

Even among the anthropoid apes, and in spite of a venerable history of
sexually selected males being bigger than females, Ralls's "big mother
hypothesis" helps to explain the emergence around 1.7 million years ago of a
hominid species with females closer in size to that of males. There was a dramatic *decrease* in the degree of sexual dimorphism (size difference between
males and females) in these animals. *Homo erectus* males and females were
about the same size as humans today, and were just embarking on the lifestyle
characterized by a division of labor between male hunters and female gatherers. To understand why *Homo erectus* females, as well as females belonging to
the closely related genus *Homo ergaster,* grew twice as large as australo-

Fig. 2.8 In *Homo ergaster,* considered by some paleontologists to be synonymous with *Homo erectus,* males were around 18 percent larger than females—far less dimorphic than australopithecines but in the same ballpark as modern humans. *(J. Beckett; Denis Finnin; Department of Library Services, American Museum of Natural History, NewYork, Image no. 2A22690)*

pithecine females—of which the famous fossil known as "Lucy" is the best known—we need to consider selection pressures *on mothers.*

Whereas Lucy's mate would have been 50 percent again larger than she was, Missus Erectus's fellow was a mere 20 percent bigger than she was, around the same order of magnitude as the 15 percent or so difference in body size that characterizes men and women in modern populations. In other words, both males and females grew larger; but by 1.7 million years ago, with the emergence of *Homo erectus,* selection pressures favoring larger body size became more important for females than for males. Why was this so? For a hunter there may be a ceiling on just how large he can grow and still be effective in the pursuit of game. Ultimately, though, speculates University of California, Davis, paleontologist Henry McHenry, the decline in size difference between the two sexes had to do with big moms making better moms.

Once hominid mothers became more terrestrial and traveled farther afield from their usual escape routes (into trees), were big moms better able to defend themselves and their babies? Were they superior foragers, able to push aside big boulders to get at underground tubers? Better able to accommodate larger babies passing through the birth canal, and, after birth, to carry large, slow-maturing babies long distances? The bigger a mother is, for example, the more efficiently she manages a heavy burden while striding along two-legged.[34]

Listening to paleontologists ponder these questions today, I think back to the early 1970s, when in order to hear Katherine Ralls lecture I would trek to the other side of Harvard Square, far from the Biology Labs (where mainstream evolution was taught), to the Radcliffe Institute—at that time a unique forum for women scholars. Ralls's enthusiasm for the "big mother hypothesis" was infectious. I can still recall the undisguised glee with which Ralls (who like me is tall) used to rattle off World Health Organization statistics on the correlation between height and easier, safer childbirth. Other times, a Jeremiah-like touch of exasperation would enter her voice. "It's so obvious, why don't they see . . ." how much more to the story there is than male-male competition and sexual selection!

Flo's Metamorphosis from Martyr to Dynast

By the end of the twentieth century, the role of Flo, Jane Goodall's most endearing mother chimp, was expanded and recast. Flo's evident tenderness and patience were only part of the story about her success as a mother. If in this book I fail to stress sufficiently this nurturing component, the reason is that I assume it is already well known, widely described, and commonly assumed. But there are secrets to Flo's reproductive success that are less well known, less often noted. These include Flo's ability to carve out for herself a secure and productive territory deep within the boundaries patrolled by the Gombe males. Many of these males were former sexual consorts; others were her own sons who had risen to a high rank in the fluctuating local hierarchy. Flo was as secure as a female chimp could be from outside males who from time to time would raid her community and, if they could, kill not just unrelated infants but adult males and older females as well.

But Flo did more than commandeer a productive larder and keep her offspring safe. She supported her offspring politically, permitting Fifi to translate her mother's advantages into her own. At Flo's death, Fifi parlayed her mother's local connections into the inestimable privilege of philopatry, remaining in her natal place. Philopatry (which means literally "loving one's home country") meant that instead of migrating away to find a new place to

Fig. 2.9 Growing bigger always involves trade-offs—some odder than others. Female hyenas often weigh just as much or more than males do, and have evolved to be even more aggressive. Being big helps a female to compete at carcasses and to discourage other hyenas from eating her babies. The high levels of circulating androgens that make this possible have led to masculinization of the female's genitalia. Her clitoris looks like a long penis— through which she gives birth. A typical mammalian birth canal extends from the cervix through the pelvis to the vagina. The hyena's is twice the usual length for a mammal of her size and makes a 180-degree turn. Because the clitoris must stretch to accommodate a four-pound fetus, labor takes hours. Mortality is very high among primiparous (first-time) hyena mothers. Up to 60 percent of infants born to primaparae suffocate while passing through the eye of this needle.[35] *(Drawing by Christine Drea. Courtesy of Larry Frank)*

live, Fifi—like half of all females born at Gombe—managed to stay where she was born. Fifi continued to use her mother's rich, familiar larder, and enjoy the protection of male kin.[36]

Make no mistake: reproductively, nothing becomes a female more than remaining among kin. Thus advantaged, Fifi began breeding at an unusually early age, and so far has produced seven successive offspring, six surviving— the all-time record for lifetime reproductive success in a wild Great Ape female. She also holds the record for shortest interval between surviving births ever reported in wild chimps. Her second-born son, Frodo, has grown into the largest male on record at Gombe and ranks in the status hierarchy just below the current alpha male, Fifi's firstborn son, Freud, while Fanni, Fifi's third-born, holds the record for the earliest ever anogenital swelling, at 8.5 years. Thus does Flo's family prosper.[37]

———

Early on, Goodall and her students noticed that when Flo approached other females, they gave nervous pant-grunts and moved out of her way. Females could be divided into those that held sway and those that gave way. What

Goodall did not immediately grasp, however, was why female rank was so important. We now know that, *given the opportunity,* a more dominant female chimp will kill and eat babies born to other females.

Over the decades that records were kept at Gombe, at least four, possibly as many as ten, newborn infants were killed by females. When Goodall reported the first two cases of infant killing and cannibalism by another mother in 1977, the so-called crimes of a female named Passion, she, like most people, assumed that the female killing these infants must be deranged. A few sociobiologists suspected otherwise and suggested that females from a more dominant lineage were "eliminating a competitor while the infant was still sufficiently vulnerable to be dispatched with impunity."

From the 1970s onward, isolated cases of infanticide by rival mothers continued to be reported for other species of social mammals—ground squirrels, prairie dogs, wild dogs, marmosets, some fifty species in all. Most cases were attributed to either a mineral deficiency or protein lust by a hungry female (since in some cases victims were eaten) or to mothers clearing out a niche and thereby making resources available for her own breeding efforts—a model first proposed by sociobiologist Paul Sherman at Cornell University. As more evidence became available, "the crimes of passion" were looking more deliberate than anomalous, and in species like chimps, other females were a hazard that mothers had to watch out for.[38]

Nevertheless, chimpanzees breed so slowly that it was 1997 before Goodall and zoologist Anne Pusey had collected enough data to show a statistically significant correlation between female rank and a mother's ability to keep her infants alive. This finding caused them to reevaluate their long-standing diagnosis of Passion's "pathological" behavior. When, two decades after the first cases were reported, Fifi's daughter attacked the daughter of a subordinate female, Pusey assumed it was a failed attempt at infanticide.

Mother chimps like Flo, then, were not simply doting nurturers but entrepreneurial dynasts as well. A female's quest for status—her ambition, if you will—has become inseparable from her ability to keep her offspring and grand-offspring alive. Far from conflicting with maternity, such a female's "ambitious" tendencies are part and parcel of maternal success.

Paradigms Widened

Darwin had set up a revolutionary new framework for understanding the behavior unfolding before us in the natural world, but it took another century to expand that paradigm to include the full range of selection pressures on both sexes. One reason it took so long to fully assess the female side of the equation was that competition between females tends to be more subtle than the boisterous, often violent, roaring and bellowing of males. Female mammals tend to confine overt competition to the spheres that actually matter in terms of status and their ability to produce high-quality offspring.

Several changes contributed to the new awareness of reproductive variation among mothers. In addition to the theoretical shift to focus on individuals, field studies were lasting longer, decades rather than months. Also, more women were doing field research. In 1875 Antoinette Brown Blackwell had lamented that "Only a woman can approach [evolution] from a feminine standpoint and there are none but beginners among us in this class of investigations."[39] A century later, 37 percent of Ph.D.s in biology in the United States were being awarded to women, and the proportion in the field of animal behavior was about the same. Although male and female researchers do science the same way, they may be attracted to different problems. The upshot of all these factors was that this time, when distaff Darwinians tapped male evolutionists on the shoulder, many of the latter were primed to respond.[40]

By the late 1980s, prominent male biologists were joining their women colleagues in pointing out the need to correct "inadvertent machismo" in their respective fields. Some of them made points similar to those Eliot, Blackwell, and Royer had tried to communicate more than a century earlier. "Research in biology," renowned entomologist William Eberhard noted, "has traditionally been carried out mainly by men rather than by women, and it is possible that, as has happened in the social sciences, research may sometimes be inadvertently influenced by male-centered outlooks."[41]

Wherever the evolution of reproductive strategies was studied, the importance of taking into account the reproductive interests of all players involved—female or male, adult or immature—was increasingly recognized. Whether in entomology, primatology, ornithology, or human behavioral ecology, researchers rushed—like prospectors in a gold rush—to seek the mother lode of new insight to be had from incorporating females' as well as males' perspectives into their research.

Scientific observation of animals living in their natural environments during the last decades of the twentieth century yielded a far more dynamic and multifaceted portrait of female nature than anything previously imagined.

Most surprising were all the ways that mothers influence their offspring's development through both genetic (including female choice) and nongenetic effects. The updated image of old Flo, for example, allows us a glimpse into the significance of such "maternal effects." Fifi, Flo's daughter, entered the world advantaged by her mother's rank, a maternal effect that pointed to ever more subtle ways—beyond genetically inherited attributes and succor—through which mothers influence the fates of their offspring. By the end of the twentieth century, the spotlight shifted so as to begin to illuminate in rigorous and controlled studies how organisms develop in specific contexts. Development would turn out to be the critical missing link in evolutionary thinking.

Underlying
Mysteries of Development

To me the Development Theory [Spencer's term for evolution] and all other explanations of processes by which things came to be produce a feeble impression compared with the mystery that lies under the processes.

—George Eliot, letter to a friend on reading Darwin's *Origin,* 1859

I suspect that many sophisticated biologists remain skeptical about selection . . . because of mysteries such as how ontogenies work.

—Richard Alexander, 1997

One reason for our fascination with Princess Diana is her Cinderella-like life story: unknown ingenue transformed into a future queen. Beekeepers routinely make such fairy tales come true just by arranging for the eggs or young larvae (less than three days old) to be fed a substance called "royal jelly."

As an egg or larva, females are totipotent, able to develop into several different forms. In the honeybee world, in which "you are what you eat," a female's lot in life—what one might think of as her class (strictly speaking, her "caste")—is determined not by her genes but by what her nurses *feed her* and by the reproductive oppression of dominant individuals. Ditto for what we might call her gender—whether or not she becomes an imperious mother or servile spinster sister.

At oviposition, the egg that will be queen is placed in a special compartment. She spends her privileged larvahood being fed a chemical concoction—royal jelly—prepared in the salivary glands of her nurses. The body of the immature, specially fed individual matures so as to differ from the ordinary worker in fifty-three different morphological and behavioral respects. Instead of becoming a sterile worker who will never produce an offspring, she blossoms into a fecund queen who will produce several million of them.[1] Two females with virtually identical *genotypes* (genetic composition at conception) look forward to two utterly different destinies. Intervening events resulting in these different outcomes constitute the underlying mysteries.

The Importance of Development

Even prior to merging of sperm with egg, even before there is anything that could be thought of as a conceptus, an embryo, or an "organism," future possibilities are being shaped by the ambient surroundings of the germ cells. In mammals, these surroundings are influenced by the mother's internal state, by nutrients in the protoplasm a mother adds to her eggs, or, as in the case of honeybees, by nutrients provided by other members of the colony.

The mysterious development of individuals, or ontogeny, includes all those complex and opportunistic emergent processes that affect how each genotype develops into the *phenotype,* the tangible properties of the organism that are influenced but never entirely determined by genes.[2] Phenotype is one of those awkward umbrella terms that began narrow, then opened up through time to cover a larger area. Today the term is still used in the original way, to describe specific ways that genes are expressed (as in a particular eye color or blood type); but phenotype is also used to refer to an entire organism, or its behavior.

The important point here is that all anyone ever sees, touches, or directly experiences is phenotypes, never genes. It is phenotypes that interface with the world and interact with others in it. Only phenotypes are directly exposed to natural selection. This is why, evolutionarily speaking, and especially for those like me who study behavior, phenotypes are what matters.

Phenotypes are produced by interactions between genes and other environmental or parental influences. They can be affected by all kinds of variables—how much cytoplasm the mother delivers in the egg, what other chemicals she adds, what time of year it is, what the mother is eating at the time, diseases she might have, even her own recent social history. This is why sociobiologist Mary Jane West-Eberhard can state so adamantly: "Nothing is genetically determined in the sense of determined by genes alone. No gene is expressed except under particular circumstances. . . . It's a kind of biological illiteracy to talk about a gene *for* anything other than a particular protein molecule."[3]

West-Eberhard is not saying that genes don't matter but rather that their powers are inseparable from context, including both external context and the developmental context, since genes act by influencing a responsive structure that is already there. This is true at every level, from immune-system defenses at the cellular level to character at the personality level. It is as absurd to talk about behavior being "genetically determined" as it is to claim that genes have nothing to do with behavior.

It is profoundly incorrect to equate "genetic" with "biological," a term that covers far more than just genetic processes. It is also incorrect to treat nature and nurture as separable entities, as in saying "The genes interact with the environment," or "Nurture does not matter." This is why it is unfortunate to hear the label "biological mother" applied to a woman who has given birth to a child and given it up for adoption, or, worse, just provided the donor egg. Such a woman is more nearly the *genetic* or *gestational* mother. By contrast to a genetic donor, the *biological* mother nourishes, nurtures, and provides the environment in which the infant develops both physically and psychologically.[4]

It is clear that genes are not puppeteers directing behavior. A range of nongenetic factors, such as mother's physical condition or social status, the season when she conceived, her own diet or the one she provided her baby, the presence or absence of father—all contribute to individualization. Parental effects encompass all the nongenetically transmitted attributes that pass from parent to offspring. Practically speaking, the mediators of such effects are often mothers. Not hereditary in any genetic sense, maternal effects can nevertheless influence the speed and course of evolutionary change, trends that sooner or later lead to changes in gene frequency—the stuff of evolution.

The dynamics of genetic and maternal effects are relatively better understood in the mother-centered worlds of hymenopteran social insects—honeybees, wasps, and ants—than they are in other animals. A finite number of chemical signals chart an individual's life course, thus permitting scientists to carry out rigorously controlled experiments showing how a specific treatment (such as feeding royal jelly) plays out during development. One of the ironies of the charge "genetic determinism" so often leveled at sociobiologists is that so many of its earliest practitioners—Edward O. Wilson, Mary Jane West-Eberhard, William Hamilton, and Richard Alexander—were also entomologists. They were acutely aware that genetics does not equal biology. They didn't call the new field sociogenetics; they called it sociobiology—for a reason.

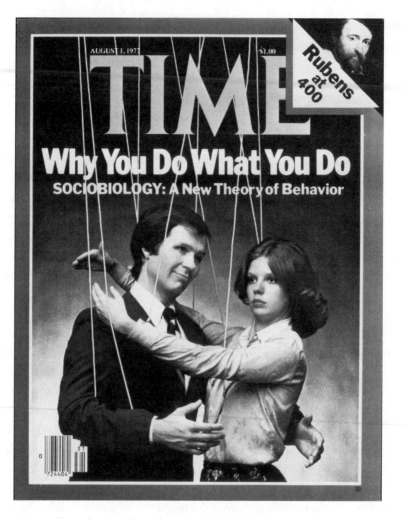

Fig. 3.1 Genetically orchestrated couple on strings dance woodenly across the August 1, 1977, cover of *Time*. The magazine promises to tell how "a new theory would explain" just "Why You Do What You Do." Images of genes controlling people like puppets are more often invoked by critics of sociobiology than its practitioners. (© 1977 Time Inc.; *reprinted by permission*)

Mother-Centered Worlds

Nineteenth-century evolutionists knew that traits were inherited, but they did not know of the existence of genes or understand how they worked. They lacked a way of conceptualizing the complicated relationship between inherited traits and alternative outcomes, or phenotypes. As late as the 1950s most animal behaviorists still took it for granted that relatively brainless, culture-less creatures like honeybees were born to function instinctively in a nar-rowly specified, or species-typical, way. Workers were predestined from birth to serve the queen and maintain the efficiency of the hive.

In 1894, Darwin's associate Thomas Henry Huxley could write confi-dently—in what he considered a progressive statement—that the "vast and fundamental difference between bee society and human society" was that bees "are each organically predestined to the performance of one particular class of functions only," while among men "there is no such predestination." Among men "it cannot be said that one is fitted by his organization to be an agricultural laborer and nothing else, and another to be a landowner and nothing else."[5]

No modern sociobiologist would disagree with Huxley's assessment that each human individual is born with variable potential. But most would emphatically disagree with Huxley's assumption that the lot of a hymenop-teran insect was quite so narrowly predestined. Far from strict destiny—a direct equation of genotype with phenotype—a honeybee's gender is merely a potential. Even in an organism born so mindless as a bee, a creature who learns remarkably little in the course of her life, a female has the potential to become *either* a worker *or* a queen, depending on the type of nurture she receives. Even whether or not a worker remains sterile or takes a stab at lay-ing eggs turns out to be negotiable.

Gender, Relatedness, and Caste

The reproductive subservience of worker castes is not quite so voluntary as believed. The honeybee queen manufactures a special "queen substance" in her mandibular gland that broadcasts an imperious olfactory message inform-ing workers of the hive: "Develop your ovaries and you're dead!" The hor-monal signals (or *pheromones*) that the queen uses to broadcast this message are derived from ancient hormones emitted by one insect to threaten another in the course of female-female competition.[6] In response to peremptory pheromonal signals passed bee-to-bee during food exchange, the workers' ovaries shut down. Yet occasionally—in spite of all this propaganda—a worker may attempt to lay eggs. But her efforts are usually in vain. Her eggs will most likely be cannibalized by other females who detect them.

Ovarian despotism by dominant females has been especially well studied in the genus *Polistes*. These hornet-like wasps range throughout North America and down to Central America. They sting like fire. Fortunately, though, many species are easy to spot due to conspicuous black, yellow, and burnt sienna body bands. If it is summer outside, paper wasps are probably, at this moment, busily constructing parchment-like nests of chewed wood pulp in the eaves of your building.

In an ingeniously simple manipulation, Mary Jane West-Eberhard—who for many years has studied the tropical paper wasps near her homes in Colombia and Costa Rica—tethered a reproductive female some distance from where her eggs had been laid by tying a slender nylon thread around the wasp's waist. As soon as this dominant female was prevented from aggressively defending exclusive access to the nest, the previously suppressed ovaries of her daughters revved up and they began laying eggs.[7]

Seemingly utopian, the paper wasps' society is more nearly an ovarian police state. This does not necessarily mean there is no future to unauthorized fecundity. Some Argentinean ants give destiny a helpful nudge by assassinating the dominant female, usurping her breeding prerogatives for themselves.[8] More often, however, the better part of valor for a worker in these mother-centered, mother-dominated societies turns out to be helping their foundress—or, once the colony gets going, their sister—to rear her offspring.

Even in honeybees, which, most would agree, do approximate buzzing automatons, genes do not determine outcome in life decisions as major as whether to become a mother. Rather, genes set limits on a range of developmental outcomes, which are very few compared with the situation in humans, where the range of outcomes is enormous—albeit still not infinite.[9]

Genes, with all their limitations, nevertheless play a very special role in the puzzle posed by highly cooperative breeding colonies of social insects. If all living things strive to reproduce, as Darwin theorized, how could one explain the dedication of the altruistic worker bees who will never reproduce and transmit genes to future generations? This challenge to Darwinian theory yielded to an ingenious solution proposed in 1963 by British geneticist William Hamilton. This reserved and self-effacing young scientist came up with a bold idea—selection at the level of kin—to explain the altruism of the queen's sterile attendants.[10]

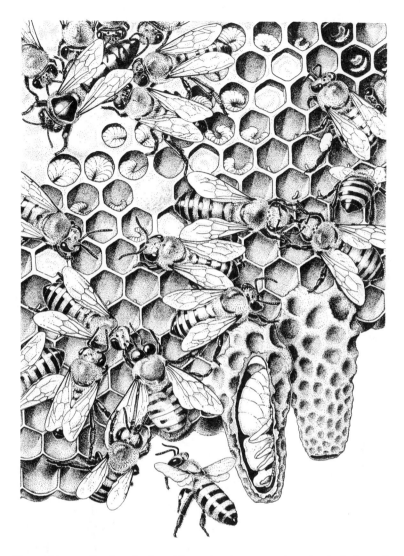

Fig. 3.2 Eusocial insects live in colonies with overlapping generations that include sterile, nonreproductive castes. Here a honeybee queen is surrounded by her worker-bee daughters who forage for pollen and then regurgitate it as nectar into the mouths of other workers. These workers add special enzymes to produce honey before storing it in the hexagonal wax cells of the honeycomb. The queen lays up to 2,000 eggs a day, tended by these workers.[11] Only one in tens of thousands of females ever becomes a mother herself. *Lower left:* One of the queen's daughters drags away a drone by his wings. Males in this world are the odd men out, disadvantaged because they are more distantly related. If entomologists got their degrees in humanities departments, perhaps we would have libraries full of dissertations analyzing "gender, relatedness, and caste"—honeybee style. *(Reprinted by permission of the publisher from* Insect Societies *by E. O. Wilson [illustration by Sarah Landry], Harvard University Press, Copyright © 1971 by the President and Fellows of Harvard College)*

Hamilton's Rule

The civic-mindedness of sterile workers earns honeybees the utopian designation *eusocial* (or, "perfectly social"), which applies to any society with overlapping generations devoted to the cooperative care of immatures and characterized by specialized reproductive and nonreproductive castes. Although young are produced in great quantity, it is the quality of the care that enables so many to survive. Busy workers spend the first three weeks of their short lives in the hive, tending their sister's young, and their next (and final) three weeks foraging for nectar in the riskier world outside.

To explain this world, Hamilton drew on his knowledge of the special reproductive attributes of social insects. He proposed that hymenopteran social insects so often put the colony's interests ahead of their own because of an especially close degree of genetic relatedness between the workers and the queen. This comes about because of an odd biological circumstance by which males have just one set of chromosomes (haploid), while females have two sets (diploid), so that organisms such as wasps engage in "haplodiploid" reproduction. In haplodiploid organisms, two sisters with the same father will share more genes in common than a mother shares with her own offspring.[12]

Primitive as they may seem, even insects have hidden zones of ovarian decision-making. Once a honeybee queen or a reproductive wasp mates, she stores the sperm in a special pouch called a *spermatheca*. When she lays an egg, she has the option of opening a valve, permitting sperm to fertilize it as it passes through her reproductive tract. A fertilized egg with two sets of chromosomes (diploid) develops into a daughter. As with most sexually reproducing animals, the resulting daughter receives half of her chromosomes from her mother, the other half from the male with whom the queen mated. But if the queen withholds sperm, something unusual happens. The unfertilized egg develops anyway, but it develops into a haploid individual with only one set of chromosomes, derived entirely from her. Haploid eggs always develop into sons. Since any male the queen mates with is haploid, this creates a peculiar skew for her female offspring, such that sisters are especially closely related. This is why the genetic payoff for a worker investing in the queen's offspring is greater than if she produced her own. Male honeybees don't have this same especially close relationship to the queen's offspring and also don't meet this same test of citizenship.

These drones, or "winged sperm dispensers" (as Ed Wilson terms them), live only long enough to mate and then die. After reaching adulthood, they spend a few days in the nest before taking off for their big (also final) moment on the mating flight. Nonmating males are either driven out of the colony or killed.[13]

Instead of focusing on the sterile worker's genetic representation in the next generation—which would be zero—Hamilton expanded the concept of an individual's lifetime reproductive success (or *fitness*) to include the *inclusive fitness* of the individual. By inclusive fitness Hamilton meant the effect that the female worker's behavior has on her own fitness *plus* the effects her behavior has on the fitness of close kin who share genes by common descent. Using this principle, Hamilton derived simple mathematical expressions predicting that altruism should evolve whenever the cost to the giver (which he designated C) was less than the fitness benefits (B) obtained by helping another individual who was related by r, a term designating the proportion of genes these two individuals shared by common descent.[14]

Hamilton's deceptively simple-looking equation $C < Br$ underlies the evolution of helping behavior in all social creatures. The rule together with the general theory behind kin selection were almost immediately confirmed by West-Eberhard for wasps,[15] and soon after for many other animals.[16] At an ultimate level, kin selection explains the universal human pattern of favoring kin. In humans different beliefs and customs underlie these patterns, but the outcome is everywhere the same: kin preferred to nonkin.[17] Indeed, as we will see, many unexpected features of maternal behavior can be understood as special cases of Hamilton's rule.

No gene or set of genes, or even any one mechanism influencing people to favor kin, has been identified. We do not know even a fraction of the ways that kin selection works. Yet wherever biologists or anthropologists have looked, animals, including people, behave *as if* there were such genes. One way or another (and, as I say, nobody understands how) all social creatures have through evolutionary time—probably in different ways—internalized Hamilton's rule.[18] In humans we can only assume that our powerful predisposition to prefer our own kin derives from very ancient emotional and cognitive systems, such as learning to recognize people familiar from a very early age and having a lower threshold for altruism in our behavior toward them. This is the simplest explanation for our similarities with other social creatures in this respect.

As Hamilton expressed it:

> [In theory] a gene causing altruistic behaviour towards brothers or sisters will be selected only if the behaviour and the circumstances are generally such that the gain is more than twice the loss. . . . To put the matter more vividly, an animal acting on this principle would sacrifice its life if it could thereby save more than two brothers, but not for less.[19]

And this is where the matter has stood for many years, the emphasis in "kin selection" on the close relatedness of the actors.

However, not all social insects with remarkably cooperative breeding systems have this kind of special haplodiploid reproduction. (Termites, for example, do not.) For this reason, attention has begun to shift to the other components of Hamilton's initial equation: the ratio of costs and benefits to actors. The honeybee queen, recall, grows up to be a specialist in egg-laying. She is a super-mother in a class by herself, a female of enlarged ovaries, able to lay an egg a minute, day and night, for up to five years. Her worker sister, on the other hand, even if she manages to produce some eggs, has severely limited prospects of rearing them. How much, then, does a sterile worker actually give up by altruistically helping her mother or her sister reproduce, accepting a fractional interest in millions of eggs instead of laying a few ill-fated ones herself? What are the costs in relation to benefits, given the females' degree of relatedness?

By themselves, the peculiarities of haplodiploid genetic systems do not fully explain why ants, wasps, bees, termites, and other eusocial insects must be counted among nature's longest-lived and most fecund success stories. Something else is needed to explain 140 million years of eusocial prosperity. We need to keep in mind Mother Nature's cardinal rule for mothers: It's not enough to produce offspring; to succeed through evolutionary time mothers must produce offspring who will survive and prosper. In short, we need to consider the importance of what I think of as "the daycare factor."

In an unrivaled reproductive success story, expeditions of leaf-cutting and harvester ants blaze trails across the forest floor, while battalions of army ants terrorize mammals in their path. Bees and wasps dot trees with their nests, and termites infest rotting wood. One-third of the animal biomass of the Amazonian rain forest teems, climbs, and swarms with billions upon billions of these social insects.[20]

The secret to their success is, quite simply, the most dedicated and efficient daycare in the biosphere. So what if some army-ant queens can lay up to two million eggs? A woman starts out her life with more than three times that many egg cells. It's not the insect queen's fecundity that is so special, it's her success rate translating eggs into adult survivors. What makes social insects so amazing is the dedicated assistance of all those allomothers. Even if the mother dies, so long as the colony persists, her progeny will be cared for.[21] It is a mother-centered world geared toward one aim: the survival of progeny.

Controlling Mothers?

She's a real Queen Bee! We use the term, often with a tinge of disapprobation, to describe a despot, a figure in charge. It's one of those metaphors that on closer inspection is more apt than people realize. But even without a queendom, some solitary wasp mothers who do not found large breeding cooperatives—like the fig wasp mother who breeds alone—nevertheless manage to exercise remarkable control over their posterity. Their power derives from their ability to predetermine the sex of each offspring.

William Hamilton showed how a solitary mother fig wasp ruthlessly manipulates her progeny in ways that suit her long-term reproductive interests. As the female lays each egg, she either fertilizes it or not, thus determining the exact configuration of daughters and sons, which she can translate into the greatest number of grand-offspring. Out of a batch of 257 eggs, one mother produced 235 daughters and just 22 sons. To explain this wildly female-biased sex ratio, Hamilton devised a theory based on local competition for mates, generally referred to as "local mate competition."

Local mate competition? What could a mother's production of sons versus daughters possibly have to do with competition to breed? Normally not much, not in outbred creatures like ourselves who avoid mating with full siblings. But in the incestuous world of the fig wasp, the number of daughters for every son matters a great deal. The wasp mother's brood will be born, and breed, right there within the fleshy pink confines of the fig. "*Local* mate competition" is an understatement. Brothers born just a hairsbreadth away from one another wait outside the nursery until the sisters hatch, then use their

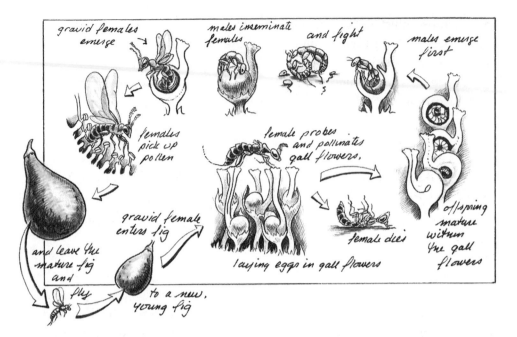

Fig. 3.3 Some Hymenoptera are social, but most are solitary. Here a gravid wasp tunnels into a fig and lays her eggs in gall flowers, where they will mature. Such males as the mother wasp produces emerge first and fight among themselves to gain access to females, who emerge shortly after. The newly inseminated females then struggle clear of the fig where they were born and fly off to find a new one. In the process, pollen from the first fig is transported to a new fig that will serve first as a nursery and next as a seraglio to her progeny, then as a tomb for the mother and her few sons, and finally as an embarkation point for the gravid daughters, who will decamp to begin the cycle anew. It is a female-centered world demographically structured to suit maternal reproductive interests. *(Drawing by Sarah Landry from Hrdy and Bennett 1979, reproduced by permission of* Harvard Magazine*)*

enormous jaws to dismember one another. The victorious male gets to mate with his emerging sisters. "Their fighting," Hamilton recalls,

> looks at once vicious—cowardly would be the word except that, on reflection, this seems unfair in a situation that can only be likened in human terms to a darkened room full of jostling people among whom, or else lurking in cupboards and recesses which open on all sides, are a dozen or so maniacal homicides armed with knives. One bite is easily lethal.[22]

When Hamilton spent a year in Brazil doing fieldwork on fig wasps, he kept a special vial labeled C-A-R-N-A-G-E to collect the body parts of males. He estimates that a million sons are murdered each time a fig tree fruits. Such is the stage the mother sets. And for what?

Within this lusty microcosm, custom-made for incest, the mother's chosen sex ratio is the one that makes the most efficient use of all the bodily resources she has to allocate to reproduction. She produces only as many sons as are needed to fertilize her daughters. Readers of A. S. Byatt's allegorical fantasy *Morpho Eugenia,* about a Victorian country estate eerily reminiscent of a social insect queendom, will be relieved to know that the theme of brother-sister incest does not derive from any special information she possesses about the sordid underside of domestic life in Victorian households (although, who knows?) as much as from the author's knowledge of Hymenoptera. In the book's film version, *Angels and Insects,* the matriarch's daughters wear fabulous Hymenoptera-styled ball gowns, complete with wasp-waists and flamboyant yellow and black stripes.

––––––

By the end of the 1970s, manipulative mother wasps had upended the stereotype of a passive, nonstrategizing, "egg-laying" machine—at least among the entomological cognoscenti. Previously, the idea of mothers manipulating the sex of their offspring, or controlling the reproductive careers of other females, seemed more like science fiction than science. But natural history was once again proving stranger than fiction.

An arcane subfield known as "sex ratio theory" arose within sociobiology to deal with the complexities of mothers who bias production of offspring toward either sons or daughters, and of parents who bias investment toward offspring of one or the other sex—a preoccupation among human parents as well (see chapter 13, below). The study of biased investment in sons and daughters illustrates what Hamilton refers to as his own "perverse, unsexy, yet fundamental (geneticist's) angle" on reproduction.[23]

Confirmation from the Jewel Wasp

Within decades, what at first seemed to some almost delusional speculations about adaptive control of sex ratios by mothers yielded spectacularly precise science. The organism whose behavior would confirm the validity of Hamil-

Fig. 3.4 William D. Hamilton describes the incestuous microcosm inside a fig, while evolutionary theorist Robert Trivers looks on. Hamilton's 1967 article "Extraordinary Sex Ratios" explained why the mother fig wasp produces mostly daughters. I used to attend their seminars, bringing my infant daughter with me. Since she was asleep in a canvas carry-all, my hands were free to take this photograph. *(Sarah Blaffer Hrdy / Anthro-Photo)*

ton's theory of local mate competition in every detail, the jewel in this crown, was an unlikely candidate in all but name. *Nasonia vitripennis* is a tiny parasitoid wasp, smaller than a fruit fly, with the unsavory habit of laying eggs on the pupae of blowflies laid under carcasses and in birds' nests. This parasite upon a parasite is commonly known as the jewel wasp.

The parasitic jewel wasp, turns out to be, in the words of biologist John Werren, "a consummate artist at controlling the sex of [her] offspring." Similar to Hamilton's fig wasps, jewel wasp mothers locate a blowfly pupa and lay eggs, most of which will hatch as daughters, with just enough sons—perhaps 15 percent of the eggs—to inseminate them. Unattractive as their housing requirements happen to be, space is in short supply. What happens, John Werren wondered, if the mother arrives at her host, injects her stinger (which is also a sensory organ) into the mush, only to detect chemically that another mother got there first and had already deposited *her* eggs? At that point, this family-planner par excellence inserts only a single, unfertilized (and therefore male) egg. Her son will hatch into a world full of opportunities: he will join the fray with sons of the first female, competing to copulate with her daughters.

Yet even a mother so much in control as the jewel wasp rarely has the last word. Werren discovered a "parasite" upon this calculating parasite. About 10 percent of jewel wasps carry a particular virus-like gene known as "the paternal sex ratio element." If the male the mother mates with carries it, that mate transmits it to her in his sperm. This parasitic gene destroys the paternal chromosomes in all the eggs that she fertilizes, converting all diploid eggs into haploid ones. The fertilized eggs that normally would have developed into daughters become sons, the only sex host capable of transmitting the parasitic gene. This parasite upon a parasite upon a parasite could theoretically cause jewel wasps to become extinct by artificially producing an all-male population.

But Werren, with the geneticist's optimism that every dilemma is only a mutation away from some sort of solution, chose to look on the bright side. Instead of predicting extinction, he quotes Jonathan Swift:

> So, Nat'ralists observe, a Flea
> Hath smaller Fleas that on him prey;
> And these have smaller fleas to bite 'em
> And so proceed ad infinitum.[24]

———

By the 1970s, then, entomologists exploring cooperative infant rearing, maternal manipulation of sex ratios, and suppression of ovulation were not just discovering new dimensions to being female; they were uncovering new dimensions to *individuality* that had to do with development. Hamilton's rule provided sociobiologists with a universal truth: it applied to all social organisms, *all other things being equal.* But when are all other things ever equal? Especially in a formula that has built into it functions like "cost to an organism" and "benefit." It's impossible to consider these without reference to the environment in which organisms develop, the age and condition of the individual, and constraints imposed by others in that environment.

Maternal Effects

For species such as primates, the mother *is* the environment, or at least the most important feature in it during the most perilous phase in any individual's existence. Her luck, plus how well she copes with her world—its scarcities, its predators, its pathogens, along with her conspecifics in it—are what determine whether or not a fertilization ever counts.

What mothers are and do can facilitate or impede adaptation to new conditions, impart to immatures a mother's own immunological defenses (through lactation) or otherwise give youngsters a boost. These head-start programs can begin even before fertilization (see Plate 1).

During the late seventeenth century, scientists thought they saw a miniature man, a little "homunculus," through their microscopes, folded up inside a human sperm, waiting to be deposited inside the womb. Even after 1827, when embryologist Karl Ernst von Baer provided a more accurate description of the mammalian egg and convinced his colleagues that miniature humans were not planted ready-made into the uteruses of women,[25] it continued to be assumed for another century that males alone directed the course of evolution. Even though mothers contributed egg cells, they were viewed as passive vessels, awaiting the life force conveyed by males.

But this, too, was not quite right. Rather than being penetrated by a sperm, the egg (or oocyte) more nearly engulfs it, quite possibly selecting which sperm to accept, and producing specific chemicals that are necessary for fertilization to take place. The sperm cell is almost pure nucleus; the oocyte contains several ingredients—nucleus and cytoplasm. Once the sperm is inside the egg, maternally transmitted instructions go to work. Nutrients stockpiled prior to fertilization supply the needs of the developing embryo. In particular, the mother's oocyte is derived from cells that, even prior to fertilization, have begun dividing. Prior to any contact with the sperm, the maternal germ cell has divided four times, into sixteen cells. One of these continues on as the oocyte. The others become "nurse cells," which manufacture nutrients and other materials that will be transmitted through the cytoplasm.[26]

This means that early embryonic development is under maternal control before the father's genes, carried by the sperm, are even activated. At the outset, the egg's acceptance of a sperm launches maternal effects. Protoplasm from the mother sets up the embryo for development, prelude to many possible maternal effects.

One of the strangest and least anticipated maternal effects ever described has to do with just such special ingredients transmitted by the mother to the cytoplasm in her eggs. It is a case that belies all stereotypical expectations about maternal virtue, defying the conventional expectation that a

Fig. 3.5 Drawing of "homunculus" from Nicolas Hartsoeker's
Essay de Dioptrique, 1694.

"madonna" ought to make a more suitable mother
than a "whore." In this instance, it is the *femmes fatales*
who make the best mothers.

Imagine flashing lights blinking on a sultry night.
But these lights are not inviting summer vacationers
to visit discos. The strobe effect emanates from lumi-
nous, phosphorescent organs on the abdomens of
Photuris fireflies. These female fireflies emit chemically produced flashes of light that mimic the mate-
attracting signal of another species, a type of firefly
belonging to the related genus *Photinus,* in which
females really did evolve to signal readiness to mate
by flashing and males evolved through sexual selection to seek them out when they did. But when an
eager *Photinus* suitor shows up, the alluring *Photuris* female eats him instead of
mating with him.

The *Photuris*-mother-to-be gets more than a meal out of this male. She also
gets his armor, since her victim has the unusual capacity to manufacture
defensive steroids that make him unpalatable to birds and predatory spiders.
The mother promptly passes this chemical protection on to the eggs she is
laying, endowing them with her chemical booty.[27]

———

Such cases are the stock-in-trade of those sociobiologists like Mary Jane
West-Eberhard who focus on development. To her, individualization begins
as a maternal effect. "An animal egg or a plant seed is already a highly organized and active phenotype before it is fertilized." She entreats us to consider
the beginning of a frog's life. Hours after fertilization, with the fast-dividing
blastula (the early development phase of an animal) already 4,000 cells
strong, none of the embryo's own genes have been activated. The only
instructions to be had are from hormones and proteins circulating in the

Fig. 3.6 Female *Photuris* fireflies mimic the sexually selected mating signal of another species,
Photinus. When unsuspecting *Photinus* males arrive
to mate, the deceptive females eat them, ingesting
their defensive chemicals, which are passed along
to their offspring. Thus, through the trickery of
their mothers, *Photuris* offspring enjoy an
increased chance of surviving to adulthood.
(Courtesy of Thomas Eisner)

cytoplasm. Far from genetically determined, initial development of this new
individual, with its "hand-me-down phenotype," is very much influenced by
maternal condition, her nutritional status or life history. This is what West-
Eberhard means when she scoffs that "The bare genes are among the most
impotent and useless materials imaginable." Thus the phenotype of the early
embryo is determined by the mother alone. This represents a maternal effect
undreamed of before the closing years of the twentieth century.[28]

West-Eberhard has been foremost among those working to integrate
behavioral plasticity in both sexes into evolutionary theory. What fascinates
this wasp specialist is the extent to which genetically similar individuals can
be shunted into different pathways of development according to conditions
encountered early in life. The identical genotype (or at least genotypes that
are very similar, as in full siblings) could develop into an organism that looks
or behaves very differently (that is, exhibits a different phenotype).[29]

The phenomenon of environmentally cued alternative phenotypes within
the same population is known as *polyphenism* (i.e., same genotype produces
more than one phenotype). Long overlooked, polyphenism, the outcome of
so many underlying mysteries, is assuming greater importance in the thinking
of geneticists. Anyone tempted by cascading research that identifies genes
"for" particular traits would do well to keep these cases in mind, as reminders
of how much context still matters.

Catkins or Twigs

The reason all the best examples of polyphenism derive from plants and
insects rather than vertebrates is purely practical. To obtain unambiguous
experimental results requires the experimenter to rear identical individuals

under different conditions. Distinctive life-forms (or morphs) found in easy-to-manipulate insects, together with their short lifespans, means that study subjects can grow up, breed, die, and yield definitive results quickly—before funding to study them runs out.

My favorite example comes from caterpillars belonging to a species of geometrid moths (*Nemoria arizonaria*) that breed in oak woodlands across the American Southwest. Entomologist Erick Greene used these caterpillars to demonstrate how different diets early in life produce utterly different morphs—organisms as different as two species. In the process, Greene showed how peculiar contingencies of a mother's existence—whether she gave birth early in the season or later—factor into the shapes her offspring must assume to survive.

In the case of the geometrid moths, mothers hatch two broods of caterpillars each year. In nature, spring broods feed on the protein-rich pollen of the oak's drooping flowers, called catkins. Long after these kittens' tails (their name derives from the Dutch diminutive, *katte,* precisely because of this resemblance) have dropped from the trees, the second (summer) brood of caterpillars hatches. Since the catkins are gone, all that is left for summer caterpillars are tough, mature oak leaves, laden with tannins, which are poisonous compounds produced by oaks to discourage nibblers. But in a world where caterpillars are what they eat, these tough leaves are just the ticket.

Whereas pollen-eating grubs metamorphose into knobby, wrinkled caterpillars that resemble oak stamens, looking to all the world (especially to hungry birds that prey on insect larvae but not plants) like drooping catkins, later-born morphs are gray-green, less knobby, and utterly twiglike, blending in with their leafy dinner and once again fooling predators. High levels of tannin from the leaves (or something associated with them) trigger the development of this twiglike morph.

Greene's elegant experiments showed that the pathway taken by the genetically coded developmental program is triggered by what the caterpillar eats in the first three days. If early broods eat fibery leaves instead of pollen, they, too, come to resemble twigs.[30]

The nutritionally superior catkin diet permits spring broods to attain a larger size by the time they pupate, to mature faster, survive better, and (once they become moths) to be more fecund breeders. Despite the disadvantages of being born late, caterpillar lines that failed to produce summer broods miss out on the opportunity to breed twice in the same year.

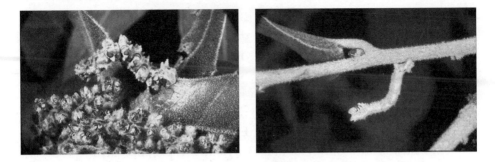

Figs. 3.7a and b When Erick Greene experimentally fed full sibs of the caterpillar *Nemoria arizonaria* different diets, two different morphs developed. Spring and summer broods look the same when they first hatch, but subsequently the early-born (spring) broods feed on oak catkins and grow up to look like drooping flowers. Later-born (summer) broods subsist on leaves and develop into alternative morphs camouflaged as twigs. If summer broods were artificially fed out-of-season catkin meals, they would stand out like solitary kitten's tails within an inland sea of twigs and leaves, easily spotted by predators. *(Reprinted with permission from* Science *243:644.* © *1989 American Association for the Advancement of Science)*

Alternative Outcomes of Development

Genetically identical individuals can grow up to be very different—that is, to have very different phenotypes—depending on circumstances encountered early in development. These flexible phenotypes result in different "morphs" or types of individuals. Simply put, in varied and unpredictable worlds there will be more than one way to survive and reproduce. Through the course of development, individuals adopt alternative strategies, manifested either in their morphology and physical appearance or in their behavior. Resulting phenotypes depend on circumstances, on which genes or receptors are switched on, which cellular and bodily responses triggered. Alternate phenotypes, or ways of being, are coded right into the genetic constitution (or genome) of the same individual.

Polyphenism, with its multiple developmental courses, is too useful a concept to confine to "simple" creatures like wasps and caterpillars. Increasingly, biologists are aware that mammals—including primates like ourselves—can develop along different pathways, even assume different forms or exhibit quite different behavioral profiles, depending on what developmental track they find themselves on. However, the underlying mysteries in large-bodied, socially complex, and long-lived organisms are far harder to pin down experimentally, and none of the cases could be so well documented as in the honey-bees and caterpillars. Consider the "Peter Pan" orangutans.

Researchers engaged in long-term studies of orangutans in the wild have long been puzzled by the curious case of males who never seem to grow up. The "Peter Pans" are so different from full adult males that the legendary naturalist Alfred Russell Wallace (the codiscoverer of Darwin's theory of natural selection), on encountering one, assumed he belonged to a different species. Various biologists since have made the same mistake.

The two orangutan body types (or morphs) are characterized by utterly different patterns of growth and reproduction; year after year, the same males get classified as adolescents—in some cases, for as long as twenty years. But if, one day, the dominant male disappears, the Peter Pan male undergoes a transformation: within months his face fills out, his hair grows, and he accumulates bulk. Abandoning his low profile for the life of a bully, it is Peter Pan's turn to patrol the forest like a quarrelsome troubadour in quest of a maiden, uttering deep roars and fighting any other adult male he meets.[31]

Primatologists Peter Rodman and Biruté Galdikas, who study orangs in the forests of Borneo, have described the low-cost, low-benefit mating strategy pursued by undersized, adolescent-looking males who skulk about females and copulate with them even though they are not sexually receptive. (Galdikas labels this the "sneak/rape" strategy, the only thing approaching rape in a primate other than humans.) Such males are seemingly unselective, attempting to copulate as often as possible, even at times of her cycle when the female is unlikely to conceive.

By contrast, a full adult male is more discriminating and concentrates on ovulating females. Such a male fiercely defends access to one, and fights to the death to drive rival males from her vicinity, thus maximizing his chance of being the father of her next offspring. This "combat/consort" strategy is far more costly than the sneaker's tactics in terms of risk to the male from combat. Furthermore, the adult male's discriminating standards mean that big males copulate only rarely (ovulating female orangs being an exceedingly scarce commodity in these highly dispersed and slow-breeding apes). Nevertheless, such copulations as the consorting big males do obtain are more likely to culminate in conception.

West-Eberhard was so impressed by the evolutionary possibilities of this kind of variation—far more common than generally realized—that she sug-

Figs. 3.8a and b The orang on the left is a "developmentally arrested" Peter Pan male. On the right is a full adult male with beard and full cheek flanges. He has much higher testosterone levels, and exudes a musky odor. Developmentally arrested males maintain a low profile and attract less aggression from dominant males. But as soon as the locally dominant male is removed, the Peter Pan male grows up, develops protruding cheek flanges, and emits long calls, turning into the very model of a Darwinian male who (as described in the *Descent of Man*) "expends much force in fierce contests with his rivals, in wandering about in search of the female, in exerting his voice, pouring out odoriferous secretions, etc." *(Photos by Jessie Cohen, National Zoological Park,* © *Smithsonian Institution)*

gested organisms may use multiple morphologies or lifestyles (say, eating one food rather than another) to "experiment" with new niches. If the trial run proves successful, and animals pursuing this new lifestyle survive and reproduce better, then new evolutionary opportunities are opened up. For example, a population of caterpillars could conceivably evolve to specialize in eating leaves high in tannins all the time. Or (to really engage in science fiction), if forest fires continued to burn in Indonesia and food was chronically short, selection might favor a Peter Pan morph who was inclined to never grow big.

Multiple phenotypes provide natural selection an opportunity to either favor or penalize genetic combinations that predispose animals to live some novel way. Such phenotypic flexibility means that evolution and speciation can occur at a faster pace than would otherwise be possible.[32]

Memes and Other Special Maternal Effects
In terms of evolution, some of the most stunning maternal effects are produced by information about the world communicated by a mother to her infant. Such information can be transmitted chemically (experiments with rats show that food choices later in life are influenced by molecules in

mother's milk) or through cultural concepts, which is possible only in species endowed with language and symbolic reasoning. Though there may have been other hominids so endowed in the past, *Homo sapiens* is the unique possessor of these capacities today.[33]

The hand that rocks the cradle rarely controls the world. But the voice that sings the lullabies and barks cautionary messages in the first years of life provides critical information about the social niche into which the child has been born. Such experiences can have a lasting effect upon his mental and emotional outlook. Through her example and direct teaching, a mother shapes critical assumptions about how the world works, what there is to eat, who there is to be afraid of, who is likely to be well-disposed, and so forth— myriad units of culturally transmitted information, or "memes."[34]

Human self-images and beliefs are not frozen and continue to change through life as individuals (active agents in their own right) encounter new social opportunities and constraints. But the fact that immature humans are so impressionable has evolutionary consequences out of proportion to the brief time period when immatures are intimately exposed to their mothers and to her immediate circumstances, or "local history."

A distinguished roster of evolutionists (including Ernst Mayr, John Emlen, George Williams, Edward O. Wilson, and Richard Dawkins) have all commented on the extraordinary gullibility of our species, especially when we are young. Call children gullible, or "learning ready," but their spongelike aptitudes function to spare small and vulnerable creatures the fatal costs of learning through trial and error. "Don't go near the water," and especially "Don't tease the saber-tooth tiger," are the examples that came to the grand-fatherly mind of George Williams.[35] One reason television is such a perilous medium is that even infants less than two years old imitate what they see on the screen, yet what appears there is determined by what happens to appeal or to sell rather than by what behavior helped individuals in a particular past environment to survive or prosper.

Few geneticists question the importance of maternal effects or early learning since they know that the course of evolution (used here to mean changes in gene frequency) can be altered by nothing more substantial than a powerful idea acquired early. A Hutterite daughter who imbibes Anabaptist doctrine along with her mother's milk is more likely to grow up to bear ten children (the average for her group) and be the least likely of any woman in any population ever studied to die without surviving offspring.[36] Meanwhile, another little girl down the way, who grows up convinced of Christ's imminent second coming, and who as a consequence joins a religious community such as that of the celibate Shaking Quakers, decreases her odds of bearing any children at all.

In part II, I will return to what is the most important of all maternal effects in terms of infant survival: a mother's decisions about how much to invest in her offspring, and in some cases even whether to nurture her infant at all. In part III, I speculate about the significance of maternal commitment for what the developing human infant learns about its social environment.

––––––––

From 1975 onward, sociobiologists began to incorporate situation-dependent phenotypes and maternal effects, along with natural selection, kin selection, and sexual selection, into our understanding of evolution. "Looking to the animals" in this new way made it inevitable that sooner or later mothers would be recognized as playing active and variable roles on the evolutionary stage. But other factors, including new protagonists among the theory builders, sped up the revision. An explosion of field studies by an increasingly diverse group of researchers in animal behavior and human behavioral ecology unveiled previously unimaginable variation in the natural history of mothers.

Unimaginable Variation

*If there were one level of feminine incompetence as strict as the ability
to count three and no more, the social lot of women might be treated
with scientific certitude. Meanwhile the indefiniteness remains
and the limits of variation are really much wider than anyone would
imagine from the sameness of women's coiffure and the favourite
love-stories in prose and verse.*

—George Eliot, 1871–72

*The most significant impact of this new [evolutionary ecological]
thinking was in its focus on variability . . . in how parents behave
and how children fare. . . .*

—Jane Lancaster, 1997

Every female who becomes a mother does it her way. From an evolutionary perspective, what mothers have in common is their high and quite certain degree of relatedness to each infant. What varies are the costs that caring for a particular infant will impose and the potential payoff in terms of that offspring's prospects of translating her investment into subsequent reproductive success. So far as natural selection is concerned, mothering is anything and everything a female does to ensure genetic representation in subsequent generations. Narrower prescriptions implying that every mother would be a fully committed, "loving" mother were just somebody's wishful thinking.

When sociobiologists followed the advice of early moralists by looking "to the animals," they did so not in search of moral guidance but to learn *why* creatures behave as they do. Instead of natural laws demonstrating how mothers *should* behave, nature yielded a series of contingent statements. Whether or not a female produces offspring depends on her age, status, and physical condition. Whether or not, and how much, she commits to such offspring as she bears depends on her circumstances, and—in cooperative breeders like humans—on who else is around to help her.

Incorporating Mothers into the Evolutionary Process

Through the choice of the males with whom she mates, the bodily resources she provides her developing young, and the social microcosm she creates for them, each mother's legacy is twofold. It includes the intertwined non-genetic and genetic endowments, which are very difficult to tease apart.

Back in the earliest days of evolutionary thinking, long before the Austrian monk and botanist Gregor Mendel showed how genes work, and long before Darwinian thinking merged with population genetics to produce the "new synthesis" of the mid-twentieth century, one of the very first evolutionists, the great eighteenth-century French naturalist Jean Baptiste Lamarck, had proposed that a mother passed on the traits she acquired during her lifetime. The oft-cited example was the giraffe's neck, supposedly stretched long by years of high browsing, and passed from mother to offspring. With the rise of Darwinian, and especially genetic, thinking, Lamarck's ideas were set aside as impossibly quaint. Now biologists realize that there *are* important respects in which nongenetic, acquired attributes—like immunities, templates for rec-ognizing relatives, social networks—pass between generations as "maternal effects," attributes that are inherited, but not as genes.

Tandem transmission of genetic and acquired attributes means that modern evolutionary theorists are having to combine Darwin, Mendel, and Lamarck into one interactive and cumbersome maternal situation. Com-pared with maternal effects, understanding paternal contributions would be relatively simple were it not for the fact that male reproductive strategies can rarely be understood apart from what mothers are doing.

All through the 1970s, in the years before sociobiology fully incorporated the new view of mothers as complex, variable creatures, there was a wide-spread presumption that "Most adult females in most animal populations are likely to be breeding at or close to the theoretical limit of their capacity to produce or rear young . . ." while, by contrast, with regard to males, it was assumed that "there is always the possibility of doing better."[1]

Researchers fixated on simple measures like "counting cops"—primatolo-gists' slang for counting the number of times each male copulates. But in species where maternal reproductive success varies a great deal, the number of matings provides only a crude and unreliable measure of any given male's reproductive success. Whether a copulation results in surviving offspring will depend on a whole range of contingencies having to do with which female a male mates with. Unless mating results in production of offspring *who them-selves survive infancy and the juvenile years and position themselves so as to reproduce,* sex is only so much sound and undulation signifying nothing.[2]

Consideration of maternal effects and other underlying mysteries takes evolutionists beyond the habitual questions raised by sexual selection theory, and the staple of so many sociobiological studies: "Will she or won't she?" "Can he or can't he?" More recently, questions like "Which mother?" and "Under what circumstances?" have become more important.

Whereas males would be under heavy selection pressure to best rival after rival just to gain opportunities to copulate one more time, females have no need to compete for mates in this way. This correct generalization was often misunderstood to mean that females lacked any "preadaptation for competi-tion" or any "genetic predisposition toward the creation of hierarchy," which was rarely true.[3] It was certainly not true if one takes into account those aspects of mothers' lives where competition matters.

Darwin's ingenious theory of sexual selection promoted a blinding hubris. If evolutionists could explain male strategies for out-competing other males and inseminating the most females, they could explain the different natures of males and females. The trouble was that this crown jewel of evolutionary theory, tailor-made as it was to explain competition between males, was poorly suited to explain the many preoccupations of females. Important sources of variance in the reproductive success of one female compared to another were overlooked.

Factors that were routinely overlooked in those early days included the female's age at first birth, the duration of the intervals between her births, social factors influencing whether her infants live or die, or even whether she reproduces at all. Nor did it always register that unless mothers gauged their reproductive effort in line with fluctuating resources and other prevailing conditions, few would manage to rear infants that survived. The poorly adapted or unlucky would die trying.

Viewing mothers the old way, no one had paid much attention to these sources of variation.[4] For example, when Jeanne Altmann first showed that high- and low-ranking baboon mothers at Amboseli differed in their proba-bilities of giving birth to a son versus a daughter, few knew what to make of it. Many found it hard to believe, because in order to understand what was going on one also had to take into account the social and ecological context in which each mother was operating, and to understand that baboon daughters born to low-ranking females were less likely to survive than sons were. Why? Studies of captive macaques with a similar social system provide one reason. Higher-ranking females in the same group harass mothers with daughters (the sex of offspring that will remain in the natal group and compete with her own daughters) but leave low-ranking mothers with sons alone. As a consequence, infant daughters suffer higher mortality than would sons born to mothers of the same low rank.[5]

With the support of their mothers and other matrilineal kin, daughters born to high-ranking baboon females rise in the hierarchy and, in turn, pass on the advantages of their acquired rank (along with such perks as early reproductive maturity, and greater offspring survival) to daughters. The female baboon, like most social mammals, introduces her baby into the network of social relationships she has forged. Daughters who grow up surrounded by high-ranking kin give birth at an earlier age to offspring more likely to survive. Since baboon daughters inherit their rank from their mother, these social advantages are transmitted across generations as maternal effects, and the reproductive advantages accumulate through time in her matriline. But this strange bias in production of progeny only made sense in the light of variation between females.[6]

Prior to sociobiology, females had been lumped together as a class, as if to see one was to see them all. By obscuring variation between one female and another, researchers also inadvertently obscured the extent to which natural selection has operated on this sex. In retrospect, it seems absurd that half the species could have been lumped into a homogeneous group this way—until we take into account the nineteenth-century contexts in which evolutionary theory emerged. Once again, the idea goes back to Victorian evolutionists like Spencer who observed that with the exception of fecund insect queens, a successful male can produce more surviving offspring, or have greater reproductive success, than females.

From fruit flies to humankind, it was taken as axiomatic that all females mate and become mothers, while among males only the luckiest or most competitive manage to become fathers.[7] It seemed to follow that the reproductive success of different mothers would be more or less equivalent, while that among males would vary tremendously. The greater reproductive potential of males was one reason that biologists focused on their behavior. Another reason was that competition between males was so conspicuous and exciting.

Females, by contrast, were viewed as plodding constants whose steady performance could be taken for granted.

Convinced that the most important variation occurs between males, some twentieth-century biologists—like Spencer—still assumed that females were less evolved than males. Why? Because *variation in the reproductive success of one individual relative to another is essential for natural selection to occur.* No variation, no selection. No selection, no evolution.

Variation and Selection

The old premise that selection acts more strongly on males than females was uncritically carried forward into modern evolutionary thought. Indisputably, sexual selection weighs heavily on male traits that affect their access to mates. But competition for mates is not the only sphere where Mother Nature is at work. The theory brilliantly illuminates the obsessive concerns of Othello, but casts little light on Desdemona beyond clarifying why it would be necessary for her to counter such detrimental effects as being suspected, chased, herded, dominated, sequestered, punished for straying, or (switching now to another Shakespeare play, *Titus Andronicus*) having offspring sired by a male other than her mate killed.

Males and females pursue different reproductive strategies. Theoretically, males compete for fertilizations, trying to inseminate as many females as possible. There is a strict limit, on the other hand, to how many times a female benefits from insemination. Her reproductive success depends not on number of fertilizations but on the contingencies of her life, the qualities of the mates she chooses, and, above all, *how successful she is at keeping alive such infants as she does produce.*

By the last quarter of the twentieth century, a previously undreamed of variation in reproductive success from one female relative to another was being documented. At the same time, unexpected—even previously unimaginable—sources of this variation were being unveiled.[8]

Not So Coy Females

Darwin assumed that females were "coy," holding themselves in reserve for the one best male. Yet field studies for primates suggested that, once again, the behavior of females was more variable than expected. Females could be "promiscuous," like males, if by that term we mean attempting to mate with many partners. But to what end? Given that a female could support only one

Fig. 4.1 Lucky Moulay Ismail the Bloodthirsty of Morocco (1646–1727) has earned a double immortality. His first derives from an astonishing reproductive success, supposedly 40 sons born in the space of one three-month period in 1704, leading to an impressive lifetime total of 888 offspring born to his many wives and concubines. Moulay's second source of immortality derives from his frequent citation by sociobiologists who used him as a chestnut to illustrate the enormous gulf separating a "big loser" (zero offspring) and a "big winner" like him. The world's records for maternity pales by comparison.

Pity poor Madalena Carnauba in Brazil who married at thirteen and gave birth to thirty-two children (fates unknown). A woman, even if fed ad libitum like a Strasbourg goose and provided ample wet nurses, would still be constrained by inter-birth intervals that inevitably spread out with the rest of her as she approaches menopause. Theoretically, a man's life-long sperm production means that only the duration of refractory periods after ejaculation, declining sperm counts, or access to ovulating women limit his fertility. Hence he always has a *chance* of doing better.

Fine, so far as this scoring system goes. But even the potency of a potentate should not be considered in a vacuum. How many offspring actually survive? For balance, shouldn't we also be discussing Moulay Ismail's mother, about whom almost nothing is known? Was she anything like Moulay's own scheming empress, Zidana? Based on a few details about Zidana that survived, she was quite effective at discrediting rivals and eliminating their sons from the line of succession.

(From Blunt 1951)

pregnancy at a time, why would she do this? Such solicitations not only take time and energy, but render females vulnerable to attacks from other males, and expose them to sexually transmitted diseases.

Consider the case of chimpanzees. A female chimp mates on average 138 times with some thirteen different males for every infant she gives birth to. Female bonobos also mate many more times than is necessary for conception, as do other species of primates living in multimale groups, such as barbary macaques and baboons.[9]

In 1997 the first-ever paternity tests from a population of West African chimps were analyzed, with unanticipated results. Samples of hair from infants, their mothers, and from the males in their community were obtained, and DNA markers from the hair were compared. The researchers had assumed that chimps live in more or less discrete communities whose boundaries were patrolled by bands of related males who share access to females in the community. These bands of males were not just defending food resources within their territorial boundaries but also the breeding females who live there. When Pascal Gagneux, David Woodruff, and Christophe Boesch analyzed the genetic data, however, they found that just over half the infants born in this community (seven out of thirteen births) were sired by *outside* males. The fathers not only lived outside the study sample, but included males that the observers had never even seen the female traveling with, much less mating with.[10]

Undetected by observers, female chimps were slipping away to solicit outsiders in spite of appalling risks. Lone chimps caught trespassing in territories patrolled by other chimps may be viciously attacked, bitten to death. Even if males tolerate a "foreign" female sporting the bright-pink swellings in her anogenital region that signal ovulation, resident females may not honor that particular passport. Wandering through unfamiliar terrain also makes a female more vulnerable to predation, not to mention the risk of disease. (It is almost certainly no coincidence that the virus causing AIDS evolved in chimps. Promiscuous habits provide the perfect niche for a sexually-transmitted virus.)

Why, then, in spite of all these drawbacks did female chimps furtively leave their home communities to breed? Were all the resident males simply too familiar? Were the females behaving so as to avoid inbreeding? Was it because males next-door seemed genetically superior in some way that was communicated to females? Were females manipulating the information available to males about paternity as insurance lest one or more of these males invade their community one day?

Mother's Sexual History as a Maternal Effect

In the wild, baby chimps are born to mothers who, on average, have mated more than a hundred times with as many as a dozen or more different males. The mother's frenetic (if transient) libido would make it impossible for any male to be certain of paternity. Nevertheless, so long as the female mates only with males in her local community, those males will still tend to be related as uncle, cousin, or grandfather to offspring sired by their relations. The possibility that an unrelated male might grow up undetected in the local breeding fraternity would be a serious threat to the breeding integrity of these "brotherhoods." Community males would not be expected to welcome another community's son in their midst—and apparently they do not.

Offspring born to a mother like Flo, secure in her feeding range deep inside the territorial boundaries of the community, would be safe enough. There, she and her young are less exposed to incursions by males from other communities than are mothers on the margins. If Flo ever bred with an outside male, it would have been on her own terms—perhaps during a furtive visit undetected by her community's resident males. But females living in home ranges at the edge of the community are less fortunate. According to Japanese primatologist Mariko Hiraiwa-Hasegawa, offspring of these mothers are at double jeopardy, likely to be killed by males in neighboring communities as well as by males in their own who suspect their mothers of having consorted with the enemy. When offspring of these mothers at community margins are killed, Hiraiwa-Hasegawa reports, sons are more likely to be the victims than daughters are.[11]

Langur males, however, differ from male chimps in this respect. They virtually *never* attack any infant born to females with whom they have mated, even if she has mated with other males as well. That is, unlike chimps, langur males err on the conservative side of the margin of error that surrounds paternity. They attack only if they can be certain the offspring is not theirs—unlike chimps (for example), who tolerate only those infants they can be confident were sired by members of their fraternal interest group.

Among the langurs studied by German primatologist Carola Borries in lowland forests of Nepal, infanticide accounts for 30 to 60 percent of all

Fig. 4.2 "No case," Darwin confessed, "interested and perplexed me so much as the brightly colored hinder ends and adjoining parts of certain primates." In about one-fifth of all primate species, tissue in the anogenital region fills with fluid and turns bright pink around the time a female ovulates. Normally swellings last just days—a week at most. But bonobo females like these remain swollen and eager to initiate sex for up to three weeks at a stretch. "Sexual swellings" evolved at least three different times in the Primate order, almost always in species where females reside in groups with multiple adult males. If females evolved to be sexually discriminating and "coy," as Darwin assumed, why do these species so blatantly advertise their receptivity and not only solicit but copulate with many partners, many times? *(Courtesy of Amy Parish)*

infant mortality. DNA evidence indicates that *none* of the infants killed could have been sired by the males who killed them. Even a completely unrelated infant will be accepted in the group, so long as it is being carried by a familiar female. It is the mother and her past relationship with the male that provide the cue for a male to either tolerate or attack a particular infant. As with the *femmes fatales* fireflies, the mother's recent sexual history would be an example of a maternal effect with life-or-death consequences for her progeny.[12]

This new awareness of female reproductive interests is transforming our understanding of animal mating systems. Wherever males attempt to constrain female reproductive options, we can expect selection for traits that help females to evade them. What are we to make of such far-flung solicitations and enterprising sexuality as are being documented for creatures as diverse as fireflies, langurs, and chimps? After all, applied to females, pejorative-sounding words like "promiscuous" only make sense from the per-

spective of the males who had been attempting to control them—no doubt
the origin of such famous dichotomies as that between "madonna" and
"whore." From the perspective of the female, however, her behavior is better
understood as "assiduously maternal." For this is a mother doing all that she
can to secure the survival of her offspring.

Whatever else these apes and monkeys are up to, it is obvious that selecting the one best male from available suitors—as Darwin imagined female
choice would work—is scarcely the whole story. Females are also actively
manipulating information available to males about paternity.

Is a male animal capable of remembering whether he mated with a particular female? The best experimental evidence testing the proposition that he
can derives not from the langur monkeys, who first inspired the hypothesis
about confusing paternity, but from European sparrows called dunnocks. A
male dunnock acts as if he can not only recall *which* females he mated with,
but how likely copulations at a particular time were to result in conception.

Female dunnocks live in cooperative breeding groups in which a female
solicits multiple males. These males, in turn, help provision the chicks more
or less in proportion to how much opportunity they had to inseminate the
mother when she was last fertile. According to Nick Davies of Cambridge
University, both alpha and subordinate males were significantly more likely
to bring food to young they fathered, or even young they *might* have fathered.
And just as the "several possible fathers" hypothesis would predict, DNA
fingerprinting (which pins down paternity more precisely than human
observers possibly could) revealed that males were often *but not always* accurate in their guesstimates.

Pro-Choice Mammals

Mothers in cooperatively breeding species are especially sensitive to who in
their vicinity is likely either to help or hinder their reproductive endeavors.
Mothers calibrate their reproductive effort according to which males, with
which intentions, and which females are also present. After birth, how much
a given mother invests may depend on particular attributes of her litter, its
size, the ratio of sons to daughters in the litter, or even the qualities of particular offspring. Deteriorating social conditions, loss of helpful kin or a mate,
or the presence of dangerous strangers can have a profound effect on maternal commitment.

The California mouse is an unusual rodent, not just because it is socially monogamous but because both members of the monogamous pair are completely faithful to their partners, and never mate with others. This is because mothers with the help of mates rear four times as many pups to weaning age as do single mothers, so both sexes are better off staying rather than straying. Outside of captivity, the presence of a male is absolutely essential to rear a large litter, to keep the pups warm and fed. Mothers who lose their mates may kill their pups rather than attempt to rear the litter alone.[13]

In the California mouse, mothers eliminate offspring after birth if they lack *assistance* from a male. A far more common pattern among other species of rodents, however, is for a mother to terminate investment prior to birth because she has reason to fear male *interference*. In house mice, deer mice, Djungerian hamsters, collared lemmings, and some species of voles, pregnant females respond to the arrival of a strange, potentially infanticidal male in their territory by reabsorbing their embryos.[14] With this efficient form of early-stage abortion, the female avoids the even greater misfortune of losing a full-term litter later on. Early-pregnancy reabsorption triggered by the smell of unfamiliar males is known as the "Bruce effect," after biologist Hilda Bruce, who in 1959 first reported it, even though the phenomenon was not then understood.[15]

Even in strains of mice that experimenters already knew to be infanticidal, not all individuals are equally likely to kill young. In some strains, almost all males are infanticidal; in others, only males of a certain "type" kill infants. For example, only socially dominant males may be infanticidal, or only males that could not possibly be the father of any infants they encounter because they have not ejaculated in the past twenty-one days, the equivalent of one mouse gestation period.[16]

In 1994, biologists Glenn Perrigo and Frederick vom Saal described a unique neural safeguard system in mice that ensures males do not kill their own infants. An internal "clock" starts up in response to ejaculation and thereafter keeps track of light-dark cycles for a period of two months. This unusual timer adaptively schedules a male mouse's transformation from cad into dad and back to cad. The male who has ejaculated becomes noninfanticidal long enough for any pups he might have sired to be gestated, born, weaned, and out of the way. After that, males revert to their infanticidal ways until they mate again. It was the "switch in time that saves mine," quipped

Irish biologist Robert Elwood, describing the transformation from the perspective of a male rodent.

As with most really critical functions, Mother Nature has retained redundant safeguards. In addition to the "switches in time," there are more primitive fallback systems so males can be absolutely certain not to kill their own
young. Hence, a male in most mouse strains can be induced to keep right on
being tolerant toward infants long after the first litter is weaned, so long as he
remains in contact with his mate, or with her smell. Somehow (probably
through androgen-mediated pheromones), females can assess which males
are and which aren't infanticidal. Pregnancies were significantly more likely
to be reabsorbed when the mother encountered a mouse that the observers
already knew from other evidence was likely to be an infanticidal male.[17]

Reabsorption of embryos—the most efficient way to terminate a pregnancy—is not physically possible in primates. However, pregnant monkeys
(baboons, langurs, geladas, and other monkeys) whose social groups have
been recently usurped by a new male have been reported to spontaneously
abort.[18]

At first, it seems counterintuitive that any female would ever produce
fewer offspring than she is capable of, much less terminate investment in a
fetus or an infant in whom she has already invested so much. But the art of
iteroparity (or, breeding more than once over a lifetime) involves knowing
when to cut your losses and weather poor conditions, the sooner to breed
again under better ones.

In species where survival of young requires extensive care, the single most
important source of variation in female reproductive success is not how many
young are born; what matters is how many survive and grow up to reproduce
themselves.[19] For such creatures, survival of at least some young requires
reproductive discretion. This is why being pro-life means being pro-choice.

The Importance of Allomothers

In most animals (reptiles, fish, insects), the mother lays her eggs and takes
off. When a mother does care for her young, she does so alone. In many
species of birds, and in about 10 percent of mammals, including a tiny fraction of primates (humans and a few species of monkeys and prosimians that
bear multiple young), infant survival depends on the mother being assisted
by others—the father and/or various individuals other than the parents—

alloparents. Sometimes, as in social insects, or the rare case of naked mole rats, such individuals provide *more* care than the mother does.[20]

Ornithologists used to call such assistants "helpers"; primatologists called them "aunts" (after the British "auntie," to designate a female relative or trusted family friend). In 1975, Edward O. Wilson decided it was time for a more dignified designation: *allo-* (from the Greek for "other than") plus *parent.* Since it is only the wise animal behaviorist with access to DNA fingerprinting who knows for sure who the father is, it would be more precise to confine ourselves to the term *allomothers*—meaning all the caretakers other than the mother (whose identity we are likely to know for sure) who help care for or provision young.

Although it may seem odd to refer to a male caretaker as an allomother, all this means is that he is an individual *other than the mother* helping the mother care for her infant. However we define them, alloparents play critical roles in all cooperative-breeding species and in many primate societies where such assistance allows mothers to breed at a much faster rate than would otherwise be possible. Among humans living in foraging societies, a helpful mate and/or alloparents were usually essential for a mother to rear *any* infant at all. In a surprisingly broad range of creatures, indispensable alloparents provide many of the same forms of care a mother might, protecting and provisioning, even suckling another female's infant in cases where the alloparent is lactating.

Communal suckling is most often reported where mothers live among their matrilineal kin, as do elephants, dwarf mongooses, prairie dogs, lions, ruffed lemurs, cebus monkeys, and bats.[21] Occasionally, however, dominant mothers force unrelated females to provide milk, as in the case of some wolves or wild dogs. The hired wet-nursing that Linnaeus and Gilibert so objected to was essentially a case of our highly inventive species consciously converging on a solution similar to that already found in nature.

Just as the labor of legions of larvae-minders has earned social insects the prize for greatest biomass, alloparental assistance permits some mammals to rear young under difficult conditions, to breed fast, or to rear young that are especially costly because they are large, quite numerous, or slow-growing. As we will see in part II, alloparenting is particularly well developed and played a special role in the evolution of our own large-brained, slow-growing, and extraordinarily inventive species.

Those Who Can, Breed; Kin Who Can't, Help Out

With their big litters and fast-growing, costly young, the dwarf mongoose of East Africa provides a classic example of cooperative breeding. Typically only the oldest female in a large group breeds, but all help her rear her young. Each animal takes a turn as baby-sitter or sentinel, standing upright on the vast arid savanna, vigilantly scanning the horizon for predators. Why do other females, themselves sufficiently mature to breed, forgo ovulating and help their kin rear super-litters instead?

Research by Purdue University behavioral ecologist Scott Creel illuminates the physiological underpinnings of this odd breeding system. The rule among dwarf mongooses, as among honeybees and other cooperative breeders, is clear: Those who can, breed, while those who can't, help out. Assistance to mothers rearing these super-litters is provided primarily by subordinate females who act so much like real mothers that, although they have never been pregnant, they spontaneously lactate to feed their charges. Harassed and often underweight, these subordinates could not hope to provide for a whole litter of their own. Under the circumstances, helping rear kin is their next best option, better than trying and failing, or challenging the alpha female and being wounded or driven away.[22]

Mongoose nursemaids respond to the presence of dominant females by a downward adjustment of estrogen levels, which temporarily suppresses ovulation. But why not breed anyway, on the off chance that a subordinate can successfully raise just a small litter? Bonn University's Anne Rasa may have discovered the answer in a related species of African dwarf mongoose. As in marmoset monkeys and wild dogs, dominant breeding females ensure the availability of allomothers by killing such infants as subordinate females have the audacity to produce.[23] The alpha female makes giving birth an option not worth the cost. At the same time, the subordinate's prudent postponement of ovulation means more care available for the alpha female's big litter, to whom the allomothers are related as well.

Contingent Commitment

Clearly, for cooperatively breeding mammals, a female's physical capacity to conceive and bear viable young is only a small part of the equation. She must also be able to carry through with the enterprise—rearing offspring to independence. Over sixteen years of monitoring a population of black-tailed

Fig. 4.3 Among cooperatively breeding tamarins, former mates and prereproductive group members carry the infants when the mother is not suckling. This allomaternal assistance sustains a rapid pace of reproduction—some mothers produce twins twice a year. Her helpers offer crickets and other tidbits as the twins make the transition from mother's milk to feeding themselves. *(Drawing by Sarah Landry)*

prairie dogs, biologist John Hoogland found that the vast majority of mothers (91 percent) are satisfied that, having committed themselves to gestation and birth, they have a chance to pull these infants through. The remaining 9 percent of mothers terminate investment at birth and make no effort to keep pups alive. Unprotected, their young are eaten by other females in the group, sometimes with the mother joining in. In other words, the endocrinological changes that accompany pregnancy and parturition do not, by themselves, guarantee that a prairie dog mother will behave in what most people think of as a motherly way.

These "nonmaternally acting" mothers, as Hoogland called them, weighed less than the others (though they were not necessarily younger). Hoogland hypothesized that these females had become pregnant "on spec," gambling on a hawk carrying off a dominant female or some other improvement in their fortunes before their due date. But when the small female's time comes, if her lot has still not improved, the odds are that a larger mother is going to kill her offspring. Nearly a quarter of all prairie dog litters are destroyed by other lactating females. At that point, the mother who gambled and failed gives it up, sooner rather than later, within the first day or so of birth: better that than to throw good calories after bad.[24]

The Real Self-Sacrificing Mother

A survey of the natural world from a life-historical perspective reveals just how special a creature the self-sacrificing mother envisioned by Gilibert must be. Such mothers exist. Women do, for example, risk their lives to save their children. After three marriages, two miscarriages, and the stillbirth of a daughter, a forty-one-year-old woman learned she was pregnant, and also that she had leukemia. Treatment could save her, but the medications would harm the fetus. She opted to postpone treatment. The baby survived. The mother died soon after giving birth.[25]

At issue is not whether some mothers value the survival of one of their offspring more than their own lives. The question is whether such mothers evolve as *species-typical universals of the female sex*. The answer is yes, but under narrowly defined circumstances. Typically, self-sacrificing mothers are found in highly inbred groups, or when mothers are near the end of their reproductive careers. The forty-one-year-old mother who gives her life for her only child is not the same individual who decades earlier might well have aborted her first.

Mothers as Amalgams of the Past

Pregnancy and motherhood forever change a woman. I do not merely mean depletion of maternal resources like calcium, a stretching and redistribution of her tissues, or alterations in her hormone profile. There are innumerable small ways. For starters, a fetus—half of whose genetic material comes from an alien organism—secretes enzymes that block a key component of the mother's immune response, providing a protected zone in which the embryo develops and the pregnancy can proceed. Such processes may or may not end with delivery. Fetal cells have been known to linger on in the mother's body for as long as *twenty-seven years*. In some instances, scleroderma, a disease that involves hardening of the connective tissue, may be triggered by autoimmune reactions to lingering fetal cells.[26] Beyond changes at the level of cells and tissue, becoming a mother is a turning point in a female's life history, altering prospects, opportunities, and, especially, a woman's priorities.

Pregnancy, labor, and delivery alter the brain. They lead to new neural pathways and the accentuation of certain sensory capacities, such as smell and hearing. Most research on these transformations has been done on laboratory animals, but almost certainly similar changes occur in women as well. Many new mothers feel their baby is so much a part of themselves that seconds before the infant begins to whimper, needle-like sensations can be felt in the nipples, and warm, wet milk leaks out. When a new mother says (as I did) that the birth of her first baby transformed her, she is not speaking just metaphorically.

A mother's body merges into synchrony with her baby's needs, and the baby's well-being becomes her pressing concern. Parts of these responses are incredibly old. Prolactin, the same hormone that coordinates maternal responses to infant demands for milk, was already orchestrating metamorphoses in amphibians and controlling water balance in the tissues of bony freshwater fish millions of years before any mammal existed (see chapter 6 for more on the role of this versatile hormone). Every aspect of our neurochemistry and emotions has a rich and convoluted history, bearing witness to multiple long-running legacies that we share with earthworms, amphibians, small mammals, and other primates.

Many of the emotions we feel today, many of our autonomic responses, first evolved in environments inhabited by ancient ancestors. Many of these conditions no longer pertain or have long since disappeared, yet, as we shall see in the following chapter, their legacy remains relevant to what we are.

The Variable Environments of Evolutionary Relevance

The tabula *of human nature was never* rasa *and it is now being read.
The inscription found is no dogma or world system and it bids to
build no empire whose later painful collapse will sweep it away. . . .
Thirty years ago I had no idea that a critique I had a hand in
could reach so far into the human sphere
and explain so much. . . .*

—William D. Hamilton, 1997

My mother, like most college-educated women in the 1940s, believed that babies came into this world with essentially blank slates waiting to be filled in. We now know that isn't the case. Far from being a blank slate, or tabula rasa, a human infant, like all apes, is born with its own agenda, preprogrammed to want to be close to whichever soft, warm creature cares for it after birth, more than likely its mother. With a repertoire of inborn "fixed action patterns," babies "root" with their lips for a nipple, suck, grasp, and nestle close, traits that have been crucial for the survival of primate infants over tens of millions of years. By crying out, signaling, clasping tight, and in emotional terms, by caring desperately, baby primates do whatever it takes to feel secure. They achieve what John Bowlby termed the "set-goal of proximity to mother."[1]

Innate behavioral systems are activated and reinforced by the same types of stimuli—touch, sounds, tastes, smells—that primate infants would have encountered in the past.[2] In the human case, the mother's voice and intonation would be among the stimuli babies are most sensitive to; they are preprogrammed to learn the language that she speaks, as well as to learn more readily at certain life phases than at others. But the most critical stimulus for infant development would be the more or less continuous presence of a sympathetic and responsive caretaker.

Bowlby's "Environment of Evolutionary Adaptedness"

John Bowlby was the first modern psychologist to follow Darwin's lead, exploring the implications of humankind's primate heritage for our earliest desires, fears, needs, and capacities. For more than thirty-five million years, primate infants stayed safe by remaining close to their mothers day and night. To lose touch was death. This explains why, even today, separation from a familiar caretaker provokes first unease, then desperation, followed by rage, and finally despair.

An infant safe inside a nursery is still well within his or her rights to feel distressed at being left alone. Under pressure, and with tough conditioning, infants learn to cope with the unnatural expectations of modern parents, but few can be imagined to look forward to spending the night in a dark room by themselves. The sensory and cognitive makeup of modern infants, the panic they still feel at separation, is distilled from innumerable past lives in which the infants most likely to survive were those who could prevent separation from their mothers. Even children's fantastical fears that something bad is lurking under the bed are assumed by some psychologists to date from a time when our ancestors spent nights in trees and predators threatened from below.[3]

The next time you are so frightened that your skin tingles and the hairs on your arms stand up, remember that once upon a time, bristling fur made a threatened mammal seem more formidable. Bowlby's own favorite reminder of our hirsute birthright was the *Umklammerungs-Reflex* (or, embrace-reflex) first described in 1918 by German pediatrician E. Moro, who accidentally jostled a baby he had just delivered. Spontaneously, the newborn's arms flung out in a symmetrical spasm, and then arced together again in closure. This same spasmodic startle occurs whenever a baby hears a loud noise or experiences some sudden change in position.

The Moro reflex causes infants to flap their arms and then clutch inward. If the infant's hands are already touching something soft, they simply cling tighter, hopefully to mother; but even a tree limb would be better than unchecked free fall. If nothing is there to catch on to, a desperate wail ensues.[4] Of less utility to smooth-skinned humans, the Moro reflex, like other grasping reflexes of the hands and feet, persists as a relict from past times, when survival depended on hanging on to a hairy protectress who might suddenly stand up and push off.[5]

In the decades just before and just after World War II, Bowlby was trying to understand the psychological development of human infants by imagining the environments in which their emotions evolved. He believed that the *basic* survival kit of ape babies would remain the same in infants born to our Pleistocene ancestors.[6] Not much was then known about how these early humans lived, but Bowlby assumed (correctly) that they were nomadic hunters and gatherers.

Borrowing the late 1930s concept of "man's ordinary expectable environment" from a German psychologist, Bowlby recast it into something he could define "more rigorously in terms of evolution . . . [with] a new term that [will] make explicit that organisms are adapted to particular environments." Bowlby's new term was "the Environment of Evolutionary Adaptedness" (EEA)—the millions of years during which "the behavioral equipment which is still man's today" would have evolved.[7]

Bowlby understood that the EEA would vary according to exactly which trait or "system" was being considered, but in the case of the human infant's emotional attachment to its mother, this Environment of Evolutionary Adaptedness has changed relatively little over the course of hominoid evolution. Compared to other behaviors that Bowlby might have focused on (such as how mothers care for their infants), the infant's powerful desire to be held close by its caretaker has changed remarkably little over the ten million years since humans, chimpanzees, and gorillas last shared a common ancestor. Bowlby's efforts to reconstruct the EEA were bounded on either side by what he knew about hunter-gatherers' maternal styles, and those of surviving Great Apes.

The reason a human infant's attachment to its mother still resembles an African ape's is that the most immediate environment of evolutionary relevance for infants was the *mother herself,* not the physical or social world Pleistocene humans and Pliocene apes inhabited. It was the mother who continuously carried the infant in skin-to-skin contact—stomach to stomach, chest to breast. Soothed by her heartbeat, nestled in the heat of her body, rocked by her movements, the infant's entire world was its mother. It was she who kept it warm, fed, and safe. Essentially the mother was the infant's niche, and for those baby apes that survived to breed, the boundaries of that niche were fairly constant.

Since 1969, Bowlby's concept of an Environment of Evolutionary Adapt-

Fig. 5.1 The model mother Bowlby had in mind when he envisioned humankind's Pleistocene Environment of Evolutionary Adaptedness was a woman like this !Kung San mother in nearly continuous skin-to-skin contact with her infant during the first two to four years of life.

(Mel Konner / Anthro-Photo)

edness has been enormously valuable for explaining infant attachment to a caretaker. But by the 1990s, the EEA had become synonymous with a specific period, the Pleistocene.[8] This more narrowly circumscribed million-and-a-half-year phase in our evolution was then used by evolutionary psychologists to explain the totality of human nature, everything from human motivations to enhance inclusive fitness (which humans share with many other social creatures) to more recent and quite specific adaptations like language. Some evolutionary psychologists even claim that we can use contemporary human behaviors (such as our mate preferences) to reconstruct this ancestral Pleistocene environment.[9] It is an ambitious goal, but it ignores just how much of what humans are, think about, care about, notice, and feel predates the Pleistocene, as well as giving short shrift to how much all human social systems depend on local history.

Expanding the EEA in Both Directions

Body and mind, for better or worse, we passed through the Pleistocene crucible. As is frequently noted, our ancestors lived as hunters and gatherers for 99 percent of the time the genus *Homo* has been on Earth. This is why I rely so heavily in this book on evidence from parents who still lived as hunter-gatherers when they were first studied by anthropologists. Nevertheless, there are important respects in which a fixation with the Pleistocene is limiting. Many traits that affect infant survival and women's reproductive success are far older than the Pleistocene, and some are more recent.

At present, no one knows which types of environment characterized the tiny population of ten thousand or so anatomically modern humans that

Mother Nature allowed to pass through the eye of the needle to migrate out of Africa between 50,000 and 150,000 years ago. Most people assume they were hunters of the African savannas. However, one might just as well guess that these were the survivors of some ecological perturbation of the time. Perhaps they were coastal-dwellers depending more on shellfish than big game, making their way along the water's edge out of Africa. But the point is, we don't know.

The twilight of the twentieth century brought with it last-ditch efforts by anthropologists to chronicle the vanishing lifeways of foragers. Fieldwork by Richard Lee, Mel Konner, Hillard Kaplan, Kim Hill, Eric Alden Smith, Kristen Hawkes, James O'Connell and others revealed greater variation among people pursuing this ancient lifestyle than can comfortably be accommodated in a one-size-fits-all concept like *the* EEA.

Given how variable human subsistence styles and family compositions are—even those of people who still live as foragers—we should not be surprised to learn that their social arrangements are very flexible, even more so than those of other primates. And some nonhuman primates, especially other widespread "weedy" and very adaptable species (like savanna baboons or langur monkeys) are very variable indeed, living in multi-male groups one place, "harems" in another, aggressive in one locale, peaceful someplace else. All primates are social. But the only specific social relationship every group can be said to share is the prolonged relationship between a mother and her infant during the first years of its life. Even the duration (although not the intensity) of this universal relationship fluctuates drastically, especially in humans.[10]

There is good reason to suppose that those foraging mothers living in the most arduous habitats, such as the !Kung San of the Kalahari Desert, gave birth only after very long, four- to five-year intervals. By the time a toddler was three years old, he would have been carried by his mother some four thousand miles. Under the conditions in which these nomadic foragers lived, infants had to suckle frequently just to keep hydrated as well as nourished. Even though a few solid foods would have been introduced by six months of age, survival depended on nursing well into the fourth year. To such an infant, its mother was cradle, protection, mobility, breakfast, midmorning juice, lunch, and dinner.

The !Kung are an extreme case. Still, there is little doubt that over the last million years or so, infants have always striven to remain in continuous contact with their mothers for at least the first few years of life. This was the infant's first choice, but living up to this "Pleistocene ideal" of mothering may have been tough for a mother to do—even in the Pleistocene. The human mother in continuous contact with her infant for four or five years more nearly represents a primate infant's favored scenario, the scenario most compatible with its well-being. But this preferred scenario is not the only one mothers employed. Wherever reasonably safe alloparental options were available, human mothers made use of them, as many mothers in foraging societies in Central Africa and South America still do. Ethnographic evidence from foraging peoples like the Aka and the Efé, as well as new evidence from other primates, suggests that alloparents were more important alternatives to continuous one-on-one contact with the mother than Bowlby had realized.

In the human case, the extended half-decade of physical closeness between a mother and her infant so typical of other apes tells us more about the harshness of local conditions and the mother's lack of safe alternatives than the "natural state" of *all* Pleistocene mothers. As Emory University anthropologist and nutritionist Daniel Sellen joked recently, the only people in the world who nurse their babies for five years are the !Kung and women anthropologists.

Rather than turning the EEA into boilerplate for a host of unverifiable assumptions about the lifestyles of humans in the last several hundred thousand years, behavioral ecologists recognize that a number of possible environments might be evolutionarily relevant, depending on which trait is at issue. Invoking environments of evolutionary relevance is an acknowledgment of accumulated past effects, without necessarily specifying when and where.

So let's be clear: humans today are an amalgam of past selection pressures on ancestors who were mammals, primates, and, most recently, hominids who lived as foragers in a range of ecological and social settings. Like baboons, langurs, and other particularly adaptable primates, humans are found in a broad range of climates, at different altitudes, over a broad array of habitats. Humans readily adapt to different habitats as other "weedy" species do; but being culture-bearing, technologically clever primates, they have even more scope to change their environment to suit their needs.

True, there are some things we can be quite certain of. Human mothers, like all members of the family Hominidae, lived in social communities. Offspring learned to recognize individuals likely to be kin by using cues provided by their mother, particularly by her patterns of association. As they matured, youngsters learned to discriminate the smell and appearance of close kin and of their mothers' close associates, and to behave more altruistically toward them. It is also certain that unlike other Great Apes, women must have lived in families and relied on other group members to help provision children who took unusually long to become independent. But beyond these points, relatively little about these early social environments is certain. We cannot know for example how long bonds between mates lasted (though one suspects that duration of marriages varied) or what kind of families women lived in (whose "in-laws," for example, they lived nearest to).

One on one, there is little doubt that a man would almost always be able to dominate his mate, as is true for virtually all simian primates. But beyond that, men's ability to control where women in their group went and what they did would have varied a great deal, depending on who else was there to back the women up. Who else was there would depend on local subsistence patterns and history, for apart from a universal tendency for primate females to avoid mating with close kin, women exhibit no clear and consistent predispositions either to leave or to remain near kin.

People who live by foraging may move daily, or almost never. They live in desolate terrain at population densities lower than one person per 250 square kilometers, in groups that rarely meet and defend no territories, or in densely occupied habitats with one person per square kilometer. Hunters can contribute anywhere from 100 percent of daily calories (as among the Eskimos) to 20 percent, with the remainder being made up by tubers, nuts, seeds, and other foods gathered by women. When people rely on shellfish or hunt with nets, both sexes often participate.[11]

Large maternal contributions to subsistence would have meant that mothers had more freedom to come and go, with important implications for whether females stayed among kin or left home at marriage, all critical, as we shall see, for how much autonomy a woman retains over reproductive decisions, how free she is to choose when and with whom she will mate. Female autonomy depends on the availability of support from her kin. This is true for all well-studied primates as well as most human societies in which one sex is more likely to move at marriage. When mothers remain among kin, they retain more autonomy than when they travel far from their natal place to live among their mate's kin.[12]

Long Before the Pleistocene

Foragers, primates, mammals—human legacies spiral backward through time, like the coils of DNA that connect us, linking us to long-ago life-forms. Evolutionarily, humans are a "mixed bag." A line from an old nursery rhyme, "Snips, snails, and puppy-dog tails," isn't too far off the mark—not just for boys but for everybody. A thrifty matron and inveterate recycler, Mother Nature is slow to discard leftovers. Conservative retention of useful molecules explains why the same endorphins, the natural morphine that made the pain of my children's births bearable, are also released in an earthworm when my garden spade accidentally severs it. The innate immune system that protects my body from bacteria makes use of the same kind of proteins that perform this function in fruit flies. Confronted with the necessity of solving a new problem, Mother Nature's first, and typically only, recourse is to use what she has on hand.[13]

Just for fun, imagine that all human traits really did evolve in the Pleistocene. What might a made-to-order Pleistocene baby be like? This ideally designed baby would fit the bill for any mother who works nine to five, but he or she would resemble no human baby ever born. Adapted for twice-daily feedings, morning and night, this PPB ("Perfect Pleistocene Baby") would be capable of digesting high-protein, high-fat milk, just like baby tree shrews and other mammal infants whose mothers leave them in nests for hours on end. Mothers would produce this rich fare and then trot off every morning with their digging sticks (or briefcases), leaving behind babies that required almost no one-on-one attention. Whichever individuals had gone out foraging or hunting the day before would take a turn resting at camp, protecting the nursery till the foragers returned.

Because learning would be postponed, the Perfect Pleistocene Baby would spend its day in a frozen or hypnotic state, curious about nothing, emotionally uninvolved, conserving energy, never crying or calling attention to itself, and requiring little in the way of monitoring. In the evening, when the mother returned, she would take the baby to a defecation site some distance from camp, so that all excretions could take place before returning to settle down for a good feed and a restful night. With this carefully engineered suite of adaptations, mothers could forage more efficiently, bring back more provender, breed much faster.

Of course it never happened that way. And if it had, the end product would

not be human. As Bowlby realized, the reason foraging mothers never produced such accommodating youngsters is that our species evolved as primates,
already committed to mothers who produce dilute low-fat, low-protein milk,
and to babies who suckle semicontinuously throughout the day and night.

Every living organism, every organ of every organism, not to mention tissues and molecules, whether or not they are still in use, bears the accumulated imprints of multiple past lives. Never permitted the luxury of starting
from scratch to produce the perfect solution, natural selection recycles
workable solutions for a "good-enough" fit, meaning simply: better than the
competition.

Consider melatonin, a hormone the body relies upon to regulate internal
clocks. Melatonin has become fashionably known as the "miracle hormone"
among jet-setters, who take it in pill form just before bedtime in a new time
zone to convince their body that it really is nighttime. Scientists are more
likely to dub melatonin the "Dracula hormone" because its production is
stimulated by darkness and inhibited by bright light.

An ancient, pre-Pleistocene compound, melatonin is found in the skin of
amphibians. In humans it is produced by the pineal gland, a pea-sized organ
once mistakenly assumed to be a functionless vestige. Far from it. One of the
pineal gland's products, this venerable light-sensitive secretion plays a key (if
still poorly understood) role in regulating body rhythms. Melatonin levels
rise at night during the hours most primates sleep and fall during the day,
permitting a pregnant mother to chemically communicate information about
day length to her fetus.[14]

We primates are, by and large, diurnal beings, adapted for daylight (a fact
that makes night shifts particularly dangerous and subject to accidents).
Instead of bungling about in the dark, risking an encounter with predators
like leopards, who see better in dim light than we do, primates spend nights
safely up trees, high on cliffs, in arboreal nests (like chimps), or swinging gently in a hammock by the fire. We do not just sleep, perchance to dream, but so
as to not to be eaten.

Ape bodies take advantage of a mother's enforced respite to do their
endocrinological equivalent of paperwork. The frequency of an infant's nighttime suckling serves the mother's body as an index of how much milk the
baby is consuming. Complex feedback loops then act like a master control,
regulating how long the mother should delay before ovulating again and conceiving her next baby (explained below, in chapter 8). Nighttime feedings

turn out to be the key ledger entry for a mother's reproductive budgeting. Surges in plasma prolactin levels produced in response to her baby's sucking are four to six times greater between the hours of midnight and four A.M. (when melatonin levels are highest) than when a mother nurses during daytime. For this reason, three or four breast-feeds during the night may have greater impact on delaying the next pregnancy than six daytime feeds.[15]

The next time you hear a nursing mother who unexpectedly finds herself pregnant grumble that breast-feeding did not suppress *her* ovulation, remind her of her primate past and all the ancient evenings her ancestors spent up in trees. In her eagerness to get a good night's rest, she probably overlooked the importance of breast-feeding during the night. She forgot, or more probably never considered, that for seventy million years, as mothers dozed on and off till dawn, infants whiled away those sunless hours by alternating between right nipple and left. As they suckled, they triggered the release of ancient compounds dating from amphibian, mammalian, and primate past lives that delay the next conception.

Every detail of our bodies has its history, and many of them have consequences. Accustomed to the beat of the mother's heart in utero, the infant prefers left nipple to right. Not surprisingly, today 83 percent of right-handed women, and 78 percent of left-handed mothers, still cradle their baby on the same side as their heart. Presumably, this explains why most Renaissance Madonnas are depicted holding babies on the left side, and why one ear is better at recognizing music and the melodic aspects of language than the other ear.[16]

From fretting babies to unplanned pregnancies, these examples suggest that unless acknowledged and understood, ancient legacies can prove inconvenient to our efforts to chart the future. But natural selection didn't stop dead in its tracks when the Pleistocene ended. We are kidding ourselves if we think so. A mother today, whether in New York, Tokyo, or Dacca, is not just a gatherer caught in a shopping mall without her digging stick. We are subtly and not so subtly different from our Pleistocene ancestors in ways that have been transmitted genetically as well as through multiple parental effects between generations.

Evolution Since the Pleistocene

There is a widespread assumption that evolution only occurs very slowly, over vast geological time spans. By and large this is true. But too many exceptions to this rule have been documented, both in lab and field, to allow anyone to take for granted that the pace of evolution is *necessarily* slow. Genetically produced life-history changes in mother fish can be documented after only forty generations in a new environment, with mothers evolving to produce bigger babies after longer intervals when predation pressures are reduced.[17]

Humans are unusual in this respect. Cultures can change much faster than bodies evolve. This puts humans in a special category, but it does not place us outside the reach of evolutionary change. We are at once relicts of Pleistocene foragers *and* altered specimens. Many people descend from ancestors lucky enough to have survived past epidemics of common post-Neolithic diseases like plague and cholera because they possessed versions of genes that conferred some degree of protection against these diseases.[18] Some are unlucky enough to have genes that interact with our new environment to produce nearsightedness or diabetes. Whenever humans find themselves living under novel conditions, there are new opportunities for selection to act. An example would be ongoing selection for heavier birth weights in babies whose mothers have recently migrated to higher elevations of the Himalayas of Kashmir.[19]

For most of us alive today, the environments of our ancestors underwent massive changes in the recent past. Almost no one you know traces his or her ancestry directly to hunters and gatherers. Rather, the vast majority of us are descended from peasants. Many were born to mothers who kept their babies in cradles or swaddled them, leaving babies with a kinswoman or other temporary caretaker while they engaged in seasonal work, planting or harvesting. These more recent legacies also leave their traces.

In mammals, natural selection will almost always have its greatest impact on the vulnerable life phases: in utero, during infancy, and just post-weaning. Assuming twenty-five years per generation, there have been about four hundred generations from the start of the Neolithic* until now. This represents four hundred opportunities for natural selection to act. A selective differential of just 2 percent can boost a gene from rarity to near fixation (that is, from a genetic frequency of less than 2 percent to more than 98 percent) in ten thousand years or less. Theoretically, changes in humankind's biological

---

* For purposes of this example, say ten thousand years ago.

and social environments since the Neolithic should be reflected in the genome of modern humans, and many are.[20]

Not surprisingly, the best documented cases have to do with disease, diet, and infant mortality—three areas where the Neolithic would be expected to expose humans to novel selection pressures. The end of the Pleistocene and the dawn of agriculture brought new types and greater abundance of food, as well as settled living, increased population, more sewage and water pollution, and the compression of people into smaller areas. Crowded conditions presented many new opportunities for water- and mosquito-borne diseases to infect people. Dense populations with open sewers allowed typhoid, cholera, and other gastrointestinal diseases to spread quickly through a population. Irrigation ditches brought mosquito-transmitted diseases to drier parts of the world where such infections had previously been rare. Respiratory diseases like tuberculosis spread from person to person in populations large enough and crowded enough to supply the pathogens with host after host. Several old diseases previously confined to other mammals, such as bubonic plague and AIDS, found new opportunities to jump across species boundaries and infect human hosts.

As a consequence of heavy selection pressure from these diseases, any genetic trait that conferred resistance to them was selected for—even against strong counterselective pressures. For example, 4 to 5 percent of people of European descent have a single copy of a gene that causes cystic fibrosis when inherited from both parents. In these populations, about one in 2,500 babies will be born with this lethal double dose and will develop the disease, which (until medical treatments became available) killed by age two.

Normally such lethal genes would have long since been selected out of the human population. Yet cystic fibrosis has been around for at least 50,000 years, and, according to microbiologist Gerald Pier at Harvard Medical School, there is a reason. In its single form, the cystic fibrosis gene protects people against typhoid fever, just as in the better-known case of the sickle-cell gene, which protects its possessor against malaria. When inherited from *both* mother and father, the bearer suffers from sickle-cell anemia.

Rapid selection could also be going on currently with respect to AIDS. There are areas in Africa today where as many as 25 percent of women of childbearing age are HIV-positive. Such HIV-positive women are thought to have a reduced chance of becoming pregnant.[21] There is also a decline in the survival chances of their offspring. Such babies are exposed to maternally transmitted infection either from mother's blood during birth or from mother's milk afterward. Even if babies escape infection, they are likely to be orphaned at a young age. Yet there are already hints that some African women are resistant to the human immunodeficiency virus (HIV) that causes AIDS. We would expect any genetic resistance to be disproportionately represented among the survivors. As selection works against those who are infected, it will favor the AIDS-resistant mothers and offspring.[22]

A surprising amount of epidemiological history can be read into the genes of surviving human populations. Genes can also tell us something about what foods people were adapted to eat. Consider the genes for lactose tolerance.

All baby mammals are born with the digestive equipment to synthesize the enzyme lactase. This enzyme enables them to break down and digest lactose, the carbohydrates in milk. Among many humans around the world the ability to digest milk sugars does not persist into adulthood. After all, being able to digest milk would be completely useless to adult foragers, who do not herd animals. Individuals who lack the appropriate enzyme to drink unprocessed milk may suffer from gas or diarrhea. This is why Western aid, which typically included powdered milk, got such a bad name in large areas of sub-Saharan Africa in the 1970s. Instead of helping, the donated milk powder made people sick.

Today, the main explanation for why so much diversity in milk tolerance exists is that since the end of the Pleistocene, some human populations began herding cows and consuming dairy products; others did not. In another example of rapid evolution, just in the last ten thousand years, the genes that promote lactase synthesis past infancy spread in populations where milk was fed long past weaning, and were lost where it was not. Fewer than 2 percent of adults in a population with a horticultural history, such as the Bantu of Central Africa, test positive for lactose digestion, and no !Kung do. By contrast, 90 to 100 percent of Tutsi populations in Rwanda and the Congo—all descended from milk-dependent pastoralists—retain the capacity to digest milk sugars throughout their lives.[23]

Some legacies from our hominid past may be adaptive in our new crowded, high-speed, high-tech, twenty-four-hour-a-day environments. We are, indeed, highly adaptable creatures who adjust quickly to new habitats, acquire new tastes, and learn new tasks—especially when young. Yet other traits, such as our lust for fats and sweets (left over from a time when it was impossible to get too much of either), can be extremely maladaptive when they lead to obesity and clogged arteries.[24] For mothers and infants, the biggest clash between ancient predispositions and modern lifestyle is the physical separation required by modern workplaces. Mothers find it stressful to go off to work and leave their infants behind. Babies find it even more so.

Balancing Motherhood and Work

Working mothers are not new. For most of human existence, and for millions of years before that, primate mothers have combined productive lives with reproduction. This combination of work with motherhood has always entailed tradeoffs. Mothers either sustained energetic costs and lost efficiency by toting babies everywhere (the way baboons and !Kung mothers do) or else located an alloparent to take on the task. What is new for modern mothers, though, is the compartmentalization of their productive and reproductive lives. The factories, laboratories, and offices where women in post-industrial societies go to "forage" are even less compatible with childcare than jaguar-infested forests and distant groves of mongongo nuts reached by trekking across desert.

The economic reality of most people's lives today is that families require more than one wage-earner—or forager. Single parents are especially hard-pressed to make ends meet. Only brief periods of prosperity or isolated blips of elite privilege have made this untrue for some people during a few periods in human existence. An expansion of the U.S. economy after World War II, for example, meant that many married women could afford to stay home with their babies. But no more. Most mothers, even if they want to, do not have the option of staying home to care for their babies.

And that's the modern rub. During the Pleistocene, women could carry their babies as they foraged or gathered firewood. Dual-career mothers *still* strive to balance their subsistence needs against the time, energy, and resources needed to rear their children. But the physical (if not always the emotional) environment in which these compromises must be made is considerably different from the workplace of our ancestors. In some respects, omnipresent conflicts create even more tension today than in the past, because the incentives to fix them strike mothers as optional. Outcomes are measured in terms of the personal toll—insecurities among infants, stress in their mothers—rather than increased mortality. Simply put, the pressures to change are less intense when children can (literally) live with the consequences.

If infants feel stressed by the separation, so do millions of working mothers. At the same time, the evolutionarily novel modern workaday world opened the door for untrammeled expression of another ancient female motivation—striving for status, or, in the case of a forager, what one might think of as "local clout."

Maternity and Ambition

It is widely assumed that competitiveness, status-striving, and ambition, qualities that are essential for success in demanding careers, are incompatible with being a "good mother," who is expected to be selfless and nurturing. "There is no getting around the fact that ambition is not a maternal trait. Motherhood and ambition are still largely seen as opposing forces," states Shari Thurer, a prominent contemporary psychologist. Sociologists can document at length the "cultural contradictions" produced by women combining motherhood with jobs in the American workplace.[25]

Under conditions of the modern world, and if we assume the old definition of mothering as an innately charitable and selfless pursuit, the point is well taken. But as I described in chapter 2, mothering in the natural world is different from the Victorian image of mothers. Mothers' work has not always been so compartmentalized from child-rearing as it is today, nor her status so separate from the prospects that a mother's offspring would survive and prosper.

Modern women may think of status as the icing on their economic cake. But once the significance of social rank is understood for such vital functions as a mother keeping another female from eating her baby (as in the case of chimps), or for keeping another female from monopolizing resources needed by her own offspring (as in the case of other cooperatively breeding mammals), the struggle for status seems more nearly a foothold on posterity than a frill. "Ambition" was an integral part of producing offspring who survived and prospered.

Establishing an advantageous niche for herself was how Flo, the chimpanzee female that Jane Goodall studied for so many years, stayed fed, guaranteed access to food for her offspring, and kept them safe from interference by other mothers. Eventually Flo's high status made it possible for her daughter Fifi to be among the few females who would remain in her natal place to breed—in Fifi's case, inheriting her mother's territory. Even more impressive data documenting the connection between female status and all sorts of reproductive parameters—age of menarche, infant-survival rates, and even sex ratios of offspring—have been compiled for Old World cercopithecine monkeys like macaques and baboons. These data strongly suggest that generalized striving for local clout was genetically programmed into the psyches of female primates during a distant past when status and motherhood were totally convergent.[26]

Evidence for human primates is less clear, in part because husbands figure so prominently in the social status of most mothers. Yet both fiction and ethnography provide multiple examples. For example, Nisa, the !Kung woman, tells what happened when her first husband, Tashay, brought home a second wife. "I chased her away and she went back to her parents," Nisa says simply. Nisa's own mother had done the same thing a generation before. The children of this new wife would have competed with Nisa's for food provided by her husband and other community members. Nisa acted so as to maintain her status as her husband's only, primary wife. Her actions were in keeping with being a "good mother." Such women do not compete for status and reputation in the spheres that matter to men (for example, being known as a great hunter or warrior); they compete in the spheres that actually matter to mothers.[27]

Occasionally, we can detect bizarre manifestations of these old connections, as in the case of the Texas mother who hired a hit-man to murder the mother of her daughter's cheerleading rival in order to derail her emotionally.[28] But for the most part, status-striving by mothers seeking to enhance the prospects of their children is more subtle. Think of the womanly rivalries chronicled for early-nineteenth-century England by Jane Austen, or by Edith Wharton for the early-twentieth-century tribal life of "Old New York." In subtle, private, and scarcely perceptible ways, both mothers and their relations close ranks so as to promote and protect the marriage opportunities (which in that world meant access to resources) of young kinswomen, while locking out other young women. We tend to think of these mothers as "controlling," "pushy," "interfering"—and I don't disagree—but the venerable ancestry of such traits is worth considering. In their environments of evolutionary relevance, these women would have been behaving like successful mothers.

Far from "opposing forces," maternity and ambition are inseparably linked. The circumstances of modern life tend, however, to obscure the connection. This is because jobs, status, and resource defense occur in separate domains from child-rearing. At the same time, civilized mores and laws mean that mothers do not have to rely on intimidation to drive off rival mothers and keep their offspring safe from competing interests. Most mothers reading this book worry far less about famine, tigers, and infanticidal conspecifics than they worry over job promotions, health benefits, and finding adequate daycare.

For the most part, mothers striving for status in the modern workplace do so outside the home. Often working mothers are driven to pursue status interests for long hours, far from home, in ways just as likely to harm as to help their baby cope with life. The conflict, however, is not between maternity and ambition, but between the needs of infants and the way a woman's ambition plays out in modern workplaces.[29]

In the modern world, status (whether socioeconomic or professional) is, if anything, *inversely* correlated with reproductive success. This is especially true for women who *earn* their status. Not long ago, sociobiologist Susan Essock-Vitale looked at the reproductive success of people listed on *Forbes Magazine*'s annual listing of the four hundred wealthiest Americans. Those women who had inherited wealth had significantly more children on average than successful businesswomen who had acquired their wealth through their own efforts. This should not come as a surprise. When given the opportunity, many women value upward mobility over time devoted to rearing a family. We need only look at the grueling hours that working mothers put into jobs as lawyers, doctors, and research scientists, careers with demands as insatiable as those of children. But if our evolutionary heritage has any relevance to what we are, how can this be?

The answer is simple. In worlds where there was no birth control, and where no female was ever celibate, there was no possibility that female rank and maternal reproductive success could be *other* than correlated. Nature built in no safeguards to ambition run awry, as it were, to energies diverted to status ends that were not linked to the production, survival, and prosperity of offspring. Now that status and the survival of offspring have been decoupled, will there be selection against women who are especially inclined or driven to achieve? Probably, if our species survives long enough, and if circumstances in the workplace don't change.

Torn between two ancient, pressing, and now incompatible urges, women are forced to make new tradeoffs. Forging workable compromises between infant needs and maternal ambition requires considerable ingenuity, self-understanding, and common sense. This is especially true in highly competitive and demanding fields. Science provides the case studies with which I am most familiar.

Novel Compromises

In 1976, the year after I completed my Ph.D., an article appeared titled "The high price of success in science," written by a young molecular biologist, Nancy Hopkins, who would go on to become a leader in her field. She argued that it was not possible for a woman in such a competitive profession—demanding, in her case, a minimum of seventy hours a week in the lab—to "be a successful wife and mother as well as a successful scientist." Her words were sobering and, looking back on that era of what Hopkins called "the bionic woman hard sell of the '70s," unusually honest. About the time Hopkins wrote that article, there were ten tenured women professors at Harvard Medical School; nine of the ten had no children.

Yet there were women who managed to successfully combine science with motherhood. None I know took ordinary routes. Mary Jane West-Eberhard, for example, whose ideas about the role of development in evolutionary processes I discussed in chapter 3, is legendary among women field biologists for the way she managed to combine her family and professional lives. "It's what we all do," she told me once, apropos of the extra-tough compromises field biology requires of mothers. "We each construct our own idiosyncratic life." In West-Eberhard's case, she opted to forgo a conventional teaching position at a major university to take a research job. The position allowed the Eberhards to live in Central America, where they could afford housekeeping help, and, more importantly, where they could do research while keeping an eye on their three children—literally, since the wasps they studied were on the roof of their house.

My own compromises took me in the opposite direction. I switched from tramping around forested hillsides, following monkeys in India, to doing research on human parents in archives in the United States, where I used part-time daycare, and along with my husband took full advantage of emerging opportunities to work less than full-time and to use fax machines, and eventually the Internet, to work at home. When my third child was born, I hired a generous-hearted allomother on a very long-term basis. She lives with us still, though my youngest child is twelve, and the allomother now pursues a part-time profession of her own.

Pretty obviously, not one of us was living the same way as our ancestors. Yet we were required to resolve similar dilemmas, and so we forged new solutions for doing so—a theme that recurs throughout this book.

Why the Past Matters

Many of us at different stages of our lives desperately desire a child. Others, out of commitment to career or for other reasons, are determined to have none. Many women are certain they will never want a child, and then change their mind. Still others have babies by accident. Those who consciously decide are often making pragmatic decisions with a watchful eye on the effects upon their career, existing children, or the overall well-being of their family. Few people give much thought to the evolutionary origins of the emotions that inform such conscious or unconscious "decisions." But I am convinced we are more tightly linked to our past than most people imagine.

Whether we think about it or act "on impulse," each of us constantly makes myriad small decisions on a daily basis that in ancestral environments would have been correlated with reproductive success. Like it or not, each of us lives with the emotional legacy and decision-making equipment of mothers who *acted so as to ensure that at least one offspring survived to reproduce.* Prudent allocation of reproductive effort and the construction of an advantageous social niche in which her offspring could survive and prosper were linked to ultimate reproductive success.

Women—and men—today have an unprecedented range of choices. So far environments and the availability of birth control make it possible to have a few children, be confident that they will survive, and invest heavily in each one. Yet certain constraints are more or less immutable. For example, although women in postindustrial societies live longer and experience menopause later than was true for most of our foraging ancestors, the outer limits are still fixed. Few women conceive past fifty. Any woman who waits until forty to bear her first child is taking a chance that she will never conceive one.

Because we are primates, adoption and the rearing of genetically unrelated babies come easily. Unlike herd-dwelling ungulates, we do not have a critical period minutes after birth when a mother must imprint on her baby's smell and bond with it then or never. If we were sheep, we would not have to worry about babies getting mixed up, switched at birth in maternity wards.

Fig. 5.2 Mary Jane West-Eberhard is legendary among women biologists for the practicality of her research choices. Here, she has climbed a ladder to study a colony of rare *Metapolybia* wasps on the roof of her house. While she worked, she had a full view of her children playing in the enclosed patio below, as if, she said, "I was god [watching over a] large tropical playpen." Like a true deus ex machina, she would from time to time intervene, calling down to reprimand squabbling sibs. As her children fledged and went to college, West-Eberhard increasingly combined doing research with work on the National Academy of Science's Committee on Human Rights, finding there an even wider scope in which to express what George Eliot termed the "maternal component" to men's and women's nature. *(Courtesy of Mary Jane West-Eberhard)*

But we are primates, and primate females in the right frame of mind find all babies fascinating and attractive. For such females, the most important ingredient for eliciting love is not the molecules producing a particular scent, or genetic relatedness, but physical proximity over time. Whether a new mother will be willing and able to keep her baby close long enough for this old primate magic to work depends on her psychological state, as well as her physical and social circumstances.

Human mothers learn to recognize their own babies in the days right after birth, and gradually "fall in love." Since babies return the favor, the baby's attachment to the mother further reinforces her commitment. This is why when babies are adopted, the younger the better. But just as surely as the forty- to fifty-year lifespan of a standard-issue ape ovary is fixed and more or less immutable, fixed also are the developmental needs of infants. The same processes by which babies attach to and learn to love their caretakers impose a tremendous cost when such attachments never form, or when sequential attachments are ruptured. Below a certain level of nurturing, the developmental outcome is disastrous. If a baby does not perceive that he or she is growing up among committed kin, even a measure of care considered adequate for development within normal limits may not be enough to produce an adult who realizes his or her full human potential of empathy for others.

By itself, giving birth does not guarantee that a mother will care for every baby she bears. A woman predisposed to be a mother can learn to love any baby, while a mother not so disposed does not even learn to love her own. This is what it means to live with the emotional legacy of a human who evolved in a hominid context where mothers relied on assistance from others to help rear offspring.

We are a clever and highly innovative species, but not infinitely so. Our past matters, not just on the physical, but on an emotional front. Does this mean we have no conscious choice over how we lead our lives? Not at all. People exercise free will all the time—but only in those areas where Mother Nature cut them some slack. A woman can choose which baby she will adopt, but falling in love with that child will not be automatic. Nor can a woman just will herself to love a child, or respond to legal prescriptions that she do so. This is one reason why a certain number of children placed in foster homes or adopted end up being returned—a painful outcome for all concerned that few adoptive parents wish to discuss. This book will make clear why efforts to legislate a mother's love—by telling a mother with an unwanted pregnancy, for example, that she must carry it to term—are so often destined to end badly.[30]

Many biological constraints—especially those having to do with disease, and with increasing or decreasing fertility—have been removed by medical innovations. Today, we have more options than humans have ever had before. But such options only crop up in the interstices we carved for ourselves with new technologies. Yet these new technologies, as often as not, bring with them new constraints. Consider just one. It takes longer, and requires more investment from parents than ever before, for children to acquire the education they need to negotiate effectively in an increasingly complex world. Such costs in turn alter emotional equations in the lives of parents, rendering some less willing to have children.

By placing human mothers and infants in a broader comparative and evolutionary framework, I offer a new slant on what babies need from their mothers, what mothers need from others in order to provide it, and explain *why*. Explanations do little to solve such practical problems as how to enlist more infant care from fathers and alloparents. I offer no plan for safe and affordable daycare. But I at least provide rough outlines for what adequate allomaternal care has to include, and why it would be worth any community's while to make such care a priority. Knowing more about the processes and history—both deep and more recent—that underlie maternal emotions and infant needs has to be the first step in meeting them.

PART TWO

Mothers and Allomothers

*Literature is mostly about having sex
and not much about having children.
Life is the other way around.*
—David Lodge, 1965

Priming an alloparent *(Courtesy of Geert Van den Broeck)*

The Milky Way

Our deeds determine us, as much as we determine our deeds.
—George Eliot, 1859

"I s sex destiny?" When this question is posed, it's a safe bet that the under-lying agenda has to do with what women *should* be doing. Should they be home caring for their children or off pursuing other interests? A comparative look at other creatures that (like humans) breed cooperatively and share responsibilities for rearing young with other group members reveals that sex *per se* is not the issue. Lactation is.

Caretakers of both sexes, wet-nurses, even "daycare"—none of these are uniquely human, nor particularly new. They are standard features of many cooperatively breeding species. As we saw, cooperative breeding is exquisitely well developed in insects such as honeybees and wasps. Shared provisioning is also common among birds such as acorn woodpeckers, bee-eaters, dunnocks, and scrub jays. Although cooperative breeding is uncommon among mammals generally, it is richly developed in species such as wolves, wild dogs, dwarf mongooses, elephants, tamarins, marmosets, and humans. In all these animals, individuals other than the mother ("allomothers") help her provision or otherwise care for her young. Typically, allomothers will include the mother's mate (often but not necessarily the genetic progenitor). Individuals other than either parent ("alloparents") also help. These helpers are most often recruited from kin who are not yet ready to reproduce themselves, or from subordinates who do not currently—or may never have—better options. In the human case, the most important alloparents are often older, post-reproductive relatives who have already reproduced.

Among mammals, the trend toward having young who require costly long-term care began modestly enough. It probably began with an egg-laying brooding reptile that started to secrete something milklike. Such egg-layers gradually developed glands especially equipped for milk production. Only among mammals did one sex come to specialize in manufacturing custom-made baby formula, to provide something critical for infant survival that

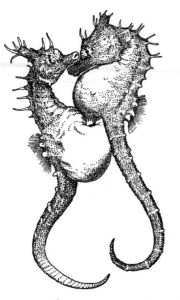

Fig. 6.1 Female sea horse depositing her eggs in the male's pouch, "impregnating" him.
(Drawing by Dafila Scott)

the other sex could not. This peculiarity has had many ramifications, especially as infants became dependent for longer periods in the primate line.

The ante was upped substantially when primate mothers, instead of bearing litters, began focusing care on one baby at a time. These singletons were born mature enough to cling to their mother's fur, to be carried by her right from birth and for months thereafter. Whether or not this intimate and prolonged association is the mother's destiny, *sex* is not the issue. Lactation is.

What Is Lactation About?

Other forms of caretaking—fathers brooding eggs, bringing food, protecting babies—are not nearly so sex-specific. Even gestation is a function that in rare cases (for example, the sea horse) a male takes on. But not lactation. Why—with the sole exception of one rare fruit bat—does lactation appear to be exclusively female? How did these curious secretions get started?

At first glance, a mother sea horse might seem to have a sweet deal. She sallies up to her mate, injects her eggs into his belly pouch, and then, carefree, propels herself off to feed and make more eggs. Meanwhile, back at the male's pouch, the sea-mare's last batch is fertilized, toted, and kept safe in the ballooning brood chamber of the now extremely pregnant male. At birth, as many as 1,500 fully formed but still minuscule and defenseless sea-foals are sprayed out into the open ocean. The sea around them teems with predators and competitors, many bigger than they are. Forced to fend for themselves immediately after birth, almost all starve.

Viviparity means keeping infants safe inside some sealed chamber within the parent's body till they can be born alive, as opposed to protected in an egg till hatching. But by itself, viviparity offers tiny, still helpless creatures only

Fig. 6.2 Parental care is rare in reptiles. But there are exceptions, such as maternal protection of young by mother crocodiles, or the reconstructed behavior of this duck-billed dinosaur affectionately named *Maiasaura*, "the good mother reptile." Her nest contained ten to twenty altricial eggs which hatched so early and so helpless that family members were required to provision them. *(Drawing by Marianne Collins; reproduced with permission from Random House, London)*

the slimmest toehold on posterity. Why not linger in the womb longer, and grow bigger before venturing into the world?

The parent so encumbered becomes less efficient at foraging, eats less well, is less able to evade predators. Lactation evolved because it beat available alternatives for fueling rapid growth in warm-blooded babies while allowing a foraging mother to safely stash immatures who will be somewhat safer in a nest or burrow than if they must fend for themselves.

Merits of Lactation

The currently favored hypothesis for why dinosaurs disappeared is that global climate fluctuations led to mass starvation among immatures. Mammals pursuing "the milky way" had an obvious advantage. Maternal provisioning through lactation spares immatures the hazards of foraging and competing with more mature animals to stay fed.

Being able to rely on the maternal larder long after birth buffers immatures from local scarcity. Remaining with a lactating mother provides a stable and viable environment for immatures who would otherwise be unable to survive severe climate fluctuations. As the dinosaurs died out, lactators came into their own.

Mammalian mothers were alchemists, able to transform available fodder—shellfish, grass, insects, other mammals, even toxic plants—into biological white gold: a blend of highly digestible nutrients and antibiotics that fuels and protects immatures during the hazardous days just after birth. A mother stockpiles energy, protein, and minerals as fat deposits on her body and doles out these repackaged nutrients on a mutually beneficial, often quite flexible, schedule.

Like the young of a cold-blooded reptile, the young of a mother bear enter the world just a fraction of their mother's bulk. But thereafter baby reptile and baby mammal grow differently. Take the American black bear. During the abundance of summer, the mother mates. Copulating induces her to ovulate, but the fertilized ova do not immediately implant. They enter a state of suspended development. As winter approaches, the mother grows drowsy and retreats to her cave to save body fuel by hibernating for the winter. Even then, continuing her pregnancy is not automatic.

If the bear has managed to store up enough fat to sustain lactation, the fertilized eggs (or blastocysts) implant, gestation and birth ensue, and the mother sleepily suckles her babies till spring. Not enough fat, and the earliest abortion nature offers takes place: implantation never occurs. The next conception is postponed pending improvement in the mother's circumstances.

If there is enough fat on board to sustain lactation, implantation proceeds and the mother gives birth to two to four cubs. After birth, cubs have months to grow before their mother brings them out in the open, and years more before they must forage on their own. If just one is born, the mother occasionally abandons it rather than allow a singleton to monopolize such a long interval between births. That way she gives birth sooner to the full complement of which she is capable.

By the time the three-year-old bears launch out on their own, they are respectably burly versions of their mother, able to eat what she eats and at least to threaten whomever she scares off. By contrast, at independence, their reptile counterparts are but twinkles in the eyes of competitors and predators alike.[1]

Even without such gimmicks as hibernation or delayed implantation, flexible lactation schedules permit mammals that originated in the tropics to adapt to freezing climes. Macaques in northern Japan give birth and spend the summer feeding and suckling their new babies. As fall approaches, the young are weaned. Both mother and weanling forage furiously to lay down fat

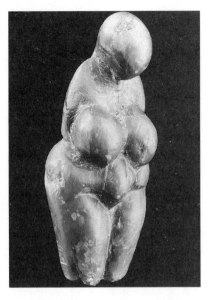

Fig. 6.3 Fat may or may not be a feminist issue, but it is definitely a reproductive one. Human imaginations have long celebrated links between fat and fertility as in this 25,000-year-old statue called the "Grimaldi venus." *(© Photo RMN—J. G. Berizzi, Musées des Antiquités Nationales, St.-Germain-en-Laye)*

stores before winter. By midwinter, all other fare is buried under a blanket of snow, and the monkeys must subsist on tree bark far too formidable for immature teeth to gnaw. At this nadir in food availability, months after their young were first weaned, mothers miraculously resume lactation.[2]

Storing Fat

It is advantageous for mother mammals to be able to stockpile fat in advance, on layaway, to be drawn upon later as needed. Still, timing remains crucial. Fat is stored out of the way in tails, buttocks, or humps on the body, where mobility will not be impeded. Deposition is under tight physiological control, and often—as in the case of the Japanese macaques—postponed as long as possible.

Consider our own species. With sufficient fat on board, some fat cells start to secrete a hormone called leptin, which triggers endocrinological transformations leading to menarche. Some time after that, a young woman actually becomes fertile. By then she will have laid down sufficient fat to help carry her through pregnancy and lactation, what some anthropologists term "reproductive fat."[3] A woman's reproductive fat is concentrated in the buttocks and upper thighs, or around the abdomen. Inadvertently, modern women relive these golden years of fat deposition when they take synthetic estrogen—a reproductive hormone—after menopause.

No comparable fat deposition occurs in men. Prior to puberty, little boys and little girls have equivalent layers of fat. In the two years after menarche, however, the proportion of fat on a girl's body increases by 214 percent, in preparation for reproduction.[4] Recognizing this, many societies give pubescent girls special foods and lessen their workloads. In many areas of village

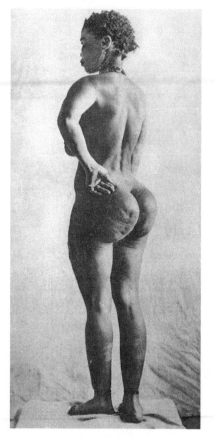

Fig. 6.4 Steatopygia, or extreme accumulation of fat on the buttocks, probably evolved as an adaptation to unpredictable food resources. A woman needs approximately 74,000 calories beyond maintenance costs to sustain pregnancy, and thereafter on the order of 600–700 calories a day to sustain lactation. Pregnant women may also respond to recurring seasons of scarcity by reducing activity levels or lowering their basal metabolic rate, reducing the total number of calories needed to sustain pregnancy—though at an unknown cost in terms of infant mortality or to themselves in terms of "maternal depletion" as critical bodily reserves like calcium are used up.[5] *(From Schultze 1928)*

India, for example, boys may be generally preferred and better fed, but when girls begin to menstruate, they are given sweets and other special treats, like eggs.[6]

A cottage industry has grown up busily generating hypotheses to explain the special buildup of fatty tissue around an adolescent girl's mammary glands. These range from the idea that breasts, like buttocks or a camel's hump, are a convenient place to store fat, to the proposal that large breasts helped hominid adolescents compete with other females for a good husband by advertising (either "deceptively" or "honestly") that she has stockpiled sufficient fat to sustain lactation. Unless the mother is starving, however, fat stored in the breasts is not normally metabolized to make milk, nor are large breasts per se (as opposed to generalized body fat) correlated with being able to produce more milk.

What is unusual about womanly breasts is how early they appear. Breast development begins at puberty, even before menarche, years before a woman is capable of conceiving or needs breasts to suckle babies. Other primates have prominent breasts, but this tissue builds up only prior to and during lactation. After weaning, monkey mothers revert to the flat-chested, button-nippled look of females who have never had a baby.

The disconnect between permanently enlarged breasts and lactation leads some biologists to speculate that breasts evolved as advertisements, but not just of their nutritional stores. Symmetrical spheres of fatty tissue might show off a woman's phenotypic quality, demonstrating how resistant she has been to disease and the various other developmental insults life dishes out.[7] The primary function of breasts, however, remains milk production.

Lactation and Lifestyles

Mother's milk—how lean or fat it is, how long lactation lasts—can reveal a surprising amount about lifestyle. Among small mammals, such as tree shrews or hares, high metabolic rates mean mothers must constantly forage to support themselves and their litters. Mothers are away from their nests hours or days at a stretch. Only unusually rich, high-fat milk tides the infants over these long absences. By comparison, infants born to early hominids— whose mothers carried them—were in constant access to their mother's nipples. Hence, like all primates, they could survive on dilute milk with moderate amounts of protein and fat and high levels of sugars. This milk, composed of 88 percent water and, like cow's milk, 3 to 4 percent fat, is specially adapted to the needs of an infant who will be able to nurse for a few minutes several times an hour and go on nursing for many months.[8]

Lactation is a perpetual buyer's market, in the sense that intensity of suckling adjusts supply to demand from the consumer, except that in some mammals the seller (meaning the mother) determines the size of the consumer to begin with. Mothers in good condition produce larger infants who require more milk sooner. Depending on her own condition, a mother may wean sooner if she is finding the burden too much to bear, or later, if she has a small litter or they are growing slowly.

———

It is easy for humans to fall into the anthropocentric trap of assuming that there is something inherently superior about placental mammals like ourselves who gestate embryos till they become babies. In fact, whether or not our own slower, more deliberate mode of reproduction is actually superior depends on the environment. In terms of the art of iteroparity as practiced by mammals who breed sequentially in totally unpredictable, extreme environments, none surpasses that of marsupials. Marsupial mammals give birth

Fig. 6.5 Mammary tissue builds up during pregnancy. In other apes, like this gorilla mother, enlarged breasts are invariably a sign that a female is lactating. Only among humans do women develop prominent breasts prior to first birth, during adolescence. *(Alexander Harcourt/Anthro-Photo)*

after short pregnancies to immatures the size of thumbnails, virtual fetuses that develop outside the womb.

Adapted to the unpredictable rainfall of the Australian outback, kangaroos have evolved a veritable breast-milk cafeteria to cater to infants of different ages. At one nipple, the mother produces low-fat "growth formula" for the tiny joey latched on in her pouch. Simultaneously, a distended nipple beside it is producing a high-fat "activity formula" for the older joey hopping beside her who comes back for an occasional drink.[9]

The kangaroo's ovarian assembly line is specifically geared for high turnover. Should either the joey in her pouch or the joey at foot for any reason cease to suckle, levels of the nursing hormone prolactin—her body's work order for more milk—fall precipitously. At this signal the tiny blastocyst in waiting (a nearly hollow globe of cells produced by the fertilized egg, inside of which an embryo will develop) is activated. The spare blastocyst emerges from developmental dormancy (diapause) to serve as a replacement.

Mortality in the outback is so high that a reproductively less flexible species would by now be as extinct as dinosaurs. But a kangaroo is a mother for any season. She simultaneously juggles progeny at three different phases of development: blastocyst in the pipeline, exterogestate fetus on the teat, nursing joey at foot. She can abort this process at any phase, with minimal risk, and without breaking her reproductive stride. A mother kangaroo closely pursued by a predator can jettison, or just allow to topple out, the joey in her pouch.[10] It's safe (for her), and very quick. Instead of stopping to scoop it up, the mother—her load suddenly lighter and her pursuer temporarily distracted—escapes with plan B already under way, since cessation of suckling activates the dormant blastocyst.

————

High-fat milk does not always mean the manufacturer is an absentee mother like a tree shrew. Sometimes rich milk is delivered in a hurry: in worlds where mother and infant are likely to be separated soon, it needs to be. Hooded seals, although relatively large mammals, have one of the shortest periods of lactation known—as brief as a week. Mothers stockpile blubber in advance, then give birth atop ice floes to pups who imbibe pure cream—a high-protein formula containing 60 percent fat, approximately 1,400 calories in an eight-ounce glass.

A seal pup can gain fifty pounds in a matter of days. It needs to. If brittle nursery platforms crack apart, weaning can befall them at any moment, leaving a pup alone in a frigid world where only the plump survive.[11]

————

With rare exceptions like the Dyak fruit bat, lactation is a female specialty. Males do not normally produce milk.[12] This may have to do with the fact that in animals with internal fertilization and gestation, the mother is the only parent who can be absolutely certain the infant is hers; males, by contrast, should evolve to be more wary about investing in offspring possibly not their own. Taking into account the resources she has already committed to gestation, it is her infant's dependence on milk and her ability to provide it (not her own sex per se) that seals her fate.

Given that infant provisioning has not always been sex-specific, and that it is not necessarily so, how did lactation come to be this utterly critical, uniquely female specialization? What left mothers holding the teat?

Prolactin and Caretaking

No one knows how lactation first evolved, but among the "accessories after the fact," the lactation-promoting hormone called prolactin is an obvious suspect. This simple protein, clearly implicated in lactation, and named in honor of its pro-lactating function, is so ancient, versatile, multipurpose, and widespread, that its fingerprints are *everywhere*—at the "crime scene" and everyplace else as well. Prolactin could just as well be called the "stress hormone," or, even more aptly, the "parenting hormone."

Unquestionably, prolactin was around, and a player, when lactation got under way. But unfortunately (for those seeking simple answers), prolactin was also around and a player in bird and fish species, among which lactation never got started. Secreted by the pituitary, prolactin molecules crop up in both sexes of a broad array of animals and are implicated in hair maturation, puberty, fat metabolism, and coping with stress. For several hundred million years before mammals ever appeared, prolactin was regulating metamorphosis in amphibians and water balance in the tissues of bony freshwater fish. Yet it is a curious feature of this hormone that wherever mothers or allomothers are motivated to protect or provision young, prolactin can be detected at elevated levels.

The role of prolactin is not easy to interpret. When Cambridge University primatologist Alan Dixson first reported levels of circulating prolactin five times higher in male marmosets carrying offspring around than in males not carrying any, his report was greeted with skepticism by experienced endocrinologists who knew that prolactin also goes up when animals are stressed. They hypothesized that the testing procedure itself was the source of the disparity—specifically, that a male on childcare duty would be far more stressed by having his blood sampled than would a male on his own.

In an ideal form of this experiment, scientists would somehow block the effect of prolactin, then artificially replace the hormone and monitor what effect these molecules have on caretaking. But technically, this is hard to do. We have only tantalizing hints about the function of this protean and ubiquitous hormone, nothing definitive. Even without the perfect experiment, it is already clear that *stress* was not causing Dixson's results. Experimenters also found high levels of prolactin in male caretakers when samples were obtained from *unrestrained* donors who were not stressed. Later, similar patterns were also found in the California mouse, where monogamously mated males are heavily involved in caretaking from birth to weaning.[13]

Prolactin levels spike when male sea horses gestate.[14] They are elevated in both sexes whenever scrub jay alloparents (typically year-old nonbreeders of both sexes) fly off and bring food back to nestlings. Among tamarins, prolactin levels go up in males right after their mates give birth. Prolactin levels rise higher in experienced than in inexperienced, first-time fathers, which is also true for mothers in species where mothers do most of the caring.[15]

Curiously, prolactin is also unusually high in contexts that seem more nearly offensive or defensive than nurturing, such as aggression *in defense of infants,* or when birds engage in tactics of disinformation to fool predators. In a wide variety of birds, especially those that nest on the ground (like kildeer, mallards, or gadwalls), the mother (often both parents) will flounder conspicuously on the ground, seemingly helpless, pretending to have a broken wing or other incapacitating injury to distract predators who have gotten too close to her nest. The nearer her eggs are to hatching, and especially if she has chicks, the greater the risks this daredevil takes, letting the coyote or dog get heart-stoppingly close before she takes evasive action. I've seen mother gadwalls in the grass, flushed by a dog, just make it to the water's edge, flapping and dragging one wing all the way, and then lead the dog swimming in circles in pursuit of what looks to all the world like a hopelessly crippled bird, before she dives one last time, surfaces, and miraculously flies away, her chicks hidden in the grass on the other side of the pond. Either as instigator for this behavior, or as a consequence of it, prolactin levels in birds engaging in these displays shoot up.[16]

Heroics aside, the higher the prolactin levels, the more attentive to infant needs both males *and* females, both parents *and* alloparents, are. Somehow high prolactin levels are implicated when mothers lactate *and when alloparents just help out.* Possibly they interact with other hormones in ways that make prolactin more nearly accomplice than ringleader. Engaging in nurturing behaviors, in turn, seems to make the pituitary secrete more prolactin. (As Eliot put it, "Our deeds determine us, as much as we determine our deeds.")

The Ambassador of Caretaking

Hormones like prolactin specialize in cell-to-cell communication. They are ambassadors, not so much empowered to *cause* results as equipped to alter the probability that, once activated, signals are passed on. (The word *hormone* derives from the Greek "to urge on.")

Neurons in the brain are the actual orchestrators of behavioral acts, and the integrators of behavioral states. Hormones are just the instigators, and their effectiveness depends on how receptive to their message the target tissues are. Hormones have an effect only where tissues are predisposed to listen. The behavioral endocrinologist John Wingfield sums up current knowledge on this subject thusly: it's the neuronal circuits that "cause behavior"; hormones merely "influence the rate at which a behavioral trait may be expressed under appropriate circumstances."

The cowbirds Wingfield studies provide a curious example. These justifiably unpopular birds build no nests, nor do they care for young. Rather, cowbirds are brood parasites who lay their eggs in the nests of other birds. Cuckolded parents find themselves in the position of bringing food back to feed an alien chick who has grown to twice the size of the legitimate denizens of the nest, and easily lords it over them, out-competing the host's own chicks for food.*

Yet even among these nest parasites, levels of prolactin circulating in the blood go up after egg-laying.[17] The difference is, cowbird mothers do not then caretake. One suggestion is that in cowbirds the neural receptors to prolactin are insensitive to its message. Or perhaps other stimuli must also be present to elicit caring. Even while experiments with cowbirds continue, what is already clear is that, once targeted tissue starts to "listen," cells become increasingly sensitized to the message and may induce changes in production of the message elsewhere. This is why when endocrinologists are asked "Does the hormone cause behavior or does the behavior cause the hormone?" they smile wryly and answer, "Yes."[18]

———

There are very few good ideas that are really new; there are even fewer new molecules. Rather than resorting to spanking new products, which might require waiting eons for the right mutations, natural selection relies on what is already available in the larder. After more than a hundred million years of selection operating on both "producer" and "consumer," however, the concoction that mammal mothers manufacture is so perfectly suited to both as to

———

* Life-history tradeoffs provide the best explanation for why cowbird babies are so big and grow so fast: their freeloading mothers can devote more resources to egg production than their hosts can, since the brood parasites won't need to allocate resources to caretaking.

appear to have been designed. It is easy to forget that the original recipe for lactation was slapped together from leftovers.

Origin of Lactation

Prolactin is a true endocrinological jack-of-all-trades, implicated in a broad range of physiological activities, from maintaining water balance to seemingly bizarre displays undertaken to lure predators away from nests. True to its eclectic heritage, this versatile hormone best known for inducing milk production—and later named for that connection—was not even first discovered in a mammal.

Prolactin was first identified when endocrinologists injected a mystery substance into *birds* and noticed that, whatever it was, these molecules caused birds of both sexes to develop brood patches on their breasts, areas of heavily vascularized naked skin that brooding birds apply, like heating pads, against incubating eggs.

In 1935 Oscar Riddle identified the substance as prolactin and discovered that an injection of it induces broodiness in birds. Whether in hens or castrated males, increased levels of prolactin are associated with a bird's urge to hover over, cover, and keep either eggs or young warm and safe. Brooding urges can be so strong that they are extended indiscriminately to young of other species.

Among such birds as pigeons, doves, emperor penguins, and flamingos, prolactin also stimulates males and females to produce "crop milk," a cheesy concoction of partially digested food diluted with mucus sloughed off from cells lining their throats. These ingredients may sound unappetizing to people accustomed to our own fare of lipid-rich excretions from cows' udders, but this fatty stuff has an avid following among just-hatched squabs, pigeon fanciers, and intellectuals intrigued by unisex roles in nurturing young.[19]

"Though a man does not brood like a pigeon . . . ," lamented Jean-Jacques Rousseau in the eighteenth century, monogamous father birds among mourning doves regularly regurgitate more crop milk into the beckoning beaks of babes than the mothers do.[20] This makes one wonder what a shot of prolactin might do for the sensibilities of indifferent fathers. (For more on this father of the Enlightenment's own fathering, see p. 310.)

Here and there parentally manufactured baby foods can be found: for example, the protein provided her young by the breed-and-then-die spider mothers, whose dissolving bodies are consumed, or the milky substance

Fig. 6.6 Among wolves, subordinate females sometimes undergo pregnancy-like hormonal changes, including elevated prolactin levels and milk production, even though not actually pregnant. Pseudopregnancy can lead to fascinating, if maladaptive, behaviors among their descendants now living under very different conditions. The mother of these kittens was no doubt puzzled when a pseudopregnant Jack Russell terrier bitch drove her away, adopted her kittens, and then settled down to suckle them. *(Sarah Blaffer Hrdy/Anthro-Photo)*

cichlid fish secrete on their scales, a mixture of skin and mucus that is nibbled off by fingerlings. But compared to crop milk and these other substances, lactation stands out because it involves specialized equipment (mammary glands) developed in only one sex.

The fact that mammary glands developed in just one sex indicates that lactation originally evolved in some animal where eggs were tended primarily by mothers, and newly hatched infants were already in close proximity to their mothers. Otherwise, there seems to be no reason why both sexes wouldn't lactate, just as both sexes in birds produce crop milk.

The best extant facsimiles for ancient mammals are the monotremes. Monotremes are egg-layers, like birds, but after the eggs hatch *only* mothers ever care for them. The female duck-billed platypus, for example, incubates her eggs safe inside a grass-lined burrow. Sealed up with their mother, the hatchlings feed on milk dripping from hairs surrounding teatless milk-producing glands on her front. Suppose, Darwin had speculated, a sweat gland of an egg-layer got plugged up in some reptilian precursor to mammals. Suppose the resulting secretions just happened to enhance the survival

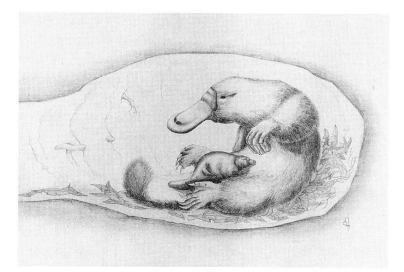

Fig. 6.7 Drawing of platypus with young feeding on nippleless mammary glands. *(Drawing by Dominic Fanning)*

of a hatchling who imbibed them. Monotremes make Darwin's accidental-provisioning hypothesis seem quite plausible.

The main alternative theory to explain how lactation got started is similar in outline. The disinfectant hypothesis posits antibacterial defense of eggs rather than nutritional benefits to infants as the initial benefit. Biologists Daniel Blackburn and Virginia Hayssen propose that a particular component of bodily secretions called lysozymes—enzymes present in human tears and blood that digest bacteria—was present in secretions accidentally applied to eggs. If this serendipitous secretion happened to protect eggs from fungi and bacteria during incubation, then leaking mothers would have higher hatching success than mothers who ran tighter ships. (And what better facilitator for a leak than prolactin, a venerable regulator of water balance.)

If eggs were being incubated someplace as damp and dark as a marsupial's pouch, all the more reason why dousing them with some lysozyme-rich disinfectant might enhance maternal reproductive success. In support of the hypothesis, Hayssen points out that a protein specific to mother's milk (alpha-lactalbumin) evolved from lysozymes.[21]

Newly hatched babies who lapped up this protein-rich antibiotic would have gotten a nutritional boost along with their immunological dose. If the antibacterial hypothesis is correct, colostrum, the thick yellowish fluid present in the breasts before and for several days after birth, may be the closest analogue to ancient mother's milk. Colostrum is packed with antibodies but, among humans at least, contains only a sixth of the calories found in the thinner, blue-tinged real thing.[22]

Across cultures and through time, opinions about colostrum have been remarkably variable. For reasons that no one understands, people from populations as disparate as the foraging !Kung, peasants in contemporary rural Haiti, and medical doctors in seventeenth-century England have all deemed colostrum harmful and advised mothers to express it, rather than risk their newborns ingesting it. Within days of birth, colostrum is replaced by real milk, and suckling is postponed till then. Given colostrum's beneficial consequences, this delay and waste seem strange. When historian of medicine Valerie Fildes took a closer look at changes in British breast-feeding practices over time, she discovered a dramatic decline in infant death rates in the first month after birth right at the time doctors changed their mind and mothers were recommended to once again put babies to their breasts immediately after birth.[23] History books don't normally have much to say about colostrum, but it is increasingly evident that this one is an aperitif with consequences.

Cellular Equivalent of a Pharmacy

Although parents in some cultures dispose of colostrum, in many others they incorporate colostrum, with its special properties, into customary childcare. An American nurse collecting a milk sample in a Swedish clinic was surprised when a new father requested a dab of the fluid. He immediately smeared it on the baby's rump. "Why?" she asked. "Oh, to prevent diaper rash. A drop of milk can also be applied in the baby's eyes, to prevent infections." Like colostrum, milk too provides some immunological protection.

Laboratory experiments corroborate such folk wisdom. Fresh mother's milk in a test tube kills one of the main dysentery-causing amoebas, *Entamoeba histolytica,* along with another common diarrhea-causing parasite, *Giardia lamblia.* A particular glycoprotein in mother's milk (lactadherin) has been shown to protect against rotavirus, one of the major causes of infantile diarrhea. I loved the heroic gentle-David-slays-dangerous-Goliath ring to the title of the first antiamoeba experiments published in the journal *Science:* "Human Milk Kills Parasitic Intestinal Protozoa," it announced.[24]

Immunological defenses are transmitted from mother to fetus through the placenta, and after birth they are imbibed by the infant from mother's colostrum and milk. Immunological protection ranks as one of the most important of all "maternal effects." These antibodies are tailor-made to protect the infant from precisely those bacteria, viruses, and intestinal parasites that are present in the environment and that the baby is in greatest danger of contracting. Like the cellular equivalent of a pharmacy, the mother's mammary glands deliver secretory immunoglobulins that work like specialized prescriptions.

Lest anything so essential for survival be left to chance, the mother is reexposed every time she ingests her baby's excretions—another good reason in addition to hygiene why mother mammals (probably no one in your family) spend so much time licking their babies. Exposed to the same pathogens her infants are, she manufactures just the right antibodies. By the fourth month of lactation, a human mother secretes up to half a gram of antibody into her milk daily.[25] Thus are mammalian young permitted the inestimable luxury of an immune system that revs up slowly, over a period of weeks and months.

Elixirs of Contentment

Breast-feeding meant prolonged proximity between mother and offspring. In addition to providing a new delivery route for antibacterial agents like those in colostrum, breast-feeding also led to the evolution of brand-new chemical agents with the unprecedented task of promoting intimacy.

I have been told that the feeling of euphoria that I experienced after the birth of my first child was a sign that one of these, the peptide hormone oxytocin, was present in large amounts. Even more fascinating is the likelihood that when my daughter breast-fed, some of the circulating "peace and bonding" hormones in me were transmitted in my milk to her, acting as a mild sedative, leaving us both with the impression that being near each other was deeply satisfying. Unlike prolactin, oxytocin does not have much of a track record outside of the mammals. Apparently, this kindest of natural opiates is a relative newcomer on the endocrinological scene, a true mammalian specialty.

Oxytocin is manufactured in both the brain and the body (in the ovaries or

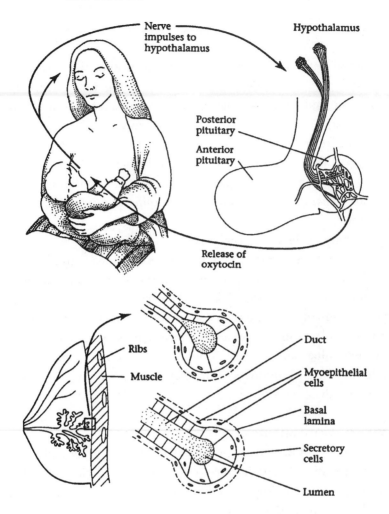

Fig. 6.8 Upon stimulation of the nipple by the baby, oxytocin is released which causes cells around the mammary glands to contract. This response is subject to classical conditioning, so that just the sound of her baby's cry can produce a mother's "let-down" of milk. *(Courtesy of Randy Nelson and Sinauer Associates)*

testes). All during pregnancy, my pituitary had been expanding, and extra receptors to oxytocin had set up shop in my brain and also in my uterus, where oxytocin would soon have a special job to do.[26] As the birth process began, pulses of oxytocin set off contractions in the smooth muscles of my uterus, causing my daughter to be expelled in a matter of hours—hence the name oxytocin, from the Greek *okus* for "swift" and *tokos* for "birth."

Two days later, as my milk came in, my daughter's sucking mouth stimulated my nipples and signaled the pituitary to release more oxytocin, which traveled in the blood to the mammary glands and caused the smooth muscles around cavities where milk is stored to contract and reply to her request with a thin stream of milk. As I became more conditioned to nursing, just the thought of her, or her cry, was enough to trigger this "let-down reflex" and cause a patch of wetness on my blouse.[27]

In spite of its no-nonsense official function, the hormone for swift birth is profoundly implicated in more subjective responses as well—the rush of warm sensations that suffuse a mother when she breast-feeds, the tapering off of inhibitions as two mammals sit companionably side by side, or groom each other, or when longtime mates nuzzle or rub each other. Oxytocin levels go up whenever a person undergoes a good massage. Manipulation of the breasts or genitals during lovemaking also stimulates release of higher levels of oxytocin, which is probably why a marked increase in oxytocin is detected during orgasm.[28]

Some women—as well as men—are unnerved by what they perceive as the "sexual" sensations mothers experience during breast-feeding. Tales abound of officials at public welfare agencies who react negatively to maternal confessions about sensual feelings from nursing their children. Some go so far as to seek removal of the offspring from so "perverting" an influence. It might be helpful for all concerned to keep in mind that maternal sensations have clear evolutionary priority in the pleasure sphere. Long before any woman found sexual foreplay or intercourse pleasurable, her ancestors were selected to respond positively to similar sensations produced by birth and suckling, because finding these activities pleasurable would help condition her in ways that kept her infant alive. It would be more nearly correct, then, to refer to the "afterglow" from climax as an ancient "maternal" rather than sexual response.

————

How susceptible an animal is to the magic of oxytocin may vary depending on its evolutionary history. This realization led neuroendocrinologist Sue Carter to predict that oxytocin would be more important in animals with long-term pair bonds, where affiliative relations between the sexes are translated into enhanced infant survival. This is precisely what her colleague Tom Insel found: many more brain receptors to oxytocin in monogamous prairie voles than in a closely related but more solitary species of vole in which males mate polygynously with several females in a love-'em-and-leave-'em pattern.[29]

Lactation and Destiny

Which came first, the chicken or the egg? The egg obviously, answers evolutionist John Hartung: "It was laid by a bird one mutation away from being a chicken." Nothing about the therapsid reptile that laid the egg that became the first milk-secreting creature could possibly have foretold how far-ranging the repercussions of this mutation would be. Out of what initially may have been little more than a bead of sweat, a new prolactin-mediated, medicated growth formula would emerge. There was nothing about the origin of lactation to indicate that it would ever have anything to do with the evolution of intelligence, but it did.

Three things about lactation were new. Undoubtedly some mothers had previously been sole caretakers among reptiles or birds, but both males and females would have been equivalently equipped to protect the nest and, when it increased their reproductive success, to provision young. Lactation was a caretaking adaptation linked to a specific sex as never before.

Second, this uniquely valuable formula for keeping infants alive would make the individuals who produced milk an even more valuable, even more limited resource than egg producers already were. As usual, males would compete for females, but with new ramifications to the old mating game, as females producing young then also nursed them. Each offspring that survives past weaning would have extracted enormous somatic investment from the mother, not to mention the opportunity costs—time out of her life spent nursing offspring she already has instead of making more. Mothers who could not sustain lactation long enough to make all that it cost her worthwhile would be selected against. Hence female ovarian functioning became more sensitive than ever to maternal condition and local circumstances. The longer the intervals between conceptions, the fewer fertile, ovulating females will be available to males. Lactation became one more factor intensifying competition between males for access to fertile females. Furthermore, other mothers, as well as males, would sometimes enter the fray to control the producers of this biological white gold.

The third innovation was the prolonged intimacy between mothers and infants that the availability of mother's milk permitted, and eventually required. Of the three corollaries accompanying lactation, this last—prolonged intimacy—would be the most significant for the evolution of social relationships and eventually new parts of the brain, and new attributes (especially intelligence linked to empathy) that, because we possess them, are of special interest to human beings.

Parenting behaviors were old hat, but intimate social relationships, and the neuroendocrinological underpinnings that evolved in the brain to maintain them, were very new. Where ecological conditions permitted daughters to remain near mothers, they did so, enjoying the protection and social support from birth such as only a mother and her matrilineal kin routinely offer.

The Beginnings of Social Intelligence

The parts of the brain devoted to biological regulation (heart rate, breathing) and the most basic drives and instincts (hunger, libido) and reflexes ("fight or flight," primal fears) are linked to the limbic system, which is located near the brain stem and hypothalamus (just behind the knobby part below your ear). Atop this older, "reptilian" brain, mammals evolved the newer, cortical portions of the brain, in particular the neocortex, which increases its possessor's ability to evaluate situations, weigh costs and benefits, plan and make decisions contingent on circumstances, including (and most especially) *social* circumstances. Though evolutionarily more recent than the subcortical portions, this neocortex is not entirely independent of them.

Evaluations made at the level of the neocortex can override those of the older parts of the brain, but these determinations are not free of limbic input. As persuasively argued by neuroscientist Antonio Damasio, rational decisions are continuously informed by old emotions. It is as if, without this primitive input, the brain ceases to plan in ways that enhance the survival and long-term fitness of the organism.

Not surprisingly, the more complicated the situations that an animal needs to evaluate, the larger (relative to body size) the neocortex needs to be. All mammals with complex social behavior have an expanded executive neocortex compared with the rest of the forebrain. Across the Primate order, species with larger group sizes also have relatively larger neocortices.[30] Add to this one of the strangest findings to emerge so far from genetics and neuroscience.

Ever since Mendel, scientists have understood that offspring inherit traits from their mother and from their father, depending on the genes they receive from each parent. Yes, but not exactly. Sometimes equivalent genes are expressed differently *depending on which parent they come from.* At least some of the time, this is due to a process called "genetic imprinting," which silences or inactivates information carried on the gene originating from the other parent. One of the most fascinating albeit still poorly understood cases of parent-specific gene expression has to do with a gene that affects a person's ability to adjust to social situations.

Like many extraordinary genetic discoveries, this finding began with an investigation of a genetic defect. It involved individuals born with only one X chromosome. Neither XY (a boy) nor XX (a normal girl), XO girls never develop ovaries, and they exhibit a suite of other traits known as "Turner's syndrome." D. H. Skuse and colleagues at the Institute of Child Health in London noticed a curious difference among girls with Turner's syndrome depending upon whether they received their lone X chromosome from their mother or their father. Although most people assume that women tend to be more socially adept than men are, girls whose X chromosome derived from their mother were *less* socially competent and less able to weed out distracting impulses than were Turner's syndrome girls who received their X chromosome from their father. These results led Skuse to speculate that many of the developmental disorders in speech and reading ability, as well as more severe conditions such as autism, to which boys are especially vulnerable, might have to do with silencing of a gene influencing social cognition and language that rides on the X chromosome.

It seems odd and counterintuitive that a Turner's syndrome daughter with her only X from her mother would be *less* sociable than a Turner's syndrome daughter who receives her lone X chromosome from her father. One possible explanation for this bizarre outcome emerges as soon as we think about it in the evolutionary context in which mammalian mothers ordinarily gave birth to sons and daughters.

Genetic anomalies aside, under what circumstances does an individual normally receive a single X from its mother? This happens every time a normal male is conceived. For under normal circumstances, a son's lone X chromosome always derives from his mother. That son's lone X chromosome would ordinarily be the only X chromosome ever unmatched and unbuffered by any corresponding chromosome apart from the characteristic little Y that male mammals have. If disinclination to form social relationships, or ineptitude at doing so, resulted from an unbuffered maternal X, all the better. It would reduce the incentive for a son to stay home. For among most mammals, it is females who remain among their kin (or are "female philopatric" or "matrilocal") and males who depart to go fight and breed someplace else, usually among strangers and away from their matrilineal kin. By contrast, daughters endowed with a "friendly" X chromosome from their dad to balance out the "irascible" X from their mom will be the ones most likely to stick around. This of course is pure speculation on my part. Cambridge University neurophysiologist Eric Keverne's explanation for the evolution of self-conscious intelligence in mammals is even wilder.

The Origins of Compassion

Almost all zoologists agree that if mammals are going to live in social groups, by and large it is the daughters who remain near their mothers, among their matrilineal kin. This is true for almost all prosimians and monkeys as well as for mammals generally. With this in mind, Keverne proposed that prolonged association among female relatives was the social environment of evolutionary relevance in which the "executive" regions of the brain evolved. These are the areas of the cortex that enable animals to size up social situations and devise coherent strategies to cope with them. Selection pressures favoring evolution of the executive brain cannot be understood without taking into account these matrilineal groupings.[31]

Extraordinary experimental evidence coming out of Keverne's lab provides indirect support for the idea that social intelligence originated in a matrilineal context. Keverne and coworkers showed that genes inherited from the mother have more influence on the newer, "executive" portions of the brain than do those inherited from the father. Keverne used specially bred mice to investigate this parent-specific gene expression. When he blocked out maternal input, babies were born brawny but small-brained. When paternal input was blocked, however, bodies were smaller, but brains were enormous. Whereas in the hypothesis about irascible sons and affiliative daughters from the XO example, sons were rendered more likely to leave, in Keverne's model the mothers render daughters more sociable, more likely to stay and be intelligent in their dealings with others. And why not? Gregarious, socially skilled daughters will remain nearby to interact with her and with hers.

But a mother's contributions are not just physiological. She also passes on a social legacy. The mother, more than any other individual, determines which other animals (for example, litter mates or matrilineal kin) infants

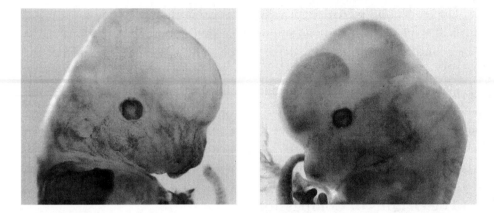

Fig. 6.9 Keverne and coworkers specially bred these mice so that genetic contributions from the father were silenced and contributions from the mother were enhanced, leading to the development of super-large brains (left). Fetal mice with the mother's genetic input reduced or silenced, and the father's genetic contribution fully expressed, developed smaller brains (right) even though bodies were larger. Braininess derived from mom, brawn from dad. *(Courtesy of E. B. Keverne)*

associate with. The mother, through her behavior, determines which other individuals her developing infants learn to identify as "familiar conspecifics" and therefore come to identify and treat like "kin."[32]

If corroborated, Keverne's work will qualify mammalian social attributes and the relationships that go with them as the most stunning maternal effects on record. Every creature with the social intelligence to respond to others as individuals and form enduring relationships shares this common mammalian heritage. All descended from distant ancestresses whose life cycle was characterized by long-lasting and intimate relationships between a lactating mother and her offspring.[33]

———

It is early days still in this remarkable research going on at the interface of genes, neuroscience, and behavior. Yet already it is clear that the "milky way" initiated the evolution of a charmed relationship between dependent immatures and their hostess. Many organisms, from coral reefs to wasps, are social; even a tadpole or a sea squirt has sufficient neurons to discriminate kin from nonkin. All kinds of other animals respond to stimuli and signals by reflexively providing a service to another—say, a father bird popping food into a beckoning beak. But only among the followers of the milky way did that old opportunist Mother Nature get to try out different neuroendocrine combinations and select for the ones that promoted the *social relationships* most conducive to infant survival and to the mother's long-term reproductive success. Sex may not be destiny in the sense that it is necessarily a female that cares for offspring, but lactation requires a female to stay near her young. Prolonged association between mother and suckling young provided both the chance *and* necessity for "social intelligence" to evolve.

Lactation turns out to be a key player in the evolution of animals who were both social *and* intelligent. In this sense lactation was a shaper of destinies not just for mothers, but for all individuals who would evolve a capacity for compassion.

From Here to Maternity

Mother's love is unconditional, it is all-protective, all enveloping; because it is unconditional it can not be controlled or acquired . . . because they are all children of Mother Earth.

—Erich Fromm, 1956

In the beginning there is some manageable rodent. Laboratory rats and mice are docile, have short generation times, and breed readily in plastic boxes, which is why so much of what we know about the genetics and physiology of maternal behavior comes from studies of these small mammals.

Shortly before she gives birth, a wild mouse transports bits of straw, feathers, fur, or, at my house, cotton stuffing from our sofa, into some safe-seeming nook. After generations of artificial breeding, a white mouse in a cage still feverishly piles sawdust into a soft mound before settling into the warm indentation her body creates.

No prior practice is necessary. Within minutes of giving birth, the mother mouse bites off the amniotic sac, eats the placenta, and sets about doing the needful. The same female who just weeks before would have either ignored or killed any newborn she encountered now gently picks up the pups in her mouth, one by one, and places them in a warm heap in her nest. If a pup tumbles out, the mother picks up the rubbery pink sausage and gently puts it back. Lost pups are located by the ultrasonic vocalizations they emit as chill sets in.

Right after giving birth, the mother is especially attuned to the smells and sounds of pups. She may also be attracted to their warmth, as they certainly are to hers. No more outlying squeaks, the mother settles her body over the huddled pups while they latch on to her nipples and nurse. The specificity of mammalian mothering boggles the mind. Even in a first-time mother, there is amazingly little trial, virtually no error. Postpartum housekeeping is direct, stereotyped, reliable. How can her efficiency in biting off the amniotic sac be other than "preprogrammed"? Yet we also know that there is sufficient variation among mothers that a fraction of them never care for pups at all.

What Is Meant by "Instinctive"

The deliberate way a mother meets infant needs, and the way infants evoke the necessary responses, testify to a highly integrated, coevolved system. The mother purposefully licks amniotic fluid and remnant blood off a helpless morsel that at another time or place she might well eat. She pays special attention to the anus and genital region. Like most nest- or burrow-dwelling mammals, stimulation from the mother's licking is essential for urination. Unless the mother performs this function, pups die.

Attractive chemicals in pup urine reward the mother for good housekeeping. She replaces her own much-needed electrolytes at the same time she keeps the burrow from turning into a growth medium for unwanted parasites and an odoriferous beacon to predators. In some kinds of rodents, mothers lavish special attention on the genitals of sons, licking them significantly more than those of daughters. When experimental psychologist Celia Moore blocked the ability of mothers to tell sons and daughters apart (which rodents do by smell), she found that sons deprived of this extra stimulation never developed neuronal pathways essential for adult sexual functioning.

––––––––

Is a mother born instinctively nurturing? ("She is a motherly type," I've sometimes heard it said.) Does something inside her change during pregnancy that makes her maternal? ("Before the baby was born, her nesting instinct really got going.") Is increased responsiveness due to stimulation from the infant? ("She just fell in love with her new baby.") Is a female gradually primed to be a mother by experiences?

For mice at least, the answer to these questions is: all of the above. "Instinctive" is a reasonable way to describe her maternal behavior, as long as it is understood that mother mammals do not necessarily exhibit automatic, full-blown commitment to infants immediately after birth. Rather, her "maternal instinct" unfolds gradually, in "baby steps" in which infants, too, are implicated.

Nature cannot be compartmentalized from nurture, yet something about human imaginations predisposes us to dichotomize the world that way. Nature versus Nurture, innate or acquired. The persistence, decade after decade, of a nonexistent dichotomy puzzles me. The recent discovery that

Fig. 7.1 In ancient China and Japan strains of "dancing mice" were as popular as the latest video game is today. In China, albino mice were used to divine the future, and the trade in them can be dated back as early as A.D. 307, small fortunes being paid for the rarest mutants. These illustrations come from an eighteenth-century Japanese text providing instructions for breeding piebald and dwarf varieties. In Darwin's time British traders carried "fancy mice" back to Europe. By the early twentieth century, scientists at places like Harvard were using their knowledge about dominant and recessive genes for coat color to study inheritance and by the end of the century to bio-engineer mice with certain genes "knocked out" in order to learn more about the role of genes in behavior. *(Figures are a composite from Chobei Zeniya's 1787 Chingan-sodategusa; "The breeding of curious varieties of mice," Kyoto. Reproduced in Tokuda 1935.)*

mother mice lacking a particular gene fail to care for their babies inevitably led to news stories about the gene responsible for "the very essence of mothering," as if with the gene she had it, without it she did not. The "nature of nurture" (as the headlines put it) may not be innate, but the need to organize information into neat binary oppositions like Nature versus Nurture may well be.[1]

Genotype to Phenotype

Scientists have long suspected that a specific area of a mammal's brain (the preoptic area of the hypothalamus) plays a critical role in sending instructions for behaviors related to nurturing. It is in the hypothalamus that a family of four genes known as "fos genes" are typically expressed. Fos genes work like master switches in an electrical system; they activate, or "turn on," other genes.

In order to study how switch genes work, Jennifer Brown, working with Michael Greenberg at Harvard Medical School, engineered a new genetic strain of mice lacking just one gene, known as fosB. These "knockout" mice seemed otherwise normal. It ought to be possible to compare mice without the fosB gene with those that had one, and in that way learn what the gene does.

The first step was to get fosB-less mice to reproduce so that sufficient numbers of knockout mice would be on hand for large-scale studies. Mice missing the fosB gene were slightly smaller, but otherwise healthy, able to mate, become pregnant, and deliver viable young. When the next generation of mutants was born, however, it became apparent that something was very wrong: almost all pups born to mothers lacking a fosB gene died within two days of birth. Were they infected with something, or defective, Brown wondered. Closer examination revealed that the problems were not physiological. Furthermore, if taken away and placed with a foster mother, they did fine. The problem was not with the babies, but what appeared to be an absence of maternal care. Within a day or two after birth, pups died of cold and starvation.[2]

The mother's hearing and sense of smell seemed fine; maternal hormones like oxytocin and progesterone were within the normal range; mammary glands were producing milk. So why was she neglecting her pups? If anyone was looking for an evolutionary sine qua non, so far as mammalian posterity was concerned, the fosB gene in mice sure looked like it. But can a gene be the "essence of mothering"?

No. All a gene can do is code for particular proteins. Getting from molecules to complex behaviors is a more complicated and dynamic process. In mammals such processes tend to be sensitive to social and environmental conditions, responses that depend on external stimuli (like the presence of pups) and are altered by the mother's own past experiences. It is not that fosB

Fig. 7.2 The research being done in Michael Greenberg's lab was focused on how switch genes worked, not on mothering. Yet it seemed strange that almost all of the pups born to the geneti-cally engineered "knockout" mothers were left outside the nest, cold, hungry and sending des-perate ultrasonic signals—to no avail. As Jennifer Brown put it later, at first the infant mortality "was enough to pique my interest and provoke me to do a larger experiment but not enough to make me think there was that much going on."[3] *(Reproduced by permission of Jennifer Brown and Cell)*

genes cause mothers to mother, but that the absence of fosB genes means something essential in the sequence of interactions that elicit maternal behav-ior in a hormonally primed female never happens—rather like being on a treasure hunt but missing *one* clue.

The fosB gene is responsible for one link in a cascade of signals from the mother's brain to other parts of her body. But it is a critical link. Brown sug-gests the following, still tentative, explanation: The first time a mother sniffs pups, their smell triggers a nerve signal that activates fosB genes in her brain. FosB genes in turn switch on other genes, which sensitize her hypothalamus to this cascade of signals. One eventual target of the signals may be nerves in the mouth and jaw that facilitate behavior essential for the retrieval of the pups, since to gather the pups together, the mother must pick up each one in her mouth. If the caretaker fails to gather pups into a squirming squeaking clump radiating pup smells, none of the other sequences of behavior that are elicited in response to such audio-olfactory cues ever get going.[4]

Whether or not Brown's scenario is the one researchers still use a decade hence, her point is well taken. Rather than the "essence" of mothering, what is missing here is activation by a gene in the preoptic area of the hypothalamus signaling nerves in the mother when she picks a pup up. Similar activation may or may not be relevant to animals (monkeys, for example) where newborns cling to their mothers. When pressed to spell out the connection between fosB and mothering in humans, the researchers themselves, wisely and warily, replied: "There is a fosB gene in humans. From there it's up to your imagination."[5]

————

What happens if a mother mouse lacking a gene for a critical step in the emergence of nurturing has that link in the cascade of responses artificially performed for her? We don't know yet. But imagine that the pups of a fosB-less mother were automatically placed in a position where she would nurse them. Would other maternal behaviors then unfold as usual? We already know that Mother Nature's designs are riddled with fail-safes, redundancies, and alternative routes.

Being pregnant and giving birth is not the only route to becoming maternal. Psychologists discovered long ago the phenomenon called "sensitization." A virgin female rat, for example, will either ignore or devour a pup she happens upon. But if she is repeatedly exposed to pups, this inexperienced "au pair from hell" becomes quite nurturing—without undergoing the hormone changes specific to pregnancy. When experimenters place pups in her cage again and again, eventually she stops killing and begins to care for them.

Most rodent females (as well as many males) can be conditioned to lick babies, crouch over them protectively, tenderly pick them up in their teeth and carry them back when they stray. Once the message starts to get through, the mere presence of pups produces the urge to care for them. Clearly, hormonal changes during pregnancy are not absolutely essential. Yet the same female, once she becomes pregnant, responds to pups faster.[6]

In a now classic experiment, blood from a rat who had just given birth was injected into a virgin female. The transfusion caused a dramatic reduction in the amount of time it took this virgin to retrieve babies. Within fifteen hours, virgin females spontaneously gathered up babies without requiring long, often gory, prior exposure. They bunched the pups together, crouched over them, and licked them just as a parous mother who had just given birth would.[7]

Lactational Aggression

Virgin mice can be induced to perform many of the same caretaking tasks as a real mother by repeatedly introducing them to pups. But this process (sensitization) has one marked deficiency: no allomother ever defends infants with anything like the ferocity of a real mother.

Within days after birth, as milk production gets under way in response to pupular demand, mothers metamorphose from mild-mannered mice into relentless biting fur-balls. I have seen a postpartum wild mouse slash the intruding snout of an eighteen-inch-long snake. Others inflict life-threatening injuries on members of their own species, male or female.[8]

We are talking about female gladiators who fully deserve Erich Fromm's epithet "all protective." In the face of such ferocity only the most determined intruders persevere. Maternal conditioning of others to stay away from her young pertains in many mammals. If you want to see a large male dog back slowly away, hold a tiny puppy in front of his nose. Dreading retaliation from the mother, he avoids the puppy as if it were the wrong end of a furry magnet.

The technical term for this Jekyll-to-Hyde transformation is "lactational aggression." But why should a female be especially aggressive when she is lactating? Even prior to giving birth, life is tough, full of predators and competitors. Partly, of course, an animal intent on defending a nest full of helpless babies has to be more ready to stand her ground. But Italian sociobiologist Stefano Parmigiani has proposed a different explanation. A mother with suckling pups has something special to protect against: conspecifics, in particular infanticidal males belonging to her own species.[9]

Lactational aggression sets in just as the period of really heavy investment in her young is under way. It takes a tremendous toll for such a tiny, active, warm-blooded creature to provide fat-rich mouse milk to seven or eight babies. Over the three weeks till she weans her pups, a mother mouse weighing just 20 or so grams herself provides them on the order of 100 grams of milk—five times her own body weight in protein, fat, and sugars. The vast majority of pups killed by marauding males are killed in the first three days. From that point on, at least, the mother has reached the point of no return; she risks all to defend her litter against trespassing conspecifics.[10]

A series of gruesome studies done twenty years ago demonstrated that suckling triggers this transformation. Mothers whose nipples were surgically removed so they could not suckle did not make the shift into hyperferocity.

Virgin females treated with hormones so their nipples grew enough for experimenters to stimulate them (an experimental procedure stimulating *both* pregnancy and suckling) *did* turn feisty.[11]

The initial explanation for the aggressiveness of new mothers was predator defense. Not very much infanticide had been observed, and what had been seen was assumed to be an artifact of captivity. But more information was coming in, and Parmigiani pointed out that without the protection provided by lactational aggression, infanticide would be even more common.[12] His argument sounded a bit like the White Knight who explains to Alice that his horse wears anklets to ward off shark bites. When Alice retorts that there are no sharks, he tells her that that just proves the ankle-chaps work. Experiments in which the mother was removed confirmed suspicions that anti-infanticide protection was warranted. (Today, such attacks are so widely documented for so many birds and mammals that many animal behaviorists consider it unethical to do experiments that involve removing a protective parent in species reported to be infanticidal.)[13]

Maternal Hormones

In real life, a new mother only gets one try. If she ignores her pups, they die. But not to worry. Shortly after conception, as soon as implantation of the embryos occurs, the placenta (which operates as an extension of embryos, and on their behalf) starts to manufacture extra estrogen and progesterone to sustain the pregnancy, and also, rather like a love potion, to put the mother in the mood.

A cascade of endocrinological events during the last third of pregnancy prepares a mother mouse for love on their first blind date. But there's a hitch. The pups' supply line, the placenta, which is also the agent serving up this cocktail, is delivered right along with the babies. Expulsion of the placenta causes levels of progesterone and estrogen to plummet—seemingly at the very moment babies most need their mother to imbibe them. Mothers get one last dose when they eat the hormone-rich, opiate-laced placenta.[14]

Although they helped to get the mother in a nurturing mood, high levels of progesterone and estrogen can scarcely account for mom's continued attentiveness. Enter prolactin and oxytocin, the hormones described in the previous chapter and most often identified with milk production and the let-down reflex that delivers it. During pregnancy, and especially just prior to birth, oxytocin receptors build up in the brain. At birth, when pulses of this hormone trigger birth contractions, these natural opiates ensure that the mother, the creature whose behavior determines the fate of the pups, greets them in a broody, mellow mood.[15]

From rodents to primates, oxytocin promotes affiliative feelings. A monkey mother whose brain receptors to these natural opiates are blocked makes fewer overtures toward her infant, is less likely to put her face near the baby's and reassuringly smack her lips. Without the warm glow from this most quintessentially mammalian of all hormones—the endocrinological equivalent of candlelight, soft music, and a glass of wine—monkey mothers tolerate offspring, but it's left to the infant to cling for dear life.[16]

Cues from Pups

Placentally produced steroids (estrogen and progesterone) prime the mother and put her on the emotional alert to detect infantile scents and sounds. Once pups slip into the world, dripping in amniotic fluids—which mothers find irresistibly attractive and potent—and the placenta is consumed, the mother is primed for pups to continue eliciting her nurturing responses. And her responses to the pups stimulate production of more maternal hormones.

As the mother instinctively hovers over her babies, it's up to them to latch on to her nipples. Their tugging and sucking, in turn, stimulates production of still more prolactin, intensifying her reactions to the infantile signals these pups emit. She continues to respond till the pups themselves outgrow the infantile traits that so turned her on. If the experimenter replaces grown pups with younger ones, however, a mother can be artificially induced to keep on nursing.[17]

Experiencing the sound and smell of pups, their stimulation of her nipples, and other maternal reactions to them all cause new neural pathways to form, leading to reorganization in the mother's brain. A mother who has given birth, or nursed babies, literally develops a different "mindset." Past experience with pups is one reason why parous mothers respond more quickly the second time around. A mother's deeds in the form of her past history do indeed determine her almost as much as she determines them.

Of Mice and Monkeys

Less is known about how hormones produced during pregnancy affect primates. Pregnant primates experience hormone changes similar to those in mice but respond in a much less stereotyped and automatic way. You could say they are more "thoughtful." For example, if the baby does not at first catch hold, an experienced mother compensates, using her hand to push the baby up to, and close against, her breast. The main source of this greater flexibility in primates is the activity in the newer parts of their mammalian brains, known as the neocortex. Greater reliance on newer portions of the brain allows primate mothers to respond with greater flexibility than mice. The flip side of this escape from rigid, stereotypical responses is that *practice* and learning become more important—even essential.

Caged monkeys deprived of opportunities to socialize and interact with babies make neglectful, sometimes abusive, first-time mothers, yet they improve with successive infants.[18] Watching these first-time mothers give birth can be hair-raising. One primiparous, socially inexperienced rhesus macaque mother was so frightened by her new baby that she leapt to the top of her cage, the infant swinging by the umbilical cord in the air behind her. When birth by caesarean section deprives first-time mothers of the experience of giving birth, they make even worse mothers. Only 3 percent of primiparous monkeys whose first infant was delivered by caesarean section even picked the baby up off the floor of the cage. They were far more interested in lapping up the addictive chemicals from the birth fluids that keepers had smeared on the baby than in the newborn itself. With luck, though, if a baby is strong enough, luring the mother close this way does the trick. Once the mother is close (to taste the attractive molecules in the amniotic fluid she has to lick the baby), a determined and enterprising newborn monkey can catch hold of its mother's fur and, with all four limbs, like a climber scrambling up a cliff-face, inch its way over to her breasts.[19]

––––––––

It seems a miracle that infants born to inexperienced first-time mothers *ever* survive. That they do is testimony to the initiative as well as hardiness of baby monkeys. It seems unlikely that primiparae living in natural groups could be as ill-prepared as their socially deprived peers in captivity; nevertheless, for reasons still not understood, wild firstborn infants die at high rates. Mortality rates among firstborn infants can be 60 percent or higher in some populations of monkeys and apes. Some human groups are also characterized by disproportionate mortality among infants born to young mothers.[20]

In light of such dismal odds, one wonders why early conceptions occur at all in these barely mature mothers, until one notes that with physical growth *and practice* mother primates (including humans) grow not only bigger but *better* with successive infants. In mice, at least, neural pathways in the mother's brain are reconfigured just by acts of caretaking. The brain and responsiveness to maternal hormones and stimuli from pups are different the second time round; she is really not the same mouse that she was.

Changes in neural pathways mean that learning—or its physiological equivalent—has taken place. The role of learning is more important still in the much more flexible mothering of primates. The real tipoff is the unabashed lust to hold and carry babies manifested by young females. Only in a handful of monkeys do future allomothers actually get to take and carry babies to their heart's content (and I'll come to these "infant-sharers" in a moment).

"Learning to Mother"

Anywhere in the world, no matter how aloof people seem, no matter how inappropriate it would be to strike up a conversation with strangers, people smile and chat up babies. Normal social rules don't apply. Strangers don't need to be introduced. A woman's overtures are considered normal. Indeed, others might disaprove if she fails to evince such interest. Why do primate females find babies so irresistible?

Primatologists have long been aware of the magnetic appeal of babies. Young females from one troop will kidnap an infant from another; orphans may be adopted by older sisters or grandmothers, even on occasion by females belonging to different communities. Mexican primatologist Alejandro Estrada tells the story of the graceful spider monkey female, an experienced mother, who adopted an ugly duckling, a chunky, phlegmatic howler monkey. The spider monkey spontaneously lactated to suckle it.

As Estrada noted, such spontaneous nurturing of an infant *belonging to another species* was obviously not elicited by genetic relatedness, or even by any trait specific to the adopter's own species. What seemed to count most was the generalized signals advertising the baby's vulnerability and newness, or "neonativity." Features relevant to this "natal attraction" include: small size; the large, rounded head; the awkward way of moving; and, especially, the distinctive black, snow-white, or golden natal coats baby monkeys are born wearing. (Natal coats and selection pressures on infants to evolve these flamboyant signals of babyness are explained in chapter 19, "Why Be Adorable?")

Fig. 7.3 Binti Jua *(Drawing by Michelle Johnson)*

Why Primates Adopt Infants More Easily Than Sheep Do

Around the world, people were charmed and amazed by the altruism of Binti Jua, a gorilla mother at Chicago's Brookfield Zoo, who gently gathered up a little boy who fell into her cage. *Time* magazine extolled her as "the gorilla of America's dreams." This simian Good Samaritan sparked excited philosophical debate along the lines of "Does one have to be human to be humane?" (Obviously not.)[21]

Primatologists were pleased by the good press for a deserving species. That Binti Jua so readily delivered the toddler to the keeper's door and handed over her windfall was impressive. All round, the safe outcome was a great relief. But few familiar with apes were surprised that this female was attracted to a child.

All female primates find babies (the younger the better) fascinating, some more so than others. And why not? Baby-lust improves the chances a new mother will care for her own baby, and if she is young, motivates her to get the practice that will make her a more competent caretaker. Finding babies attractive is a luxury primates can afford because the chances are so slim that a mother will mistake some other female's baby for her own and nurture an alien infant at the expense of her own. By contrast, in other very social or herd-dwelling animals, where infants are born more precocial—on their feet and running within minutes—there would be an ever-present danger that a mother's milk might be diverted to a stranger while her own infant starved. Consider a baby lamb or other herd-dwelling ungulate. Minutes-old babies are quite capable of wandering over and latching on to the nipple of another mother in the same herd if she fails to fend them off. For a mother in such a species, open-ended tolerance toward babies would be evolutionarily disastrous, causing the mother's milk to be diverted to strangers while her own babies starved. As a result, mothers in such species, mother ewes for example, have been selected to imprint on the smell of their own baby within minutes of birth and thereafter reject babies that don't smell exactly like that.

In sheep, the imprinting process begins as the lamb passes through the birth canal. Dilation of the ewe's cervix induces a surge of oxytocin. For a brief window of time lasting less than five minutes, neural changes in response to this pulse of hormones make the mother sensitive to certain smells, like typing a user code into a computer.[22] Although under normal circumstances a ewe is repelled by the smell of amniotic fluid (so that a female would be amenable to this process only after she herself had given birth), during this sensitive period she finds the scent of amniotic fluid temporarily irresistible. As the mother avidly licks this slimy ambrosia off her singleton, twin, or triplet babies, she imprints on the scents in their wool. And then the door to maternal tolerance slams shut. The mother implacably rejects any lamb that does not wear the olfactory badge that she just learned to identify.

Highly discriminating sheep mothers are the opposite of communal breeders. There are certain strains of mice, for example, in which a mother will suckle any baby within ten days or so of the age of her own litter that for any reason ends up in her nest. Evolutionarily, of course, infants she found there would likely belong to her sister or some other close relative who evolved to share a nest this way because of the significant survival advantages for both mothers' pups.[23]

Mother primates fall between these two extremes: they find all babies—especially very little ones—attractive, but they learn to discriminate between their own and other infants. Learning begins shortly after birth, thus reducing the chances that a mother will ever mistake or prefer another female's infant to her own, but not precluding the possibility of accepting a foster baby.

Based primarily on what is known for humans, a mother primate can identify her own baby's smell within a day of birth. She can pick out her own baby's T-shirt from among a soiled pile. Smell and sound are apparently even more important than sight. Within forty-eight hours of birth a mother can differentiate the cry of her own baby from others.[24] Some hospitals find it worthwhile to record "cry-prints" right along with footprints.

In humans, much of this learning is happening on a subliminal level, such that when a new mother was recently informed by the hospital a few days after giving birth that a mix-up had occurred, she found it very difficult to believe and at first refused to accept her own baby. "I want the baby I brought home," she told officials at the hospital.[25] This mother presumably had imprinted on her baby immediately after birth, but to a far less definitive

Fig. 7.4 Right after birth, ewes are powerfully attracted to the smell of the viscous film coating their kids. Licking off this amniotic fluid, they learn to recognize their own infants' smell. Shepherds use a process known as "slime grafting" to induce a ewe whose lamb has been born dead to accept another baby, usually a "bummer" lamb rejected by its own mother. (Since ewes can care for only two young, and triplets are not rare, redundant, unwanted lambs are not rare either.) The motherless (bummer) lamb is hog-tied and "slimed" either by placing the bummer underneath the delivering ewe's vagina, or smearing it with the birth fluid. As the mother licks away, she imprints on the bummer lamb as her own and thereafter is more likely to allow it to suckle, and once that happens, to adopt it. *(Sarah Blaffer Hrdy / Anthro-Photo)*

degree than a sheep would have. With a DNA test and the tincture of time, the mother came around.

In the environments in which most humans evolved, there was no possibility that another woman's baby would *accidentally* be substituted for a woman's own. If a woman adopted an infant to replace her own, chances were that foster-baby would be a relative—far and away the most common circumstance for adoption in both human and nonhuman primate societies.[26] This is why adoption—especially of very young infants, or when the infant is a substitute for her own rather than in addition to it—often works so well and results in relationships between mothers and adopted infants indistinguishable from those of mothers and offspring they gave birth to.

Maternal programs are quite open-ended in primates, as compared to, say, sheep. This distinction may seem an obvious point. Yet a failure to acknowledge it led to a major misunderstanding in the interface where ani-

Fig. 7.5 This spoof from the *National Lampoon* picks up on the facility with which adoptions take place in the human species. Human mothers fall in between herd-dwelling animals, which are very selective about which babies they care for, and solitary mothers who are so indiscriminate about which babies they care for that babies belonging to other species can be slipped undetected into their nest. Human mothers learn to recognize their own babies soon after birth, yet a woman also readily "falls in love" with any infant with whom she remains in close proximity. *(Courtesy of J2 Communication/National Lampoon)*

mal behaviorists meet pediatricians and specialists on child development (discussed below, in chapter 22, "Of Human Bondage").

Infant-Sharers

Fascination with babies is universal in primates. What differs between species is the willingness of mothers to let anyone else hold her newborn. Invariably other females come up to a new mother to nuzzle or inspect the neonate. Some try to take it from her. For the most part, however, mothers respond possessively, as if to say, "Look . . . maybe even touch . . . but you are not to take this baby away." Among most baboons and macaques—which belong to the cercopithecine branch of the Old World monkeys, characterized by ranked matrilineal clans—mothers are anxious about loaning their babies, and with good reason. Most mothers steadfastly refuse to. If a subordinate female is forced to hand over a baby to a female from a more dominant matriline, the mother may not be able to get it back, with the result that the infant starves to death.[27]

Chimpanzee mothers have a special worry. These apes are unusual among primates insofar as they hunt, and mothers have to worry about the possibility of a meat-hungry chimp eating her newborn—a peculiar maternal concern that may not fit the mental image most people have about these apes. It's probably significant that chimps are more carnivorous than most primates, and also that they are among the minority of primates in which mothers do not necessarily remain to breed among close kin.

Then there are the infant-sharers, monkey species in which females are especially closely related and also have dominance hierarchies so relaxed that mothers can afford to freely give up babies to their kin—heaven, if you are an eager young allomother, and very convenient for mothers. It is to these infant-sharing monkeys that we need to look to get a clear picture of *which* of all these baby-lusting females are actually *most* motivated to get their hands on babies—to hold and carry them—and *why*.

Even though a newborn monkey clings from birth to its mother's fur, keeping it there requires constant minor adjustments, as anyone knows who has tried to use their hands to get a job done while carrying something both vulnerable and awkward. A mother who turns her baby over to a sitter is freed to pluck and eat more mouthfuls of food per minute. What primatologists call "freedom to forage" is the payoff for maternal permissiveness. In species characterized by infant sharing, babies grow faster. This is presumably because better-fed mothers can make more milk for them. Infant-sharing mothers also breed at a faster rate, after shorter intervals.[28]

Invariably the most persistent allomothers, and also the allomothers who keep the baby the longest once they get hold of it (*if mothers will let them take it*), are young females who have never had an infant of their own and who—as the anthropologist Jane Lancaster pointed out—have the most to gain from "learning to mother." This is the equivalent age-sex group that in human societies spends hours and hours playing with dolls. The species where mothers are most permissive about loaning out their babies—langurs, colobus monkeys, and other "leaf-eaters"—belong to the colobine subfamily of the Old World monkeys.

The hanuman langur monkeys of India that I studied are classic infant-sharers. Groups tend to be small on average, with only a single breeding male. Females are as closely related as sisters or cousins. Unlike baboons and other species with fixed female hierarchies, langurs have relatively fluid female hierarchies that fluctuate through time, with young females rising in rank about the time they start to breed and then declining in rank as they grow old.[29] In these relaxed matrilineal groups, mothers freely farm out babies to their kin.

Within hours of birth, baby langurs are spirited away from their mothers. They spend up to 50 percent of daylight hours passed from allomother to allomother. The newer the baby, the more irresistible allomothers find it, although the baby at the center of all this complains loudly at being taken away from its mother's breast.

I was a nulliparous female in my own right back when I studied infant-sharing in wild langurs, and found the cries of the borrowed babies enormously stressful because they made me want to intervene—which was not possible. Allomothers occasionally treated infants roughly, pulling them by one leg or, if they got bored with the baby and were trying to make it let go, pressing its face to the ground. Yet in thousands of hours of observation, I never once saw an allomother harm her charge. The younger and less experienced an allomother was, the harder she tried to keep the baby happy, patting it on the back, nuzzling its face, never pushing it off.

More experienced females who have had several infants of their own eagerly take the infant a few times, perhaps turning it upside down to inspect its genitals, but quickly lose interest. Besides the inexperienced youngsters,

Fig. 7.6 A juvenile langur monkey takes an infant from another allomother, a very pregnant female on the verge of giving birth. Both are very motivated to take and carry babies. *(Sarah Blaffer Hrdy / Anthro-Photo)*

the only females likely to keep an infant longer than five minutes are very pregnant females approaching term—probably primed to feel broody by the same maternal cocktail that induces maternal behavior in expectant mice.[30] Adult male langurs, however, have no interest in carrying babies, and immature males account for only a tiny fraction (0.2 percent) of attempts to take babies. Playing at being a mother is something only future mothers are motivated to do (which is different from saying males can't or won't caretake—see chapter 9, "Three Men and a Baby").

Since langur milk appears to be just as low-fat as the dilute milk of other monkeys in species where babies spend all their time within reach of the nipple, my guess is that consumers make up at night for time spent off their mother's nipples during the day.[31] Allomothering the langur way is one of those win-win situations from which both over-burdened mothers and eager allomothers benefit, with babies none the worse for it.

Fig. 7.7 Newborn infants exercise a magnetic appeal. Young females who have never had an infant of their own, along with very pregnant females hormonally primed to mother, are especially mesmerized, and sometimes downright obnoxious in their efforts to get hold of one. This may be one reason langur mothers seek secluded places to give birth. I crawled for seventy feet under thick bushes to photograph this mother giving birth. By the time I found her, so had her group mates. The group's alpha female (a temporary status) was very pregnant at the time and desperately eager to have a look. The new mother, unwilling to give her neonate up yet, lay on top of the neonate and crossed her paws (the newborn's drying umbilical cord can be seen just sticking out from under the silver hairs on its mother's stomach). In the foreground, an older juvenile female is also mightily interested, but, as langurs are wont to do, bashfully looks away from the neonatal object of her desires—while inching ever closer. *(Sarah Blaffer Hrdy / Anthro-Photo)*

Culture and Socialization

In some human societies, mothers pass babies around like some infant-sharing primates do, though more often only the mother holds her newborn. This degree of variation in the pattern of maternal care within the same species is one respect in which humans differ from other primates. But there is another even more important respect. Unlike other primates, women imagine ahead of time what it will be like to give birth and to be a mother. Their expectations are built not just on what they themselves have experienced, and from observing others and practicing with their babies, but from what others (especially other women) tell them it *should* be like. Anyone who observed Prissy's panic and heard Melanie's anguished screams in the near-death labor scenes from the film of *Gone With the Wind* at an impressionable age is more likely to

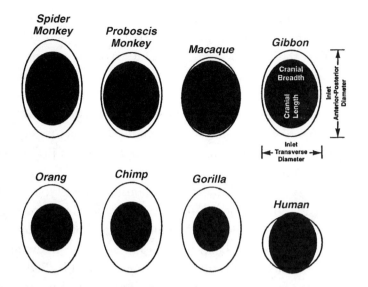

Fig. 7.8 This diagram of the size of the neonate's head (the black oval) relative to the pelvic outlet in monkeys and apes illustrates why delivering babies is so much tougher for women compared with other apes. Labor in a gorilla is short, on the order of twenty minutes, and enviably easy. By contrast, human births take far longer, and range from easy to extraordinarily difficult. It took sixty-four contractions (occurring an hour apart at first and then gradually occurring every two to four minutes) over an eight-hour period to deliver my third baby, and all present regarded it as an "easy" birth. For such ordeals, women can thank the engineering demands of a pelvis that allows upright walking. Quite simply my baby's cranium was larger than the anterior-posterior dimensions of my pelvic outlet, requiring my son's head to enter the outlet facing sideways. Evolution of larger brains around 1.5 million years ago unquestionably played a role in this tortuous process. In contrast to apes, but more like humans, monkeys also undergo a tight squeeze. It's obvious that giving birth hurts; the parturient monkey strains and may yelp. But she does not require special obstetrical assistance. *(Adapted from Rosenberg and Trevathen 1996)*

assume she will need painkillers at parturition than someone whose expectations about childbirth were shaped by Pearl Buck's *The Good Earth*. The peasant woman's baby slipped out and the mother went back to work in the rice fields the same day.

Even though nerve signals work the same way, something as obviously biological as pain in childbirth is experienced differently depending on cultural expectations. Women develop expectations not just about how they should respond but about how they should experience their own sensations and emotions. A human mother experiences particular emotions quite specific to her infant even before it is born, something I doubt other animals do.

An Initial Reserve

In some cultures, like my own, the birth experience is seen as "labor." The "work product" is the baby, whose successful arrival is supposed to—and often does—evoke spontaneous joy. To this day the word that comes to mind when I recall my own first birth is *euphoric*. My husband describes the moment our first daughter was handed to him as the happiest in his life. Yet Wenda Trevathan, an anthropologist who studies birth across cultures, was surprised to learn how atypical this "immediate reaction of joy" actually is in the context of general human experience.

In many cultures a guarded maternal response is more typical, what Trevathan describes as "a period of indifference while the woman recovers from the exertion of delivery." Among the Machiguenga and other peoples of lowland South America, the newborn is set aside after the midwife cuts the umbilical cord, almost ignored, till the mother has been bathed. Only after the mother begins to nurse the child, hours later, or perhaps the next day, does the mother become concerned about the infant.[32]

In many cultures, formal acceptance of the infant as a new member of the community is delayed, and only celebrated some time after birth. Expressions of ebullience at birth would be regarded as unseemly, like tempting fate. In others, emotions are put on hold until people can be assured all is well. There is a matter-of-factness about giving birth. People simply do what they need to do. There is also an awareness that this is a time of real danger, spiritual as well as physical (see chapter 20).

Traditionally, most anthropological fieldworkers were men. Rarely would they be eligible to attend births. Remarkably few detailed and objective accounts of births from different cultures are available. Brigitte Jordan provides one of the few detailed published accounts for Mayan-speaking people in Yucatan, where she noted that there was "*no* reaction. No smiles. No talk or exclamations." This account is very similar to that recorded by psychiatrists in a sample of mostly married, well-off and healthy British mothers. The psychiatrists were struck by the seeming "indifference" of mothers holding their babies for the first time, just after birth—particularly first-time mothers, 40 percent of whom reportedly felt no particular affection for their babies initially, compared with 25 percent of experienced mothers (who had given birth before). Strong feelings of attachment to the baby emerged in the days and weeks following birth.[33]

Among other things, the Mayan bystanders were waiting for what Westerners call "the afterbirth." With no remedy available against postpartum hemorrhaging, safe delivery of the placenta is critical to a safe outcome. Among the best predictors of whether a mother seems indifferent or actively pleased, suggests Trevathan, is how much social support she has.[34]

A Shortage of "Fixed Action Patterns"

If only women started to scratch furiously in the dirt just prior to giving birth! Or if after giving birth women felt an irresistible compulsion to lick the newborn all over its body, or eat the placenta like other apes do.[35] If only, like dogs (whose wolf ancestors also reared their pups in nursery dens), we cleaned up our infants' excrement by eating it. If only mothers engaged in such "fixed action patterns," their stereotyped responses would put to rest the debate over whether or not there is an innate component to maternal behavior in humans. But we don't.

Primatologist Kelly Stewart, one of the few people who has observed a wild gorilla give birth, describes that mother's compulsion to eat the placenta. "She left her newborn lying on the ground, picked up the placenta and ate it with two hands, like pizza," Kelly told me. Only after that did the mother turn her attention to the baby.[36]

But there are no "fixed action patterns" universally exhibited by new mothers in *Homo sapiens* comparable to mammalian mothers licking babies and biting off the amniotic sac. Humans respond to placentas very differently and quite variably—disposing of them as waste, as we do, or (as in many other cultures) burying them, either privately or with some public ritual. Eating placentas seems to be more common among back-to-nature cults in California than in any bona fide tribal society I know of.

Apart from some Tibetan mothers,[37] postparturient women rarely lick their babies. Instead, Bisayan women cleanse them with coconut oil and dust newborns with powder; Mixtecans grease them with almond oil; Marshelles with pure coconut oil; the ancient Greeks with olive oil; the Tewa bathe and then rub babies with cornmeal; the Caraja apply red dye; the Tiwi of north Australia cover neonates in charcoal; Hottentots rub newborns all over with moist cow dung, followed by the juice of a fig tree, then smear them with sheep fat and finally dust them with fine powder.[38]

The myriad ways mothers respond to birth, and especially the broad range of human customs surrounding birth (which depend on what specific substances, symbols, social rituals, and incantations they have available) makes it

Fig. 7.9 When a !Kung woman like Nisa gave birth for the first time, she went off from camp alone, gave birth, buried the placenta, and brought the baby back to camp, where her husband's mother cut the umbilical cord and "shaped" the infant's head.[39] *(Marjorie Shostak/Anthro-Photo)*

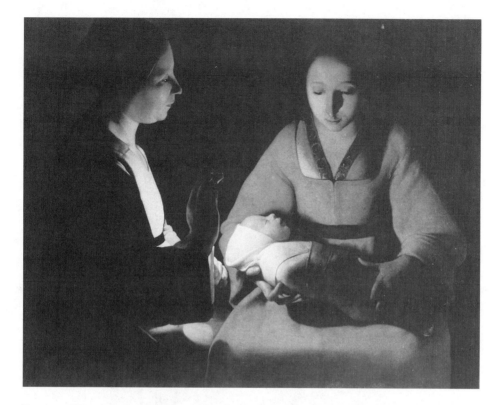

Fig. 7.10 Wrapping the newborn in bands of cloth was considered essential in seventeenth-century France, not just to keep the baby warm, but to protect the fontanel. People believed that cold vapors might enter the brain case through this soft spot and damage the brain. Swaddling also kept the body straight and prevented an infant from reverting to a fetal position, and thus was regarded as part of the socialization process, transforming an infant from a more animal-like creature into a civilized human-being who will walk upright. *(Le Nouveau Né by Georges de la Tour, 1593–1652, courtesy of Musée des beaux-arts de Rennes)*

impossible to come up with a list of species-typical universal traits that is any more than obvious (the baby emerges from the birth canal covered with fluids that need to be cleaned off somehow). The handful of behaviors that can be reliably considered cross-cultural universals are all quite practical. They include cleansing and immediate inspection of the baby, genitals first, followed by touching the baby all over and perhaps long looks into the baby's face. Even this inspection is skipped over by some inexperienced young mothers. In my own case, I felt a powerful urge to hold the baby close and feel its skin and body next to mine, and especially to smell the baby, which is

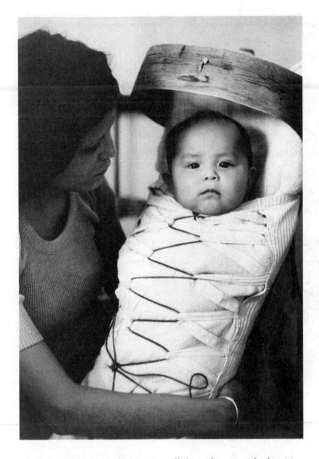

Fig. 7.11 Like swaddling, strapping infants to cradleboards, as with this Navajo baby, prevents tactile contact with the mother but may have a calming effect on the infant. *(James Chisholm/ Anthro-Photo)*

what many women do. But I cannot be sure my own urges were "universal," since according to those ethnographic accounts we do have, many mothers delay holding the baby.[40]

Postpartum Depression

Approximately half of all new mothers experience "postpartum blues" or weepiness a few days after birth. A smaller portion (around 10 to 15 percent of women in Britain, North America, and Uganda, for example) go on to experience serious depression several weeks after birth.[41] The most commonly cited symptoms are anxiety, sleep disturbances, concern for the baby, depression, irritability, and hostility.

That the arrival of a "bundle of joy" should have such counterintuitive effects seems strange. Most people assume that being depressed right after giving birth is not particularly adaptive. A few, however, go a step further and ask why, then, are postpartum "blues" so common?

One popular psychoanalytic interpretation in the old days was that mothers became depressed following birth due to their "sense of loss" at no longer being pregnant.[42] (Remember, most psychoanalysts in those days had never experienced the very memorable last week of pregnancy, the onset of which can be pinpointed, as a friend once told me, to "when you can't stand it another moment and burst into tears. Then you know you will give birth in one week.")

Few contemporary psychiatrists, and even fewer mothers, are persuaded that women are anything other than relieved when pregnancy ends. Today, there are three main explanations, none mutually exclusive, all evolutionary.

As discussed above, there is considerable evidence for mammals that neurochemical changes during pregnancy and the postpartum period evolved to ensure proximity of mothers and infants. Hence, some medical experts have come to see the new mother's depression as a pathological response to modern procedures that interfere with her following this natural mammalian program of mother-infant proximity, or else depression is simply a by-product of these changes, period.

Detailed naturalistic observations of mothers with different amounts of exposure to their babies might allow testing of this proposition. Unfortunately, most information derives from routine medical exams. "Well, Mrs. Jones, how are you feeling today?" followed by a weak, unconvincing reply: "Fine, I guess . . ."—an interchange that provides far less information than could be obtained by actually observing how many minutes of each hour Mrs. Jones and her new baby are in contact, whether she breast-feeds, and how often, compared with a control sample of cheerful moms.

The second hypothesis, favored by evolutionary psychologist Stephen Pinker and others, assumes that the postpartum depression is specifically human in origin (rather than mammalian or primate). They trace it to a conflict between a mother who, in some earlier nomadic hunter-gatherer phase of human evolution, would have chosen to invest no further in an infant if it were defective or if she lacked a mate, and civilized circumstances, which prevent such a mother from bailing out. Contemporary constraints include both custom and stringent laws that make infanticide ill-advised or impossible. "The depression is most severe," Pinker writes, "in the circumstances that lead mothers elsewhere in the world to commit infanticide."[43]

The fact that depressed mothers were less affectionate with their babies, and that half of them were depressed while pregnant, even before they gave birth,[44] are both consistent with Pinker's hypothesis. However, medical researchers counter by citing cases of women suffering postpartum depression whose children are healthy, social circumstances good, father present and willing to invest[45]—in short, mothers who do not obviously seem to resemble that small fraction of hunter-gatherer mothers who would have committed infanticide (the best estimates for nomadic African hunter-gatherers like the !Kung are on the order of one in a hundred births,[46] far lower than the fraction of women who suffer from postpartum depression). Furthermore, whereas most women who abandon or kill their own infants tend to be young, women suffering postpartum depression and postpartum psychoses seem to represent the average age of mothers generally.[47]

I have talked to women suffering postpartum depression. Often they are mystified as to why they should feel so "sad" or "so angry," and quite certain that they want to keep their babies. Why would a vestigial emotion that applied to only a fraction of women (say, one in a hundred) in the evolutionary past apply to one in ten or more today?

This brings me to the third of these evolutionary hypotheses, first proposed by psychiatrist I. Mastrodiacomo and associates in Italy. Their explanation relies on older mammalian responses, not necessarily rooted in the past hundred thousand years of human evolution. It applies to a wider range of mothers—those with "old" as well as "new" brains. According to this third hypothesis, which might be termed the "vestigial lactational aggression" hypothesis, postpartum depression is an endocrinological by-product or left-over from an intense intolerance of others that was once adaptive among mothers who might need to protect infants from either predators or conspecific members of their same species. The root of her depression derives not from the mother's suppressed desire to abandon her infant, but from a fierce compulsion to protect it that fills her with hostility toward others. The worse off she is, or the more potentially threatened the mother feels, the more defensive she should be.[48]

Such maternal emotions would not be nearly so inappropriate as feeling infanticidal. Nevertheless, such hostile emotions are rarely, in any human culture, considered exemplary in a woman. In many cultures (especially patriarchal ones) women are taught that they should never feel or behave aggressively, that they should cheerfully accommodate those around them. Mothers (mammal, primate, or human) with minimal support, those who felt the most stressed, would include among their numbers a tiny fraction who might benefit from abandoning their babies, but many more who would benefit from being able to protect them. Is "unexpressable" hostility being registered as postpartum "depression"?

All during pregnancy, a woman's anterior pituitary (which along with the placenta is a major producer of prolactin) increases in size by about 40 to 50 percent, preparing the mother's body for lactation. Prolactin may also be implicated in responses that make her behave more defensively to protect her young. Whether she decides to breast-feed or not, a woman right after birth has high circulating levels of the same hormones that in other mammals are implicated in "lactational aggression."

Hormone profiles are consistent with the hypothesis that hostile feelings in these recent mothers represent vestiges of lactational aggression. Prolactin levels remain high at a time when estrogen and progesterone precipitously drop. Women tested seven days postpartum had higher hostility ratings than female hospital employees who were asked to volunteer to serve as controls. When postpartum women were matched with patients known (for other reasons) to have elevated prolactin levels (hyperprolactineia), the two groups scored the same on hostility measures.[49] Even women with no particular depression undergo a postpartum decline in "positive feelings" toward husbands in the couple of months immediately after birth, which seems very odd—except as an artifact of lactational aggression.[50]

Little Steps to Maternity

All mammalian females have innate maternal responses, or "maternal instincts." The question is, what do we mean by such terms? With fosB, we learned just how important one gene—just one gene out of a total genome of 100,000 or so—can be. One step at a time, something as simple and discrete as a build-up of hormone receptors in the brain can determine whether a particular signal from a pup fails to register, or becomes magnified into a network of responses. Small nudges favoring distance by the mother, or affiliation, can make an enormous difference in whether a bond between mother and infant is forged. Many of these tiny differences can contribute to maternal decisions of life-or-death significance for the newborn.

But this unfolding of genetic instructions does not conform to the common view of instinctive maternal devotion summed up in Fromm's epigraph, love that is spontaneously expressed and "unconditional." It is not true that women instinctively love their babies, in the sense that they automatically nurture each baby born. Neither do other mammals—although when they *do* care it is hard to explain their behavior as anything other than instinctive. In other words, there is probably no mammal in which maternal commitment does not emerge piecemeal and chronically sensitive to external cues. Nurturing has to be teased out, reinforced, maintained. Nurturing itself needs to be nurtured.

Instead of old dichotomies about nature versus nurture, attention needs to be focused on the complicated interactions among genes, tissue, glands, past experiences, and environmental cues, including sensory cues provided by infants themselves and by other individuals in the vicinity. Complex behaviors like nurturing, especially when tied to even more complex emotions like "love," are never either genetically predetermined or environmentally produced.

Family Planning
Primate-Style

Nature has so wisely ordered things that did women suckle their children,
they would preserve their own health, and there would be such an interval
between the birth of each child, that we should seldom see a houseful of babes.

—Mary Wollstonecraft, 1792

"*Nine months, no charge . . .*" croons the earthy, caressing voice of the songstress. Who is she trying to fool? Perhaps she is forgetting, or was generously not counting what it costs to carry a fetus to term. The mother's gift of life can be measured in calories, minerals, opportunities to invest in other children present and future, charged against her own continued survival in times of famine—or in teeth. In practical terms, the German proverb *Ein Kind, ein Zahn* (One child, one tooth) comes closer to the mark.

This old saying (which exists also in Danish, Russian, and Japanese) is consistent with a recent study of Danish fraternal twins ages seventy-three years and older: the more children a woman had, the fewer teeth she retained in old age. Mothers of low social status lost about one tooth per child, while those better off more nearly lost a tooth for every two children.[1]

Nine months, and there definitely *is* a charge. Pregnancy lasts over eight months in gorillas, and closer to nine in orangutans and humans, making the 7.6-month gestation of chimps a bargain by comparison. Our hominoid ancestresses topped off marathon pregnancies with a long, calcium-depleting three to five years or more of lactation, which probably cost them more than a tooth per child.

Lactation, the most exorbitant phase of reproduction, lasts longer in primates than in any other mammal of comparable body size. Furthermore, a primate mother must not only provision but also tote her infant.[2] If she succumbs in the process, her infant most likely dies with her. Thus mother and infant had to coevolve some way to space births that was calibrated to the needs and condition of each. But how? Just what are the implications of

175

widely spaced, singleton births, and of the staggering maternal commitment to each infant born that such scheduling permits? Why is maternal commitment toward babies so much more contingent on circumstances in human mothers than is the case in most other primates? I am convinced that part of the answer lies in the past: not in the distant past, but in the last 50,000 years or so, a period for which we know almost nothing about child-rearing practices. Once again, comparative evidence (from other primates and people living nomadic, foraging lives) offers the best hope for insights.

Life in the Slow Lane

It's no wonder wild populations of apes have so much trouble bouncing back after environmental disasters and in many areas of the world today hover on the brink of extinction. Nor is it surprising that it took so long for the earliest humans to amble out of Africa and only very, very gradually people the globe. A mother ape's commitment to each infant born is part and parcel of having evolved for "life in the slow lane."

Seventy million years ago the earliest squirrel-like, scurrying ancestor of these apes lived differently. She bore multiple young; and most likely she stashed litters in nests the way the rare ruffed lemur (who typically has twins) does today. Ruffed lemur females still sport three pairs of nipples as a memento of more prolific times. Under pressure to reproduce before something ate them, primitive primates weaned fast-growing babies and bred again quickly, compressing as many litters as possible into this short, busy, vulnerable lifespan.

Their modern descendants, however, have evolved to breed at a more deliberate pace. Birth rates, growth rates, and death rates among apes are a quarter to a half those found in other comparably sized mammals. In a life lasting half a century, ape mothers rarely produce more than five widely spaced offspring. Taking years to reach her mature body size, an ape breeds late and spaces births four to eight years apart—although no other ape takes quite so long to mature as immatures in humans typically do.[3]

Costly Offspring

Maternal investment begins at conception, extends through gestation and lactation, and then, in most mammals, terminates at weaning. For many, weaning marks the end of further contact. A mother who meets her offspring years later may fail to show any sign of recognition, may even attack, as if

Fig. 8.1 "Human beings are said to have a longer period of immaturity than other animals; but it is not prolonged childhood which distinguishes us so much as prolonged parenthood" Charlotte Perkins Gilman pointed out in *Concerning Children* (1901).[5]

(© Flying Fish, Inglewood, California)

the latter were a stranger—something dog breeders know very well. But not so with primates and mammals with prolonged mother-offspring relationships.

A critical distinction between humans and other animals, even other cooperative breeders, is the sheer duration and extent of parental investment, along with the diverse forms parental invest-ment takes. Weaning (to quote again from a song, this time a lullaby) is "only the beginning." The provisioning of human young continues on and on, past childhood into adolescence.[4] Behavioral ecologist Hillard Kaplan estimates the provisioning costs among South American foragers for one human from birth to independence at 10 to 13 million calories, not counting the food an older child collects for himself. No foraging woman can meet such outlays by herself.[5]

Long after weaning, human parents continue to invest in offspring; they even help a grown daughter at her first birth. Parents and other kin help set up progeny to breed, negotiate settlements of the property that customarily changes hands when a son obtains a wife ("bride price"), or care for and pro-vision grandchildren. In complex, stratified human societies, parents provide substantial dowries, goods delivered to the groom's family by the bride's. The present $133,000 plus cradle-to-college price tag of a middle-class North American child merely represents the extreme end of a trajectory of greater investment in offspring embarked upon by primates seventy million years ago.[6]

Unconditional Commitment of Mother Monkeys

In other mammals, a mother who bears a particularly large or costly litter can be discriminating about *which* young she cares for. A mother hamster or prairie dog, or even a domestic dog, will sometimes cope with a large litter by pruning, nudging aside the smallest or weakest pups so as to concentrate on the strongest. But such pruning is not an option for a primate mother. She gives birth to a single infant, born after an unusually long gestation. This baby arrives with prioritizing built in. Its mother has already opted for quality over quantity.

Commitment of a monkey mother to her singleton, high-cost baby almost never depends on *attributes of the infant itself,* traits like weight, sex, birth order, or physical conditions that affect survival prospects. Like all mammals, primate mothers are primed by potent hormonal cocktails during pregnancy, and at birth are exposed to the newborn's irresistibly infantile signals—its shape, smell, sounds, and natal coat. The big difference between these primates and, say, prairie dogs is the extraordinarily low incidence among primates of maternal failure to attach to their babies. In fairness, however, it must be noted that primate neonates themselves deserve some credit for this, since they often clinch the deal by grasping hold of the mother's fur right after birth. The neonate quite literally *attaches* to his mother.

––––––

The fact that humans are primates makes the degree of maternal ambivalence found in our species seem all the more curious. Mother monkeys and apes stand out among mammals for the extent to which they almost never discriminate among offspring. Monkeys born blind or suffering from gross locomotor deficits like spastic cerebral palsy are assiduously cared for, *so long as they can initially cling.*[7]

Being able to hold tight seems to be the only viability test newborn monkeys are subjected to. Pass that test and the mother soon after learns to recognize and love the baby as her own. Thereafter she holds him even if the baby becomes too weak to cling.

Once emotionally attached, a monkey or ape mother will carry for days the limp, even decomposing, body of an infant that has died. Ever so gently, the mother lays the corpse on the ground while she feeds, fetching it when she is ready to move on. Gradually the distance between the mother and the object of false hopes extends. She moves farther and farther away to feed. Elapsed time between visits to the now desiccated corpse grows longer until, reluctantly, one day, with obvious ambivalence, the mother leaves behind the flattened strip of fur.

Even as she distances herself, a mother may continue to defend the corpse. Once, years ago, before I knew better, I attempted to examine the desiccated corpse of an infant langur. I was mobbed, not only by the mother and her female kin but also by the burly father. Langurs, *Presbytis entellus* (named for the boxer Entellus lionized in the *Aeneid*), are in fact normally not at all aggressive toward humans. On this occasion, however, it was as if the ancient Trojans were battling to win back the body of Hector or some other fallen comrade. Dropping my quarry and shielding my face, ignominious as Paris, I beat my retreat.

Such passionate maternal devotion in the service of a lost cause—for an infant who has died—persists in the monkey mother's repertoire presumably because comatose infants revive just often enough to make this energetically costly allegiance to an immobile infant worthwhile in evolutionary terms. Once in a blue moon, pale eyelids flicker open again.

This unconditional commitment to her infant, irrespective of its sex or other physical attributes, is, as we will see, one of the key differences between monkey and ape mothers and human ones. Across human societies, murder rates among men are always higher than those among women. Not surprisingly, men tend to murder people they are not related to. But when women *cause* someone else's death (through sins of omission as well as commission) that person is most likely to be her own newborn baby.[8] In this respect, women are utterly different from other primates that, like us, produce one baby at a time.

Infanticide is hardly unique to humans, and is widely documented among primates, both human and nonhuman. But in other primates, the killer is almost always an unrelated individual, never the mother. Even when nonhuman primate females are implicated in infanticide, mothers don't harm their *own* infants, they kill someone else's. Only under the direst circumstances does a mother cease to care for her infant or actually abandon it.

It is not that unusual for a mother monkey to treat her baby roughly, to briefly drag it, or even punish it with a slap or threaten it with a toothy grimace—especially when she is trying to wean. But no wild monkey or ape mother has ever been observed to deliberately harm her own baby.[9]

Singleton versus Multiple Young

Does this mean that primate mothers are by nature more nurturing than other mammals? Does giving birth to one baby at a time somehow sharpen maternal commitment? Were it not for the thick dossier implicating human mothers in abandonment and outright infanticide, it would certainly appear so.

Let's consider the evidence then for monkey and ape mothers that do give birth to multiple young. Twins are rarer than in humans, and unless conditions are very favorable, mothers have trouble rearing them. When monitored, each twin gains weight more slowly and is weaned later than a singleton baby. Often, survival of both twins is compromised, and cases of twins provide several of the very few recorded instances when mother monkeys do appear to discriminate between infants based on viability.

When a twin situation is artificially produced by a particularly zealous mother adopting a second, "foster" infant *in addition to her own,* the mother's own infant manages to spend more time on her breast. That infant also grows faster than the fostered baby, presumably because the mother is able to distinguish between them and discriminates in favor of her own. Similarly, a wild gorilla mother who gave birth to twins abandoned the weaker of the two.[10]

The most telling evidence that multiple births is a factor in maternal abandonment comes from such unusual species as marmosets and tamarins, who routinely give birth to multiple young. The commitment of these mothers to their babies is far more conditional than that of other monkeys. Cotton-top tamarins illustrate the point. These little South American monkeys are archetypal "colonizers," adapted to reproduce quickly when new foraging opportunities become available. They are among the several exceptions to the rule that most primates are slow breeders.

Tamarin mothers routinely give birth to twins, occasionally to triplets, and rely on the father and other group members to help rear them. When faced with a shortage of allomaternal assistance, mothers reject one or more babies, reducing their litter to a manageable size. If helpers are not on hand, the majority of mothers refuse to undertake the daunting task of rearing multiple young. The babies are abandoned, usually within the first seventy-two hours after birth.[11]

Can Monkeys Discriminate Between Infants?

Mother monkeys come close to psychologist Erich Fromm's ideal of a loving mother: "unconditional" and "all-protective." But are they unconditional in their commitment because they lack the cognitive capacity to tell the difference between infants, between a son and a daughter, a weak baby and a strong one?

This can't be the answer. Alloparents readily discriminate between one baby and another based on age, sex, degree of relatedness, or physical condition, especially when abusing another female's infant. Other group members may attack juveniles that behave in a strange or spastic way. Macaque allomothers harass daughters born to low-ranking mothers but tolerate their sons. Infanticidal male chimps, on the other hand, discriminate the *other* way, killing male infants whose mother may have been impregnated by an outsider, while tolerating her daughters.

There can be no doubt that nonhuman primate mothers have the capacity to distinguish a healthy infant from a sick one, a son from a daughter. Yet there is no evidence that mothers use such information to discriminate among their offspring. They never withhold care *on the basis of physical attributes like sex, age,* [12] *or poor health*. Their willingness to care is not affected by infant quality. In this respect, nonhuman primate mothers differ markedly from many human ones. This is not to say that no monkey or ape ever abandons an offspring. Under extreme circumstances they do so. But abandonment depends on the mother's condition or social situation, rather than attributes of the infant itself. The circumstances are as revealing as they are unusual.

When Monkey or Ape Mothers Abandon Babies

Cases of wild monkey mothers who abandon their babies almost always involve mothers in very poor condition or faced with terrible predicaments. A case in point would be a mother confronted with the arrival in her group of a determinedly infanticidal male.

A primate mother's worst nightmare is to have an infanticidal male permanently install himself in her group. Typically, mothers fight him off as best they can, but time is on his side. Confronted with repeated attacks, some mothers "give up." But the cases of special interest here involve mothers with older, nearly weaned infants, infants who usually would be past the age when infanticidal males would target them. (For example, most infants killed by invading langur males are under six months old.) But some mothers find themselves in the terrible predicament of having to decide either to remain in their troop with a male who is attacking her infant or to leave the safe harbor of the troop she is accustomed to.

Confronted with infanticidal males, both gorilla and langur mothers have been known to take the extreme step of temporarily leaving their group. Some take their infant and try to deposit it in the custody of the ousted father or with the baby's brothers. In some cases, infants subject to such sudden weaning and separation from their mother may survive. Among the many liabilities, however, is that an infant old enough to survive on its own is also old enough to find its way through the forest alone, back to its mother. In both gorillas and langurs, abandoned youngsters were observed who eventually caught up with their mothers again, only to be attacked by the new males then accompanying them—suggesting that each mother "knew" what she was doing when she deserted her baby "for its own good." In the langur case, the little female was only wounded, not killed. The youngster (who was years younger than the age of menarche) then did something amazing. She presented to the male and shuddered her head in the stereotypic pattern of an estrous female sexually soliciting her attacker, as if to remind him "I'm too old to kill. I'm someone you could mate with."[13]

"Unconditional" versus "Contingent" Commitment

Giving birth as they do to one widely spaced infant at a time, a monkey mother's marching song is understandable: "If it clings, I will carry it." But what else would we expect from a mammal that breeds so slowly and is so cautious about embarking on pregnancy, so discriminating about which pregnancies were continued? The majority of pregnancies, after all, terminate before implantation of the embryo occurs. Considerable birth management has already been exercised by the time each primate infant is born.

The singularity and costliness of each birth is what makes the contingent nature of maternal commitment among human mothers so especially curious. Yet there is an enormous amount of evidence (reviewed below, in chapters 12 through 14) that not all women do anything like commit unconditionally to each baby they bear. The fact that some human mothers give or withhold care depending on the infant's sex, or some other specific attribute, is unexpected and curious not because mammalian mothers are *unconditionally* nurturing (they are not) but because other primates are.

No other primate discriminates between offspring on the basis of infant attributes. My question, then, is: What happened in the course of human evolution and history to make mothers in our own species so much more discriminating than other primates? No other primate, ever, scrutinizes an infant for defects, or buries a newborn alive because it is born the "wrong" sex. Does our extremely contingent nature of maternal commitment mean that human mothers are born with less developed instincts, with blanker slates in this regard than other primates?

A Population in Check

Paleontologists, demographers, and anthropologists all agree that for most of the Pleistocene, humans lived at extremely low population densities with an infinitesimally slow rate of growth. Anthropologists studying modern genetic diversity in order to reconstruct ancient demographic history have hypothesized that their effective breeding population never expanded much beyond ten thousand people.[14] These early humans could have traversed vast distances without ever encountering as many new faces as one of us sees in an evening at the movies.

Demographers estimate for the beginning of the Paleolithic that it would have taken *fifty thousand years* for a ragtag group of one hundred survivors to multiply to ten thousand. Confronted with hard times, many small local populations died out. Blessed with periods of abundance, others expanded, split, and spread. Even so, the estimated doubling time was probably around once every 15,000 years, contrasted with a human population currently doubling every fifty.[15]

An astonishing fact from primate population genetics underscores how newly fecund our species is. There is more variation in the genomes of vanishing populations of East and West African chimpanzees than exists in all of humanity put together. Why? Because all humans on earth today probably descended from the same small population of hunter-gatherers last in the same place roughly 100,000 years ago. After an exceedingly slow start, the carpet of humanity covering the globe today was laid down over such a short period that *Homo sapiens* as a species did not have time for the extensive local evolution and microdifferentiation at a genetic level that characterized the many different ancestral populations of chimpanzees.

So how did our ancestors maintain their slow reproductive pace for so long? What limited early hominid populations? And what changed?

Explaining Slow Growth

The very fact that populations were so dispersed made it unlikely that epidemics or warfare, which are essentially local phenomena, could have sufficed to keep the lid on population growth over such long periods.

Unsatisfied by Malthusian explanations—murder, pestilence, and famine—the generation of evolutionary anthropologists that preceded my own (most prominently J. B. Birdsell) postulated that hunter-gatherer parents kept family sizes within manageable bounds by practicing infanticide. Darwin had noted that "Barbarians find it difficult to support themselves and their children, and it is a simple plan to kill their infants." Birdsell went further, hypothesizing infanticide rates as high as 15 to 50 percent of live births.[16]

By the 1970s, fieldworkers began to replace these guesses with real numbers. Infanticide rates as high as 40 percent could be documented, but primarily in more settled and patriarchal societies with preferences for sons so strong that they killed daughters, not in nomadic hunter-gatherers considered representative of Pleistocene foragers. In her classic demographic analysis of !Kung San nomads, still making their living in an arduous old-fashioned hunter-gatherer way, anthropological demographer Nancy Howell reported only six cases of infanticide in 500 live births. Infanticide was culturally sanctioned, even regarded as a duty under certain circumstances. But circumstances requiring infanticide—birth defects or a birth so ill-timed that it would jeopardize survival chances for an older child—were uncommon.[17]

Practiced at such low rates, infanticide could not account for such slow population growth. For one thing, infanticide right after birth (which !Kung infanticide always was) unless practiced again and again is not an effective way to reduce population growth, since a human (or ape) mother who loses a neonate becomes pregnant and gives birth soon again, perhaps within the year. Furthermore, birth intervals reported for hunters and gatherers were if anything *shorter* than those in other hominoid apes among whom mothers *never* killed, and very rarely abandoned, their own infants. Hence, although infanticide was almost certainly practiced, it was a poor candidate to explain slow growth of populations like the !Kung. Something else had to be responsible for the very low birth rates and slow population growth. What? The answer resides in the ovaries and the lifestyles of foraging ape females, and can be divided into three categories: delayed maturation; a long delay between first menstruation and the point where a woman actually becomes fertile (a period known as "adolescent subfertility"); and long intervals between births.

Delayed Maturation and "Adolescent Subfertility"

By the standards of most mammalian life histories, all apes take a long time to mature to breeding age. Humans, however, are off the scale (see below, chapter 11). Within the age range possible for their species, however, individuals mature at different times. In general, better-nourished human females reach menarche sooner, conceive earlier, and reach full body size sooner. A chimp in a productive territory blessed with supportive and influential kin (Flo's eldest daughter, Fifi, comes to mind) reaches menarche as early as eight, several years before other females in her community do. In the 1970s, nutritionist Rose Frisch confirmed nineteenth-century suspicions that women resembled apes and other mammals in this respect.[18]

With reproductive fat beginning to accumulate, some of the fat cells secrete the hormone leptin. Soon, every ninety minutes the girl's hypothalamus will release regular pulses of gonadotropin-releasing hormone.[19] These pulses are the critical event in puberty, stimulating the pituitary gland to secrete hormones addressed to long-dormant ovarian follicles—follicle-stimulating hormone and luteinizing hormone, which essentially tell this egg-incubating organ to switch on.

But the first menstrual cycles in humans and other apes have a special twist. Whether females mature early or late, cycling begins long before pregnancy is possible. In apes, this period of adolescent subfertility lasts anywhere from six months to three years, protecting an adolescent who is sexually active from conceiving.[20] In some species (chimpanzees and bonobos come to mind), adolescent females are sexually very active indeed.

A female chimpanzee, for example, may reach menarche at eight. But it will be several years before her sexual swellings (as a result of rising estrogen levels in the middle of her cycle) are full-sized, and grown males pay attention and mating begins in earnest. But even then, a female chimpanzee may copulate on the order of 3,600 times during successive subfertile cycles before she conceives for the first time, around age fourteen, and gives birth.[21]

An adolescent's sexual swellings are especially conspicuous. Young females use them like "diplomatic passports" that permit safe passage through hostile territories. This way the wandering female can check out competitors and local resources in foreign communities while she decides where to eventually settle to breed. In the meantime, young females are free to explore and to continue growing.

By fifteen, a chimpanzee female has already borne her first baby and attained full body size. From then on, she is far more fertile than younger females. Furthermore, because of her greater maturity and experience, an infant born to such a mature female is more likely to survive. It is as if male chimps are programmed to know this and respond accordingly. Given a choice between two sexually swollen females, male chimps invariably choose the older one.[22] It is interesting to speculate why men in some societies differ from other primates in this respect, placing so much emphasis on youth. One reason, I suspect, is that these men are in a position to monopolize access to their mate and to literally *possess* her long-term. The mate he acquires may end up living in his social unit for a long time, whereas a male chimpanzee is merely seeking to fertilize a female and have done. Even though males of their respective species may have different criteria for sexual attractiveness, women and female chimps are nevertheless characterized by the same pattern of fertility, shaped like a lopsided bowler hat.

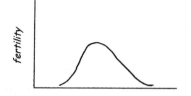

Early adulthood is the time women are most fertile, their pregnancies most likely to have a good outcome, and their infants most likely to survive. For most women, fertility peaks between the early and mid-twenties, although among some poorly nourished West African and Nepali groups peak fertility is much later, around twenty-six to twenty-nine years.[23] Fertility is lower and the probability of miscarriage higher among the very young and very old.[24] After peaking in early adulthood, fertility declines with age, falling to zero at menopause, sometime in the forties for chimps and women in most foraging societies, later for women in postindustrial societies—and around thirty in a captive macaque.

———

Late menarche and adolescent subfertility protect a young female from a dangerous reproductive enterprise unlikely to yield surviving offspring. The same stockpiled resources could be put to far better use later, when offspring would be more likely to be born healthy and to survive. The importance of such delays may be even more critical for human mothers than other primates because conscious planning plays such an important role in our ability

Fig. 8.2 At fourteen this !Kung girl is just reaching puberty. Under the harsh conditions of the Kalahari, she is still several years away from menstruating for the first time. By contrast, a rich diet and a far more sedentary lifestyle mean that girls in the United States mature more rapidly. The majority reach menarche by age twelve and a half, four years earlier than this nomadic gatherer. Adolescent subfertility, not chastity, ensures that she will be nearly twenty before she gives birth. *(Irv DeVore/Anthro-Photo)*

to cope. Yet faculties critical for emotional control and for organizing and
carrying through with plans of action are still maturing at the same time
her ovaries are. This is why some psychiatrists refer to early reproductive
maturation as "starting a car engine without a skilled driver."[25]

Since the Neolithic, and especially in the past several centuries, better-
nourished girls have begun to mature earlier, and are capable of conceiving
earlier than ever before in human existence—closer to twelve than twenty.
In the United States in 1996, a half-million babies were born to girls between
ages fifteen and nineteen, 11,000 to girls fourteen and under—the highest
rate of teenage births of any industrialized country.[26]

People refer to this as the "problem" of teenage pregnancy, yet it is more
nearly a problem of failed contraception, an undermining of evolved safeguards that under conditions more typical of human existence protected very
young girls from inopportune conceptions. Any adolescent girl living under
foraging conditions who found herself in the unusual situation of being
plump enough to trigger ovulation in her early teens would almost certainly
be in an unusually productive habitat. She also—and this is important—
would have to be surrounded by well-disposed adults helping to provision
her (see chapter 11). In modern societies, however, adolescents can be terribly disadvantaged, lack all manner of social and economic support, yet still be
so hypernourished that they reach menarche at twelve and conceive by fifteen. The amount of fat a girl has on board has become a dangerously mislead-
ing signal telling this young mammal that it is a good time to go ahead and
reproduce, when it is anything *but*. In the United States today early childbearing and large numbers of closely spaced births are the two greatest risk factors for child abuse and infanticide.[27] I return to this topic in chapter 12.

The way the hypothalamus and the pituitary respond to signals secreted by
fat cells is another one of those clues suggesting that evolution did not end
with the Pleistocene, nor stop when radiation out of Africa got under way.
Depending on the geographic origin of her ancestors, different girls' thresholds for responding to the hormones that trigger menarche are set slightly
differently. Selection pressures on her not-so-distant ancestors have become
encapsulated in her genome. Hence, age of menarche varies not only with a

girl's immediate circumstances but also according to where her ancestors came from.

Even fed the same diet of high-fat, high-carbohydrate Big Mac equivalents, daughters of mothers born in the warmer climes of southeastern Europe reach menarche earlier on average than daughters whose ancestors came from northwestern Europe. Just why this is so (cold winters? taller bodies? greater likelihood of famine?) is not known. Nevertheless, recent European migrants to Australia, all currently living in the same environment, start to menstruate at different ages depending on whether their mothers came from southern or northern Europe—strong evidence that some genetic component influences age of menarche.[28]

Absent Fathers and Early Menarche

In all primates, menarche can be speeded up or slowed down by various factors. Primate females respond to social stress, mother's low status, or unpredictable resources, by delaying puberty. Curiously, one of the first findings about human mating systems to catch the attention of sociobiologists was an apparent exception to this pattern. Girls growing up in households where the father was absent reached menarche earlier on average than those with the father present.[29]

Struck by this correlation, evolutionary-minded anthropologists Pat Draper and Henry Harpending proposed that girls growing up in "father-absent" households are conditioned to expect men to provide little and to be unpredictable ("cads" was their memorable term, as opposed to more nurturing and reliable "dads"), and respond accordingly by pursuing alternative, more promiscuous reproductive strategies. Having written off the likelihood of being able to forge an enduring pair bond, the girl opts for a more opportunistic strategy aimed at short-term gains.[30]

In line with this model, girls growing up without fathers engage in sex earlier, and with more different partners. Some researchers hypothesize that sexual maturation might be speeded up by a higher level of social stress in the home that "triggers an early increase in gonadal and adrenal hormones facilitating early sexual activity among girls." Essentially, then, early menarche is assumed to be an effect produced by ecological *and* social circumstances, including the daughter's perception of her mother's plight.[31]

It is an intriguing hypothesis. In some circles it is already accepted as fact.

It is, however, a hypothesis that makes primatologists uneasy, because the underlying reproductive physiology is so "unprimatelike." "The less predictable the resource base, the greater the social stress—the earlier the female matures?" a primatologist asks incredulously. Completely the opposite is true for resource-deprived baboons and low-ranking female chimps. In other primates, stressed females *delay* rather than speed up maturation.

The main competing hypothesis is that the correlation is an artifact of something else. This alternative hypothesis focuses on a girl's genetic inheritance from a mother who matured early rather than some nonheritable "maternal effect." If mothers who reach menarche earlier, engage in sex earlier, and become pregnant at an earlier age, are, as a consequence of this personal history, less likely to end up in a stable relationship, *then* their early-maturing daughters would find themselves growing up in a household without a father. This would make early menarche more nearly an incidental correlate of absent fathers than an adaptation to cope with them.[32] At this writing, the puzzle remains unresolved.

When Primates Give Birth Young

No matter how or why she happened to conceive so young, there is a general (and reasonable) perception that girls in their early teens make poor mothers. Young mothers are disproportionately represented among neglectful, abandoning, or even infanticidal mothers in both police records for North America and ethnographic accounts for tribal societies.[33]

It is difficult to tease apart complicating variables—heredity, the girl's own childhood experiences, prenatal care, social support, marital status, available resources. Is it really giving birth at an early age that produces inadequate mothering, or is it the correlation in our society between giving birth early and low socioeconomic status?[34] In monkeys, however, the correlation between giving birth in adolescence and substandard caretaking is less complicated.

Like many an American girl who gives birth in her early teens, some sedentary and unusually well-provisioned monkeys also breed early. Consider the vervet monkeys kept by primatologist Lynn Fairbanks in a large outdoor colony near Los Angeles. Fed ad libitum, hypernourished three-year-olds gave birth a year or more sooner than they would have in the mixed African woodland-savanna habitats where the species evolved.

To field primatologists, accustomed to the virtues of mother monkeys, the behavior of these "teenage" mothers came as a shock. Compared with physically mature vervets, these "young" mothers were negligent to a degree not seen in the wild. Some failed even to pick their neonates up off the ground. Once again, as with the twin-bearing tamarins, abandonment almost always occurred within the first seventy-two hours (see above, page 180). It's as though something critical for maternal responsiveness never quite clicked on. Their infants died at twice the rate of infants born to full-sized mothers in the "prime" breeding years of early adulthood.[35]

It is meaningless to point out that such lapses in maternal solicitude are never reported in the wild, because vervets that young never conceive in the wild. There is an obvious temptation in the case of the hyperfertile vervets to compare the consequences of their artificial, provisioned lifestyle with the plight of human teenagers. Settled living and plentiful food have removed constraints on fertility that for tens of millions of years protected anthropoid primates from giving birth at such young ages.

From a purely biological perspective, whenever chastity among teenagers is an unrealistic expectation, providing them with birth control can be viewed as a reasonable way to compensate for recent and drastic transformations in the circumstances of human existence—rather like pressurizing the cabin for people flying on airplanes, to compensate for hurtling them through an environment they were never adapted to. If a primatologist instead of a politician were president, it's likely that Joycelyn Elders, ex–Surgeon General of the United States, would have been given a medal rather than fired. The immediate impetus for the surgeon general's dismissal was her public comments to the effect that sex education and alternative outlets for libido (such as masturbation) would be preferable to early, unprotected intercourse by teenage girls who are fertile at the unprecedented age of thirteen.

Extra Support at First Birth

For any primate, first birth is an especially vulnerable time. A wide range of adaptations help a young female to negotiate it. Delayed maturation and adolescent subfertility are only part of the story; there are also patterns of social behavior that provide extra support to first-time mothers. Among the best-documented such pattern is the one found among matrilineally organized societies of monkeys where mothers help to ensure that daughters are better fed, and buffered from attacks at just the point in life when they begin to reproduce.

Few females subscribe to a social etiquette more exquisitely subtle, orderly, or predictable than Old World cercopithecine monkeys. As each female baboon, macaque, or vervet matures, she develops a set of status expectations about who belongs where. Cercopithecine societies are composed of ranked matrilineal clans, with each clan in turn organized into a rigid female hierarchy. It is mothers and grandmothers who shape it, by perpetually monitoring the competition. Matrilineal clans stick together, rallying to suppress or exclude social-climbing daughters of lower-ranking lineages, while orchestrating advantageous social positions for their own, exercising all the tribal "discipline of a small society," as Edith Wharton termed the social hierarchies produced by society matrons in early-twentieth-century New York.

The two rules that prevail in cercopithecine societies are: (1) daughters fit into the female social hierarchy just under their mothers; and (2) if there is more than one daughter, the younger, just about the time she matures and peaks in her reproductive potential, rises in rank above her older sister. But given that she is smaller, younger, and weighs less, how? The younger sister rises above her older sister with the help of an indomitable ally who ranks above both of them—their mother.[36]

As her younger daughter approaches breeding age, the baboon or macaque mother takes her younger daughter's side in confrontations with her bigger, heavier, previously dominant older sister. The likeliest explanation for this seemingly whimsical maternal preference is that she boosts her younger daughter's fortunes at a critical juncture.[37]

Worlds away, hunter-gatherer girls giving birth for the first time act as if they, too, feel the need for special support. Young couples in nomadic free-form societies like the !Kung could live almost anywhere with anyone they chose. But the most common pattern is for the new couple to remain near the girl's parents till after the birth of her first child.[38]

A man who catches the fancy of a girl, and persuades her (or her family) to have him, typically settles for a time near her folks. If the girl has not yet begun to menstruate, sex may be postponed. Meat provided to the bride's family constitutes a form of payment for reproductive rights to the daughter by the prospective son-in-law.

It is a way of life that passed out of favor with the Neolithic. In most pastoralist and agricultural societies, the bride decamps to go live with her husband's family at marriage. But among foraging people, living as our ancestors most likely did (that is, those hunter-gatherers who did not rely on boats or horses), daughters are about as likely to stay near kin when they first marry as they are to leave. Anthropologists Kim Hill and Magdalena Hurtado report that of twenty-one young foraging women among the Aché women of Paraguay, sixteen continued to hang their hammocks within an easy holler of their parents—at least for her first marriage (the average number of marriages in this society is ten).[39]

This means that a young woman pregnant for the first time enjoys the support of kin at a time that will prove critical for her continued survival as well as for her lifetime reproductive success. Furthermore, if the marriage does not work out, as is common, parents are on hand to support her.

———

A curious reminder of how adaptive it once was for a pregnant woman to be near kin when pregnant persists in a modern woman's sense of smell. As has been experimentally demonstrated for other mammals, ovulating women prefer the scent of males who have genetically produced immunological attributes (or, "major histocompatibility complexes") *different* from their own. Presumably this is to decrease the chances that a woman will mate with close kin. Mice manage to avoid incestuous matings this way, by sniffing the male's urine. Women use body odors instead. They can distinguish between different men's smelly T-shirts, and rank them as either "attractive" or "unattractive." Instead of preferring alien smells, very different from their own, however, women taking birth control pills that simulate pregnancy exhibit the reverse preference: they *prefer* the smell of those with immune systems genetically most similar to their own, men most likely to be brothers or even their father.[40] Perhaps women whose bodies have been artificially induced to simulate pregnancy subconsciously gravitate toward kin.

"Got Milk?"

Once a woman is mature and bears a child, what maintains the characteristically long three-to-five-year birth spacing typical of hunter-gatherers? Specifications for a foraging woman's ideal birth-spacing mechanism would include:

1. the mother's nutritional condition;
2. her workload;
3. how much milk any infant currently suckling still needs; and
4. whether environmental conditions (especially resource availability) seem to be getting better or worse.

These are exacting specifications. Amazingly, the "ecology" of a woman's ovaries takes into account all four.

In the eighteenth century, some women, among them Mary Wollstonecraft (an early advocate of women's rights), were well aware that if mothers would only breast-feed, "there would be such an interval between the birth of each child, that we should seldom see a houseful of babes."[41]

The understanding that nursing delays the next conception emerges in the diaries of some aristocratic women of this period who deliberately contrived to switch from wet nurses to breast-feeding so as to delay becoming pregnant again.[42] So far as medical science was concerned, however, breast-feeding continued to be viewed as a notoriously unreliable form of birth control. The "breast connection" was long dismissed as an old wives' tale.

Not until the 1970s did scientists appreciate that breast-feeding was the foundation of family planning in primates, including people in traditional societies. The "puzzle" of long birth intervals among hunter-gatherers was officially solved when anthropologists studying the !Kung recognized just how *often* babies were sucking on their mother's nipples. Mel Konner and Carol Worthman reported that babies who were carried everywhere by their mothers nursed *several times an hour* during daytime and slept with their mothers, nursing on and off through the night. Konner and Worthman demonstrated that frequent stimulation of the nipples was associated with specific hormonal responses in foraging mothers.

Right after birth, a rapid rise in circulating levels of prolactin signals onset of milk production. Thereafter, each time the baby's mouth closes around a nipple, stimulation from the baby's sucking signals the hypothalamus. This small region at the base of the brain, in turn, lowers secretion of dopamine, which triggers the anterior pituitary (a small gland near the hypothalamus) to pump out more prolactin. Continued tugging on the nipple causes prolactin levels to spike higher, increasing fifteen-fold above prenursing levels before returning to the base level in about three hours—unless the baby sucks again.

In women whose infants nurse frequently, prolactin levels remain continually elevated. These high prolactin levels are correlated with some as yet unknown function that suppresses ovulation. Whatever it is, this thermostat is set at night (see discussion of melatonin and body rhythms above, in chapter 5).

Sucking on one nipple, perhaps kneading and pulling on the other with its fingers, the infant does its baby-best to stimulate prolactin levels. From the perspective of an infant striving to survive, high prolactin levels not only sustain lactation, they help hold at bay the conception of a new sibling, whose impending arrival would encourage the mother to wean at once the babe at breast. So far as the mother is concerned, the link between intensity of suckling and postpartum infertility prevents an organism already burdened by metabolizing for two from being saddled with another pregnancy and the even more daunting task of metabolizing for three.[43]

Environmentally Sensitive Feedback Loops

Fashion-conscious readers of *Harper's Bazaar* who are also interested in avoiding conception may have noticed an interesting new product being advertised in its pages: a high-powered mini-microscope that allows the user to detect changes in her saliva that are indicative of cycle state. To say that this snazzy assay of steroid levels in saliva is an improvement over the rhythm method is to put it mildly. Instead of groping groggily for her thermometer, wrestling with doubts about whether the slight dip in her early A.M. temperature means she is about to ovulate, or getting over being sick, smart women can "Simply anoint the lipstick-size scope's reusable slide with a thin layer of saliva and examine it through lens."[44]

Celebratory bubbles form in nonfertile times, when it is safe (so far as contraception is concerned) to have unprotected sex. During fertile times, straight lines with cross-hatching will appear. The dissemination of knowledge that led to this gadget ("used by European women for decades" and only $39.95) provided an even bigger windfall for anthropologists studying the ecology of the ovaries.

No longer faced with the need to persuade women to volunteer to have their veins jabbed for a blood sample, Harvard University's Peter Ellison and his coworkers fanned out around the world with stopwatches, notebooks, and small glass vials. Trekking to pygmies living deep in the forests of Zaire, to highland tribes in Papua New Guinea, to peasants in the Himalayas of

Nepal and the Andes of South America, or merely ringing the doorbells of housewives in Eastern Europe and Massachusetts, anthropologists observed how often mothers suckled their babies, noted how much exercise the mothers got, how much energy they had to expend to do their work, and then requested them to please spit here, into this vial.

Ellison and company were helping to unravel an incredibly complicated system of interconnected, environmentally sensitive feedback loops involving the hypothalamus, pituitary, and the ovaries (technically referred to as the "hypothalamo-hypophyseal-gonadal axis"). The reason breast-feeding turns out to be so quirky and unreliable as a means of birth control is precisely because this dynamic system is so keenly responsive to maternal condition.[45]

On average, eighty minutes of suckling a day spread over a minimum of six bouts should suppress menstrual cycling for eighteen months.[46] But it takes more suckling to suppress ovulation if the mother is sedentary, less if she has to walk long distances carrying heavy loads, or if she is jogging ten miles a day. A professional ballerina, accustomed to grueling workouts, would have been unlikely to conceive in the first place. Exercise notwithstanding, the direct cue for how close to nutritional independence the mother's baby is will be how hard or how often the baby sucks. But frequency of suckling can be affected by the mother's work schedule, how far she has to commute, and whether she carries the infant with her or leaves him behind. (Of course, when a mother uses a breast pump, her body goes by how frequently she expresses milk, not by the infant's feeding schedule.) When there is no suitable substitute for milk, no safe water, or no safe place apart from the mother's breast for the baby to be, then weaning must of necessity occur late. A mother who neither breast-feeds nor works out may begin menstruating again as soon as two months after giving birth. However, women nutritionally depleted by either recent food shortages or from breast-feeding her last baby may cycle without necessarily ovulating.

To many women, it is scarcely news that nutritional depletion makes them less fertile. As one pygmy woman living in the Ituri Forest of Zaire confided to Ellison, "Hunger times are not birth times."[47] Elsewhere in Africa, among agro-pastoralists like the Kipsigis of Kenya, families seeking wives for sons pay higher "bride-wealth" for young women who are plump and who reach menarche earlier because, hypothesizes anthropologist Monique Borgerhoff Mulder, the Kipsigis recognize that plump wives are the most fertile.

Back-Load Model

Like a modern working mother agonizing over when to go back to work, or whether or not to leave her baby at the nearby daycare center, whose staff she does not quite trust, mothers have always found themselves calculating whether it's safe to leave baby with an allomother. Can she return to camp in time to suckle it? How hungry or dehydrated will it be? If she takes the child with her foraging, will she be able to carry back enough food to make the arduous trek worthwhile? Most such choices in the Pleistocene, however, were even harsher. The baby of a mother who miscalculated the time to wean probably died.

Some environments were safer than others. Anthropologist Nick Blurton-Jones and others have called attention to the harshness of the climate in a desert locale like the Kalahari, the ease with which a wandering child might become lost in the vast, featureless flatlands; the sheer distances involved. A gathering woman could not but take her infant with her. Yet she also paid dearly when she did.

Women frequently had to travel to distant groves of mongongo trees to collect protein-rich nuts that are a staple food in the !Kung diet. The trip could be daunting, a dehydrating six miles each way. Given small group sizes, even if another woman in her band was lactating, this woman might not be willing to volunteer as a wet nurse, or might have to leave camp herself.

The availability of palatable, alternative foods had to be factored in to age of weaning. Infants born into worlds without safe water or soft, easily digestible foods for weaning had to wait. Instead of Gerber's baby food, babies made do on milk and insect grubs or fibrous vegetables masticated in their mother's mouth, or, like lucky Aché weanlings, sucked on armadillo fat. For a foraging mother to remain in close enough proximity to nurse could require carrying babies—plus supplies and gathered provender—back-breaking distances. Birth of another baby too soon could prove disastrous. Hence Blurton Jones proposed that, in fact, far from limiting population growth, endocrinological feedback loops that spaced babies at long intervals actually worked to ensure that mothers replaced themselves, by *optimizing* the survival of such infants as they did produce. Babies born at shorter intervals might destabilize a mother's precarious juggling, contributing to her demise and/or their own. When tested against data for the !Kung, the "back-load hypothesis" held fairly well.[48]

As important as the extensive research on the !Kung has been for fleshing

Fig. 8.3 "Women who have one birth after another like an animal have a permanent backache!" warns a !Kung proverb. This woman in the last months of pregnancy returns to camp with her four-year-old son riding on her shoulders. Her leather sling, or kaross, permits her to carry nuts or gathered food weighing twenty-five pounds or more, together with another five pounds of personal possessions and water for the trip in an ostrich egg canteen, *and* her child, who weighs almost thirty pounds. A child this age would already have been carried by his mother some 4,900 miles. *(Richard B. Lee / Anthro-Photo)*

out our understanding of Pleistocene lifestyles, it is important to remember that the Kalahari was just one of many possible permutations for hunter-gatherers. Far to the north of Botswana, in the rock-strewn hillsides of Tanzania, Hadza foragers collect all the tubers and baobab pods they need without having to travel more than two miles from camp. Mothers are rarely gone longer than an hour (average travel time is twenty-five minutes). Infants are left behind at a much younger age (closer to age two than four, as is characteristic among the !Kung), often with subadult caretakers. Because they have the option to leave babies with an allomother, Hadza mothers can produce infants after shorter intervals than !Kung mothers without compromising

infant survival. As a consequence, the Hadza population is growing by 1 percent a year rather than holding steady. Similarly, abundant game in the South
American forests, where Aché foragers live, means more protein and fat for
women, birth intervals even shorter than among the Hadza, and higher population growth rates. The "sleeper" here, the nonobvious but very important
determiner of population regulation, is the little infant, whose suckling
transforms ecological differences into demographic ones.[49]

Division of Labor

With the adoption of slings, like the leather kaross !Kung women still use,
mothers could transport both infants and quantities of gathered food. The
technology of such carryalls is so unassuming that it is all too easy to overlook
the significance of this early technological revolution. It was the beginning—
perhaps as early as 50,000 years ago—of an economically significant division
of labor based on sharing.

A rudimentary chimpanzee-style division of labor whereby males obtained meat from hunting while females specialized in vegetable foods and
small prey items has characterized hominoids for more than five million
years. But with mothers' capacity to carry provender *in addition to babies,* they
would have had stores of food to share. The new division of labor meant that
men could go hunting confident in the knowledge that if they failed, women
would have sufficient gathered food back at camp to tide everyone over.
Anthropologist Jane Lancaster notes that new technologies for carrying
babies and also for obtaining food more efficiently (spears and sharpened digging sticks) meant that mothers who were better fed could give birth after
shorter intervals. Improved efficiency in food-getting contributed to shorter
birth intervals, and to the expanding human populations that were edging
their way out of Africa.[50]

The Real Neolithic Revolutions

As the modern world closed in, bringing outsiders to once remote locales to
claim land once freely used by peoples like the !Kung, the last vestiges of
Pleistocene lifestyles came to a standstill. Former hunters and gatherers
spent more and more time in one place. Many kept livestock, or a garden, or
relied on handouts, or otherwise traded the freedom (and the uncertainty) of
a nomadic life for greater security. In the case of the !Kung, former foragers

Fig. 8.4 Leather slings, nets, or woven baskets made it possible for mothers to travel long distances carrying several offspring, or food plus offspring. These innovations left no trace in the fossil record. The balancing baskets like these used by this Japanese mother are still widely employed. *(Courtesy of Andy McCarthy)*

indentured themselves to Herrero pastoralists in exchange for milk. Instead of making long treks to find food, carrying dependent children with them, women stayed close to home sites.

Whenever people cease wandering, intervals between births grow shorter. For ethnographer Richard Lee, observing the transformation of the !Kung San was like fast-forwarding through the Neolithic transformation of early humans.

"This sudden embarrassment of riches in terms of births," he noted, "is already imposing hardships on !Kung mothers and children alike, a degree of stress that reveals the existence of a third system interlocked with production and reproduction—a system I will call the emotional economy of the San." The Neolithic revolution did not so much mean a radical overhaul in an economic sense; what was radically different was the abrupt arrival of the next infant.

No one had to bear the brunt of these changes more acutely than the infants themselves. For a child in a village weaned after just a year and a half, "The effects are striking," wrote Lee.

> The most miserable kids I have seen among the !Kung are some of the 1.5- to 2-year-old youngsters with younger siblings on the way. Their misery begins at weaning and continues to the birth of the sibling 6 to 8 months later and beyond. The mother, for her part, has not only a demanding newborn to care for but the constant intrusions of an angry, sullen 2-year-old. A grandmother or aunt may do her best to feed and cheer up the older child and to give the overworked mother some relief, but it is clear to the observer that something is out of kilter.

The scene contrasted markedly with Lee's own earlier descriptions of loving, infinitely tolerant !Kung mothers attending deftly to the cooing and crying of each infant born. Gradual transitions from foraging to settled living—what some refer to as the Neolithic revolution—were experienced by the infants as a series of neonatal crises.[51]

————

The prelude to the Neolithic was long and slow, dating back a million years or more. People gradually became already more efficient at extracting food, using sticks to dig tubers, traps or nets to catch fish and game. Furthermore, as I will discuss in chapter 11, a new type of allomother became available, helping to provision offspring and making it possible for mothers to wean children sooner and space births more closely. Some early humans, especially those living in rich riparian or lakeside habitats, would not have needed to forage quite so widely. As new food sources became available, and as people spent more time in fewer places, birth intervals shortened and populations began to grow.

Later, with the introduction of agriculture, these effects were greatly magnified. By 11,500 years ago in central China, by 10,000 years ago in the Middle East, Mexico, and parts of highland South America, and closer to *fifty* years ago among such remnant foragers as the Aché and the !Kung, lifestyles changed completely. Foragers lingered in a few locales, literally putting down roots, depending more and more on cultivated strains of wild rice, emmer, einkorn, oats, barley, wheat, millet, or, in the New World, squash and maize.[52] Woman the forager adjusted to the more sedentary routine of woman the grinder. The availability of ground grain, and fired pottery to cook it in, meant gruel was available year-round as a weaning food, so that infants could be weaned as early as around six months and still survive.[53]

Population after population, independently, each according to its own schedule, traded the freedom of a nomadic lifestyle for short-term security.[54] The long-term costs could be measured in the classic combination of higher birth rates coupled with higher rates of infant mortality in the face of recurrent famines, epidemics, and war.

For humans, the long birth intervals typical of chimpanzees and other apes grew shorter.[55] Human infants were, if anything, *more* costly than those of other apes, yet these slow-maturing, big-brained "ape" babies were arriving as often as every two years or less. Large-bodied ape babies were being born at monkey-like intervals. In areas where food was plentiful, selection actually once again favored multiple births. In parts of the world where famines were rare, the incidence of twinning increased.

Scientists at the University of Turku in Finland have shown that women living on islands, with plentiful and constant supplies of fish, had higher rates of twinning than women on the mainland. Island mothers who gave birth to twins had a higher lifetime reproductive success, but the same was not true on the mainland, perhaps because crop failures and recurrent famines meant that both twins and the mothers who bore them were more likely to die.[56] Indeed, in parts of the world where food has traditionally been less plentiful, the only twins ever born are identical twins. Such twins represent an accident of early cell division rather than an inherited tendency toward multiple ovulations.

Replacing Singleton Births with Clutches

Long birth intervals were a staple feature of the coevolution between ape mothers and their infants. However, over tens or even hundreds of thousands

Fig. 8.5 Archeologist Theya Molleson has found bone wear in the big toes, vertebrae, and knees of 7,000-year-old skeletons from Abu Hureyra in northern Syria that indicate women spent long hours on their knees grinding domestic cereals against a stone base, much as this Central African Bemba mother is doing. While a mother is occupied by this daily grind, her baby might be held by an allomother, cradled nearby, or wrapped on to her mother's back using a sling arrangement like this one. *(Photo by Audrey Richards, African Institute)*

of years, hominid mothers were beginning to produce infants after shorter intervals. Like the island-living Finns, some mothers were even reproductively favored for producing two at a time. Increasingly, periods of heavy investment in successive infants overlapped, creating the functional equivalent of litters (or the asynchronous hatching of broods found in some birds).

At some point in human evolution, ape mothers with reproductive physiologies and temperaments adapted to rearing singleton young found themselves simultaneously nurturing several young of different ages. Multiple dependents became the "facts on the ground" that mothers had to adjust to and eventually even adapt to.

What humans, ruffed lemurs, and a handful of cooperatively breeding monkeys like tamarins have in common with many birds is that, unlike other apes, they live in families where mothers simultaneously care for multiple young. Closer birth spacing over the course of human evolution exacerbated dilemmas confronted by mothers who must then decide how to allocate

Fig. 8.6 A wicker basket on wheels permits this Japanese mother to take her nursing infant to work with her in flooded rice fields, with important consequences for both birth spacing and infant survival. Baby carriages, strollers, and infant car-seats that strap into motorized vehicles are modern innovations that—like the Pleistocene sling—permit mothers to bring infants with them as they forage, or just run errands. *(Courtesy of Kawai Takahi)*

resources among dependent young with competing needs. These pressures increased as the Neolithic brought about an unprecedented level of fertility, but at a potential cost to particular infants who may not have been just *what* (in terms of sex or other attributes) or been born *when* the stork ordered.

A mother's genes would continue to be represented in the population or disappear, depending on her ability to assess which offspring would best translate her investment into long-term reproductive success, and on how much assistance she managed to elicit from fathers and alloparents. I turn now to the means by which mothers sought such help. First on the list: how to elicit help from males in caring for infants.

Three Men and a Baby

A capon will sit upon eggs, as well as, and often better than a female.—this is
full of interest; for [there are] latent instincts *even in* brain *of male.—Every*
animal surely hermaphrodite—

—Charles Darwin's notebook, 1838

Mother-infant relations are particularly intimate and long-lasting in primates. Great Ape mothers carry their infants wherever they go. Fathers, by comparison, are rarely in direct contact with babies. Even when, as is often the case in humans, men help provision the mother, one-on-one care of infants by the father is unusual. Observing this, the young George Eliot pondered the question of "a class of sensations and emotions—the maternal ones—which must remain unknown to man. . . ."[1] Later she changed her mind. Like Darwin puzzling over a capon he saw brooding eggs, Eliot became fascinated by the possibility that males *could* express the tenderness and compassion we normally associate with mothers.

How, she wondered, would "a lone man . . . manage with a two-year-old child. . . ." What would happen, the novelist asked, if a motherless two-year-old toddled through a blizzard and ended up at the door of a lonely man? She titled her thought experiment *Silas Marner,* a novel that is still required reading for English majors. In it, Eliot intuited what primatologists have since documented. Care is most readily elicited from a male primate under three conditions:

1. long-standing familiarity with the immature;
2. the nearby infant is urgently in need of rescue; and, especially
3. the male has a relationship with the mother.

In Silas Marner's case, solicitude was elicited by little Effie's obvious need, combined with a perception of kinship (Eliot tells us Marner identified the golden-haired orphan with his long-lost "little sister"). Once triggered, Marner's tenderness came naturally as he lifted the toddler and she

Fig. 9.1 Peasant father spoon-feeds gruel to child. *(Etching by A. von Ostade, 1648, courtesy of Wellcome Institute Library, London)*

clung around his neck, and burst louder and louder into that mingling of inarticulate cries with "mammy" by which little children express the bewilderment of waking. Silas pressed it to him, and almost unconsciously uttered sounds of hushing tenderness.[2]

I know of no better human case history for exploring what Darwin termed the "latent *instincts*" for nurturing that lurk "even in *brain* of male."

What does it take to get a man to care for infants? And—turning to a question much on the minds of mothers today—why don't fathers help more? I'll first explain the immediate, or *proximate,* reasons why, in general, primate mothers hold babies more than fathers do, then turn to the *ultimate* causes for why male primates are not more inclined to care for infants.

In all creatures where fertilization takes place inside the mother many weeks or months before the infant is born, males cannot be certain of paternity. A male's best clue to whether or not he is the father will be whether he had sexual relations with the mother, and if so, their timing, and their frequency. Most female primates have evolved to mate over a period of days, and (when feasible) with a range of male partners. There are few species in the order Primates in which males can be certain of paternity.

This evolutionary history can still be detected in the patterning of sexual behavior in women today, and in the psyches of men who are obsessed with the chastity of their mates. No matter that females did not evolve a flexible and assertive sexuality in a vacuum. (It was an essential tactic for ensuring well-being of their infants that would scarcely have been necessary if females could choose an acceptable partner and count on him.) Given the situation as we find it, females mate with more than one male. This leaves males little choice. They must mate with as many females as they can, or else find themselves at a relative disadvantage vis-à-vis their rivals' efforts to transmit their own genes to the next generation. Like mothers, males make tradeoffs of their own. Males must choose between parenting offspring they *may* have sired, and seeking to mate with additional females and possibly siring more. Often such tradeoffs (along with uncertain paternity) make it disadvantageous for males to respond to the magic of infantile signals with the same alacrity that mothers do. Mother Nature has set male thresholds for responding high. Yet it is not the case (as an egg-sitting capon led Darwin to suspect) that males never care for infants. "Instincts" to care slumber in the hearts of primate males, including men. Under what circumstances are they activated?

Godfather Gorillas

Far stranger transformations than Silas Marner's abound in the natural history of other primates. Binti Jua, the mother gorilla who rescued the little boy at the Chicago Zoo, got all the press, but a decade earlier another little boy on the Isle of Jersey fell onto the floor of the gorilla enclosure at the zoo there. That time the simian good Samaritan was a silverback male named Jambo, who stroked the boy's back and kept other apes away.[3]

Primatological lore is rich in anecdotes about aloof males transformed into instant heroes by endangered infants. Adoptions of ape infants whose mothers have disappeared is almost always by kin, typically brothers or possi-

Fig. 9.2 Babies elicit the tenderest responses from the toughest guys. The Greek hero Heracles (better known as Hercules), draped in the skin of the lion he has just battered with the club he carries in one hand, holds the baby Telephos ever so gently in the other. According to legend, Heracles had a brief affair with a (supposedly chaste) priestess of Athena who abandoned their baby at birth. The baby, Telephos, was being suckled by a wild deer when Heracles found him. *(Statue from the Villa d'Este in Tivoli, Italy, 2nd c. A.D. Courtesy of the Louvre)*

ble fathers. The same males notorious for killing infants unlikely to be theirs will, at other times, look out for those that might be.[4] Picture the extremely domineering hamadryas baboon male, a snout the color of raw beefsteak with the intrusive habit of biting the neck of females who stray from his harem. Yet when one of "his" females (this is the only nonhuman primate species for which the term "harem" actually applies) went into labor too close to a cliff, this tyrant responded with deft tenderness to catch the baby. A similar intervention was witnessed in captivity, but this time—even more amazingly—the midwife was a male orang—an event almost beyond the imaginings of fieldworkers, for in the wild male orangutans are solitary. Shaggy hundred-kilo titans crash through the jungle oblivious to mothers and infants, passing females by like ships in the night, meeting with them only to mate.

No less bizarre are the multiple cases of care-in-a-contingency offered by notoriously nonnurturing langur males, best known for stalking babies, not caring for them. Yet in a pinch, a brother or former consort of the mother takes custody of an infant deposited with him by the mother. In a plot reminiscent of the film *Three Men and a Baby,* a mother langur leaves a nearly weaned infant with a group of males recently driven out of her breeding group by an usurping male. She will return alone to face the potentially infanticidal "stepfather" who now resides there. The same langur male that as a juvenile showed no interest in joining his sisters as they scampered about with borrowed babies becomes transformed into a solicitous custodian by the overtures of a needy youngster.

Why Males Don't Mother More (Proximate Causes)

If the circumstances are conducive, almost any primate male can be induced to behave in a nurturing way. How is it, then, that it is almost always females who end up holding babies? In only a tiny minority of species do males care for infants even remotely as much as mothers do. Confronted with such overwhelming evidence, it just seems "natural" to conclude that since mothers provide the womb, develop the mammae, and energetically invest the most, they therefore are the sex selected to nurture babies. End of story.

But what if we don't end the story there? What if we look beyond the obvious to inquire further: *Why is it* that, even if they are not hungry, baby primates prefer mothers? Or, to bring the question closer to home: Why, even among bottle-fed babies with both parents working outside the home, does the traditional division of labor between father and maternal caretaker so often emerge?

In many mammals, retrieving, licking, "brooding," or protecting pups are unisex *potentials.* What Darwin termed the "latent instincts," the basic wiring, for such behavior seems to be there.[5] The underpinnings for caretaking are there; it's just not expressed under ordinary circumstances. Why not?

Fig. 9.3 The film *Three Men and a Baby* is the ethologically correct chronicle of a few weeks in the lives of three incorrigible, field-playing bachelors with no previous space in their brain for commitment of any kind until they find a baby outside their door, dropped off by a former girlfriend. Like a male rat sensitized to caretake, the bachelors become not only competent but committed. The cues eliciting care are both the infant's obvious need and the males' past relationship with the mother. *(© Touchstone Pictures, courtesy Disney Publishing Group, reproduced with permission)*

Gender ideology is no help. How can culture and socialization explain sex roles in mammals lacking language and symbolic thought? There must be evolved emotional differences between males and females, differences that go beyond the two major physical differences, birth and lactation. What besides ideology produces this seemingly unbridgeable chasm of difference between aloof fathers and "instinctively" caring mothers? Initial differences turn out to be surprisingly minor—tiny compared with the magnitude of the eventual dichotomy.

Fig. 9.4 Even when fathers are obviously devoted to their offspring, fatherly love is rarely translated into direct care of infants. During the first six months of his daughter's life, this doting !Kung San father will hold her less than 2 percent of the time.[6] *(Mel Konner / Anthro-Photo)*

Small Differences Much Magnified

What magnifies small differences into major divisions of labor? The simplest answer is that *people* do, by following the path of least resistance. As Ed Wilson put it, "At birth the twig is already bent a little bit."[7] Where natural inclinations lead depends on how much effort is expended bending them back. Among humans, conscious effort can minimize preexisting differences. More often, small initial differences in responsiveness are exaggerated by life experiences and then blown out of all proportion by cultural customs and norms.

Imagine two working parents who start out with the noble intention of an equal partnership caring for their new baby. The benefits of breast milk are too important to forgo, so the mother uses a pump and stores bottles in the freezer for the father to heat up when she can't be there. Within weeks, they notice the baby has developed a preference for the mother. Soon it is painfully obvious that the baby *wants* its mother. Aware that her husband is hurt and her baby is unhappy, the mother quits her job and stays home with the baby. "After all," she sighs, "you can't fool Mother Nature."

Why did this young couple's good intentions go for naught? Recent findings by anthropologists Joy Stallings and Carol Worthman at Emory University, collaborating with developmental psychologist Alison Fleming and coworkers at the University of Toronto, provide relevant clues. The researchers asked new parents to listen to two recordings. One was the sound of a day-old infant crying first thing in the morning when he wanted to be fed. The second tape contained the more jagged and alarming cries of a baby being circumcised. Reactions of mothers and fathers were carefully monitored, and hormone levels (of cortisol, testosterone, and prolactin) measured. At the first signal of real distress, both mothers and fathers responded with equal alacrity. But if the infant merely sounded uncomfortable but not in extremis, if the cry was merely "I want" rather than "Help! Help!" the mother was the quicker to respond. It's possible that the mother's greater responsiveness and the physiological reactions that accompany it are learned. More probably her lower threshold for responding to infant signals is innate.

————

So the mother is more sensitive to infant needs than the father. So what? Who cares? And that's just the point. The act of caring has its own consequences—habits of mind and emotion. When we get down to the "underlying mysteries" that George Eliot called attention to, the causes of difference can be just that simple. No doubt, other things are going on. The point is, consequences are magnified out of proportion to initial causes.

Just because the mother is more readily galvanized to respond to infant demands does *not* mean that fathers are not able to do so, or that they cannot become adequate caretakers, "good enough" caretakers, or that baby primates cannot form primary attachments to a male. Rather, a seemingly insignificant difference in thresholds for responding to infant cues gradually, insidiously, step by step, without invoking a single other cause, produces a marked division of labor by sex.

Recall our imaginary couple, just home from the hospital. From the nursery wafts the first sputtering sound of fretful wakening. Mother is already out of her chair. She soothes baby, who never even reaches full cry. All is quiet. Baby coos contentedly. Further intervention is superfluous. "I would only be needlessly intrusive," the father tells himself. There is no reason to move. Mother has baby. Why disturb the peace by taking him from her? The result, of course, is that the baby's attachment to his mother intensifies. The baby starts to complain whenever he is transferred from her to someone else.

Yet all along there were alternatives. She could leave her baby alone with her husband more. He could request that his wife wear earplugs, or, like Odysseus, bind herself to the mast so she will not be able to respond to the irresistible call of her little siren. The neural equivalent of earplugs is what Mother Nature opted for in the case of titi monkeys, rendering mothers indifferent to the allure of infant cues. The result? Infants strongly prefer fathers, and the males "just naturally," without conscious determination or outside intervention, do most of the childcare.

Eliciting Paternal Devotion

Titi monkeys are as monogamous as primates get. The mother is so attached to her mate that she borders on indifference toward her infant. After birth, the father carries his newborn 93 percent of the time. This unusual monkey is the exception that proves a larger rule: when a male primate's reproductive success is substantially enhanced by assisting his mate rear offspring, and when he has no better reproductive alternatives (the female titi makes sure of that by driving away any female who enters her territory), he helps.[8]

Sure, either titi partner may occasionally stray. They are primates, after all, never so blindered that given an opportunity they won't at least lust in their hearts or, should his or her partner's vigilance lapse, copulate with an outsider. (Extra-pair copulations were first reported in titi monkeys in the 1960s, long before reports of female philandering in monogamous birds caused a sensation in ornithological circles.)[9] Nevertheless, the primary commitment of these lovebirds is to each other.

Fig. 9.5 In togetherness titi-style, monogamous pairs are often seen side by side, tails entwined. *(Drawing by Virginia Savage)*

The threat that his adultery may lead to alienation of his care from her infant explains why a titi mother is more attached to her mate than to her baby. When psychologists William Mason and Sally Mendoza at the California Primate Center used various endocrine measures to monitor responses to different circumstances, they found that titi mothers were far more stressed by a temporary separation from their mates than from their babies.

Once the infant has suckled, the titi mother wearies of her heavy baby. She pushes it off. Since the mated pair typically sit side by side on the same branch, tails entwined, the father is nearby. Rejected by its mother, the baby climbs aboard its father till it's time to suckle again. Thus does the baby form a primary attachment with him, not her.

This is not really so strange. On the rare occasions when human fathers or allofathers end up the sole caretaker, infants form primary attachments to them. As the kindly neighbor who volunteers to teach Silas Marner how to care for a child observes: "See there, she's fondest of you. She wants to go to your lap, I'll be bound. Go then: Take her Master Marner. . . ."

When Care Is Neither Exclusive nor Costly

Any male who provides exclusive care to an unrelated baby, and thereby fails to sire children of his own, may find many satisfactions. But he does not increase his reproductive success. This is why Mother Nature sets the threshold for direct and exclusive care of young higher in fathers than in mothers. If primate mothers respond to infant needs right after birth, they are unlikely to ever misdirect their care. Males cannot be so sure.

But what about assistance divisible among a number of babies? What if care is only intermittently required? Far more common than the exclusive, all-consuming care given by titi monkey fathers to offspring almost certain to be their own is the flashier brand of male care, the quick intervention of a

Robin Hood who shows up every so often, behaves like a hero, and then fades away. Such fathering is divisible (or "partible" in the lingo of primatologists) among many recipients. The important point is that a male need not be certain of paternity to proffer paternal-like assistance.

Between the "dad" who is devoted to his kids and the "cad" who deserts them lies a broad intermediate zone filled with occasional dads, and "temporary heroes." Such fathers may divide their time between many possible offspring. If an infant is *possibly* theirs, and intervention is not too risky, these scattershot fathers can afford to be less than certain. If protection of an infant from a predator or another male is essential for its survival, males can scarcely afford to withhold it from an infant that might be theirs. Although there is a tendency to assume that early hominids lived in nuclear families the way married people do today, or perhaps in small "harems," no one knows how they lived. Hence, this "temporary hero" type of father must be included as one of the various possible alternatives for how early hominid males interacted with their kids.[10]

———

Savanna baboon mothers breed in multi-male troops and stay in touch with favorite former consorts. These special male friends look out for progeny possibly their own. Hence, their relationships might provide insights into the formation of the kind of pair-bonds we find in human parents committed to the same child. However, it took the equivalent of a primatological "sting operation" among the infanticide-prone baboons at Moremi to prove that some males had the well-being of the infant in mind.

Ryne Palombit, Robert Seyfarth, and Dorothy Cheney from the University of Pennsylvania set up the sting by discreetly placing speakers where they knew the baboons would pass. As each target male passed, he would hear the taped sounds of a newly immigrant male harassing one of the recent mothers he had a special relationship with. The adult male reacted immediately, rushing to the defense of mother and infant. However, if the tape recording indicated that the animal harassing her was just another female, the same male exhibited little interest. If the mother's infant was no longer alive, the "strange-male-hassling-your-girlfriend" tape elicited no response whatsoever from him.[11] The researchers concluded that protection of infants from strange males was indeed high on the possible father's list of priorities.

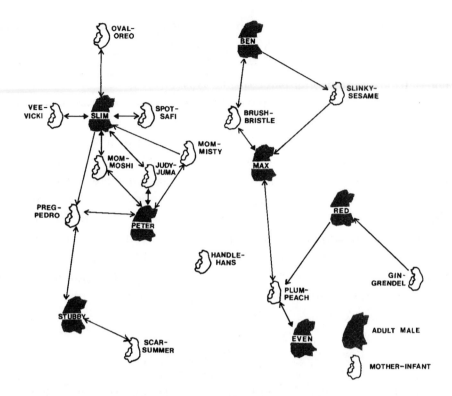

Fig. 9.6 Jeanne Altmann's 1980 diagram shows the special relationships forged between savanna baboon mothers and adult males they have mated with. During the first three months after the birth of her infant Moshi, his mother, Mom, stays closest to male Slim, but also engages in frequent grooming with Peter. These "godfathers," of the genus *Papio,* defend the mother and her infant from other baboons. This may not be "quantity time" but it is "quality time" in the most elemental sense. Occasional intervention by former consorts can mean the difference between an infant's life or death. *(Courtesy of Jeanne Altmann)*

Several primatologists, most notably Carel van Schaik and Robin Dunbar, have proposed that one of the main selection pressures on males to remain near their mates after breeding—an essential precursor to monogamy and any form of the nuclear family—was to guard mothers and offspring from other males—especially infanticidal ones. As novelist Alison Lurie put it in *The War Between the Tates,* her savagely funny dissection of American marriage, "We need [men] sometimes, if only to protect us from other men."

A primate mother's best bet in such a system would be to find a mate strong enough to protect her. However, if her mate dies, or is driven off by another male, she would be better off if she had associated with several males, thereby improving the odds that there will be several candidates in the neighborhood who classify her offspring as possibly "kin." At issue here is not whether males *can* help protect or care for infants, but how mothers can motivate them to do so. Next to ensuring that she conceive with an appropriate mate, the most important selection pressures shaping female sexuality in primates have to do with forging relationships that promote tolerant, even protective, relations between a past or present consort and her infant.

The Trouble with Being a Male Primate

A mother mammal relies on proximity right after birth to learn to recognize the smells and sounds of her newborn baby. A male has only his past relationship with the mother to go on. If he has managed to control access to the mother during every moment when she was last fertile, his paternity is guaranteed. This contrasts with the situation for, say, pronghorn antelopes: when it's time to breed, the pronghorn female surveys the scene, selects the most vigorous male, and mates with him—just once. Then the matter is closed till next season.

Among such prosimians as the little African galago, the female may mate with more than one male, but only during a brief period, lasting a few hours. The rest of the time an epithelial membrane seals the galago's vagina shut, and intromission is impossible. But for simian primates, the window of opportunity during which copulations can occur is more flexible. Worse (from the male's point of view), in most primates there is no conspicuous midcycle signal advertising ovulation, such as the red swelling on a baboon or chimpanzee. Under some circumstances (for example, when strange males enter her troop) a monkey female may solicit males although she is not ovulating, something neither Saint Augustine nor the Catholic Church was aware of when they mistakenly assumed that intercourse without possibility of conception was unnatural.[12]

Exactly why monkey and ape females are so sexually flexible, or why they go to so much trouble to solicit matings from extra males, is much debated. Among the various explanations proposed is that females are ensuring that conception occurs. Or they might be increasing the odds of bearing offspring sired by males with superior genes. Perhaps they are reducing the chances of inbred offspring sired by close relatives.[13] One consequence of female wanderlust is that only a male who completely excludes other males from his

Fig. 9.7 Relatively few primate species advertise ovulation with conspicuous sexual swellings. In most species, females are more subtle. A langur around the time of ovulation solicits males by presenting her rump and frenetically shuddering her head. She exhibits no other visible sign. Females vary in the intensity of their head shudder, so only a resident male has reliable information about her actual cycle phase and probable fertility. *(Daniel B. Hrdy / Anthro-Photo)*

mate's vicinity can afford to regard his recent sexual history with the mother as a reliable cue for paternity. By the same token, the margin of uncertainty surrounding paternity could forestall males who recently mated with the mother from harming her subsequent offspring—an infant that just *might* be theirs. This is one of several benefits mothers derive from being able to manipulate the information available to males concerning paternity.

Whatever the explanation for this polyandrous component to female mating habits, there are obvious repercussions for males. To remain competitive with other males in his vicinity, a male primate must grow large enough to dominate and control females, and to exclude rival males (the way dominant male gorillas do). Or else he must evolve large testicles and ejaculate plentiful, high-quality sperm in order to compete in a different arena, inside the reproductive tract of a female he will never manage to monopolize.

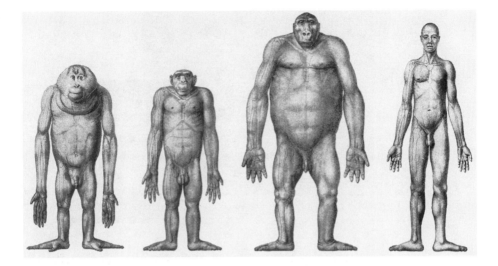

Fig. 9.8 Comparisons of the genitalia of Great Apes conform with the general rule. Males in one-male breeding units, like gorillas, have smaller testes relative to their body weight than do males in species where a number of males mate on the same day with a female who conspicuously advertises ovulation. Hence, a 170-kilogram male gorilla who is able to prevent other males from getting anywhere near ovulating females in his "harem" has testes weighing just 27 grams, compared to the enormous 140-gram testes of a considerably smaller, 45-kilo chimp. Because a chimp does not have the luxury of excluding competing males, his sperm competes inside the female's vaginal tract with the sperm of rival males. Humans have testes proportionally larger than those of the underendowed gorilla, but considerably smaller than those of chimps. Men fall with orangs among primarily monandrous (or one-male-at-a-time) breeders. This might be interpreted as showing that our ancestors lived in one-male breeding systems, but with just enough lapses to maintain continuing selection for moderately large testicles. Alternatively, it could mean that selection favoring large testes was once important but no longer is, and men are slowly evolving smaller testes. With no advantage to maintain the extra sperm-producing capacity of testes, through time the average testes size in humans might become smaller still, sperm counts lower. *(Courtesy of the A. H. Schultz-Stiftung, Anthropological Institute, University of Zurich)*

Larger testes give a competitive edge to the male who produces the most (or most competitive) sperm. But without DNA tests, who knows which male that is?[14]

Woman's Sexual Legacy

From galagos to bonobos, female primates range from being sexually receptive for just a few hours right around midcycle to being *able* to copulate (although not necessarily desirous of doing so) across extended periods of the

A

B

Fig. 9.9 a and b. Early reports about bonobos sounded like flights of feminist fancy. In addition to their free-wheeling sexual ways, females have feeding priority, and may share food they obtain with the offspring of female friends. Bonobo swellings can last weeks at a stretch. Prolonged sexual attractiveness helps females to cement alliances and to exchange sexual favors for food. (a) This bonobo engages in face-to-face sex, rubbing her genitals against those of the female beneath her. As in common chimps, the bonobo's clitoris (insert b) is both absolutely and relatively larger than that of a woman. In bonobos it is frontally placed, presumably to facilitate the achievement of orgasm during genital-to-genital rubbing and to provide a proximate reward for alliance building. Some evolutionists have argued that the clitoris is nothing more than a vestigial penis, present in females because it is necessary in males. More likely, however, selection has also operated independently on the clitoris and the penis. This would explain why the chimp and bonobo clitoris is larger than that found in humans, while the opposite is the case for the penis, which is larger in humans than in either bonobos or chimps.[15] *(Courtesy of Amy Parish)*

cycle. Human females fall at the extreme end of this continuum, at the far edge of flexible, situation-dependent receptivity. Although in most cultures people avoid intercourse during menstruation and for long periods post-partum, women are capable of mating on any day of the menstrual cycle. Yet vestiges of "estrus" (or cyclical sexual urges) persist.

Like other primates, women experience a mild to pronounced increase in libido between menstruation and ovulation, during the phase of the cycle when the follicle ripens in preparation for releasing the egg. The exact time of ovulation can be speeded up or slowed down depending on circumstances. If, for example, a woman is exposed to pheromones from the armpits of another woman who is just about to ovulate, it may cause her to ovulate sooner. In 1998 Martha McClintock, one of the pioneers of biosocial psychology, experimentally confirmed the existence of these long-suspected airborne chemicals wafting from one woman to another. She was able to entrain one woman to another's cycle by attaching cotton swabs from the armpit of the donor to the upper lip of a woman in another cycle phase.[16]

This olfactory component to female cycling has to be very old. The same may be said for behavioral predispositions linked to ovulation, such as a woman's increased sexual yearnings at midcycle. These responses serve as reminders that even though our rear ends do not swell up and turn bright red like a chimp's, and although we rationalize our actions more than baboons and langurs do, the origins of our sexual urges predate the Pleistocene. Modern hominids have not entirely lost cyclical estrus, although its manifestations have become much modified and subdued.

A broad range of field and laboratory studies now confirm that women around the time of ovulation—when a monkey or ape would be in estrus—are more likely to feel self-confident, to experience erotic fantasies, and to initiate sex. This pattern became apparent when researchers excluded sexual behavior initiated by husbands or lovers and just focused on female-initiated sex. A woman also has a somewhat lower threshold for experiencing orgasm around ovulation—a patterning to female libido found in both heterosexual and lesbian pairings, and during both intercourse and masturbation.[17]

Women at midcycle are more restless and move around more. They have enhanced motor capability, possibly a heightened sense of smell,[18] and are probably better able to discriminate healthy from unhealthy people. All in all, women at midcycle test higher, including in an academic sense—all the better, one assumes, to use their "wanderlust" to greatest advantage by assaying the available males and choosing well, and perhaps also using heightened sensitivity to danger to avoid punishment by possessive mates. Recall (p. 85) the West African chimps who slipped away to the neighboring community and back without anyone (apparently other chimps as well as human observers) the wiser.

So far the only in-depth study of hunter-gatherers that combines personal interviews with information on a woman's cycle state (determined from blood samples) is the one done in the 1970s among eight women belonging to the same community as Nisa. Anthropologists Carol Worthman and Mel Konner documented statistically significant increases in "sexual desire" around the time of ovulation. These women were more likely to report having had sex with their husband during the follicular phase of their cycle than during the postovulatory, luteal phase. They were also at that time more likely to have sex with men other than their husbands. Midcycle also brought with it a slightly higher probability (not statistically significant) of experiencing orgasm.[19]

––––––––

For a long time, it was assumed that the female orgasm was a uniquely human trait. Some even suggested that female orgasms evolved in the course of human evolution to make women satisfied with one male. But this is unlikely. For one thing, we now know that female orgasms occur in at least some other primates, although in a different context than in humans.[20]

It is possible that in baboons and chimps the pleasurable sensations of sexual climax once functioned to condition females to seek sustained clitoral stimulation by mating with successive partners, one right after the other, and that orgasms have since become secondarily enlisted by humans to serve other ends (such as enhancing pair-bonds). Anyone who notices how erratic the occurrence of female orgasm is, from intercourse alone, compared with male orgasm, which invariably accompanies ejaculation, might well wonder whether this curious psychophysiological response currently has any adaptive function at all.

Recently, British biologists Robin Baker and Mark Bellis suggested that female orgasms do currently have a function: to ensure that a mother's egg is fertilized by the best male. Their hypothesis rests on three as yet unproven assumptions:

1. that orgasm produces an "upsuck" response that increases sperm retention;
2. that orgasm increases the probability of impregnation; and
3. that women are more likely to experience orgasm when mating with males who have superior genes.

So far, all that has been conclusively demonstrated is that given sufficient clitoral stimulation, any female achieves orgasm—with a male partner, a female partner, or masturbating alone, although females may experience orgasms more readily at midcycle—which may also be when women are more likely not only to become pregnant but to seek extra-pair copulations.[21]

A persuasive case can be made that orgasms (whether in women or bonobos) dispel tension and strengthen bonds between partners. But such observations are not sufficient to argue that orgasms evolved for this purpose, or that orgasms currently have any effect on a woman's fertility or ability to keep her infants alive. If anything, the opposite seems more nearly true, given that the vast majority of women today across large areas of the world (much of Asia, North Africa, and the Middle East) live in coercively patriarchal societies. Female libido and sexual assertiveness are dangerous predispositions in such contexts, more likely to get a woman beaten, disfigured, or killed than to increase her reproductive success.[22]

For compared to other primates, men have many more sources of information about where their mates have been, and what they have been doing (not the least of which is gossip). Knowing how risky extra-pair sex can often be for females in my own species has led me to wonder if female orgasms may be a once adaptive retention now no longer selected for, like the grasp of a just-born baby for maternal fur that no longer exists, a reflex gradually fading out of the human repertoire. If so, our descendants living on starships eons from now may find themselves wondering what all the fuss was about.

———

Much has been made of the fact that men have stronger libidos than women do. In Darwin's time, experts were convinced that women were sexually passive and that "the appetite never asserts itself." Today it is assumed that women *are* interested, albeit not nearly so much so as men.

When asked by a psychotherapist how often she and her husband have sex, Diane Keaton in the title role of the film *Annie Hall* complains, "Constantly! I'd say three times a week." By contrast, Woody Allen, playing her husband, answers, "Hardly ever. Maybe three times a week." To make this same point using quantitative data, evolutionary psychologists have a favorite experiment. They send out student shills to proposition opposite-sexed "research subjects" encountered about campus. In line with both the old stereotypes and the widely accepted predictions from evolutionary theory, 75 percent of the male students approached agreed to "go to bed" with the experimenter (no information available on what happens next), while none of the coeds did.[23] The dichotomous results conform to the widespread expectation that ardent and lustful males pursue discriminating, chaste, or even, as Darwin termed them, "coy" females.

The results of the solicitation experiment are taken as powerful confirmation of Robert Trivers's truism: since mammalian mothers invest so much more in offspring than fathers do, they cannot afford to be indiscriminate; they need to be much more selective about who they mate with than do males.[24] But the tricky part comes when the interpretation of these results is taken a step further, cited as proof that there has been far stronger selection on male sexual desire than on that of females, or even to show that there has been no selection in women for sexual desire. Such a black-and-white view of the sexes is seriously at odds with field observations of female chimpanzees and barbary macaques in estrus soliciting and copulating with multiple partners with a lust and avidity that evolutionary theorists traditionally reserve for males.

In their extreme forms, such stereotypes derive not only from wishful thinking on the part of Victorian men but from an invalid comparison. The experiments ignore social context. It is both physically dangerous for a woman to go home with a strange man and dangerous to her reputation for her to be perceived as interested in doing so, especially when the individual soliciting her behaves so peculiarly and indiscreetly as the experimenter. An even bigger problem is that one of the subjects being compared is a continuously potent male, the other is an only intermittently fertile female.

In men, production of sperm in the testes goes on all the time. Males ejaculate, and then, after a refractory period, can, given the opportunity, ejaculate again. Their sperm count may go down in successive ejaculations, but it is soon replenished. Indeed, a male's batteries may be recharged sooner rather than later if a novel and desirable partner is introduced, or if a male reencounters his own partner after a long separation. In short, this is a gamete-making machine almost continuously charged and ready to reproduce, not just ever-ready to mate but ever-ready to fertilize. Compare Mr. Ever-Ready to a woman.

Women, like men (indeed, even more so than men, since no erection need be sustained) are continuously able to copulate, even though they may not, and often do not, desire to. However, if we confine the comparison of men's and women's libidos to just the period when women are actually reproductive, comparing Mr. Ever-Ready with Ms. Intermittently Fertile, the much-remarked dichotomy between ardent males and coy females begins to pale. Given the boorish behavior of the shills, the coed still ought to say no; but it would be informative if the experimenters had more sensitive methods to detect whether or not she was aroused by the proposal.

Why Smart Women Make Foolish Choices

Adaptive or not, holding steady or fading out, it would be a mistake to underestimate the power of libidinous retentions. Consider the mysterious process we call "falling in love." Ancient legacies figure here as well.

No one has been more innovative in examining the subtleties of women's cyclicity than Dutch sexologist Koos Slob. Slob was among the first scientists to measure uterine contractions and increased heart rate in female macaques during sexual climax, thus helping to confirm that female orgasms occur in other primates and were almost certainly part of the package of traits the earliest hominids brought to the human experiment. When Slob used sophisticated new techniques to measure sexual arousal in women, he was amazed not just by how profoundly cyclical women's sexual responses were, but by how far-ranging the implications of that cyclicity could be.

During the first half (or, follicular phase) of their cycle, women were more aroused by erotic films than during their postovulatory (luteal) phase—not a result one would expect in an animal whose sexual responses were culturally constructed, or among whom love and romance depend only on context. But Slob's work also carries a warning for those who would reduce human mate choices to simple chemistry. A woman who happened to be in the follicular phase of her cycle the first time she saw the film footage was more likely to be sexually aroused by it than were controls who saw the same film at other times. But the second time the test subject saw the same film, she responded to it the same way she had the first time, regardless of her cycle state. It was as if recollections of her past libidinous feelings took precedence over how fertile she was on that day.[25]

Along with all the sensible things a prospective mother should be attending to (such as a potential mate's health, body size, strength, ability to protect her and her children, his intelligence, and indicators of "good genes"), women may also be influenced by such seemingly extraneous variables as the cycle day on which they first met.

Imagine: a woman at midcycle looks at a man more intently than she might otherwise do. Chronically alert to just such opportunities, the man stares back. Subconsciously, he notes her body language, the dilation of her pupils, and gauges her interest and responds accordingly. One thing leads to another . . . but, when, later in her cycle, she remembers that man, her emotions will be colored not just by his merits, but by how she felt when first aroused. It is a comparative sexologist's angle on why smart women so often make foolish choices. An awareness of our primate heritage carries with it a warning to women of good sense to keep an eye on their calendars. Ovulation can be hazardous to your judgment.

Some Fathers Who Can Afford to Care

Human mothers need male investment as never before. No other primate produces babies quite so needy, or dependent, for quite so long. Yet theoretically only a male with a high certainty of paternity should be willing to provide it. Of the 40 percent of primate species that exhibit some form of male care, only monogamously mated males like titi monkeys with a high probability of being the father provide *direct* and *extensive* care. Ethnographic information for different human societies (these data all referring to the *same* species) similarly suggests that paternal care is most intensive where monogamously mated men have a high certainty of paternity.[26]

Consider the Aka pygmies of Central Africa. During the first six months of life, the average father holds his infant more than 20 percent of the time and remains within arm's reach of the baby an unheard-of 50 percent of the time. Aka fathers mostly marry just one wife. They spend long periods at camp and appear to have invented "flex-time." Help from fathers sustains relatively high fertility (Aka women average 6.3 live births) without causing child mortality to rise any higher than among the !Kung, who reproduce far more slowly.[27]

Aka togetherness is facilitated by a family-friendly workplace. Hunting—mostly for small game—is done using nets, so that men, women, and children can all safely and practically participate. When men go off to hunt, mothers and children go with them. How typical or atypical was the Aka lifestyle for forest-dwellers in the Pleistocene? Not known. Recent archaeological discoveries indicate, however, that hunting with nets is very old, dating back 30,000 years or more.[28]

In Western industrial countries like the United States, the 1990s are being hailed as the new "era of the involved father." Fifty-six percent of men interviewed in a recent study by the DuPont company indicated that they want to spend more time with their families. But is this really so new? Ethnographic and historical sources suggest that some men were always interested in being with their families. Variation among men in the past probably fell along a similar continuum ranging from the "new man," proudly and euphorically "engrossed" in his baby son, to the stereotypical self-absorbed CEOs who act as if they do not know they have children, to "deadbeat dads," totally resistant to investing in them.[29]

Hype about "new fathers" notwithstanding, the DuPont survey calls attention to a valid observation. Across cultures, the amount of time men spend caring for kids is the best predictor of how connected they feel toward them. For example, the same proximity that puts a foraging father on the spot to help out, and induces him to become emotionally involved with his kids, also brings greater certainty of paternity. An Aka father spends 46 percent of each day within view of his wife. This is one reason such peoples as the Aka have become to ethnography what titi monkeys are to primatology. Both represent ideal types—"chestnuts," as it were—useful for illustrating a correlation between pair-bonding and direct and extensive male care.

The Hearts of Men

Given the obvious advantages of male care, why don't primate fathers—human fathers in particular—always tend to their progeny, or at least devote more attention to infants? Part of the reason, as I hope I have shown, is simply opportunity and exposure. Given the right circumstances, males do care. But there are also more fundamental reasons why women have lower thresholds for responding to infant needs than men do. Males who invested in infants not their own would be genetically out-competed by males whose priority was seeking additional mates. Males are torn between impulses to protect and care for their offspring, at least a little bit, and their desire for novel partners.[30]

From a man's perspective, a mate with heightened libido at precisely the time when she is most fertile is a dangerous liability. Not long ago, feminist economist Claudia Golden assumed she was making a joke about remote fathering when she quipped that "We never ask [high-powered career] men whether they have children. I guess we assume some of them don't know. . . ."[31] But truth to tell, how could they? Golden had inadvertently stumbled on one of the ultimate causes of bad fathering, the nagging problem male primates have always faced: How to be certain of paternity? We primates are not like fish, where males come along and fertilize eggs that are already laid. Nor are we absolutely monandrous, like the pronghorn antelopes, among whom a female mates just one time per conception. It's not just "a *wise* father who knows his own child," it's either a father who runs his home like a seraglio with a eunuch at the gate, or else one who has a DNA fingerprinting lab at his disposal.

True enough, as economist Golden stresses, many men *are* far too busy striving for status (in part to impress other men and additional mates) to be anything other than oblivious to their children. But there is another reason these men ignore their infants: their thresholds for paternal investment are set high. And the reason for *this* is that female primates did not evolve so as to guarantee their mates certain paternity. For humans the consequences can be surprising.

The Show-Off Hypothesis

In 1986 a group of behavioral ecologists embarked on a re-study of the Hadza of northern Tanzania. Aware that hunting was the main source of protein, they began with the classic premise of optimal foraging theory. They predicted that men would plan their hunts so as to maximize the amount of meat they brought back to camp, consistent with acceptable risk-taking. But the researchers were surprised to discover how inefficient the hunters were. A Hadza hunter could expect a full month of failures for every successful kill. This was not because game was hard to come by, but because the hunters held out for the biggest prey they could take—creatures like the sleek, elusive, and marvelously fleshy eland. Men tracked these imposing beasts for days, even though the same effort devoted to lesser prey—rock hyraxes or guinea fowl—would yield higher returns.

Whenever the researchers could persuade the hunters to lower their sights, the Hadza invariably came out ahead in terms of protein and calories earned for effort expended. Self-inflicted inefficiency was the first surprise. The second was discovering that after a month of unsuccessful hunts, when a Hadza hunter did kill something big, he retained only a fraction of the meat for his own family. A hunter who killed an eland kept only 19 percent of the enormous 350-kilogram carcass for his own wife and children. The vast majority of it was shared with other group members who spontaneously appeared "to help" eat it. Only if a hunter repeatedly failed and was forced to resort to hunting small game did his family get to keep the entire portion.

Well, such is the price of success, one might argue. An eland is too much meat for one family. But the Hadza climate is very dry. Sliced meat quickly develops a crusty rind and could easily be stored, or traded as jerky. Yet this does not happen. As is typical of hunter-gatherers, Hadza hunters neither brag about success nor attempt to claim meat as personal property.[32] Instead, they tolerate a system in which the families of failed hunters—sometimes including families of men who have not killed prey in anyone's memory—can end up with the most meat.

According to the Hadza ethic, the more a hunter obtains, the more he gives away. This is a fairly general pattern among hunter-gatherers. From each according to his means, to everybody else. Had evolutionary storytelling come full circle, back to group selection? Were men storing up credit for another day, against an unlucky stint when the hunter would have nothing, while another, luckier, group member might have meat to reciprocate? Was meat so rich and succulent, so desired and public a good, that no man could afford to retain exclusive access?[33] Or were men choosing to hunt larger prey because by killing a creature with mythical status some of the animal's charisma rubbed off on the hunter? This is when one of the team, Kristen Hawkes, started to wonder if large game—meat that a man would have to give away—might not be worth more to a man in terms of prestige than as food in the mouths of his children.

According to Hawkes's "show-off" hypothesis, it was reputation that hunters were maximizing, not protein. Not only would other men respect him more, but women wowed by his prowess and intrigued by the prospect of gifts of meat might grant him sexual favors. What looked like parental effort was more nearly reproductive effort, as hunters exchanged food for sex in a time-honored performance characteristic of every primate in which males hunt.[34] That "women like meat" was the standard !Kung explanation for why a particularly poor hunter remains a bachelor and has no prospects of ever being anything but celibate.[35]

———

How typical are Hadza show-offs? Not known. However, a recent survey of the shopping habits of 167 British couples was eerily consistent with Hawkes's view of male foragers: "showing off" took priority over economy.

Even when men have the same amount of shopping experience as women, husbands and boyfriends shop less economically. Seventy-three percent of the time they chose different brand names than their wives, almost always more expensive ones. Men spent 10 percent more on shampoo, 6 percent more on butter, 5 percent more on coffee. Men apparently find it hard to resist the brand names and the occasional big-ticket items—the thirty-dollar bottle of burgundy that magically materializes in the shopping cart.[36] Male shoppers are more tantalized than women by the prospect of showing off with the "grand gesture."

Economic surveys from developing countries as far afield and as culturally different as India, Guatemala, and Ghana reveal that the nutritional level of children in a family does not increase in direct proportion to paternal income. Only increasing women's income has a direct effect. The show-off hypothesis is not only consistent with the man who stops off at a pub for a pint with his buddies while the kids are hungry at home, but with actual statistics from the United States Department of Health and Human Services concerning the large number of men who are more amenable to making car payments than paying child support.[37]

Monogamy as a Compromise That Children Win

Once again, Nisa's biography provides a !Kung San forager's perspective on the tensions underlying human pair-bonds. Nisa marries four times, always monogamously. When her first husband, Tashay, brings home a second wife, Nisa recalls, "I chased her away and she went back to her parents." Several of Nisa's marriages dissolved under the strain of infidelities, either her husband's or her own. In addition to her four husbands, eight lovers pass in and out of her life. Nisa is quite obviously in love with several of them. "Pairbonds" were formed, but the relationships did not last.[38]

Two of Nisa's pregnancies probably derive from affairs with men other than her husband at the time. As Nisa's daughter Twi grows up to look more and more like her husband's brother, with whom Nisa was having an affair when the child was conceived, her husband reminds her that his younger brother is the likely progenitor and therefore "will help take care of her." Whenever Nisa finds herself between husbands, when she is widowed or divorced, she sets out across the Kalahari to find her brother and live with him.[39]

Hunter-gatherer societies like the !Kung San are as egalitarian as traditional societies ever get. Nisa's husbands were physically stronger than she, able to dominate her, but if she was unhappy enough, Nisa could always vote with her feet and leave. Even when Nisa was caught by her husband in flagrante delicto with a lover and beaten and threatened with murder, others stood up for her, and life went on. In more patriarchal societies, her perpetual adulteries would have been lethal.

Since none of Nisa's children survived to adulthood, the life of this spunky woman can scarcely be said to typify success in evolutionary terms. Yet the tensions that characterized her marriages are the same ones that Nisa's mother mentions. Again and again, her predicaments crop up in women's life stories.

Nisa cherished her freedom of movement, her freedom to choose mates, and, if her husband did not provide sufficient food, her freedom to negotiate with lovers. Each husband, on the other hand, wanted multiple wives for himself but also to maintain exclusive sexual access to Nisa. There is a dynamic tug-of-war in these relationships that is at odds with conventional pipe dreams about humans having an innate tendency to form long-lasting pair-bonds, unions in which both sexes have a powerful commitment from within to adhere. Such cases make it hard to sustain the illusion that lifelong monogamous families are the natural human condition.

Monogamy in Nisa's case is more nearly a compromise than a species-typical universal. Monogamy is the most harmonious common ground she and her husband of the moment can arrive at. And when it works, children benefit. Monogamy reduces inherent conflicts of interest between the sexes. *Her* reproductive success becomes *his,* and vice versa, promoting harmonious relations between genetically distinct individuals striving toward common goals.

Sociobiology is not a field known for the encouraging news it offers either sex. Yet its most promising revelation to date has to be that over evolutionary time, lifelong monogamy turns out to be the cure for all sorts of detrimental devices that one sex uses to exploit the other. As usual, this point is most convincingly demonstrated in organisms that breed much faster than we humans do. Once again, fruit flies are the organism of choice, the current favorites for studying coevolution between the sexes. This time, the experiments have a "happy ending."

Recall that female fruit flies mate promiscuously with many partners. To counter female promiscuity and thwart their rivals, male ejaculate is laced with special components that enhance an ejaculator's reproductive success, even though their cumulative effects in females are toxic. Amazingly though, in just forty-seven generations of experimentally imposed monogamy, researchers Brett Holland and William Rice produced strains of male drosophila whose seminal fluid was no longer toxic to females. At one level, monogamy is a compromise offering something for everyone—especially immatures. All males got to breed, at least a little. Females had less choice among males, but they lived longer. Meanwhile, offspring were more viable.

So who says evolutionists are necessarily antithetical to family values? Theoretically at least, I know of no reason why Holland and Rice's experiment would not produce similar results in the temperaments of men and women, although the experimenters would need to locate a population of volunteers who could precommit for descendants, guaranteeing their absolute fidelity for fifty or more generations. The result would be a strain of men and women whose first priorities were to the well-being of their children.

———

No wonder, then, that as a mother in the modern world, where children require more parental investment than ever before over a longer time if they are to prepare for a successful life, I personally am partial to the companionate monogamous marriage. Long-term trust permits unparalleled efficiency and emotional satisfaction. I place a high priority on the benefits that two cooperating parents offer children. Precisely because I am aware of my bias, I consciously guard against distorted readings of such evidence we have for early hominid mating systems as would cause me to project upon my ancestors—who lived under different circumstances—the same choices I have made.[40]

I assume that mothers in the past were *emotionally* similar to me. But they made their decisions under different, vastly more arduous, circumstances. It would scarcely be wise, or fair, to extrapolate my self-interested priorities to them. Nevertheless, from Victorian times to the present, this is what many anthropologists and evolutionists have blithely done.

Earlier commentators failed to consider how unusual are the particular environmental and demographic conditions that make long-term monogamy

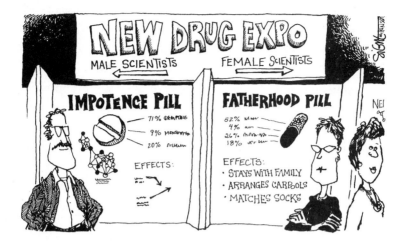

Fig. 9.10 Underlying male tradeoffs between investing in parenting effort versus additional matings have not changed all that much in the last million years. Nor have the perennial tensions between what fathers might prefer to do and what women and children need them to do. *(Signe Wilkinson/Philadelphia Daily News/Cartoonists & Writers Syndicate)*

advantageous for both sexes. For monogamy to benefit a mother, her mate must be in a position to protect her or to reliably provision her. Demographic rates and sources of mortality are also important. High adult survival rates among men make it worthwhile for them to invest in relationships. But does the protection they offer matter? This depends on what the sources of infant mortality are (pathogens, or human enemies?). Where fathers make a difference and infants also have good prospects of survival, a father can afford the emotional luxury of sharing his mate's commitment to quality over quantity. When, however, children are susceptible to sudden, unpredictable demise, it should not be surprising that men go to great lengths to sire as many as possible, in the hopes that some will by chance survive.

———

An old ditty provided by psychologist William James is a great favorite with sociobiologists. It runs: "Higamous, hogamous, woman's monogamous. Hogamus, higamus, men are polygamous." Contemporary Darwinians Donald Symons, Roger Short, and Laura Betzig have made strong cases that men in the past sought many mates because by doing so they increased their reproductive success. Women, by contrast, did better locating one good man willing and able to invest in her offspring. But there is a missing caveat to James's ditty: "Except where males are poor providers, likely to die young, or when resources are unpredictable—then mothers are far better off polyandrous"— if they can be so safely. A woman's children may be better off with several "fathers" than with one inferior or unreliable one.

Unpredictable providers confront mothers with a dilemma: Should she rely on one man for much, or on several for something? What about in patriarchal societies, where maternal choices are circumscribed from the outset. Should she seek 100 percent of a poor mate or settle for some fraction of the investment to be provided by a polygynously mated potentate? From a mother's perspective, the optimal number of "fathers" all depends.

The Optimal Number of Fathers

It seemed that [Rosamond] had no more identified herself with [her husband, Lydgate] than if they had been creatures of different species and opposing interests. . . .

—George Eliot, in *Middlemarch,* 1872

That fathers matter is obvious. But they cannot always be relied upon. When fathers die or defect; when men seduce or rape women and then decamp, or when they fail to protect their wives from the violence of other men; when husbands seek other mates, prospects are diminished for the offspring they leave behind. In the absence of effective laws protecting the person or property rights of women, early mothers had to rely on such fathers or alloparents as they could enlist—husbands, lovers, or, failing them, male kin. No matter how Apollo-like the progenitor, how skilled a hunter, how good his genes, how viable his immune system, his absence put his children at a disadvantage. Even in matrilineal societies, where descent and transmission of rights are transmitted through the female line and mothers typically have an unusual degree of reproductive autonomy, when property rights are transmitted, they still pass from the mother's brother to her son. Parents recognize that sons are more likely to be able to hold on to property than daughters are.[1]

Being Fatherless

In industrialized countries, disadvantages to fatherless families include economic hardship, reduced status, and generally declining prospects. Costs to children are measurable in poorer school performance, higher rates of delinquency for boys, and early pregnancies for girls. In foraging societies fatherless children are more likely to die.

"I am without the man I married . . ." the !Kung San woman Nisa wailed when she lost her first husband. Men were the main providers of protein. Well might she ask: "Where will I see the food that will help my children grow? Who is going to help me raise this newborn? My older brother and my younger brother are far away."[2]

Among the Aché, an infant who lost his father was four times more likely to die before the age of two. Even if the father was still alive, Aché children whose parents divorced were *three times more likely to be killed* than if the marriage endured. When a widowed or abandoned mother takes a new mate, risks to her infants shoot up. Terrible prospects are one reason why some foraging peoples bury orphans alive along with the deceased parent.[3]

Stepfathers Worse Than No Father

Westerners appalled by such barbaric treatment of the fatherless should take a look at their local newspapers. Child homicide in civilized societies is nowhere tolerated, very much against the law, and uncommon. Nevertheless, in North America when the father of offspring under two years of age no longer lives in the home and an unrelated man or stepfather lives there instead, this rare event is seventy times more likely to occur.[4]

The murder of infants by stepfathers or mothers' boyfriends resembles the circumstances under which sexually selected infanticide evolved in other primates: males from outside the breeding system increase their own chances to breed by eliminating offspring sired by rivals. The superficial similarities have sometimes led to the erroneous conclusion that child abuse as we know it today is or once was adaptive.[5] Some clarification is in order.

Canadian psychologists Martin Daly and Margo Wilson were the first to demonstrate increased risk to infants from having unrelated men in the house. They were careful to stress that in postindustrial human societies, neither child abuse nor infanticide is adaptive. More likely than not, the boyfriend goes to jail and the mother is prosecuted for neglect. More important, the attacker is not some invader entering the breeding system from outside it: he already has keys to the apartment and access to the mother's bed.

———

Imagine: the mother goes off on an errand, leaving her baby in the boyfriend's care. She may or may not have an inkling of the risk. Perhaps she senses that her boyfriend resents diversion of household resources, including her attention, to some other man's child. (Among the Aché, mothers themselves sometimes kill fatherless infants after a conscious evaluation of what the future holds.) Perhaps boyfriend and baby are already off to a bad start—all the more reason why the baby may reject such tentative comfort as this man offers. The baby cries, makes demands not willingly met by a man in no way sensitized for this task. Mother Nature has set high his threshold against altruism toward this insatiable stranger. Because of the low degree of relatedness between the man and the child, the benefits don't come close to outweighing the costs of care.[6]

But beyond his lack of solicitude for an unrelated, very vulnerable but demanding dependent, the abusive boyfriend may have little more in common with an infanticidal monkey than a certain nonspecific impatience, a general predisposition to respond violently to repeated annoyance. By contrast with the boyfriend's violence, the infanticidal langur's attack is purposeful and goal-oriented. Injury comes after hours of single-minded stalking of the infant. Such males are often in a special state of arousal, as evidenced by an erect penis, though without other indications of sexual interest. The langur utters a distinctive hacking vocalization, a "cackle bark," rarely heard in other contexts. His attack is as organized and focused as a shark's.

If relatives of the infant intervene, or if the mother eludes him, the langur male starts over, inching closer. Infants killed by infanticidal langurs typically die from puncture wounds in the skull. To use canine teeth to bite a baby that way is the simian equivalent of pulling out a hunting knife and stabbing the quarry. This behavior, when observed, cannot be considered accidental. Infant-biting by usurping male langurs bears little resemblance to the tragic blend of frustration, brutality, and terrible judgment that results in "shaken baby syndrome."

A more appropriate animal analogue for a brutal stepfather would be an alloparent of either sex compelled to invest in an infant he or she has lost interest in. The motive is not to kill the infant in order to increase reproductive access to the mother, but to rid oneself of an encumbrance. What *evolved* is not the bizarre and maladaptive alternation of solicitude with torture that we know as "child abuse." What evolved was a high threshold for responding in a solicitous way toward an offspring not likely to be genetically related—the equivalent of emotional earplugs.[7]

When Strangers Capture Mothers

Boyfriends who fatally abuse babies in their charge do not thereby enhance their own opportunities to mate. But when raiders abduct women and leave

Fig. 10.1 When an experienced, middle-aged langur uses her back foot to scrape off a baby she has borrowed and then tired of, it looks abusive. But she does not so much seek to harm the infant as to prevent it from clinging to her. Free of her encumbrance, she leaves the infant alone. Fortunately, an abandoned baby is picked up within minutes by either the mother or another allomother. *(Sarah Blaffer Hrdy / Anthro-Photo)*

behind or savagely kill unweaned infants, they often do. What then? Are these men expressing infanticidal tendencies such as those that evolved in other primates? It has the same outcome as infanticide does in other primates. Does this mean that the killing of infants by such men is homologous—that is, similar in its mechanisms and evolutionary origins—to infanticide by raiding males in other animals?

Some cultural anthropologists dismiss such behaviors as infanticide during war as pathological, produced by colonial transformations of otherwise peaceful tribal worlds. Some dismiss all forms of murder as idiosyncratic or deranged.[8] They assume evidence from other species is irrelevant because animals have no symbolic culture. They assume that infanticide, to the extent that it is really going on, must be due to culturally constructed attitudes that condone brutality.

Yet the earliest evidence we have, from the *Iliad* and other ancient narratives, suggests that raiding goes way back. New assessments of archaeological evidence (recently and cogently summarized in Lawrence Keeley's *War Before Civilization*) convinces me that men have always engaged in intergroup conflict over access to resources, just as males do in other primates using less elaborate or calculated means.

Intertribal conflict is very ancient, and fertile females are among the resources likely to be competed for.[9] Clearly, animal evidence *is* relevant. The difficulty comes in trying to be specific about just *how* it is relevant. *Just why* does the capture of mothers put their children at such horrifying risk? It is bad enough that infants are often forcibly left behind to starve. Worse still is

Fig. 10.2 "What's wrong with him? He rapes young girls, and then takes off? He fathers children secretly, then lets them die," observes a character in a play by fifth-century B.C. dramatist Euripides of Apollo's habit of raping lovely maidens. *(Detail of an Attic red-figured hydria from Capua, attributed to the Coghill painter, ca. 440 B.C., British Museum, London, Inventory no. E 170)*

the propensity of some raiders to deliberately kill a vulnerable child that is no immediate threat to them. The matter is problematic not just because the behavior is gruesome and the evidence sketchy, but also because we know so little about the mechanisms involved.

Observing phenomena like lactational suppression of ovulation, we understand fairly well how innate neuro-endocrinological feedback loops adjust birth intervals in primates. The mechanisms work the same way in a langur as in a human. So long as the mother's infant suckles frequently, ovulation is suppressed, although nutrition and energy expenditure are also relevant variables. The outcome is birth intervals compatible with both infant survival and maternal success over a long breeding career. It is far less clear what, if anything, is the same when a strange male appropriating a female kills her infant. We don't know if, at a physiological level, brutal monkeys and brutal men are motivated in anything like the same way.

Rational Actors Can Also Behave Like Brutes

One of the most reliable accounts of infanticide by tribal raiders comes from Elena Valero, a Brazilian woman kidnaped by Yanamamo warriors when she

Fig. 10.3 Today the term "trophy wife" is likely to crop up in a Gary Trudeau cartoon about the glamorous young wife of an aging, self-important CEO. But in the past such trophies were a primary objective of warfare. In this vase painting from 480 B.C. the commander-in-chief of the Greek army, Agamemnon, commandeers the most beautiful of their recent captives from his lieutenant, Achilles. Agamemnon leads Briseis away by the wrist in the time-honored gesture denoting both "taking possession of" and the marriage union. "She trailed on behind, reluctant, every step . . ." adds Homer. "Women are the constant cause of war," said Darwin, and the stronger "carries off the prize." Friedrich Engels concurred: "In Homer young women are booty and are handed over to the pleasure of the conquerors," he wrote, "the handsomest being picked off by the commanders in order of rank."[10] *(Attic red-figure skyphos attributed to Makron and signed by Hieron on one handle, courtesy of the Louvre)*

was eleven years old at a time when intertribal warfare and raiding for women was still endemic in the forested region between the Upper Orinoco River and the Upper Rio Negro. No sooner was she kidnaped than Elena Valero's captors, Kohoroshiwetari Yanamamo, were themselves attacked by rival Yanamamo, the Karawetari. Again Elena was taken captive and handed over to one of her abductors as a wife. She would spend the next twenty years

Fig. 10.4 Rumored atrocities against children outnumber accurate eyewitness accounts. Yet many wartime attacks on children are all too real. The Bible (Exodus 1:16) tells us Pharaoh commanded Shiphrah, Puah, and the other Hebrew midwives to kill all male children at birth. Much later, King Herod "sent forth, and slew all the children that were in Bethlehem, and in all its borders from two years old and under" (Matthew 2:16). Assuming it's true, why should Herod kill infants of no immediate threat to him? *(The Slaying of the Innocents, Nicholas Poussin, [1594–1665], Photographie Giraudon, courtesy of Musée Condé, Chantilly)*

among the Karawetari, marry twice with different captors and bear three children before finally escaping. She would witness, and hear about, many more raids. But none were so horrifying as the second one:

> They killed so many. I was weeping for fear and for pity, but there was nothing I could do. They snatched the children from their mothers to kill them, while the others held the mothers tightly by the arms and wrists as they stood up in a line. All the women wept.

Elena Valero and the other women fled before the raiders, taking their children with them. Were mothers with babies victimized because they were especially vulnerable, or were they specifically targeted? With ease and absolute callousness, one of the Karawetari raiders "took the baby by his feet and bashed him against the rock. His head split open and the little white brains spurted out on the stone."

When much later, anthropologist Napoleon Chagnon interviewed different Yanamamo groups, people told him about women being kidnaped, their infants merely left behind to starve. But in Elena Valero's account, the children of captured women—and especially sons—were very deliberately targeted:

> the men began to kill the children; little ones, bigger ones, they killed many of them. They tried to run away, but [the Karawetari raiders] caught them, and threw them to the ground, and stuck them with bows, which went through their bodies and rooted them to the ground. Taking the smallest by the feet, they beat them against the trees and the rocks. The children's eyes trembled. Then the men took the dead bodies and threw them among the rocks, saying, "Stay there, so that your fathers can find you and eat you."

Elena Valero goes on to describe how mothers tried in vain to engage the raiders in conversation and dissuade them from killing their offspring. One woman pleaded, "It's a little girl, you mustn't kill her." Another gambled desperately to save the life of a two-year-old snatched from her arms by telling the raider, "Don't kill him, he's your son. The mother was with you and she ran away when she was already pregnant with this child. He's one of your sons!" The man pauses as he mulls over this possibility before replying, "No, he's [another group's] child. It's too long since [that woman was] with us." The man then took the baby by his feet and bashed him against the rocks.[11]

———

Grisly recollections, whether they derive from the Yanamamo or Bosnia, raise special problems. "Rational actors" seem to behave as brutally as chimps, eliminating other males and old females, carrying off fertile females, killing immatures, and (like chimps and langurs) especially targeting immature males.[12] Some sociobiologists, like Harvard's Richard Wrangham, attribute similarities between human and simian genocide (and he uses that term for both) to common genetic attributes in both that cause males to behave in a "demonic" way. Some evolutionary psychologists propose that humans evolved distinct psychological mechanisms, or "modules in the brain for homicide."[13] But so far these are speculations.

The argument that infanticide can be attributed to the common heritage of chimps and humans, who share 98 percent of their genetic material (Wrangham's homology argument), is weakened by the fact that bonobo males also share 98 percent of their genetic material with humans yet have (so far) never been reported to behave in quite so "demonic" a manner. However, as bonobo specialists Amy Parish and Frans de Waal point out, one reason bonobos do not engage in infanticide is because even though females do not remain among kin, the strong alliances that females forge with other females make mothers too formidable.[14]

Gorilla males, like chimps, share nearly 98 percent of their genes with men, and they, too, are highly infanticidal. Yet infanticide by gorillas is a solitary activity, not the demonic work of males in groups. At the same time, equivalently "demonic" patterns (infanticide, perhaps special targeting of male infants) can be found in more distantly related monkeys like langurs, who share with humans only about 92 percent of their genetic material.[15] It is not clear what we are dealing with here.

Some suspect (and I agree) that superficially similar patterns in men's behavior derive from a much more open-ended program than in other primates. Humans endowed with similar general motivations and emotions have only so many practical options for solving similar problems they confront, so they converge on the same solution as other primates. Human raiders commit similarly brutal acts as gorilla, chimp, or langur males. But men consciously evaluate costs and benefits, as well as future consequences of their actions. They calculate contingencies: How much more slowly, for example, are mothers burdened by infants likely to travel? What are the chances that a son spared will grow up to avenge his father? Might these children be useful alive?

Yanamamo articulate their unwillingness to share resources with the offspring of others. They have learned through observation that the fiercer they are, the more other Yanamamo—enemies, comrades, and women they seek to control—fear and respect them. Yet they worry out loud about fears that other men will retaliate or engage in the same sort of atrocities. As Elena Valero witnessed, some men talk to mothers whose infants they are about to kill, rationalizing their actions. One raider even performs some quick arithmetic to assess the woman's claim that one of his comrades might be the father of the infant he is about to kill. The calculus of Hamilton's rule ($C < Br$) is ingrained not just in his genes but in his memes—in learned tribal values that are not just a matter of genetic influences, as in wasps, but are part of the way he is predisposed to look at the world.[16] Nevertheless, by killing these infants, their mothers' captors eliminate an encumbrance on the trip home, and are likely to increase their own opportunities to mate and sire offspring. In the long term, such actions will increase their group's access to resources—something else they are consciously aware of.

My own guess is that the behavior of infanticidal men is homologous to that of their primate cousins in only the most general sense. They are motivated to strive for status, to compete for access to females, to avoid investing in unrelated infants, to adopt patterns of behavior more likely to enhance than to decrease long-term inclusive fitness. The specific similarities, then, are merely analogous solutions to common problems these variously endowed animals confront.

But even as we acknowledge that we don't really know the reason, I think we can conclude that women have always had a problem with unrelated men behaving savagely toward their children. Sometimes there is nothing mothers can do about it. Other times there is.

The Importance of Him Being Earnest

Women are not peafowl, looking for peacocks who provide nothing beyond their genes. Nor are they cockroaches, nor butterflies on the prowl for packets of nutrients that males deliver along with their sperm. Human females can't afford to be so single-minded as to look just for the father with the "best genes" or even the guy who momentarily offers the most resources. Few evolutionists would argue that a mother will be indifferent to signals indicating the appropriateness or genetic quality of their mates. But, frankly, my dear, in a species like ours, maternal decisions have to go beyond genes. If a man inseminates and decamps, and the resulting infant dies prior to weaning, well before any heritable traits from the father have a chance to be expressed, their excellence is of no consequence.

Not long ago, in the most extensive study of its kind, women in more than thirty countries were asked to rank the traits they would most value in a husband. After "mutual attraction" (which may include subconsciously registered cues about mate quality, body size, and ability to protect her) women attached the highest priority to traits like "dependable character," "emotional stability," and "maturity"—what Nisa would call having good "sense."

Several things about how such criteria would have played out in a hunter-gatherer world like Nisa's are worth noting. First of all, just reaching adulthood offers some measure of quality control; fools rarely managed it. Furthermore, women often chose between closely related men. In Nisa's case, her lover was a full brother to her husband, both men carrying many of the same genes. Is it really so surprising, then, in a species where allomaternal care matters so much, that Nisa, like the majority of women in the international survey, ranked reliability above good looks, good health, physical performance, or other indicators linked to specific genes? Even material resources paled beside good sense—a composite of innate temperament (no doubt at some level influenced by genes), upbringing, and life experiences.

There is considerable evidence from a wide range of societies that women exchange sexual favors for material resources, and that in societies where males control access to such resources, women commonly marry for money. Yet "financial prospects" still ranked twelfth on the international list of favored attributes.[17] In the Pleistocene, hunting success was erratic, and there was no wealth to possess. Hence modern preoccupations with "how to marry a millionaire" strike me as recent, closer to ten thousand than to a million years old. Marrying for wealth is an artifact of patriarchal societies and their aftermath, where powerful men (or patrilines) monopolize access to the resources women need to rear their young.

In such societies, union with a wealthy man may indeed be a promising route to secure status and even reproductive success. But among foragers, who have no alimony, reliability in a husband was a trait worth considering. So were such contingency plans as a mother could make ahead of time, should she have reason to fear that her mate might let her down.

When Should Mothers Line Up More Than One "Father"?

When loss of a parent is a real risk, survival is hand-to-mouth, and mothers would be hard put to provide alone for progeny, why should mothers confine themselves to just one mate, if by mating with several they increase their off-spring's survival chances? If the logic here sounds out of character for humans—creatures we have reason to think spent much of their evolutionary history in primarily one-male breeding systems—consider the Western tradition of asking trusted friends and especially influential people to serve as "godparents." The goal is to line up fictive parents who, if needed, will look out for your offspring. But beyond symbolic designation of godfathers, why not offer them some tangible, real probability of paternity? In some parts of the world, sexual liaisons are part of the way a mother sets up networks of well-disposed men to help protect and provision her offspring.

————

The area of lowland Amazonia that today spans northern Brazil and Venezuela down to Paraguay has long been a patchwork of forested tracts and vast stretches of traversible savanna, unbroken by any natural geographic barrier. This land is occupied today by hunter-horticulturalists belonging to language groups that have long since diverged as tribal groups went their separate ways. In addition to fish and meat, these tribes today still live on manioc that women grow in temporary gardens. Along with a similar lifestyle, these people still share the ancient belief that a baby is sired by more than one man, a biological fiction anthropologists refer to as "partible paternity." That is, fetuses are thought to be built up over time, like the luster on pearls, by repeated applications of semen. It is a system of belief (if men buy into it) that benefits mothers and children at the expense of any one male's certainty of paternity.

No matter whom they are studying, anthropologists begin by collecting genealogies, then classify people by type of kinship system. When Kim Hill asked the Aché who their fathers were, however, he found he needed to expand his termniology. Three hundred and twenty-one Aché listed a total of 632 fathers. That is an average of almost two "fathers" each. As in many tribes—including the Canela, Mundurucu, and Mehinaku of Brazil, the Bari and Yanamamo of Venezuela—the Aché believe fetuses are a composite product of several different men with whom the mother had sexual relations. Hence the Aché have a word, *miare,* for "the father who put it in"; *peroare,* for "the men who mixed it"; *momboare,* for "the ones who spilled it out"; and *bykuare,* for "the fathers who provided the child's essence." Men who provided

Fig. 10.5 The ceremonial designation of godparents who stand beside the parents at baptism is a time-honored way for parents to improve their child's prospects in life. We often call these people "Aunt-so-and-so," or "Uncle," reinforcing the fiction of genetic relatedness. Parents may specifically discuss with godparents provisions for a child who might be orphaned. *(Courtesy of Marion Hunt)*

the mother with meat while the baby was forming are seen as especially likely to have given the child its essence.[18]

In addition to the mother's husband, who is the socially designated father, a baby can be born with additional, secondary, fathers, who share some obligation to support this child as he or she matures. The Mehinaku joke about this joint paternity, referring to it as an "all-male collective labor project."

All men with whom a woman had sex when she became pregnant, and including the period just prior to when she was detectably pregnant, are expected to provide food for her child. Hence, it is scarcely surprising that just as soon as she suspects she is pregnant, a Canela woman, like a groupie after a rock star, attempts to seduce the tribe's best hunters and fishermen. This is a slightly different way of thinking about why the best hunters might get the most lovers. Are they out-competing other men for reproductive success, or incurring more obligations? Probably both.

In any event, this extra provisioning may explain why Bari children with several fathers have measurably better survival chances. According to anthropologist Stephen Beckerman, only 64 percent of the 638 Bari children who

Fig. 10.6 An Aché woman rests during early stages of labor. According to Aché traditional knowledge, a child can be born with one or more fathers. These "godfathers" are expected to give gifts
of meat and other help to the child, and are critical for the survival of any child who is orphaned.
(Courtesy of Magdalena Hurtado and Kim Hill)

had no secondary fathers survived to age fifteen, compared with 80 percent
of the 194 children who did. Singly fathered children born to the same
mother did not survive as well as did their multiply fathered siblings. Among
the Aché, Hill observed that children with just one father received less help,
but when a mother lined up *too many* fathers, the extreme uncertainty of
paternity dissuaded all candidates from helping. Children identified as having
one primary and one secondary father had the best survival rates.[19]

───────

The optimal number of fathers under these demographic and subsistence
conditions appears to be two. Are husbands jealous? Among the Aché, men
deny it, but then later beat their wives. Not surprisingly, Aché mothers try to
convince possible fathers that the club is more exclusive than it really is.
Among the matrilineal and matrilocal Canela, where women have more say,
there seems to be less need for discretion. Anthropologist William Crocker
of the Smithsonian, who has studied the Canela for many years, is convinced

that husbands are not jealous. Whether or not Canela husbands are telling the truth about not minding, they join with other group members in encouraging wives to honor the custom. In what may be one of the more extreme cases on record, unusual for hominids, ritual sex with twenty or more men during all-community ceremonies left some Canela mothers with an array of candidates for "fatherhood."[20]

With resources unpredictable and often scarce, in environments where high adult male mortality makes divorce and widowhood likely, husbands may see merit to their wives' infidelity. There is the insurance factor and also enhanced social cohesion. Friendship is a widespread, if little remarked, rationale for wife-sharing among peoples such as the Eskimo, or among ritual age-mates in parts of eastern and southern Africa where men share wives with real or fictive brothers. Husbands turn a blind eye, appreciating that the well-being of children whom they are fond of (and also have the best chance of having fathered because they had regular sexual access to the mother) is at stake.[21]

———

In a traditional Canela marriage ceremony, the bride and groom lie down on a mat, arms under each other's heads, legs entwined. The brother of each partner's mother then comes forward. He admonishes the bride and her new husband to stay together until the last child is grown, specifically reminding them not to be jealous of each other's lovers. Survival of a mother's children takes priority over a man's exclusive sexual access to his wife. The Canela have taken the "touch of polyandry" that crops up so often in the mating behavior of other primates, justified and legitimized it through ideology, and magnified its significance many times over through ritual. But the system is a fragile one. When the Canela come into contact with outsiders, Canela men almost at once became more possessive of their wives.[22]

Mother-centered societies like the Canela fare poorly when they come into contact with more patriarchal ones. The freedom to wheel-and-deal, which allows Canela and Bari mothers considerable leeway in negotiating for child support, is the antithesis of patriarchal marriage, which, as Marxist theorist Friedrich Engels pointed out years ago, was devised with "the express purpose" of producing "children of undisputed paternity." By forcing wives to submit to husbands in order to gain access to resources under his control, patriarchal husbands translate ownership of the means of production into "ownership" of wives; "should the wife recall the old form of sexual life and attempt to revive it," Engels wrote, "she is punished." The "patriarchial con-straints" hypothesis receives support from such limited evidence we now have about misattributed paternity: rates are extremely low, less than 1 per-cent in the most patriarchal family systems and as high as 10 to 20 percent in some underclass populations or in partible paternity cultures like the Yana-mamo.[23]

Matrilineal descent systems disappear as soon as farming becomes inten-sive, when plows, livestock, and paid employment are introduced, and irriga-tion systems built. But the possibility of mother-centered lifestyles is latent, and always there, reinvented all the time when resources become unpre-dictable, adult male mortality rates go up, or whenever it becomes impru-dent for a mother to rely on protection or provisioning from a single man.

Across the shantytowns of South America, in African townships, in urban underclasses of both hemispheres, along with some tribal groups like the Canela, wherever husbands prove unreliable providers and male patrilines do not have so tight a hold as to make such tactics dangerous, mothers massage the networks of possible paternity. Socially, marriage is monogamous, but mothers form polyandrous sexual liaisons with several "fathers" to elicit such contributions as these men can provide. "Why put all your eggs in one bas-ket?" quipped one modern Zambian woman, by way of justification for what struck her interviewer as an exploitative attitude toward men. The woman had good reason to ponder the question—for, like a Hadza hunter who just killed an eland, if the father of her children encounters a sudden windfall, he may use it to cultivate ties with other women rather than provisioning her and her children.[24]

"Breakdown" of the Nuclear Family Scarcely New

There is a tendency among politicians and moralists on the right to decry the "demise" of the nuclear family, as if this were some new cataclysm brought about by the loosening of sexual mores in the wake of birth control. Some-times feminism and sexual liberation are specifically blamed, as in a recent editorial in the *Wall Street Journal* titled: "Feminism Isn't Antisex. It's Only Antifamily." The more promiscuous women are, the less interest men will have in marrying, runs the logic.[25]

Meanwhile critics from the left assume the breakdown of the family must be due to capitalist, racist, or colonialist oppression. This at least is more accurate insofar as such forces contribute to high male mortality rates and undermine the ability of a given man to be a regular provider. But both views are seriously limited, insofar as they assume that very old patterns can be explained by recent historical trends, mistaking correlations and outcomes for causes. Specific demographic and ecological conditions that have always motivated mothers to line up multiple "fathers" may be new. But the emotional calculus behind the decisions that inner-city mothers make every day is very old.

Is a Fraction of a Father Ever Preferable to a Whole One?

Of course the breeding system best suited to the goose will often look different from the one preferred by the gander. Polygynous arrangements where one male monopolizes sexual access to as many females as he can afford, sequestering them from other males, keeping them all for himself, is a common goal of many male animals. In patriarchal human societies, men use alliances with patrilineal kin—fathers and brothers—to achieve this goal. One way for patrilines to control women is to gain control over the resources that mothers need to survive and reproduce.

"Fraternal interest groups," whether based on real or fictive kin alliances, constitute the business end of patriarchy. Such alliances are particularly likely to prevail where raiding or feuding is common and where, therefore, men are essential to protect property, women, and progeny. The greater competitiveness of men in groups is the main reason why patrilocal marriage systems proliferated along with the cow, the plow, and denser populations these last ten thousand years.[26]

In such societies, male interests and priorities are given more weight than maternal ones, with the result that high fertility (quantity) often takes precedence over child well-being (quality).

––––––––

By the time late-nineteenth- and twentieth-century anthropologists began to collect information about residence patterns in different cultures, the majority of cultures were already patrilocal. That is, some 70 percent of human societies were living in male philopatric arrangements. About two-thirds of these patrilocal societies have patrilineal descent groups, and most are polygynous. Matrilineal cultures like the Canela, which reckon descent through the mother, emerge only among people living matrilocally. Such matrilineal arrangements are fragile, and they quickly disappear after contact with patrilineal herders, agriculturalists, or wage economies. A mid-twentieth-century survey revealed 15 percent of the world's cultures were matrilineal, and they were becoming scarcer.[27]

A matrilineal society is not the same as a "matriarchal" one—the inverse of a patriarchal society in which powerful matrilines maintain public and private social control. Indeed, outside of myth, I know of no evidence that any matriarchal society ever existed. However, there is general agreement among anthropologists that women living matrilocally among their kin tend to have more autonomy than when they do not. Such women are likely to have many more opportunities to take advantages of devices such as the fiction of "partible paternity." By contrast, the whole thrust of patriarchal societies is to guarantee powerful men certainty of paternity. Especially in patriarchal societies that are also polygynous, mothers have relatively little freedom of choice, and, once mated, have access to only a fraction of the paternal investment that a man divides among the offspring of multiple wives. How, then, do men convince women to accept a breeding system so far from optimal for the well-being and survival of her children?

Elaborate modes of socialization, rituals, and whole mythologies have grown up to endorse male control over the inconvenient sexual legacy that women inherited from their primate predecessors. Among the more common myths of patriarchy is the claim that patriarchal polygynous family structures actually benefit women. It is a myth that social scientists themselves occasionally help to perpetuate.

———

It was George Bernard Shaw who first popularized the notion that polygyny benefits women when he claimed that "the maternal instinct leads a woman to prefer a tenth share in a first rate man to the exclusive possession of a third rate one"—a claim reaffirmed over the years by Nobel Prize–winning economist Gary Becker in his *Treatise on the Family,* along with a host of evolution-minded social scientists. When evolutionary biologist Robert Trivers provocatively inquired of Harvard anthropologist Irven DeVore whether polygynous marriages—in which women get only a fraction of a husband instead of a whole one—weren't just a "male chauvinist fantasy in which males of strength and status give vent to their basest sexual appetites and reproductive drives?" DeVore replied:

> That is the judgement of most people in our society, and it's heavily reinforced by church, state and cultural values. But in most cultures, women would be furious if a law were passed that decreed they could not become the second, third or sixth wife of a wealthy, high-status male when the alternative was a monogamous union with a poor, low-status male.[28]

DeVore's logic—frequently echoed among those evolutionary psychologists who assume that patriarchy is the natural, species-typical state for humans—derived from a classic study of red-winged blackbirds conducted by a founding father of behavioral ecology, Gordon Orians. Polygyny in blackbirds is not maintained by males preventing other males from getting near fertile females, but rather by males first monopolizing the resources that females need to breed. Each breeding season a male drives away male competitors, establishes his own "territory," and then sits back to see which females settle there.

By contrast with males in "female defense polygyny," in this "resource defense polygyny," males vie for territories and then females come along and "decide" whether their interests are best served by being the first mate of a male on a "poor" territory, or the polygynously mated wife of a male on a "rich" one. Since females were the ones choosing between males, the existence of "female choice" seemed to confirm that a female derives more benefit from becoming the second wife (or mistress) of a resource-rich male rather than sole wife of a poor one, or a female who never gains access to any territory at all. (Resource-defense polygyny, though originally modeled for birds, was subsequently applied to humans by John Hartung, and this model was confirmed in a study of the Kipsigis people of Kenya by Monique Borgerhoff Mulder.)[29]

So is this true? Do women benefit from polygyny—and, more to the point, do mothers? The answer all depends on what a woman's options are.

Obviously, if men monopolize all the possible niches for breeding and all the sources of production, so that the only way a female gets access to the resources she needs to breed is by joining him and his other wives on his territory on his terms, polygyny is her best option.

If male resource control is a given, then yes, a mother accepting male terms is better off than if she produced young without a territory and she and her young all starved. (As nineteenth-century Darwinian feminist Eliza Burt Gamble worked it out for her own society a century ago, "the female of the human species is obliged to captivate the male in order to secure her support.")[30] Similarly, a mother surrounded by roving bands of marauders is better off with a powerful protector. But in either case, her choice is scarcely unconstrained. Hence, to argue that a woman is better off as mistress to a wealthy man than as the sole wife of a poor one ignores other alternatives, including the possibility that the woman might be protected by kin, by law, or given the same access to resources that men have through patrimonies or access to remunerative work.

No doubt, when men monopolize resources, a woman may sometimes be better off polygynously mated to the wealthiest. But for several of the best-studied polygynous societies, even when there are major disparities between husbands in terms of wealth, a mother who ends up with a fraction of a rich man is *not* better off than one who has access to all of a poor one. Part of the reason is the competition between mothers that the situation generates. The Dogon of Mali, one of the best-studied polygynous groups, illustrate just how high the costs to mothers and children can go when male interests take priority over maternal ones.

When Mothers Compete for a Fraction of a Father

The Dogon are West African millet farmers. Although unusual for Africa, a continent with many matrilineal societies, their classically patriarchal family system has many parallels with societies in Asia and, to a lesser extent, the Western world. (By *patriarchal* in this discussion I mean societies with patrilocal residence, patrilineal inheritance, and male-biased institutions, including preferential inheritance of property by sons.)

Polygyny among the Dogon—as in other patriarchal societies—occurs hand in hand with various means of monitoring female sexuality. Each woman's menstrual cycle is open to public scrutiny. By custom, as soon as she detects bleeding, a woman must relocate to a special hut. (Such practices are documented in about 12 percent of tribal societies.)

Countering millions of years of evolution, the Dogon have become a culture where ovulation cannot be concealed. Failure to comply with the rules, especially cheating (pretending to menstruate when a woman is actually pregnant), brings social reprisals in the real world, and the prospect of worse punishments in the supernatural one. In this way, a man can be confident that any woman he marries is not already pregnant by another man.[31]

In addition to menstrual monitoring, ancient female incentives for confusing paternity are countered by removing each girl's clitoris. The Dogon take it for granted that after clitorodectomy a woman will find sexual intercourse outside marriage less tantalizing, not worth the risks. In this way, older men with several young wives can be as certain of paternity as any primates in the world—Dogon certain.

Does certainty of paternity ensure that Dogon men invest more in their children? Not necessarily, especially not if they have several wives and many sons. As in most patriarchal societies, a man's attention tends to be focused on gaining and maintaining prestige, with its corollary of more wives and more children.

Typical for this area, child mortality among the Dogon is very high. According to anthropologist Beverly Strassmann, 46 percent will die before age five. What is more noteworthy, though, is that the chances of a child dying are seven to eleven times higher if the mother is living in a polygynous family, working together with and sharing meals with cowives, than if she is married monogamously. In a monogamous union, a mother's loss is equally the father's. But in the case of a man married to three wives, the polygynist comes out ahead reproductively even if more than half his children die.[32]

Unlike a mother's goal, which is generally "quality," well-spaced, healthy offspring, each one well provided for, the Dogon's father's goal is "quantity"—as many children as he can have, even if many die. Among the Dogon, this is particularly unfortunate, because land is increasingly in short supply, and men can bequeath a homestead only to one select son, or to a couple of sons. Yet men have little incentive to take fewer wives. For women work hard. Owning them confers prestige, a higher standard of living, and reproductive success.

Fig. 10.7 When statues of "mother goddesses" cropped up in archaeological remains from prehistoric Europe and the ancient Near East, some feminists optimistically interpreted their existence as evidence for ancient matriarchies. Alas, without more information all we can conclude is that fertile females were of interest then just as they are today. Similar representations ranging from great art to obvious pornography are found in societies with very different degrees of female autonomy. This statue of a young woman just past menarche comes from an intensely patriarchal society, the Dogon of West Africa, where men hold political power and use it to quite effectively monitor and control female sexuality. Statues such as this one are displayed at the funerals of powerful men to symbolize their ability to to possess such a woman. *(Courtesy of the Metropolitan Museum of Art, New York)*

Strassmann is candid about not knowing for sure what causes higher mortality among children—especially sons—born to polygynously married mothers. But Dogon mothers themselves are unequivocal: they claim that their sons are being poisoned by cowives. Strassmann was invited to attend rituals at which masked dancers intimidate women to deter wives from such nefarious pursuits. Indeed, wild-sounding accusations about children poisoned by cowives can be extensively documented in Malian court records. Occasionally women actually confess to poisoning a rival's child.[33] But why primarily sons? Because daughters leave home when they marry, and it is sons who are favored, and who remain at home to compete with their father's other sons for inheritances of scarce land.

Why don't wives married to the same man manage to cooperate more? In some societies, they do, especially if the husband marries sisters. Among Australian Aborigines, wise men often seek to marry wives who are related to each other precisely because such women are known to get along better.[34] But among the Dogon, the benefits of reduced strife do not outweigh the benefits to the patriarch of discouraging his wives from forming alliances. The patriarch's strategy is to "divide and conquer." Strassmann notes that sisters and other related women are *specifically prohibited* from marrying into the same patriline. Such strictures make it hard to sustain the functionalist argument that these polygynous families are set up for the common good, with eugenic intent to promote the well-being of all concerned. In a world where the optimal number of fathers per child is pretty obviously at least one, not some fraction of one, most Dogon men still aspire to be polygynists.

Conflicting Maternal and Paternal Interests

In literature, conflicting interests between the sexes is a theme as old as the Greek myths, yet few have zeroed in on the key issue quite so pointedly as the brilliant, troubled, and utterly misogynist Swedish playwright August Strindberg. In his 1887 play *The Father,* the wife of the patriarch (merely identified as "the captain") torments him with the possibility that their only child may not have been his. Lashing out in helpless rage, the captain exclaims, "If it's true we are descended from the apes, it must have been from two different species [males and females]. There's no likeness between us, is there?"

Writing about the same time, the Swiss-German jurist Johann Bachofen decided that some of the myths about ancient matriarchies overturned by men were not just stories but real history. His anthropological classic *Das Mutterrecht* (The Mother Right) traced successive stages of human cultural evolution. He began with an imagined primeval state when all people lived in matrilineal societies, followed by a second stage characterized by a bonobo-like interlude of "primitive promiscuity." These were mother-centered worlds, at the opposite extreme from patriarchal worlds like the Dogon's, which Bachofen viewed as the third and final stage of cultural evolution.

As anthropologists read the evidence today, however, there is no support for the notion of successive *stages* in human evolution, and even less support for the prior existence of matriarchies. Ancient myths about an age when women ruled, only to be usurped and put in their proper place by men, are not so much history as reflections of the chronic tensions between paternal and maternal interests. Themes about men overthrowing female rule are rampant in the origin myths of ancient Greece, Amazonia, and the ancient Near East. This does not mean events like those described ever happened, but rather that conflicting interests between males and females, matrilineal and patrilineal interests, are perpetually at work.

These tensions derive from the ongoing dialectic between mothers who seek the optimal breeding situation for themselves and fathers who strive to counter them and impose their own ideal conditions. Monogamy happens to be an unusually stable compromise, in part because it raises survival prospects for children. The alternative to monogamy is for matters to sort themselves out so as to favor the sex with the most leverage, usually males. When one sex is forced to accept a breeding arrangement that is grossly disadvantageous, as many women in patriarchal societies must, what prevents mothers from defecting? This is where myth and ideology come in.

Convincing Mothers to Accept Poor Terms

How might a female primate be persuaded to accept sequestered breeding arrangements that may compromise her health and nutrition and diminish her reproductive success? Why would a female primate ever give up such hard-won assets as concealed ovulation, or collaborate in the removal of the clitoris of her daughters? How to convince women to restrict their selection of mates to just one male and pressure her to breed at shorter intervals than might be optimal for her health and give birth to more children than is healthy for her? How do males manage to maintain a one-male breeding system, especially one that is not monogamous and in which females receive so little paternal investment care in return? How are mothers persuaded to make do with only a fraction of a father, and not necessarily even one of their own choosing?

One widely employed solution, found among gorillas and hamadryas baboons, is to threaten with physical violence any female who resists, or, like langurs and chimps, to make it so risky for infants of mothers short on male defenders that mothers cannot afford to refuse their offer of protection.[35] Such rudimentary versions of patriarchal mating systems probably characterized some (many? . . . most?) early hominid populations. Full-fledged patriarchy, such as the Dogon developed, is more recent. It depends upon constraints far more insidious than just out-manning daughters and wives. Patriarchal arrangements like the Dogon's or Victorian Britain's rely for their persistence upon the invisible constraints that creep inside human imaginations and insinuate themselves into the way men and women conceptualize who they are.

Humankind's unique capacity for symbolic thought brings with it language. Having the ability to talk not only means that people can gossip about what other people are doing (and often what is being discussed is the sexual conduct of women) but that girls will have the ability to internalize what others think and say. For, unique among animals, humans *imagine how others perceive them.* Many animals have templates for how they should behave in order to avoid punishment; even macaques and baboons manage this. But only among humans do these transactions take on historical and symbolic dimensions, resonating inside verbal brains as humans strive to avoid disgrace and achieve honor. Typically, proper sex roles for men include being domi-nant, or "manly"; for women, being modest, chaste, and, above all, a "good mother."

Enforcers of Womanly Modesty

In addition to physical constraints, phalanxes of supernatural creatures work overtime inside human imaginations to convince women living in the real world to comply with gender norms. Among the Maya-speaking maize farm-ers of southern Mexico and Central America, women are afraid to go out unchaperoned, especially at night, and with good reason. Among the Tzotzil Maya, women fear the *h'ik'al,* a small super-sexed demon with a several-foot-long penis modeled rather distantly on Cama Sotz, the large-testicled ancient vampire bat from the *Popul Vuh* (the central Mayan religious text) whose practice it was to zip down from the sky and decapitate people.

Today, this fearsome creature, too horrible to mention by name lest he come, confines himself to seizing women who behave immodestly or who fail to respect taboos about menstruation (there are certain activities, like cook-ing, that menstruating women are supposed to avoid). The *h'ik'al* snatches women, carrying them off to his cave, where he rapes them. A woman impregnated by the *h'ik'al* swells up and then gives birth night after night, until she dies. Not surprisingly, few women are interested in finding out whether such demons, devils, furies, or hellfire are real or not.

Did Modesty Evolve?

No one doubts that women behave more modestly than other apes. Women really are very different in this respect. Women, for example, do not rub their genitals against a partner of either sex in full view of other people, the way bonobos do. No grown man or woman is so likely as a chimp to masturbate in

Fig. 10.8 Once a year at the festival of *Carnival* held in a Mayan town in southern Mexico called Zinacantan (literally, "the place of the bat"), a man blackens his face and impersonates a legendary chastiser of sexual sins. This contemporary enforcer of sex roles publicly humiliates wives who have behaved immodestly. He also frightens children in ways they are unlikely ever to forget. *(Sarah Blaffer Hrdy / Anthro-Photo)*

public, and universally (even where people go mostly naked) there are parts of the body (especially the genitals) that people would be ashamed for others to see.[36] In every culture, no matter how casual about sex, people are embarrassed to be caught in flagrante delicto. In many societies women would blush just to be suspected of imagining a compromising situation. Being caught, or even suspected, brings a wave of mortification not unlike (and I am guessing) a chimp who has just offended a dominant animal or been caught trespassing.

We women are incomparably more inhibited than chimps. But when did this come about? And why? Is this difference due to early learning, or has there been selection through time, such that women lacking modest inclinations died young, before they could reproduce? It is unsettling to ponder the

Fig. 10.9 Fearsome "enforcers" monitored the conduct of women among the now extinct Selk'nam foragers of Patagonia. According to Selk'nam cosmology, there once was a time when women ruled. This imagined matriarchal order ended when, for the good of society, all the powerful women were massacred. Thereafter (and therefore) it was deemed essential for women to remain submissive and never again challenge male interests. To this end, once a year, men impersonated misogynist demons called "shoorts" and rampaged through camp. Any willful or wayward wife was a special target. A shoort, in one of his eight different disguises, went to her hut, shook it violently, stirred up the hearth, scattered her belongings, and if she looked out, stabbed her with a sharp stick.[37] The closest contemporary comparison would probably be terrorists rampaging through family planning clinics. *(Photograph by Martin Gusinde, from the Anthropos Museum and Institute, Germany)*

possibility that generations of male domination and claustration of women have left traces in the human genome.

However it happened, girls in every culture learn at an early age to emulate suitable behaviors, to speak and dress in appropriate ways. The earliest written records describing standards of modest conduct for women date back thousands of years. In preclassical Greece, for example, girls were brought up to be constantly aware of *aidos*, a composite concept involving respectful modesty and a self-consciousness about the ever present danger of sexual shame. *Aidos* is signaled by a girl's dress, her veil, her downcast eyes, and the way she handles herself.

Daughters were sequestered inside the home. None would leave until they were married. In his play *Tereus,* Sophocles has homesick Prokne review the consequences of male philopatry and female transfer from a woman's perspective

> When we reach puberty, we are thrust out. . . . Some go to strange men's homes, others to foreigners', some to joyless houses, some to hostile [ones] . . . and all this, once the first night has yoked us to our husbands, we are forced to praise and say that all is well. . . .[38]

In ancient Greece, aristocratic women lived their whole lives indoors, literally in the shadows (as Aristotle noted), their faces veiled from direct view, their glances averted from men, never looking directly at any man until their wedding day, when they looked straight into the eyes of their husband to signify their compliance.

————

Was all this necessary? Certainly the ancient Greeks thought so. Young girls were identified with bears, and to a lesser extent lions, common enough in late Pleistocene Greece. Both bears and lions bear multiple young, and any ancient hunter who chanced to witness a lioness mate would be unlikely to have forgotten it. Female lions mate as often as every fifteen minutes, day and night, with anywhere from one to three males, like the wild sexually uninhibited maenads of Greek legend—or Canela women.[39]

For the ancient Greeks, a woman's animal nature lurked at the core of her being. From early childhood, it was deemed necessary to "tame" her. Homer's word for "wife," *damar,* means "broken into submission." Following this train of thought, fifth-century B.C. comedic playwright Plato Comicus warned that "if you relax too much, the wife gets out of control." This was a very different view of female nature from the one that prevailed in Darwin's day. By then, patriarchy had made such deep inroads into all fields of learning, even natural history, that it was simply assumed that women "happily for them" have little libido.[40]

Virtue Transformed into a "Maternal Effect"

Patriarchal cultures, whether among the Dogon, ancient Greeks, or in Asia, differ in how they define female nature and the most sought-after virtues. Always, though, virtue comes linked both to chastity and to accepting the patriarch's assessment of what would constitute the optimal number of mates for a mother to have: one—him alone.

The trick was to convince a woman that being chaste was both in her interests and the same as being a good mother. A mother will have an obvious stake in compliance when the status of her offspring depends on her "virtue," on how well she measures up to patriarchal standards. From that point on, it becomes very difficult to tease apart a husband's interests from those of his wife and her children.

———

As was true in Victorian England, women in Ching dynasty China—the eighteenth century's Eastern renaissance, noted for producing women writers of great merit—were meant to be industrious, devoted to the care of children and to their husband's parents, and, *above all,* chaste. A mother's virtues, it was assumed, rubbed off on her children—as did her vices.

In Asia, the concept of "maternal effects" was transformed into exquisite metaphors for women tending the gardens of their virtue in ways that were transmitted to their children. Chinese mothers were considered so important to the early education of scholar sons that writing by women was actually encouraged and much admired. Historian Susan Mann has pieced together the cloistered lives and work of these Ching women writers. Their specialty was writing biographies of "exemplary women." Most admirable of all were young women who committed suicide after the death of a fiancé, or lived on as chaste widows who brought honor to their sons.

As with women's education in England and the United States before the twentieth century, the intention was for women to become better mothers, not more educated people. Accomplished or not, it was unthinkable for an upper-class Ching woman to join men, or even to be seen in public without a chaperone. Nor was it likely. The bones of her feet were crushed in childhood and then bound so as to prevent growth. When the period of excruciating pain was ended, she would only be able to walk supported by maids. Her parents would arrange her marriage, to a man who would treasure her tiny feet, as erotic to him perhaps as trim ankles to Homer, breasts to Americans, buttocks to Brazilians, only better, since they guaranteed that his love would be too crippled to act on any cyclical wanderlust.

If a daughter or wife expressed ideas judged inappropriate or rebellious, or behaved so as to attract dishonor, her father or husband had the right to kill her. Ching daughters who brought shame upon their families (as George Eliot had upon hers by living openly with an already married man, George Lewes) would be strangled or hacked to death.[41] Little was known in England at that time about these Chinese literary women, yet Eliot, with her broad-ranging intellect, felt a natural affinity with Ching women when she wrote:

> You can try . . . but . . . never imagine . . . what it is to have a man's force of genius in you, and yet to suffer the slavery of being a girl. To have a pattern cut out . . . this is what you must be; this is what you are wanted for; a woman's heart must be of such a size and no larger, else it must be pressed small, like Chinese feet. . . .[42]

Even when the goal is the survival of a family line, maternal and paternal interests do not necessarily overlap. Patriarchal societies are those in which patrilineal interests have, over time and by whatever means, come to prevail over strictly maternal ones. The goal is to produce offspring—often many of them—of undisputed paternity, no matter the cost to their mothers.

———

Surveying this dismal scene, a world in which interests of mothers and children are so often subordinated to patrilineal aims, raises a troublesome question at a theoretical level. How could a species like *Homo sapiens* ever evolve? How could females so clearly at risk of reproductive exploitation be selected to produce such large-brained, vulnerable, slow-maturing offspring? Why not instead opt for (or more likely, continue to produce) babies that a mother could rear by herself—the way chimp mothers do. Although chimp moms require males for protection from other males, they do not need mates to provision their young. To me the only possible solution is that humans did not evolve in mating systems where mothers depended entirely on husbands to provision them. Otherwise, why wouldn't females be selected for self-sufficiency—as chimp females have been? Rather unlike other apes, the earliest representatives of the genus *Homo* were cooperative breeders. It's time then to ask: just who besides mothers subsidized the long, slow development that is the hallmark of the human species, the very late maturation that distinguishes us from every other primate?

Who Cared?

Allomothah, allofathah,
Hamadryas and gelada,*
Please be careful lest you pass me
Off to someone who'll . . . harass me . . .

Keep me near my loving parent.
There's much to fear, and risks inherent
When you farm your infant out to go
With some big ape you may not know . . .

Blood is thick compared to water,
And you might never have another . . .

—Anonymous verses submitted to Professor R. L. Trivers
by a student, to be sung to the tune of Alan Sherman's
summer-camp song parody "Hello Muddah, Hello Faddah!"

If a Pleistocene father's nagging source of anxiety lay in the past (just who did sire his child?), a mother's stretched into the future. Hers was a perpetual, existential angst about who would help provide for her children. Would her mate survive, stick around, and if he did, how committed would he be to her children? As tantalizing opportunities came up or as she aged, would he become involved with other women? We romanticize the past if we imagine that a mother in the Pleistocene would have been any more certain about her mate than women are today. Yet, on her own, it was beyond her capacity to provide for herself and several offspring. Given her predicament, how much did fathers help? And when they fell short, who else helped? This inquiry is yielding a new appreciation for an eclectic assortment of alloparents who provided mothers the only "safety net" the Pleistocene had to offer.

---

* Refers to hamadryas and gelada baboons.

New data from extant foragers are yielding clues that help resolve the puzzling inconsistency. Human babies needed more allomaternal assistance than any ape ever before, and needed it far longer. How could a mother so dependent on assistance from others, yet so uncertain about whether the father would provide it, afford to put so much on the line every time she gave birth? Part of the answer lies with a very special type of alloparent—post-menopausal females, former mothers who lived on not just for a few years, but for decades after their own last birth. Significantly, these older women gathered more food than they consumed. New information on just how much of this surplus postreproductive females may have contributed to the weaned offspring of others is changing the way some anthropologists think about the evolution of extra-long human childhoods.

Contrary to common wisdom, protracted childhoods may not have evolved because they were essential for the development of big brains. Rather, long childhoods might have been a life-historical by-product of an organism adapted for a long lifespan. "Grandmaternal" clocks set slow in anticipation of a long life are looking like one of several reasons why selection for greater intelligence came at an evolutionary discount in our species.

Subsidizing Long Childhoods

All apes mature slowly, but none so slowly as apes belonging to the genus *Homo*. Humans take advantage of a long period of dependency by having brains that grow rapidly after birth, learning to talk, engaging in trial-and-error experimentation during play, acquiring such skills as tool use, and absorbing all manner of tribal wisdom. Since Darwin's time it was assumed that provisioning by increasingly skillful hunters subsidized this leisurely development. "The most able men," wrote Darwin, "succeeded best in defending and providing for themselves and for their wives and offspring," with the result that more of their children survived, leading to the evolution of "greater intellectual vigour and power of invention in man."[1]

Put another way, we had man-the-provisioner to thank for both our inventiveness and for the period of grace between birth and maturity. A century later, this basic hunter-centered scenario is still the most widely accepted framework for interpreting all incoming evidence. In 1974, the discovery of the *Australopithecus afarensis* fossil Lucy, and shortly afterward the discovery of footprints left by three australopithecines walking across volcanic ash 3.6

million years ago, proved beyond doubt that small-brained australopithecine apes walked upright. On the basis of little more evidence than this, it became widely accepted that humans became bipedal so fathers could carry food back to their mates.[2]

By the 1980s and 1990s, however, evidence from other primates and extant hunter-gatherers was challenging the notion that children took longer to mature because bipedal hunters provisioned mates and dependent young. As discussed in the last two chapters, paternal agendas were too complicated for any mother to count on her mate's help. Furthermore, such paleontological evidence as was available (mostly from teeth) provided no support for the notion that the earliest bipedal hominids, the australopithecines, took longer to mature than other apes. The earliest evidence for long childhoods seems to date from *Homo erectus,* who, according to paleontologist Holly Smith, reached adolescence later than chimps, but not so late as humans do. At this point, the most reasonable conclusion is that something other than upright walking and the hypothesized provisioning by fathers made possible hyper-long *Homo sapiens* childhoods.[3] When did this occur? And where did the extra subsidies needed to rear these especially needy offspring come from?

Among hunter-gatherers men are the main source of protein. But this does not mean that wives of the best hunters receive the most meat. Often, as in the case of the Aché or the Hadza, meat goes to the group at large, donated by fathers more interested in "showing off" than providing for their children. Data from the Aché hint at incentives for doing so: the best hunters reportedly have the most love affairs. Furthermore, in some of these groups the reason that the offspring of the best hunters are better fed is not because they obtain more meat from their fathers, but because the best hunters manage to marry the most enterprising gatherers.[4]

It was not so much that Darwin and several generations of anthropologists had been on the wrong track. Rather, there were more tracks than initially assumed. When anthropologist Frank Marlowe conducted yet another study of the Hadza, he found that even when men "showed off" by bagging the most prestigious game and thereby provisioned the camp at large, more meat somehow reached children the hunters regarded as their own than reached their stepchildren. Marlowe concluded that provisioning by men was at least partly "parenting effort," albeit not always as single-minded and dedicated as many mothers might hope. To varying degrees, fathers seek to provision their

Fig. 11.1 Reconstructions of *Australopithecus afarensis* routinely depict them in pairs, ignoring the extreme sexual dimorphism that characterized this species, a trait normally associated with polygyny. Here a four-and-a-half-foot-tall female (close kin to the famous "Lucy") is shown with her five-foot-tall mate's arm gratuitously wrapped about her shoulders as they saunter through a diorama at one of the world's great natural history museums.[5] *(Courtesy of Department of Library Services, American Museum of Natural History, New York)*

own *and* show off. Fortunately for mothers, fathers are not the only providers mothers depended on.

There is a proverb African parents dust off when a son comes of age and sets out to find a wife: "First find yourself a good mother." By "good" they don't mean having a fine character. Nor are they thinking about heritable traits that a particularly beautiful or healthy woman might pass on to progeny. By "good" these tipsters have in mind a mother with numerous hardworking and well-connected relatives. It is quintessentially good advice for a cooperative breeder.

Alloparental Safety Nets

Anyone who has ever held a new baby knows that it takes two hands and cautious dexterity to support both the baby and his or her floppy head. Compared with baby chimps or baboons, who obligingly hold tight, human babies do not help much. Yet a mother has no place safe from lurking predators or stinging insects where she can set her vulnerable charge while she digs for roots or picks fruit from branches overhead. Even wearing a sling, a new

Fig. 11.2 An Aché mother suckles her twenty-month-old daughter while simultaneously weaving a palm mat. Mothers routinely combine work with mothering. Nevertheless it is beyond the capacity of new mothers to provide for their own needs plus those of older, still partially dependent children. They must rely on others to subsidize their survival. *(Courtesy of Magdalena Hurtado and Kim Hill)*

mother is bound to be less efficient. How, then, does she make up the caloric shortfall between what she can gather and the extra calories a nursing mother needs for herself and other dependent children?

Anthropologist Magdalena Hurtado was among the first to show that deficits were being made up by other women. Among the Aché, foraging mothers are immersed in a network of casual, opportunistic, and typically reciprocal relations involving cowives, in-laws, neighbors, and blood-kin. In some societies—Andaman Islanders in the Indian Ocean, Solomon Islanders in the Pacific, Efé pygmies landlocked in Central Africa—a mother who hears another mother's baby whine may offer it her breast. In the Solomon Islands, taboos prevent a woman from nursing her baby in the tuber gardens, so she has to leave her infant with another lactating woman (often a sister-in-law) while she hoes and harvests.[6]

As in all cooperative breeders, human animals have an internalized emotional calculus predisposing them to protect, care for, and allocate resources to individuals they classify as kin, their genetic relatives, or those they think of as kin. Familiarity from an early age would be the most common cue, affection and sympathy the immediate conscious motivations. Whenever costs of helping are less than the benefits to their "kin," alloparents help. But calculating Hamilton's simple-sounding terminology of "costs" and "benefits" can be complicated. All other things being equal, with costs and benefits held constant, closest kin should help the most. But when were other things ever equal? Benefits to an infant who gets fed are fairly obvious. But how to calculate the costs to the baby's aunt who provides those benefits? Those costs will vary with her circumstances. A grandparent, an aunt, and the baby's half-sibling all share one-quarter of their genes with the infant. Based on degree of relatedness alone, they should be equally motivated to help. But what if the aunt has a new baby of her own, or the grandmother is still caring for her last child, or the half-sibling has recently had malaria and feels weak?

The most readily available assistance is not necessarily provided by the closest kin. Fathers are not the only ones with complicated agendas.

Like, More Available Than Diligent

Around the world, alloparents not yet ready to breed are the working mother's mainstay. These prereproductives are available precisely because they are either physically immature or have not yet "fledged," found a feeding ground, a "nest site," a job, a mate, or territory—though they are preparing to. This last qualification makes them dangerously distractable.[7] For such reasons, teenagers may be more available than they are diligent, proficient, or useful.

Among some foraging groups, like the Pumé of Venezuela, the average woman will be in her mid-twenties before she starts to gather all the food she herself consumes, much less the surplus food needed for the children of other women. Even when they engage in hard work—say, like digging up tubers—women between the ages of eighteen and twenty-four obtain less food per hour than senior women, age forty or more.[8] When food is carried back to camp (and this may sound all-too-familiar to postindustrial office workers and soccer moms alike), the hardy young women somehow end up carrying the lightest loads.

Typically, adolescent girls gravitate toward easier tasks, such as berry picking. Childcare is one of these lighter tasks, energetically far less demanding than hacking tubers out of rock-hard soil. But an adolescent's heart may not be in her work, not even when it's a light task. In an extreme case, !Kung kids under fifteen spend less than three minutes an hour in productive labor. It is as though teenagers are saving themselves for better things—which, reproductively, of course, they are. Childcare is a job their younger sisters, mere children themselves, may be more eager to undertake. (If baby dolls are surprisingly scarce in traditional societies compared to societies with lower birth rates, it's because little girls have the real thing near at hand.)[9]

Modern mothers joke grimly about the narcissism of the young, their self-absorption, restlessness, and preoccupation with the other sex. From their point of view, however, girls near menarche are actually hard at work of another sort, reprogramming their hypothalamus and ovaries, storing up resources as a down payment on the demanding reproductive career they are about to embark upon. Adolescents may have expectations and priorities quite different from those of parents. When the chips are down, without necessarily knowing why, they hesitate to risk their own reproductive potential on behalf of someone else's young.

"Child labor, what a joke," groans a mother of my acquaintance, worn down by her daughter's obstinance about household chores. The behaviors parents so deplore turn out to be adaptive when "the young and the nestless" light upon a niche of their own. If our language to describe such "fledglings" has a birdy, aviocentric ring to it, that is because, as ornithologist Steve Emlen delights in pointing out, cooperatively breeding birds and humans conform to similar "decision rules" in respect to "helping at the nest."

Extracting labor from adolescents and young adults is not an isolated problem. In many foraging societies, women do not hit their stride as efficient, diligent workers until well into adulthood. Often, the heaviest burdens fall on the backs of older women. Older Pumé women, for example, already depleted by a lifetime of childbearing, end up carrying the heaviest loads—weighing anywhere from 7 to 100 percent of their body weight.[10]

Fig. 11.3 This seven-year-old Pumé girl lives with her maternal grandmother and her aunt. An orphan, she is unusually hardworking. One of her jobs is to care for her year-and-a-half-old cousin while her aunt and grandmother are off foraging for wild tubers in the flooded savanna near her wet-season home. *(Courtesy of Charles Hilton)*

A New Kind of Alloparent

WANTED: EXPERIENCED CARETAKER, FULL BODY SIZE, SELFLESS, WITH NO FUTURE REPRODUCTIVE EXPECTATIONS. MUST BE GENETICALLY RELATED TO MY OFFSPRING AND PROVIDE OWN MEALS, AS WELL AS OCCASIONAL MEALS FOR OLDER CHILDREN.

Across diverse taxa, cooperatively breeding animals rely on "helpers at the nest," drawn primarily from the ranks of the prereproductive. Taken to the extreme, among honeybees and other eusocial insects, heirs to the most elaborate and efficient "daycare" system ever evolved, prereproductive recruits join the class of "sterile workers" never to become reproductives. However, in most cooperative breeders, once a female starts to breed, she becomes less available to rear the young of other mothers, and goes on producing babies until she dies.

There are some species (not necessarily cooperative breeders) in which females cease cycling and then live on for a few years after menopause (lions and baboons are examples), but these former mothers care mostly for their own last offspring. In only three species are females known to quit reproducing and then go on living for a *long* time: humans, elephants, and pilot whales. All happen to be particularly large as well as long-lived mammals.

Women reach menopause by fifty but then live on another two to three decades or more. Elephants and pilot whales breed until forty or so, then live on among their daughters and granddaughters: until sixty in the case of the whales, even longer for elephants.[11] For females this old, the costs-to-benefits ratio of helping kin is completely altered. As compared to young females, whose reproductive careers stretch out before them, older females have a low threshold for responding to signals of need from immature relatives. Much like a worker bee in a honeybee colony, who opts to become a sterile helper rather than reproduce and almost certainly fail, these postreproductives forage on behalf of needy kin. When available, this rarest of all commodities, the postreproductive female, makes an ideal alloparent, assisting other mothers to protect or provide for their weaned offspring.

Why Postmenopausal Females Exist

If an egg is just a hen's way of making sure she is represented in the next generation's gene pool, what better route to success than for a hen to live on and on, punctuating eternity with more eggs? A surprising number of creatures enjoy such "indeterminate" lifespans. Among sturgeon and other teleost (bony) fishes, the longer a mother lives, the bigger she grows. The bigger she grows, the more fecund she becomes. Her long lifespan translates into reproductive success measurable in the gold standard of abundant caviar. But not among mammals.

One by one, some of the twentieth century's greatest evolutionary theorists (J.B.S. Haldane, Sir Peter Medawar, George Williams, William Hamilton) have been drawn to the puzzle of senescence: Why do animals grow old and die instead of breeding on and on? Each emerged with some variant of the same solution: mammals have been selected for traits that help them survive in the short term even if the same traits prove detrimental in the long term. Or, as Hamilton put it: they opt to "live now, pay later."

Life is risky. Sooner or later, every organism succumbs to something—lightning, floods, enemies. Evolutionists reckon that genetic traits that benefit an organism early in life and help it reproduce before that fatal encounter will be selected for, even if they carry with them associated (or "pleiotrophic") effects detrimental later in life. Decay, then, must be viewed as a by-product of early success. Probably to avoid the risk of mutations which might otherwise arise in the dividing cells of the germline, women are born with a fixed number of ovarian follicles, all they will ever have. After fifty years of slow attrition, all the viable ones have been used up.

Inevitable aging provides a plausible explanation for why the remaining follicles decay, with a resulting decline in levels of circulating estrogen, around age fifty in healthy women and chimpanzees; the same happens at around thirty-plus years in captive macaques, and closer to twenty-four in wild baboons.[12] An extremely lucky wild baboon may manage to live as long as twenty-seven years, but she will start to experience irregular cycles by age twenty-three, about a year before she stops menstruating altogether. Similarly, wild lions stop menstruating about two years before they die. This cessation of cycling prior to death, or menopause, has led some evolutionists to wonder if perhaps there was selection on females to "stop early," for ovaries to wind down before kidneys, heart, and other organs do. According to this hypothesis, menopause is an adaptive way to cope with the inevitability of senescence: mothers quit producing new babies and live on just long enough to ensure survival of that last offspring.[13]

Jane Goodall's "old Flo" is often cited to illustrate why "stopping early" might be useful. This chimpanzee mother was already noticeably feeble when she gave birth to her last baby, which died in infancy. At that point, her next-to-last offspring, Flint, reattached himself. But when Flo lay down one day, never to rise again, Flint—who, Goodall argued, had never become properly independent—became despondent, grew ill, dying within a month of his aged mother. How much better it would have been, if Flo had never given birth to that last baby and instead devoted her remaining energy to making sure Flint was strong enough to survive on his own. One healthy offspring, this argument runs, is better than two dead ones.[14]

A series of distinguished theorists, most notably George Williams, Richard Alexander, and Paul Sherman, have argued that menopause is an adaptation to prevent aging females from making Flo's mistake. Stopping early guarantees that old mothers who give birth to one last baby will care for it long enough to see this reproductive enterprise through to a successful conclusion. If menopause also means that old females are available to help grandchildren and other kin as well, so much the better. This was the first of three quite distinct hypotheses to become known under the rubric of "the grandmother hypothesis"—although stopping early might more accurately be termed "the prudent mother hypothesis," since she who stops reproducing shortly before she dies is as likely to help her own offspring as those of her daughter.[15]

———

As with all long-lived primates, a woman's fertility gradually increases after menarche, when she first begins to cycle, peaks in early middle age, then slowly brakes to a full stop. In women who lead arduous lives, reproduction may cease closer to forty than fifty. But only a fraction of women anywhere go on reproducing beyond fifty, and none past sixty. Fifty is a venerable upper limit to the warranty on primate ovaries. It has remained constant for as long as can be known, a milestone as inevitable among women in classical Greece and Rome as among women in contemporary Greece, Italy, or California.[16]

From a comparative perspective, this scheduling is consistent with information about ovarian lifespans in other apes. "Live now, pay later," the accepted explanation for senescence, provides a perfectly good explanation for the reduction with age of a woman's reproductive potential. (The technical term is *reproductive value,* defined as the average genetic contribution to subsequent generations that an organism of a particular age is expected to contribute.)

In humans, declining fertility with age is accompanied by a greater likelihood of birth defects, as well as a higher likelihood of miscarriage. Even if the DNA remains intact, cell structures in the supporting cytoplasm may become defective over time. Resulting reproductive failures contribute to even longer intervals between births as mothers approach the end of their reproductive careers. Birth intervals may also be stretched out due to the greater indulgence accorded late babies when they do arrive.

Provided that mothers are not already overwhelmed by childcare, older women are notorious for showering extra-tender loving care on last-borns. From traditional sub-Saharan Africa to village India, to suburbs in postindustrial America, we find mothers "spoiling" the last-born child. Often, he or she is identified with a special term of endearment. Where I grew up in Texas, a last child was lovingly referred to as "the caboose." The rough equivalent among the Gusii people of Kenya is *omokogooti,* the nickname for an older woman's last-born, who will be breast-fed longer, weaned gently, and disciplined less.[17] According to Hamilton's rule, such increasing maternal altruism with age (and decreased reproductive value) is only to be expected.

Why Old Females Are So Self-Sacrificing

Decline in reproductive potential with age means that mothers near the end of their reproductive careers have less to lose by giving of themselves to their own offspring—or by assisting kin with theirs. In many species (recall the self-sacrificing mothers who allow spiderlings to eat them) females become increasingly altruistic as they decline in reproductive value.

Langur monkeys provide a particularly vivid primate example of old females who become simultaneously more self-sacrificing *and* more heroic as they age. Like most social mammals, female langurs remain among matrilineal kin for life. Instead of the rigid hierarchies formed by macaque and baboon females, a langur female's rank rises and falls over her life. Females scramble up the social ladder in their youth, occupy top ranks of the hierarchy during their prime breeding years, then gradually opt out of competition as they age, sinking back to the bottom. As a mother langur ages, intervals between her births grow longer and longer. Finally she ceases to reproduce altogether. Increasingly peripheralized, the aged langur becomes socially invisible—like some aging starlet on the fringes of Hollywood.[18]

Yet when neighboring groups trespass into her troop's territory to exploit a fruiting tree or some other finite resource to be found there, or when unrelated males invade and threaten infants belonging to her relatives, this same old langur will discard her mantle of debility and rise up to heroically defend her descendants' interests against great odds. Forget the image of an aging starlet then, and think total warrior.

Such a hero was old Sol, a langur I estimated to be at least twenty-five years old. She had ceased to menstruate, and never reproduced in the half-decade before she died. Hers was a solitary life spent on the fringes of the group. Yet when the male shown in figure 11.4 invaded her troop, it was Sol who repeatedly charged this sharp-toothed male nearly twice her weight to place herself between him and the threatened baby. When the infanticidal male seized the infant in his jaws and ran off with him, Sol pursued the attacker and wrested the wounded baby back. With danger momentarily past, and the wounded infant once again in his mother's arms, old Sol resumed her diffident attitude.

That an arthritic old female would become marginalized with age is scarcely surprising. More curious was Sol's transformation from decrepit outcast to intrepid defender. Never before or since have I seen any animal so unabashedly throw caution to the winds and take on an obviously stronger animal with such fierce and selfless determination.

Natural Selection and Postmenopausal Females

Increasing altruism with age (what fund-raisers call "donative intent") may be critically important to explain why females in some species have evolved to live on past the point where reproduction is still possible. By itself, though,

Fig. 11.4 Aged and decrepit, "Sol" avoided conflict with younger female relatives. Yet here, along with another older but still reproductive female, she charges an adult male langur to retrieve an infant that he has seized from its mother. The infant's body can be seen flying out like a rag from the male's jaws. What's oddest of all about this picture is that neither defender is the mother. That vigorous young female remained on the sidelines, out of harm's way. This was one of seven occasions when I witnessed old Sol intervene to thwart this male's attacks on infants in her troop. *(Sarah Blaffer Hrdy / Anthro-Photo)*

greater altruism may not be enough. Three conditions have to be met before there will be selection for females to live on not only past the time when their ovaries wear out, but past the time when their last child becomes independent. First, there must be donative intent toward kin. Next, there must be something beneficial an old female can do for kin, like protecting or provisioning them. Finally, the cost of having an old female around must be offset. That is, like an old langur, she may opt out of competition with kin, or else remain alive only so long as she increases the total amount of food available. The more nearly these conditions are met, the more likely selection will favor long postmenopausal lifespans.

So far, efforts to measure how much well-intentioned older kin enhance the reproductive success of offspring and grand-offspring yield mixed results. Data from wild baboons show that the presence in the group of a grandmother—whether she was menopausal or still fertile—was generally beneficial; but the fact that a grandmother was *postmenopausal* did not translate into statistically significant increases in the reproductive success of her kin.[19]

Data for captive vervet monkeys—an infant-sharing species in which females directly care for other females' young by carrying them—showed that infant survivorship was significantly higher if the infant's grandmother was present in the group; but again it did not matter whether or not she was postreproductive herself. When the maternal grandmother was experimentally removed, her daughter's reproductive success declined: intervals between her births grew longer, and when she did produce infants, they were less likely to survive than if her (still reproductive) mother was around. These ill effects were most pronounced if the daughter herself was young and lacked maternal experience.[20]

———

Fairly obviously, having a mother or other well-disposed older kin around is beneficial. But when is a female better off (in terms of genetic representation in subsequent generations) devoting herself to the offspring of her relatives rather than continuing to breed herself? There is little evidence that benefits provided by an old female to her kin ever outweigh advantages from continuing to reproduce as long as she can.[21]

Postreproductive Longevity: Adaptation or Fringe Benefit?

The "live now, pay later" argument provides a perfectly good explanation for why ovaries senesce. So far as I'm concerned, senescence combined with stopping early explains menopause. After five decades or so, women cease to have viable eggs. But this does not explain why women and a few other creatures go on living not just for a few years but for several *decades* after menopause.

Is their long postmenopausal lifespan adaptive, or just a by-product of the fact that women today live in safer, healthier environments than ever before? Once people managed to kill off their major predators, line up better and more consistently available food supplies, and invent antibiotics, it seems perfectly reasonable to suppose that like barnyard animals, people would go on living longer simply as a fringe benefit of this new domesticity.

This is a possibility. But the suggestion that living on after menopause is just "a fringe benefit of domesticity" has its own problems. What about wild elephants, another large-bodied, long-lived species living in tightly bonded social groups in which alloparental care, including suckling other females' babies, is very important? What about the various octogenarian hunter-gatherers? No forager has ever made it to anything close to Frenchwoman Jeanne Calment's record 122-and-a-half-years. But such foragers frequently survive past middle age and remain active and vigorous for three to four decades after their last child is born. Among the !Kung nomads most people who reached age sixty could expect to live another ten years.

Like elephants, humans are a species built to last. This is the conclusion of the most extensive available study of primate body weights and brain sizes as these relate to lifespan. Based on this study, human primates appear designed to live to about seventy-two, far longer than any other ape, and considerably longer than should be necessary for any mother who gives birth for the last time at around age forty-five to pull her child through (the "stopping early" hypothesis).[22] So far, the most reasonable conclusion is that postmenopausal survival is more than a by-product of modern lifestyles.

Emerging evidence for the importance of genes and underlying design factors related to longevity is even more convincing. The genealogy of Jeanne Calment is a case in point. When researchers traced this Frenchwoman's ancestors back to an eighteenth-century community of shopkeepers near Arles, they found that all thirty-two of her great-great-great-grandparents tended to be long-lived. Unless they died in accidents, her ancestors included people who apparently just by chance lacked the genes that predispose people to die of degenerative illnesses. Remarkable genetic luck converged in this durable Frenchwoman, and passed to her brother as well; he lived to be ninety-seven. Even more convincingly, in studies of twins, age of menopause is closer among identical, or "monozygotic," twins than among "dizygotic" ones.[23] Rather than a fringe benefit of domesticity, the most plausible explanation for human lives lasting into the sixth and seventh decades and beyond is that genes of people who lived longer were selected for. But why? And especially, what advantage would there be for women to live on past the age of reproduction?

————

All primates tend to be long-lived. They take about 25 percent more time to breed and die than other mammals of comparable body size. But even by this slow standard, the nineteen or so years that elapse before a hunter-gatherer woman makes the switch from growth to reproduction seems long, given

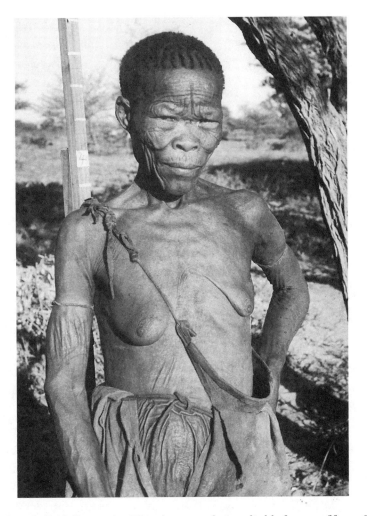

Fig. 11.5 About half of all nomadic !Kung hunter-gatherers died before age fifteen. But those who survived to adulthood had a good chance of reaching old age. About 8 percent of the population was sixty or older. A few survived into their eighties.[24] *(Irv DeVore/Anthro-Photo)*

that all apes stop breeding by fifty.[25] Women actually reproduce during relatively few decades of their long lives. Why?

We already know that in species where females remain among their kin, females become more altruistic with age. Based on data from peoples like the Hadza, older women also appear to work harder, and more effectively. They bring more gathered food back to camp than they consume. Incredibly fit and

wiry grandmothers, postreproductive women in their fifties and sixties, forage the longest hours, dig deeper for tubers, and spend more time gathering
berries and processing food than any other category of forager. In terms of
motivation and proficiency, women hit their stride sometime after thirty, by
which time they usually have weaned children. From then on, such women
keep on working, long after they quit bearing children. And if they don't have
dependent children, they give the surplus away—usually to kin.[26]

For Hadza youngsters in the vulnerable life phase just after weaning, having postreproductive kinswomen on hand—aunts or grandmothers who,
quite literally, dig up something for them to eat—allows them to maintain
their weight better, grow faster, and are more likely to survive. Grandmothers are also repositories of important tribal knowledge. In societies like
the !Kung, the elderly are revered. As they put it, "the old people give you
life."[27]

Yet such reverence is a luxury not always forthcoming. Cultural attitudes
about old people vary, just as attitudes toward infants do. Attitudes toward
the infirm or the vulnerable are sensitive to local circumstances. For example, old women have lower status in those societies where hunting is emphasized (the Eskimo, for example, or Chipewyan, or the Aché) than they do
among the !Kung or the Hadza, where gathered food is such a large part of
the diet.[28]

An extra incentive beyond inherent altruism may help explain the dili-
gence of aging foragers. Some may fear being left behind—or worse. Because
meat is so important in the Aché diet, contributions by old women are less
important and Aché attitudes toward the old far less benevolent. In the Aché
culture, as among the Eskimos, euthanasia is practiced. Kim Hill and Magdalena Hurtado recall a startling interview with an old hunter, then in his
mid-seventies, reminiscing back to a time when just the sound of his foot-
steps on the leaves of the forest floor struck terror in the hearts of old
women. For he was a societally sanctioned specialist in eliminating old
women deemed no longer useful. He described his mode of operation: coming behind an old woman unawares, he would strike her on the head with his
axe.[29]

Today we take for granted that old people will be cared for long after they
cease to contribute to their families (and I am glad we do). But the capacity of
humans to evaluate contributions of family members, and to adjust their tolerance accordingly, has to be included in any attempt to model the evolution

of postreproductive longevity through kin selection. Without such flexibility, the burden imposed may often have been greater than a family could sustain.

———

The modern stereotype of grandmothers is more nearly a smiling white-haired lady bearing Christmas presents than a wiry forager lugging large tubers critical for warding off starvation among younger kin. For ours is an androcentric culture in which women are valued and judged according to their reproductive capacity (or sexual attractiveness) and few are accustomed to thinking of postmenopausal women as having a function. Yet in the United States today, around four million children are currently being reared primarily by a grandmother. A recent headline in the *New York Times* sums up the situation: "As welfare rolls shrink, load on relatives grows: Weary Milwaukee grandmothers tell of strain."[30]

Sociologist Arline Geronimus even suggests that having a healthy grandmother available to help may increase the odds that young women with poor educational options (and worse marriage and job prospects) will neglect birth control and become pregnant at an early age. According to this counter-intuitive rationale, disadvantaged girls either consciously or unconsciously sense the merits of reproducing young, while they are still healthy and free of the worst consequences of sexually transmitted diseases, and especially while mothers and grandmothers are still around to help rear their baby. Others emphasize psychological benefits to grandmothers. Given the stigma attached to old age in women, helping to rear grandchildren may enhance an aging woman's sense of worth and outweigh the substantial costs in time and energy. Few Westerners give much thought to the antiquity of this pattern.

Although not the case in the West today, in many parts of the world, and throughout history, special roles for grandmothers were institutionalized. Often grandmothers were expected to accord priority to the reproductive interests of their progeny, as was the case in nineteenth-century Japan. For Tokugawa era Japan, a woman's reproductive life ended when her eldest son married—whether or not she had reached menopause. Even though she still lived in the same house with her husband, it would have been scandalous for her to give birth in the household now regarded as her son's. Her energy was meant to be focused on helping her son to prosper, and to produce, in turn, the finest heir that he could.

Our Grandmothers' Clocks

All apes grow up slowly, but large-brained humans take the longest. In particular, young humans are nutritionally dependent for a far, far longer time. Young chimps and gorillas who have just been weaned feed themselves. Just-weaned children do so only partially, if at all. Since processed baby foods enable women to wean offspring far sooner than other apes do, human babies may be weaned earlier and spaced just a few years apart rather than separated by half a decade or more. This leaves human mothers with a load no other ape mother undertakes: several simultaneously dependent youngsters. How could a woman afford to put herself in this predicament? Why embark on a costly enterprise with odds seemingly weighted against the likelihood that she could rear such closely spaced dependent young to maturity? These questions lead back to the big difference between humans and other apes: our much longer postmenopausal life.

As behavioral ecologists began pondering differences between women and other apes, it occurred to them that the postmenopausal life phase tacked on at the end of a woman's reproductive career might be more than a recent by-product, or fringe benefit, of domesticity. A group of them—including Kristen Hawkes, James O'Connell, and Nick Blurton-Jones—turned to Eric Charnov, a biologist who specializes in life-history theory. Using what Charnov calls "invariant rules" for how mammalian life phases are pieced together, they started to think of long lifespans as one integral component of a package. If life is likely to last longer, it is worth taking the risk of growing for a longer time before beginning to breed—like taking the trouble to carefully construct a piece of machinery that will be in use for a long time. The bigger a mammal and the longer she lives, the longer it takes her to mature. (By established convention, all life-history models are calibrated using a breeding female.) The luxury of very slow maturation is primarily available to those, like elephants, who can count on staying alive for a long time. This is why slow development is not an option for a vulnerable little creature, like a vole, who could be eaten at any moment. By this same logic, other slow-growing apes mature faster than humans because they will also die sooner.[31]

Viewing extra-long human childhoods as an integral part of a long life that includes a stint as postmenopausal grandmother helped explain the life-historical differences between humans and other apes.[32] To distinguish this model from other versions of "the grandmother hypothesis," which are aimed

at explaining menopause itself, rather than the extra-long life after menopause, I call it "the grandmother's clock hypothesis." This hypothesis presup-
poses that there has been selection for genes that permit humans to live on
past the point where reproduction is no longer possible, as well as selection
for internal metabolic adjustments that delay maturation. According to this
hypothesis, long childhoods came not as a gift from man-the-hunter, but as a
life-historical prelude to a long life in which all phases of growth, maturity,
senescence, and death are interrelated. Broadly defined this way, the long
postmenopausal stint is, in fact, an integral part of the human organism's
reproductive plan.

Accordingly, the metabolic clock that programs humans to mature slowly
could have evolved only in a primate-like species in which mothers were
already under selection pressure to live longer. This would occur if mothers
were selected first to live long enough to get their last offspring to independence and thereafter selected because their altruism produced increased survival of close kin.

Suppose, then, that once upon a time there was a dedicated forager with a
fortuitous complement of genes. She happened to live on past the point when
standard-issue ape ovaries normally wear out. Because she was a proficient
forager and already predisposed to help kin, she channeled surplus food to
recently weaned grandchildren because her own daughter (or perhaps her
son or brother's mate) had a new baby. Thus genes contributing to longevity,
shared in common among these kin, came to be disproportionately represented in succeeding generations, leading to a line of women who lived to
sixty or seventy years.

For those willing to entertain such a scenario, it is tantalizing to note that
long childhoods first appear in the fossil record about the time that *Homo erectus* females grow larger. For many reasons, "big mothers" often make better
mothers. They are able to more safely bear large babies, protect them, and
efficiently carry them long distances; and with their greater strength they can
more easily dig up and carry gathered provender. Thus, hominid females
could afford the luxury of growing for a longer period of time and getting
bigger if they were also leading "reproductively productive" lives by provisioning kin for longer.

Paleontologists have long had reason to suspect some change in hominid
economies about 500,000 to 1.7 million years ago that either made more

Fig. 11.6 A wiry Hadza woman in her sixties applies herself with skill and great industry to gathering, moving heavy boulders to get at tubers beneath. *(Courtesy of James O'Connell and Kristen Hawkes)*

food available or somehow lessened the risk of starvation during recurring famines. Without such an improvement, it becomes difficult to explain the evolution of larger body size among *Homo erectus* females. There may have been technological innovations that made food more available—superior gathering and hunting techniques as well as fire for cooking have all been suggested. The possibility I prefer, however, is that the change derived from the kind of "fitness bonanza" (Hawkes's phrase) postreproductive gatherers provided to kin.[33]

Hypotheses about origins are notoriously difficult to test, and this one is still so new that few archaeologists or ethnographers have tried. However, if this life-historical version of the grandmother hypothesis holds up, it will offer a very different slant on the evolution of wise *Homo.* Much depends on just how typical patterns of grandmaternal provisioning documented for the Hadza turn out to be among African foragers generally. Were older female relatives likely to keep young relatives from starving during the relevant phase when longer human lifespans were evolving?

According to previous models, a man's success as a hunter permitted him to obtain mates and do a better job of provisioning his offspring. Since mothers could count on food from the father, they could afford to bear slow-maturing babies, whose brains would have a chance to grow and develop during a long, costly period of dependency. But think about it. The reproductive benefits of being a little bit smarter would have had to be tremendous in order to offset the obvious costs of taking a long time to mature. What if a girl or boy died before puberty, as most probably did? What if some fast-maturing slouch out-competes a slower-developing contemporary? A long, slow period of brain growth, development, and learning becomes more feasible, however, if early members of the genus *Homo* were already committed to maturing slowly.

It is hard to imagine how being a hominid just a little smarter than others in the population could by itself offset the risks of delaying reproduction as long as two decades. By contrast, benefits to be had from surviving past reproductive age were clear and immediate—especially for females who were already dedicated to helping kin, and in a position to do something survival-enhancing for them. Being a little smarter did not have to offer big payoffs right away to compensate slow maturers for delaying reproduction: modest initial benefits would suffice if childhoods were long anyway.

Long lives, and with them long childhoods, would have altered the hominid equation. Small reproductive payoffs from being smarter would be sufficient to select for sapient brains in worlds where greater intelligence did not have to compensate completely for the enormous costs of delayed maturation. Of such lives one could confidently say, I grow slowly, therefore I think. I develop slowly, *ergo cogito.*

Unnatural Mothers

"I don't know how I felt about the baby. I seemed to hate it—it was like a heavy weight hanging around my neck; and yet its crying went through me, and I daren't look at its little hands and face. . . ."

—Hetty's confession, from George Eliot's novel *Adam Bede*, 1859

No girl in the Pleistocene could have reached menarche as early as many twelve-year-olds do today, nor become pregnant in her early teens. Even those who conceived in their mid-teens did so under unusual—spectacularly good—conditions. Such a forager would have required nutritional subsidies from others to permit her to mature so early. In the modern world, however, being fat enough to ovulate is no longer tied to having a supportive social network who will help rear her child.

What happens, then, when an unwed teenager, sufficiently well fed to ovulate but lacking social support, finds herself pregnant? The novelist George Eliot was one of the first writers to explore this topic. Her fictional case study accords with what psychiatrists and epidemiologists have learned since.

Case Study in Neonaticide

Eliot's 1859 novel *Adam Bede* traces the coming of age of a plump, pretty, and self-absorbed dairymaid, Hetty Sorrel, an orphan living with a respectable and prosperous aunt. Seduced by a local squire, Hetty becomes pregnant. Lacking kin support, she denies the pregnancy to herself, conceals it from others, and makes no preparation for the impending birth. In short, Hetty exhibits what psychiatrists today recognize as the main risk factors for neonaticide.[1]

Like the notorious New Jersey teenager who gave birth in a bathroom stall during a school dance and left her baby in the trash, Hetty gives birth, in secret and alone. She is numb, confused, and unable to feel anything about the baby except her fear that if the birth is discovered, the consequences will be bad, her future ruined. "I buried it in the wood . . . the little baby . . . and it cried . . . I heard it cry. . . . But I thought perhaps it wouldn't die—there might somebody find it. I didn't kill it—I didn't kill it myself."[2]

The baby's cries just postpartum would have been an unusually powerful auditory stimulus; they would indeed have pierced right "through" the mother. Comparisons of women in different reproductive phases reveal that even prior to stimulation of the nipples by the baby, the abandoning mother's hormone levels would be altered. Following studies of new mothers today, we can expect that Hetty's heart rate would have accelerated at the first utterance, decelerated as the cry intensified, and then risen again, with concomitant changes in pulse and perspiration—measurable through conductivity of her skin to an electric current, the same response polygraphs pick up when an anxious person tells a lie.[3]

Eliot depicts Hetty as amoral and emotionally shallow, but otherwise healthy. She is a physically lovely young human animal, not so much deranged as emotionally detached, especially from a baby she does not want to take responsibility for.

Charged Topic

George Eliot's portrayal of a pregnant teenager's plight was as realistic as it was unusual. It remains true that infants born to teenage mothers, especially those lacking social support, are most at risk.[4] Yet then as now, it was still assumed that mothers are naturally nurturing. Any woman who kills her own infant must be deranged.

Consider early-twentieth-century admissions records for Broadmoor, Britain's state asylum for the criminally insane. Forty-eight percent of the women hospitalized between 1902 and 1927 were women who had committed infanticide.[5] Today, being a devoted mother is still equated with mental health.[6] But Hetty is not crazy; she merely falls at the noninvesting end of the continuum of maternal responsiveness found among mothers who give birth at a young age, under poor circumstances, or with insufficient allomaternal support.

Because killing one's own infant is so abhorrent to us, even less comprehensible to many people than is male brutality toward an unrelated infant, there is a tendency to compartmentalize the mother's actions as the intentional killing of her infant, and to consider her behavior in isolation from her circumstances, even though they are functionally related. When we treat infanticide as an aberration, and as a crime (which, of course, in all modern

Fig. 12.1 Unlike mothers who kill neonates right after birth, mothers who kill older children are likely to be mentally ill, often suicidal, or desperate beyond reason. Such was the case with Margaret Garner, "the modern Medea," who killed her infant daughter and two older sons. She was captured as she and other slaves attempted to escape. Margaret Garner's terrible choice, between killing her children and having them reenslaved, was the subject of Toni Morrison's novel *Beloved*. *(From a wood engraving that appeared in* Harper's Weekly, *based on a photograph by the Brady studio of a painting by Thomas Noble; Library of Congress LC-USZ62-84545)*

societies, it is), we are likely to obscure underlying motivations.[7] Many people snatch at implausible straws so as to cling to the conviction that the emotional ambivalence many mothers feel about investing in infants is "unnatural," and hence very rare, and completely separate from more common, or "normal," maternal emotions. No connections are drawn between the ordinary distancing of a new mother from her infant, on the one hand, and these very extreme failures of maternal bonding that end in tragedy, on the other. Even those who accept that infanticide takes place—among heathens, somewhere else—are often reluctant to accept its natural occurrence among civilized or Christian people.

The unwillingness to acknowledge infanticide among creatures like ourselves, except as morally compromised, subhuman behavior, has a sporadic history in the West. The euphemistic catchall term *overlaying* used to be

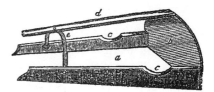

Fig. 12.2 Florentine *arcutio*. *(From Fildes 1986: pl. 7.4)*

invoked when infants were "accidentally" smothered by their caretakers. In fifteenth-century Florence, for example, 15 percent of deaths among infants sent to wet nurses were attributed to this type of mishap. Taking "overlaying" at face value, an eighteenth-century physician advised Britons to adopt a new invention, the Florentine *arcutio,* a three-foot-long wooden cage designed to prevent a woman from accidentally suffocating a baby in her bed. Italian nurses were obliged to use them under pain of excommunication.[8]

There is no evidence, however, that *arcutios* made any difference. Coroners' reports for London in the period between 1855 and 1860 (about the time George Eliot was describing Hetty Sorrel) list 3,900 deaths, mostly of newborns, from "overlaying." In subsequent inquests, 1,120 of these cases were deemed murder.[9] Then as now, one suspects, it was rare for co-sleeping to result in smothered babies. (Harder mattresses, or forbidding caretakers to drink or take opium, would have been more effective at preventing *true* accidents than prescribing this cumbersome contraption.)

Long after the *arcutio* was first invented, thousands of infant deaths in eighteenth-century England were still being attributed to overlaying. Today, doctors and coroners sweep fairly obvious cases of infanticide under a new euphemistic rug: unexplained sudden infant death syndrome (SIDS). Unquestionably, some babies (about 1 in 1,000) spontaneously stop breathing for unexplained reasons; many of these SIDS cases may involve heart-rhythm defects.[10] But not so many as otherwise cynical detectives, coroners, and pediatricians readily accept.

In England, it took a justifiable invasion of patients' privacy by a suspicious British pediatrician to end a massive and futile quest for a will-o'-the-wisp—a supposed genetic trait running in families that causes successive infants to suddenly stop breathing. Over a period of years in a London hospital, medical personnel working in conjunction with police secretly videotaped mothers attempting to smother their infants who had been hospitalized for recurrent sleep apnea. In the United States, the highly publicized 1995 murder trial of Waneta Hoyt had the same effect.[11]

Mrs. Hoyt was a New York State housewife with serious emotional and financial problems whose four newborns and one older infant successively died, their deaths attributed to SIDS. An amazing feature of the Hoyt case is how many people expressed doubts, only to have them overridden. People around Waneta Hoyt were reluctant to hurt her feelings by expressing suspicions.

The case would never have come to trial had not one unusually blunt woman medical examiner from Dallas mentioned to a Syracuse detective how statistically preposterous the familial SIDS diagnosis was. People believe what they want, she told him, and "they don't want to believe that a mother could possibly do something like this to her own child."[12]

The next-to-last of five Hoyt babies to die of sudden infant death syndrome (in 1970) was in the hospital being monitored for her "sleep apnea condition" just days before she died. By then a handful of psychiatrists were cautioning medical personnel that *some undeterminable portion* of SIDS cases might be infanticides.[13] Several nurses at the hospital became concerned by the mother's indifference to her infant, along with the fact that when alone with "Molly," Mrs. Hoyt rarely touched her, and never held the baby close even during bottle-feeding. One nurse noted in Molly Hoyt's charts that "I discussed my concern for the baby with [the doctor in charge] this a.m. . . . The interaction between mother and baby is almost nil in my opinion. . . ." Another told the director of the SIDS research project that "I don't like the way the mother is acting. . . . There's no maternal instinct there. She pulls away from the baby like she really doesn't want to be bothered."[14] To no avail.

The Hoyt case cannot be dismissed as tunnel vision by medical researchers intent on proving at any cost a pet theory about SIDS. Too many people had to ignore too much. They simply took for granted how a mother must feel about her infants. After the fifth death, Waneta Hoyt went on to apply for and adopt a two-and-a-half-year-old baby, a child she and her husband successfully reared.

The "suggestion that a proportion of children with recurrent apnea . . . have been suffocated [still] tends to produce either outrage or skepticism," Roy Meadow ruefully noted. Meadow was in a position to know. He was the pediatrician whose hidden videos documented twenty-seven different mothers "creating" infant breathing disorder. An unfortunate legacy of the misdiagnosed SIDS cases is that now parents whose infants *do* suddenly die of natural, hard-to-diagnose causes will have to cope with both that tragedy *and* the unjustified suspicion of medical personnel newly converted to a previously undreamed-of possibility.

————

While studying infanticide among *nonhuman* primates, I would learn in the most up-close-and-personal way how emotionally charged this topic can be. A year after completing my Ph.D., in 1976 I was in Washington, D.C. for the annual meeting of the American Anthropological Society. One of the grand old men of physical anthropology was sitting in the front row as I laid out the lines of evidence that led me to conclude that infanticide had been going on among primates for a long time, and that under certain conditions it was an adaptive rather than a pathological behavior. I was not even talking about mothers, but about the phenomenon of unrelated males from outside the breeding system who kill infants sired by competing males. "Offspring are vital to the continued survival of any species," I began. "At first glance, it seems surprising to find selection for behavior that does not contribute to the survival of infants, odder still to find selection for behaviors that actually decrease infant survivorship."

As I spoke, I noticed the muscles on the great man's face tighten, and his jaw set. At the end of my talk he rose abruptly and turned to face the audience. Langur monkeys elsewhere did not behave the way I had described, he announced. These monkeys were "not normal." Later he was to tell a public audience at the American Psychological Association meetings that the monkeys I studied were deranged. On that day, he walked down the main aisle and out of the room while I was completing my reply. In the pages of *The American Scientist* another anthropologist questioned whether these could possibly be "*Normal monkeys?*" (the heading under which the letter ran). Since everyone knows that animals reproduce in order to perpetuate their species (group selection still reigned supreme in those days), she argued, no normal animal would deliberately kill an infant. Such death rates as I was reporting from infanticide could only represent "destruction, not adaptation."[15] Because they did not benefit the species, they must be pathological.

Within anthropology this debate still persists,[16] long after accumulating evidence for beetles, spiders, fish, birds, mice, ground squirrels, prairie dogs, wolves, bears, lions, tigers, hippopotami, and wild dogs led biologists to take for granted that there are a range of conditions under which mothers cull their litters and abandon or cannibalize young, and an even wider range of circumstances in which unrelated males or rival mothers take advantage of the vulnerability of infants. None of these cases, even when far less well documented, met the same kind of resistance as did the initial suggestion that infanticide among primates like ourselves was an evolved behavior, an adaptive way for individuals to resolve dilemmas confronting them in the course of reproduction, and not out of the range of "normal" behavior.

————

Unwillingness among social scientists to believe that primates evolved to behave in such a repugnant way is one reason why many anthropologists are still reluctant to accept that infanticide might be adaptive under some circumstances. But there is also a more complicated reason why anthropologists hesitate to attribute infanticide to the people they study. They worry about the eagerness with which humans ascribe atrocities to other peoples, and the various ways we use labels like "inhuman actions" to dehumanize those accused of committing them.

Accusing people of depraved behavior can be the first step to viewing them as "moral inferiors" and denying them basic human rights. Sensitive to such concerns, some anthropologists are reluctant to publish data on behavior that might be used to justify social or political intervention.

Extracting sound data from propaganda and confidential sources is problematic. Even in societies where infanticide is sanctioned, it remains a painful subject. Almost all mothers are reluctant to talk about it.[17] Gathering reliable information on live births that terminate in infanticide usually requires fieldworkers to augment detailed censusing with in-depth interviews with both mothers and their associates.

In 1990, a conference on infanticide sponsored by the National Science Foundation and the Italian government[18] was convened in what seemed an appropriate venue: a mist-shrouded medieval village named Erice on a mountaintop in Sicily, well out of public view. I was preparing a technical report for the conference on ecological and cultural factors contributing to high versus low rates of infanticide in tribal societies. Why is infanticide common in some groups, virtually nonexistent in others?

Fig. 12.3 In North America "wild Indians" supposedly captured children and dragged them to death. No one knew how often, but the charge provided a rationalization for U.S. cavalry to destroy whole villages with minimal disturbance to their consciences. Recently the American public was prepared for the bombing of Baghdad by being reminded that "For seven years Saddam Hussein has murdered Iraqi children [by the] thousands. . . ."[19] *(Courtesy of Barker Texas History Center, University of Texas at Austin)*

Unusually detailed information was available for some dozen societies. At a gross level, the answer was obvious. Mothers kill their own infants where other forms of birth control are unavailable. Mothers were unwilling to commit themselves and had no way to delegate care of the unwanted infant to others—kin, strangers, or institutions. History and ecological constraints interact in complex ways to produce different solutions to unwanted births. The task of evaluating the evidence became unexpectedly complicated, however, when several of the fieldworkers responsible for collecting the most detailed case studies expressed concern that even the use of code names for particular individuals, and even for tribal affiliation (routine practice), might not be enough to disguise identities. They requested that all reference to their data be deleted. Another anthropologist who was at the conference to describe discriminatory treatment of orphans among the South American tribe he studied withdrew his paper from the published proceedings of the conference, fearing condemnation from colleagues.

From a scientific point of view, the problem was so stark that all thirty scientists from around the world who had gathered at Erice signed a statement (prepared by biologists Fred vom Saal and Stefano Parmigiani) calling for a "spirit of free inquiry and reporting of politically sensitive findings."

Of course scientific information can be misused by others for political ends. But those intent on doing so will do so whether or not their information is accurate. The !Kung, for example, have become widely known as *The Harmless People*—the title of an early ethnography about them—while the Yanamamo are best known by Napoleon Chagnon's epithet *The Fierce People*. Both stereotypes (as the authors of these works are the first to point out) are narrow characterizations of multifaceted people—oversimplifications not meant to be taken at face value. Yet ultimately such scholarly caveats are thrown to the wind when politicians turn to these publications to justify their actions. The homelands of both groups are currently being appropriated by technologically superior neighbors: the lands of the "harmless" people because they are deemed too childlike to have property rights, those of the "fierce" ones because they are viewed as too nasty, too savage, to deserve them.[20]

In the end, signers of the "Erice Statement" concluded that more harm than good is done "by attempting to suppress the truth." By 1998, medical doctors as well were calling for more open discussion of maternal ambivalence and infanticide.[21] One of the ironies of the attempt to suppress reporting on infanticide is that the more we learn about the full range of maternal behaviors, the weaker become the grounds for ethnocentric moralizing, for making distinctions between "civilized" and "savage" people, between Christians and non-Christians, and so on.

Many millions of infant deaths can be attributed directly or indirectly to maternal tactics to mitigate the high cost of rearing them. These tactics include leaving infants at foundling homes with terrible survival statistics, in addition to the everyday decisions by women in many parts of the world who decide for various reasons not to breast-feed and hence rely on powdered formulas, of necessity mixed with local water so polluted as to make baby food a prescription for dysentery. When mothers distance themselves psychologically from excess or inopportunely born infants, such decisions become easier. But it is important to be clear: odd cases aside (and the Hoyt case may or may not be one),[22] mothers don't set out to commit infanticide. Rather, abandonment is at one extreme of a continuum that ranges between termination of investment and the total commitment of a mother carrying her baby everywhere and nursing it on demand. Abandonment is, you might say, the default mode for a mother terminating investment. Infanticide occurs when circumstances (including fear of discovery) prevent a mother from abandoning it. Although legally and morally there is a difference, biologically the two phenomena are inseparable.

Unkindness of Kin and Reliance on Strangers

The late historian John Boswell is best known for his writings on same-sex unions among early Christians. How this man with no special interest in parenting came to produce a monumental study of child abandonment in the West (*The Kindness of Strangers: The Abandonment of Children in Western Europe from Late Antiquity to the Renaissance*) is a fascinating story in itself.

While researching early Christian sexual mores, Boswell came across some odd advice given by prominent early theologians. Men should be careful never to visit brothels or have recourse to prostitutes *because in doing so they might unwittingly commit incest.*

What a peculiar and oblique appeal to virtue! Boswell's curiosity was piqued. "Although the 'exposure' of children was a part of the standard litany of Roman depravities," Boswell noted, "I had understood [the babies] were left to die on hillsides. I never imagined that it was a widespread or common practice, and certainly had not thought that Christians abandoned babies."

Was the warning predicated on the remote possibility that someone's abandoned child could end up in a brothel "worth mentioning simply because it would be so dreadful if it did happen?" In the end Boswell succumbed to logic: "If Christian fathers did not abandon their children, they would hardly be dissuaded from visiting brothels by the threat of incest."[23]

Boswell's book grew out of this chance encounter with an odd admonition to virtue. From trial and church records, civil and canon law, Boswell determined that children were given away and abandoned in great numbers throughout this period by parents belonging to every social class. Parents had all sorts of reasons for not rearing a child. Perhaps they already had too many children to feed, or daughters to provide dowries for when they grew up. Perhaps the child's birth, if known, would bring shame upon the family, or perhaps the infant was defective. These early Christian parents, much like the "barbarians" Darwin and various anthropologists described, abandoned rather than killed their unwanted infants. The deeper Boswell delved, the clearer it became that very nearly the majority of women living in Rome during the first three centuries of the Christian era who had reared more than one child had also abandoned at least one. He found himself looking at rates of abandonment of around 20 to 40 percent of children born. If Romans gave to crippled beggars it was because "everyone is afraid he might say no to his own child."

Still, Boswell assured readers, most of those abandoned "were rescued and brought up either as adopted members of another household or as laborers of some sort." Abandonment, he assumed, functioned as a benign way of recycling children, transferring them from those who had too many to those who had too few. Whether they were exposed anonymously, sold, donated to monasteries as "oblates," or substituted for someone else's deceased child, Boswell remained convinced that abandoned infants died at a rate only slightly higher than the normal; "the death of *expositi* [exposed infants] does not appear to have been common at any time under the [Roman] empire." According to this very optimistic view, "the 'kindness of strangers' in every age seems to have been sufficient to rescue most abandoned children."[24]

Boswell emphasized the altruistic end of the wide range of motivations people have when they stoop along a muddy road to pick up helpless, crying, starving babies, and he seriously underestimated how difficult it would have been to keep infants who were deprived of breast milk alive without baby bottles, sterile water, or formula.

When medievalist Mary Martin McLaughlin reviewed *The Kindness of Strangers* for the *New York Times Book Review* she reminded readers that as soon as better records become available (by the early fifteenth century) and we get "substantial testimony to the fates of children, the picture is far from happy." Abandoned children were sometimes substituted for people's own children, or became valued laborers as "other sources of slaves dried up." These children fared the best. But by and large, the treatment of abandoned children was not nearly so "gentle" as Boswell hoped. Some significant portion of those who survived were being sold as slaves and prostitutes—which is what they were doing in brothels in the first place.[25]

The lasting importance of Boswell's scholarship does not lie in his ideas about the recycling of babies by strangers, but in his documentation of how widespread maternal abandonment of infants was in antiquity. Some portion of European foundlings surely survived. But that proportion varied with time and location, depending primarily on whether or not abandoned babies ended up in the arms of a suitable wet nurse. Even those who survived would have confronted an uncertain future at best.

"Unintended Consequences on a Massive Scale"

The explicit aim of foundling homes was to prevent abandoned infants from dying. The "Hospital of the Innocents," one of the earliest such institutions in the world, still stands in Florence, a stately reminder of a catastrophic experiment in social engineering. Founded in 1419, with assistance from the silk guilds, the Ospedale degli Innocenti was completed in 1445.[26] Ninety foundlings were left there the first year. By 1539 (a famine year), 961 babies were left. Eventually five thousand infants a year poured in from all corners of Tuscany.

Although the best known, the Innocenti was just one, the largest, of sixteen such foundling homes in the Grand Duchy of Tuscany. Three centuries after it opened, mortality rates were still appalling. Of 15,000 babies left at the Innocenti between 1755 and 1773, two-thirds died before reaching their first birthday.[27]

———

Elsewhere in Europe groups of citizens and governments were similarly disturbed by the large numbers of unwanted infants left along roads and in gutters. In city after city the same painful experiment was repeated. In England,

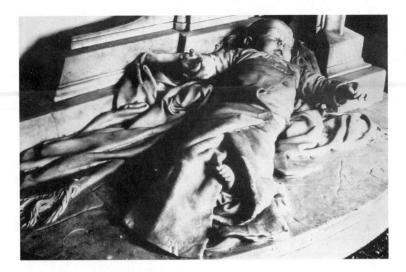

Fig. 12.4 Marble statue depicting *The Awakening of an Abandoned Infant* (Le réveil de l'abandonné, *by Robert, 1894, courtesy Musée de l'Assistance Publique, Hôpitaux de Paris*)

retired sea captain Thomas Coram got royal backing to build a home for deserted children. When it opened in 1741, it became immediately apparent that he had underestimated the magnitude of the problem. Mothers fought at the gates for admission. By 1756 the British Parliament guaranteed funds to ensure open admissions, with the result that within four years, 15,000 children had been admitted. Mortality rates soared, as there were simply not enough wet nurses to go around.

To the east, imperial foundling homes in Moscow and St. Petersburg were formally opened by Catherine II of Russia. The bountiful empress had fallen under the influence of an idealistic reformer named Ivan Betskoi who drew her attention to the children abandoned in the streets of Moscow and the "incomparably greater number" who "having managed to draw their first breath are deprived of it secretly by merciless mothers and their inhuman accomplices."[28]

Having traveled through Europe, Betskoi was familiar with the writings of reformers like Rousseau. He became convinced that with proper rearing, almost anything was possible. Betskoi persuaded Empress Catherine that foundlings could be reared to provide the trained labor pool that would allow Russia to develop along Western lines.

Fig. 12.5 The artist William Hogarth presented this drawing to Captain Coram, who used it as the letterhead for stationery to solicit funding to hire wet nurses for the foundling home. The letterhead depicts a naked baby in the bushes, a swaddled baby being deposited on the ground by its mother, and another in the arms of a foundling home official. Older foundlings are shown industriously occupied—the boys preparing to be sailors, the girls, domestics. *(Courtesy of Yale Center for British Art, Paul Mellon Collection)*

The very tangible results were the imperial foundling homes in Moscow and St. Petersburg, intended to qualify Russia as a player in the mid-eighteenth-century European Enlightenment. The doors of these grand repositories were formally opened in 1764. Both the St. Petersburg and Moscow foundling homes were soon admitting a steady stream of ill-fated applicants. Of the 523 children admitted during the first year, 81 percent died. There followed two years of improved survival prospects, culminating in the catastrophe of 1767. Ninety-nine percent of 1,089 infants admitted that year failed to survive to the next.[29]

The foundling homes became focal points of contagion for smallpox, syphilis, and dysentery. But the key problem was always how to feed infants without introducing lethal diarrhea-causing pathogens. Without nutritionally fortified baby formulas and sterile water to mix with them, the availability of breast milk has always been, and in many parts of the world continues to be, the single most important predictor of infant survival. As this reality emerged from a grim period of trial and error, administrators in Moscow and St. Petersburg developed plans to contract with peasant women to breast-feed babies from the foundling homes.

A fine plan in principle, but the administrators failed to foresee the number of parents who would seize the chance to delegate care of their children to others. By providing payments to wet nurses, the state foundling homes also created financial incentives for the *torgovki,* women peddlers who scoured the countryside for abandoned babies to deposit at the foundling homes. These babies were then transported from the foundling homes *back* to the countryside, where they generated pitiful stipends for peasant women. Many of these "wet nurses" kept the passbooks guaranteeing payment and passed their charges on to even more poorly paid—and not necessarily lactating—women. Even more desperate were the unmarried women who managed to secure for themselves skimpy sinecures by getting pregnant, depositing their own baby at a foundling home, and then qualifying to wet-nurse a foundling for pay. A tiny, lucky percentage of the hired wet nurses (if anyone in this tragic network can be called lucky) managed to bribe a foundling-home employee to assign them their own infants. In the words of historian David Ransel, the state's well-intentioned plan for caring for infants became a case study in "unintended consequences on a massive scale."[30]

Foundling homes offered ambivalent parents an easy, blameless option for delegating the costs of lactation and provisioning to someone else. Unfortunately, unless they essentially adopt the baby, unrelated alloparents can rarely be counted on for wholehearted commitment. And for mothers without access to birth control, not suckling their newborns often meant that they conceived again soon, sometimes within the year, only adding to the number of unwanted babies.

Epidemics of Foundlings

Because there were rarely enough lactating nurses to go around, foundling homes did little more than forestall death from exposure—just long enough to ensure that the baby was baptized. Without the nutritional and immunological benefits of mother's milk, most died in the first months, of infectious diseases or starvation. We know this because of just how well the staff at the foundling homes did at least part of their job. Keeping infants alive was often beyond their capacity. But an extraordinary bureaucracy grew up to record, in neat columns, detailed information on each of their charges: the exact date the infant was admitted; its sex, age, and whether or not the baby was baptized; any identifying tokens, coins, or scraps of cloth or notes parents might have left with the baby; and date of death. As historical demographers began, locality by locality, to transform neatly scripted columns of figures into rates and totals, it became clear that foundling homes were magnets for a much wider population than simply unwed mothers and poor domestics seduced by employers. Parents—often married couples—from a broad catchment area saw the orphanages as a way to delegate to others parental effort for offspring they could scarce afford to rear. Mothers poured in from the rural areas to deposit babies in the cities. What has generally been studied as a patchwork of various, discrete, local crises is really a wide-scale, demographic catastrophe of unprecedented dimensions.

I still recall the crisp autumn day in the old cathedral city of Durham, England, when at a conference on abandoned children, the full extent of a phenomenon I had been aware of for years sank in.[31] The talks were routine scientific fare. Overhead projectors flashed graphs and charts onto a screen. The black lines sprawling across the grid summarized data from European foundling homes, tracking changes in infant mortality rates over time.[32] As the morning wore on, the phenomenon of child abandonment was described, country by country, epoch by epoch, for England, Sweden, Italy, even Portugal's colony in the Azores. Gradually it dawned on me that this phenomenon affected not tens of thousands or even hundreds of thousands of infants, as I had long assumed, but millions of babies. I grew increasingly numb. I recall that I had difficulty breathing. As I distanced myself emotionally from these findings by seeking to analyze *what* they meant, I may have experienced (in a very remote and quite insulated way) the sort of surreal distancing other mothers long ago must have undergone as they adjusted so as not to see or feel what was before *them*. My subconscious, no doubt dwelling on my own children, dredged up, of all things, fantastical illustrations from Maurice Sendak's ominous classic *Outside Over There*. It was a book I had read to each of my own children. Sendak depicted columns of babies floating mysteriously over bridges, and down paths and stream beds. At that moment they clearly represented ghosts, streams of babies flowing from the city out to wet nurses in the country in one direction, and from poor peasant households in the countryside into urban foundling homes, flowing back the other way. There was nothing exotic about this heritage. It was my own. In time, I became desensitized to this information. In what follows, I treat it like any of the other material that scientists call "data."

Italy provides some of the most complete records on infant abandonment, and these data have been analyzed by a roster of distinguished historians and demographers, among them the anthropological demographer David Kertzer. By 1640, 22 percent of all children baptized in Florence were babies that had been abandoned. Between 1500 and 1700, this proportion never fell below 12 percent. In the worst years on record, during the 1840s, 43 percent of all infants baptized in Florence were abandoned. In the Grand Duchy of Tuscany around the same time, 5,000 were abandoned—practically 10 percent of all those born.

As in much of Catholic Europe, a *ruota,* or rotating barrel, was installed in 1660 to replace the old marble basin at Florence's main foundling home, the Innocenti. By 1699, however, it was necessary to place a grill across the opening to prevent parents from shoving in older children as well.

To the north, at the foundling home in Milan, 343,406 children were left between 1659 and 1900. For Milan in the year 1875, 91 percent of illegitimate children whose births managed to get recorded were abandoned. But the Italian cities were not isolated cases. Comparative data compiled by Kertzer for the period 1880–89 reveal an annual average of 15,475 infants abandoned in Moscow; 9,458 in St. Petersburg, which is comparable to figures showing 9,101 abandoned in Vienna during the 1860s, and 2,200 in Madrid between 1800 and 1809. The majority would not survive. In one of the worst sets of statistics, of 72,000 infants abandoned in Sicily between 1783 and 1809, about 20 percent survived. The scale of mortality was so appalling, and so openly acknowledged, that residents of Brescia proposed that a motto be carved over the gate of the foundling home: "Here children are killed at public expense."[33]

Putting Espositos in Perspective

In the "civilized" world, a woman who suffocates her newborn has committed a crime. She goes to jail. But a woman whose baby dies because she psychologically distances herself from her newborn and opts not to breast-feed, with the result that the baby succumbs to dysentery, is viewed merely as ignorant. Similarly, a mother who abandons her infant to a foundling home— even those where mortality rates are in the vicinity of 90 percent—is

Fig. 12.6 From medieval times onward, rotating barrels were set up in the walls of foundling homes so that parents could deposit an unwanted baby, ring a bell, and fade anonymously into the night. It would have been unusual for the father to accompany the mother, as depicted in this engraving of *l'Hospice des Enfants Trouvés* (Paris) by Henri Pottin (1820–64). *(Courtesy of Mary Evans Picture Library, London)*

regarded as unfortunate, but legally and spiritually blameless. Technically, her infant will die of malnutrition or dysentery, not neglect; she did not kill it.[34]

Even as cultural amnesia and other sleights of mind wipe clear the Western slate concerning acts and omissions responsible for more infant deaths than from several plagues combined, tangible reminders of the West's legacy of "unnatural mothers" persist in marble statues, stately Renaissance buildings where unwanted infants were warehoused, in police reports, and in crumbling ledgers from the foundling homes. Even phone books in most large metropolitan cities still bear witness.

Throughout Europe it was the practice for each foundling to be given a name as he or she was logged in, a first name and then a generic last one, names like Esposito (Italian for "exposed") or Trouvé (French for "found"). In Milan, many were given the last name Colombo, for the pigeons that alighted on the roof of the foundling home there and adorned its emblem. (This practice was abandoned in 1825, as the Milanese authorities found it awkward to

Fig. 12.7 When Napoleon decreed that every hospice in France should be equipped with a *tour,* a device similar to the Italian *ruota,* the thing came to be known as the "Napoleonic wheel." The French poet Lamartine extolled the wheel as an "ingenious invention of Christian charity, which has hands to receive but neither eyes to see nor a mouth to tell." As late as the mid-nineteenth century, Kertzer counted 1,200 such depositories for babies around Italy. By 1875, however, after revolving for two centuries, these wheels of misfortune began to be shut down. A well-intentioned system had spun out of control.[35] *(Courtesy of Mary Evans Picture Library, London)*

have tens of thousands of people with the same last name, and they worried about the stigma attached to being abandoned.) A durable fraction of these foundlings were lucky, robust, or resourceful enough to survive. They grew up and went on to have families of their own. During a recent visit to Boston, I counted in the phonebook 86 people of Italian descent in the metropolitan area whose male ancestors years ago had been given the institutionally provided name Esposito—sometimes Exposito, but always meaning "an exposed one."[36] In "Les Pages Blanches" for Paris 1996 there were 46 Espositos, 1 Esposti, 2 Espostos, 8 variants on Degli Esposti, 64 Trouvés, plus—as if to advertise just how this family's fortunes have improved—one listing for a family business that read: "Trouvé, Per et Fils."

Desperation, Destitution, Self-Delusion

When parents left infants at foundling homes, did they know what they were doing? At the outset, when an infant was left there by an unmarried woman (likely to be destitute) or sent in by desperately poor people from the countryside, it's possible that the baby would actually have a better chance of surviving at the foundling home than with its mother. Many foundlings in Renaissance Florence were illegitimate children of slaves or domestics and, as such, would have died at three times the average rate. Such abandonments could be construed to be in the child's best interests. But what about decisions made after mortality at the foundling homes rose to the catastrophic levels that they almost inevitably did, as more and more mothers made the same choice. When mothers set their infants in the barrel and rang the bell, did they have any idea how staggering the mortality rates were? Surviving documents from fifteenth-century Florence suggest that parents not only had some idea that foundling homes were dangerous places, but some even made shrewd suggestions in an attempt to tip the balance in favor of their infant's survival. Some left pathetic instructions begging personnel at the Innocenti to send this baby to some outside wet nurse, not keep it at the hospital, where chances of survival were worse.[37]

Volker Hunecke records case studies from eighteenth- and nineteenth-century Milan for tailor "Filippo A———" and his wife, who kept their first son at home and then deposited the next six (in the space of five and a half years) at the nearest *tour*. When his first wife died, he remarried "Cecilia B———" and together they deposited five more infants in five years. After a year and a half, the mother tried to retrieve them, but only two of all these children were still alive. "Francesco G——" and his wife "Amalia S——" similarly produced twelve children in thirteen years. The first died soon after birth, all the others went to foundling homes. Only one daughter survived. The point here is that these parents had the information *to know* that things at the foundling home had not gone well; and, being human, they would have communicated their misfortune to others.

After a point, parents had to have known. But they were making decisions based on immediate costs (discovery, in the case of an unwed mother; lost wages; destitution) rather than on the basis of rumored misfortunes behind distant walls. In doing so, they relied on the all-too-human gift of fantasy and self-delusion. Hetty cries that she did not mean to kill her child. Mothers who left their infants at foundling homes found solace by fantasizing fabulous destinies of upward social mobility for abandoned progeny, who like Romulus and Remus, the abandoned twins supposedly adopted by a she-wolf, might be miraculously saved and survive to found a dynasty.

Questioning Maternal Instinct

Decades before the sudden-infant-death-syndrome scandals surfaced in the 1990s, or before data from the foundling homes started to be quantified in the 1970s, psychiatrists, historians, and social scientists all noted the poor match between real-life mothers and the nineteenth- and early-twentieth-century stereotypes of instinctively nurturing mothers. Feminists in particular had long ago lost patience with Darwinian perspectives that struck them as essentialist and which patently disregarded women's felt experience. They were eager to discount biological explanations, and had little incentive to keep up with what was going on in reproductive ecology or sociobiology. They continued to project onto these fields their own worst assessments about essentialist and determinist assessments of "female nature" even after biologists themselves had abandoned these types of explanations. The result was that feminist theorists were producing models to explain what was essentially a biological phenomenon (namely, the failure of an infant to elicit nurturing responses from its mother) but without any reference to biology. They used the evidence of high numbers of non-nurturing mothers as a tool to jettison altogether the confining stereotype of the instinctively nurturing mother that had long been used to prescribe social roles for women.

———

Instead of taking a closer, critical look at the original, biologically based explanations to see if perhaps something had been left out, feminists (along with other social scientists who were trying to explain the widespread practice of abandonment by mothers) patently rejected evolutionary explanations. The biological basis for motherhood was replaced by a new environmentalism. The way a mother feels toward her infant must be solely determined by her cultural milieu.

In France, where this view of "socially constructed" mothers originated and then gradually spread, a brilliant and animated philosopher, Elisabeth Badinter, in 1980 the first woman to become a professor at Paris's prestigious École Polytechnique, and an appropriately iconoclastic descendant of Clémence Royer, could stare straight into the eyes of a reporter from the *Nouvelle Observateur* and say: "I am not questioning maternal love"—pause, for she knows what comes next will be scandalous—"I am questioning maternal instinct."[38]

Although David Lack's ideas about the tradeoffs mothers in nature make were by then built into sociobiology (discussed above, in chapter 2), most social scientists still assumed that in nature, mammal mothers instinctively and automatically care for every infant they produce. Badinter's reasoning was simple. If mother love is instinctive, all normal mothers should be loving. However, if the vast majority of mothers in eighteenth-century France had opted not to rear their own infants but to delegate their care to inadequate wet nurses instead, this was more mothers than could reasonably be dismissed as aberrations. Furthermore, Badinter knew that not all these women were unwed and destitute. It was also apparent that children related to parents to the same degree were not being treated in the same way, which to her could not be consistent with a biological basis for maternal love.

Such maternal love as Badinter could document was often discriminatory and selective. A wet nurse might be brought in from the outside to nurse an older son while the younger son was sent far away. If not spontaneous and automatic, Badinter argued, maternal love had to be a nonbiological social construction. It was a sentiment produced by a particular cultural context, peculiar to a specific historical time and place. Her best-selling book on the "myth" of a mother's instinctive love for her infants stressed the fact that although many mothers who abandoned infants or sent them to wet nurses were destitute, many others were "bourgeois" mothers whose banishment of their babies was discretionary. (I will return to this topic in chapter 14.)

Social Construction of Mother Love

The idea caught on among social historians that "good mothering," even the concept of childhood itself, must be a recent cultural invention.[39] Social historians like Philippe Aries in France and Edward Shorter in the United States hypothesized that parental emotions and the internal workings of family life derived from particular attitudes and customs. Such customs are built up and change over time. Not only do such cultural constructions take on a life of their own, they were never influenced by biology. It was a model that could safely eschew any discussion of an evolved human nature.[40]

Fired by the notion that parental attitudes change over time, the American psychohistorian Lloyd de Mause compiled masses of evidence on infanticide and abandonment in earlier times. Childhood, he announced, was actually "a nightmare from which we have only recently begun to awaken." He laid out a succession of stages, beginning with an "infanticidal mode" prior to the fourth century A.D., when parents "routinely resolved their anxieties about taking care of children by killing them." By the eighteenth century a phase of "intrusive" child-rearing ushered in by reformers led to the child being "nursed by the mother, not swaddled, not given regular enemas, toilet trained early, prayed with but not played with, hit but not regularly whipped, punished for masturbation, made to obey promptly," culminating in the "socialization mode" of the nineteenth and twentieth centuries, when newly empathetic parents became more concerned with training than conquering children, and today's modern "helping mode."[41]

Anthropologists as well were inclined to set aside Rousseau, Spencer, and the various essentialists. They followed the historians, divorcing maternal emotions from biological predispositions and situating them in particular economic and political contexts.

"Anthropologist Calls Mother Love a Bourgeois Myth," announced one headline.[42] Years of studying desperately poor mothers in Brazilian shanty-towns as they distanced themselves from doomed children and watched them die had convinced Nancy Scheper-Hughes that "Mother love is anything *other* than natural and instead represents a matrix of images, meanings, sentiments, and practices that are everywhere socially and culturally produced." She vividly describes how mothers convince themselves that their children lack the will to live, and how they subsequently draw back. The cultural art of breast-feeding, she decided, had been lost, so that the baby is fed powdered formula diluted with water teeming with diarrhea-inducing microorganisms.

When, almost inevitably, death ensues, mothers do not disguise their grief behind a stoic façade—they feel none. "The traumatized individual," writes Scheper-Hughes in *Death Without Weeping*, does not "shrug her shoulders and say cheerily, 'It's better the baby should die than either you or me.' " Rather (paraphrasing General William Westmoreland's famous observation about his allies during the Vietnam war), "They do not grieve the way we do." We err, she argues, by attributing to Third World women "a very specific cultural 'norm' [of maternal love] from our own Bourgeois society."[43]

Universality of Parental Emotions

Scholar after scholar detailed massive mortality owing to such "unnatural mothers," then sought to ascribe the absence of maternal commitment to cultural constructs—attitudes or historical factors peculiar to particular periods: the absence of a "concept of childhood";[44] the notion that "mother love" had not yet been invented;[45] mothers conditioned to expect children to die;[46] pressure put on unwed mothers by the Catholic Church,[47] together with the creation of foundling homes that serve as magnets for abandoning parents; unprecedented population growth (the doubling of the human population between 1650 and 1850);[48] colonial and capitalist oppression to explain lethal child neglect in the Third World today;[49] and so on. These circumstances were indeed highly pertinent. Historical and ecological context had important implications for how mothers assessed what their own, or this particular infant's, prospects were. Social and economic context had everything to do with what alternatives a mother had to choose from. But whereas these represent highly relevant circumstances, they are not explanations for *why* mothers were abandoning babies. And in the debates over the construction of "mother love," such variation in maternal responses by no means disproved the involvement of innate mechanisms. Indeed, understanding the biological basis of "mother love" is essential for understanding what is going on here.

———

Long before populations in western Europe burgeoned, well prior to the eighteenth-century boom in sending children away to wet nurses or abandoning them outright, European parents sought ways to cope with unwanted children. Around the world, hunter-gatherer societies have suffered high rates of child mortality without compromising close mother-infant relationships. "Their death made me feel pain," moaned Nisa, the !Kung mother, in a culture where on average 50 percent of children die before adulthood—and 100 percent of hers had. "Eh, Mother! I almost died of that pain."[50]

Hundreds of British and American diaries kept by literate parents (alas, mostly fathers) between 1500 and 1900 reveal the same continuum of emotions—from indulgent to abusive, distant to engrossed—that one finds in cross-sections of parents today.[51] Further back in time the record grows skimpy. But such accounts as exist tell the same story, albeit from a male perspective, since few mothers left written diaries. "How great a joy it was to me and his mother; and soon came his movements in the womb which I noted carefully with my hand, awaiting his birth with the greatest eagerness," wrote Giovanni Morelli, a fourteenth-century father from middle-class urban Italy. Morelli sounds like a textbook example of the modern "engrossed" father— supposedly a historical novelty.

> And then when he was born, male, sound, well-proportioned, what happiness, what joy I experienced; and then as he grew from good to better, such satisfaction, such pleasure in his childish words, pleasing to all, loving towards me his father and his mother, precocious for his age.[52]

How profoundly modern as well sound the guilty self-recriminations of the same father when his beloved firstborn, Alberto, dies in 1406 at age ten: "You had a son, intelligent, lively, and healthy so that your anguish was greater at his loss," he berates himself. "You loved him . . . but treated him more like a stranger than a son . . . you never looked approvingly at him . . . you never kissed him when he wanted it. . . . You have lost him and will never see him again in this world."[53]

In 1746, Madame d'Épinay sounds like the woman next door when she writes of her new baby: "I think of nothing but this little creature from morning till evening." Advantaged enough to hire both a governess and a tutor, she chose to breast-feed the baby herself and was so attentive to his needs that he developed "a passion for having me always near him. He cries when I leave him. . . . I sometimes think when he smiles as he looks at me . . . that there is no satisfaction equal to that of making one's fellow creature happy." This obvious bond between mother and infant blossomed during the historical

Fig. 12.8 Women like Madame d'Épinay were following the advice of Jean-Jacques Rousseau "Man of Nature" (1712–78). He admonished men who "cannot fulfill the duties of a father" not "to become one." Nothing, he wrote "can dispense [a father] from caring for his children and bringing them up himself." He offered this advice later in life, after he himself was no longer a poor writer seeking to better himself. Rousseau was rationalizing to himself how it was that all five of his children ended up in foundling homes. At the time he noted that the arrangement "seemed to me so good, so sensible, so appropriate, that if I did not boast of it publicly it was solely out of regard for their mother. . . . All things considered, I chose what was best for my children, or what I thought was best." All in all, Rousseau mused in his *Reveries,* it was quite lucky for the children, because if they had been left with their mother they would have been spoiled by her and transformed into "monsters." It was a matter of "principle" for him to do what he did. Like other early moral philosophers, Rousseau was strong on what it was biologically "natural" for mothers to do, and took for granted both that "in the animal kingdom the laws of nature reign unencumbered."[54] He also took for granted that he knew what those laws were. *(Engraving by Auguste Claude Le Grande, 1785, private collection)*

heyday of French wet-nursing. The parents of less fortunate infants sent them to languish, often to die, far from home.[55]

When Circumstances Change

Most telling perhaps are real-world case histories that allow us to ask: What happens when the same mothers, with the same social constructs, are placed in different circumstances? Among the Ayoreo Indians of Bolivia earlier in this century, following a period of grave social disruption (the Chaco War, between Bolivia and Paraguay, 1932–35) nearly every woman in one village had committed infanticide. Between them, the women in this sample had buried alive 38 percent of all infants born. One mother code-named "Asago" viewed her first three husbands as poor prospects for long-term support and buried at birth the *first six of the ten children* she bore in her lifetime. Yet infants born to women who had managed to forge stable relationships, or who, having grown older, had decided to proceed with a family no matter what, were loved and cherished. "Even when trained as an anthropologist," the ethnographers noted, "it is difficult to believe that someone known as a charming friend, devoted wife, and doting mother could do something that one's own culture deems repugnant."[56] Across cultures—in South America, New Guinea, Europe—the same mother who regretfully eliminates a poorly timed neonate will lovingly care for later ones if circumstances improve. Indeed, among both the Ayoreo and North Americans in Martin Daly and Margo Wilson's sample of infanticidal mothers, those under the age of twenty were the most likely to respond to poor circumstances by committing infanticide, while older mothers were far less likely to do so.[57]

How a mother, particularly a very young mother, treats one infant turns out to be a poor predictor of how she might treat another one born when she is older, or faced with improved circumstances. Even with culture held constant, observing modern Western women all inculcated with more or less the same post-Enlightenment values, maternal age turned out to be a better predictor of how effective a mother would be than specific personality traits or attitudes. Older women describe motherhood as more meaningful, are more likely to sacrifice themselves on behalf of a needy child, and mourn lost pregnancies more than do younger women—presumably because the latter foresee more opportunities to conceive again.[58]

Sometimes an improvement in circumstances leads to an entirely different child-care style. Nancy Scheper-Hughes describes a poor Brazilian woman who finally finds herself attached to a husband with a small but predictable income. Spontaneously, this woman reinvents the "bourgeois" concept of mother love and readopts the lost art of breast-feeding, which supposedly had disappeared from her shantytown subculture. She invests emotionally and financially in each child, and marvels at the outcome: "From then on almost all our babies lived."[59]

Unintended Experiment at La Maternité

Even when circumstances remain grim, extended contact between mother and infant (especially if the mother is breast-feeding) can elicit emotions that undermine the strongest pragmatic resolve. Social historian Rachel Fuchs, whose research deals with the effect of public policies on infant abandonment, describes an unusual experiment that resulted from an attempt at social engineering undertaken in Paris in the years 1830–69. It dates from the time when Europe's epidemic of abandonment was winding down.[60]

Exceedingly poor women who could not afford a midwife gave birth at La Maternité, the major state-run charity hospital in the Seine region. The hospital was directly across the street from the Hospice des Enfants Assistés, the only place around where infants could be abandoned legally. In an effort to reduce the numbers of abandoned babies, a group of French reformers came up with a plan. A subset of indigent women was obliged to remain with their infants for eight days after birth. What most people would consider unethical manipulation today produced remarkable results. Under this "experimental" regimen, the proportion of destitute mothers who subsequently abandoned their babies dropped from 24 to 10 percent. Neither their cultural concepts about babies nor their economic circumstances had changed. What changed was the degree to which they had become attached to their breast-feeding infants. It was as though their decision to abandon their babies and their attachment to their babies operated as two different systems.

Fuchs's analysis is consistent with this interpretation. Infants whose destitute mothers left the hospital on the day of birth had a fifty-to-one chance of being left behind. Infants whose mothers left just two days later had a six-to-one chance of being abandoned.

Underpinnings of Contingent Commitment

A mother's commitment to her infant—and in the case of humans, this is what we mean by "mother love"—is neither a myth nor a cultural construct.

As with other mammals, a mother's emotional commitment to her infant can be highly contingent on ecologically and historically produced circumstances. No one knows how the underlying mechanisms work. But it is a reasonable guess that such mechanisms involve thresholds for responding to infant cues. These would be endocrinologically and neurologically set, possibly during pregnancy and prior to birth, rendering a mother more or less likely to become engaged by infantile cues as she makes decisions about how much of herself to invest in her infant.

Most mothers remain close to their infants in the period after birth, and over a period of days, weeks, and months the attachment between mother and infant grows stronger. But some mothers are so detached at birth that their own actions ensure this never happens—a situation rendered far less common in apes and monkeys than in humans since (and this may not be the only reason) neonates are able to catch hold of the mother's fur and cling to her from birth long enough for backup systems related to maternal responsiveness to kick in. Such cases of "unnatural"-seeming abandoning mothers offer far more insight into the underlying processes than do the more usual cases where babies are born, picked up, and cared for.

Across primates, including humans, wherever mothers abandon babies, they almost invariably do so within the first seventy-two hours—as in the case of Lynn Fairbanks's "teenage" vervet mothers, or the hyperfertile captive tamarin mothers with inadequate allomaternal assistance. This does not necessarily mean that there is a critical period right after birth during which mothers must bond or else. Rather, what it suggests is that close proximity between mother and infant during this period produces feelings in the mother about her baby that make abandonment unbearable. (I discuss this in more depth below, in chapter 22).

———

Far from invalidating biological bases for maternal behavior, a closer look at the historical and ethnographic record reveals mothers who respond to a range of circumstances with a fairly predictable range of emotions. Their responses to infants remain consistent across vast spans of time and space, in the face of bewilderingly variable social histories. These consistencies remind us that we descend from creatures for whom the timing of reproduction has always made an enormous difference, and that the physiological and motivational

Fig. 12.9 A modern *tour*. This incubator was set up outside the Schopf-Merei Hospital in Budapest in 1997 to cope with a surge of unwanted births after free contraception in Hungary was discontinued. *(AP/WideWorld Photos)*

underpinnings of a quintessentially "pro-choice" mammal are not new. These consistencies in maternal nature transcend historical peculiarities, and the vagaries of local ecologies and demography. It was not the response of mothers in ancient Rome, or eighteenth-century France, or twentieth-century Brazil that was unnatural. In fact, what was unnatural was the unusually high proportion of very young females, or females under dismal circumstances, who, in the absence of other forms of birth control, conceived and carried to term babies unlikely to prosper.

In the next chapter I explore circumstances under which mothers treat offspring differently not because of when in the mother's life they happen to be born, but because the baby is born one sex or the other. The mothers involved are not necessarily immature or poor. Many have mates or kin networks to help support them, or are otherwise well-positioned to rear their baby. Indeed, in many of the best-documented cases of sex-selected infanticide, elites—those with the most resources, not the least—are implicated. Why?

Daughters or Sons?
It All Depends

"The son was alive then, and the daughter was at a discount. . . ."
—from George Eliot in *Middlemarch*, 1872

Seventy thousand calories, nine months, seventeen-plus years of room, board, and extras—and yes, there is a charge. Parents expect returns from their investment in children. How often have you overheard a parent telling a child that he or she is "a disgrace," "good for nothing," or complaining that a son or daughter "will never amount to much"? How many parents have entertained such thoughts themselves? "It's for your own good." Or especially, "I just want you to live up to your potential." Talking to young people this way is so commonplace that it has become routine fare in melodramas. What is at issue here?

There is a contract stored deep in the minds of parents: they expect those to whom they give so much to bring credit to the family name, or to translate parental investment into either cultural success or its former correlate: enhanced fitness for the lineage. Parents may justify their behavior by claiming to act in "the child's interests." Closer scrutiny often reveals parents defining those interests in line with their own.

In the West, such conflicts have tended to be over education, inheritances, career decisions, social or sexual choices. Parental preferences rarely place infants in mortal peril. Elsewhere, though, parents literally sacrifice children to family goals. Nowhere are underlying tensions more manifest than in societies where parents resort to sex-selective infanticide in order to obtain specific family configurations.

China's Missing Daughters

In 1991, results from China's massive census of every hundredth household became available, sparking worldwide comment. "Where have the girls gone?"[1] It is normal for slightly more boys than girls to be born: 104 to 106 boys per 100 girls is considered normal. But comparisons of expected sex ratios with those from China's 1990 census revealed that out of a total 1.2 billion people, millions of girls that should have been counted seemed either not to have been born, not to have been reported as born, or else eliminated so soon after birth as to escape notice in the census. Instead of the expected 106 sex ratio there were 111 boys per 100 girls. Perhaps Asians are genetically disposed to produce more sons than other people, some suggested.[2] Demographers, however, are convinced that girls are being eliminated on a massive scale, either through prenatal sex determination followed by selective abortion of female fetuses, or through neonaticide.[3]

Later-born children are most at risk. Westerners assume that the one-child-per-family policy means that Chinese families only have one or at most two children. But that is not necessarily the case, especially in rural areas. Dispensations for extra offspring can also be obtained—especially for parents who only have girls. But often a fine is imposed, and many families are reluctant to bear penalties for an extra child without getting the sex they want.[4]

Either sex may be acceptable for the first birth. This explains why the current Chinese sex ratio for first births is within the normal range—106 boys reported for every 100 girls. For higher birth orders, however, sex ratios start to climb. For families producing a fifth child, 125 male births are reported for every 100 daughters.[5]

Policy to Blame—or Parental Preferences?

Viewed in historical perspective, China's one-child policy has enhanced the well-being of *wanted* children and helped the country to catch up economically.[6] But small families also increased pressure for a son. China's "missing daughters" have become an international cause célèbre, with special condemnation reserved for the one-child policy itself.

Female infanticide, however, was practiced long before Mao's population policies were introduced in the second half of the twentieth century. In southern regions like the Lower Yangtze, where Shanghai is situated, the only plausible explanation for *so many* missing daughters is either sex-selective abortion or infanticide.[7] Infanticide rates are higher today than ten years ago, but they are lower than in centuries past. In some areas, childhood sex ratios in the eighteenth and nineteenth centuries were as high as 154:100.[8] In large cities like Beijing, wagons made scheduled rounds in the early morning to collect corpses of unwanted daughters that had been soundlessly drowned in a bucket of milk while the mother looked away.[9] One nineteenth-century woman interviewed recalled eliminating eleven newborn daughters. Another could not recall the exact number, except that she had borne more daughters than she wanted.[10]

Such anti-daughter prejudice was scarcely new. A Chinese poem recited 2,500 years ago celebrated the arrival of a son who should be dressed in finery, laid on an elaborate bed, and given a jade insignia to hold. A daughter, by contrast, would be dressed in a wrapper, laid on the ground, and given a wooden whirligig. According to a popular proverb: "More sons, the more happiness and prosperity."[11]

Whether current distortions in China's sex ratios are due to selective abortion of female fetuses or female infanticide, existing laws are not effective. Sex ratios are most skewed in remote rural areas. Labor provided by sons is more essential there, and laws harder to enforce. The strongest skews are found in southern China, where discrimination against daughters was traditionally more pronounced.[12]

The call for more and tougher laws detracts attention from the underlying problem: long-standing parental desires for a particular family composition. This means that unwanted infants, if they survive, are likely to grow into unwanted children who will be fed last and fed least, have less attention paid to their education and medical needs, and suffer physical and emotional abuse. A more effective and humane solution would focus on changing parental mindsets. But how? Ongoing propaganda campaigns—for example, the signs posted all over proclaiming "Little Boy, Little Girl, Both Okay"—have had limited impact.

Fig. 13.1 Public sign from urban China. Essentially it says: "Little Boy, Little Girl, Both Okay."
(*Courtesy of Craig Kirkpatrick*)

A Widespread and Very Ancient Bias

The first step is to understand what ancient and deep-rooted parental preferences for sons versus daughters are about. Sex ratios as high as those found in China today (116:100) can be documented for other Asian countries that *do not* have such coercive family planning.[13] Far beyond the boundaries of China, wherever preferences for offspring of one sex are so extreme that sex-selective infanticide is practiced (in about 9 percent of the world's cultures), sons are the desired sex.[14]

Outside of China, female infanticide is well documented for other parts of Asia, among tribes in highland New Guinea and South America, as well as in ancient Italy. Wherever it is found, extreme son preference and the devaluation of daughters that accompanies it go hand in hand with patriarchal ideologies. Indifference to the fates of daughters can be stunning, as evidenced by the note of a Roman soldier sent his wife in the first century B.C.:

> I ask and beg you to take good care of our baby son. . . . If you are delivered of child . . . if it is a boy keep it, if a girl discard it.[15]

In India, special mantras from the *Veda,* sacred texts of Hinduism, are still chanted when a wife becomes pregnant. If by some mischance the fetus is female, this text expresses the hope that she will be magically transformed into a son.[16]

Various well-meaning pundits have proposed letting the "mania for sons" take its course. Playwright, congresswoman, and ambassador Clare Boothe Luce was among the most outspoken of them. She correctly noted that the Chinese desire for sons motivated parents to have larger families, since those with only daughters kept trying for sons. She proposed a "male-child birth pill" as the "quickest way of peacefully slowing down the [population] clock." Furthermore, Luce suggested, as daughters became scarcer, the status of women would rise.

Laws of supply and demand, however, do not always work, especially not where odds are stacked against a sex that is not only scarce but is socially disenfranchised. In urban China, scarcity has indeed provided women with undreamed-of opportunities. In television broadcasts that fall somewhere between *The Dating Game* and talent shows, desperate bachelors make their appeals, then anxiously await a summons, as female viewers choose among potential mates. But the very same scarcity that drives urban bachelors to these extremes makes women's lives more perilous than ever in rural areas. The incidence of rape, kidnap, and even women being bought and sold, has risen along with the number of wifeless men.[17] Women may be in short supply, but as a class they are no better off. In 1995, China was the only country in the world where the suicide rate for women exceeded the rate for men.[18]

———

For parts of the world where "a daughter's birth makes even a philosophical man . . . gloomy [while] a son's birth is like sunrise in the abode of gods,"[19] prenatal sex diagnosis with the option of selective abortion arrived on the scene like a divine gift. The old proverb according to which "eighteen goddess-like daughters are not equal to one son with a hump" is taken quite literally by parents who use prenatal diagnosis not to guard against genetic defects, but against paired XX chromosomes. Of 8,000 abortions performed at a clinic in India, 7,997 eliminated fetuses parents had been told would be daughters. (Typically, mothers being tested already had one or more daughters.)[20]

Officially, such discrimination is banned. Asian countries have far stricter laws against using prenatal tests this way than do Western countries.[21] But the laws are unenforceable. In 1988 Maharashtra state in south India banned all prenatal sex determinations. India's Parliament followed suit. In 1994, nationwide penalties of three years in prison and a fine (equal to about $320) were imposed on anyone found guilty of administering or taking prenatal tests solely to determine the sex of a fetus. Korea followed suit the same year, making it a crime to abort a female fetus. Such laws notwithstanding, volunteer organizations in India estimate that around 80,000 abortions after sex tests are still performed every year (surely an underestimate). The situation is similar in Korea. Meanwhile in the poorest areas of Asia, where prenatal testing is largely unavailable (e.g., in Tamil Nadu or Rajasthan in India), female infanticide continues. Unwanted daughters may be dispatched either the traditional way (by smearing opium on the mother's nipples or by poisoning with plant extracts) or the "modern" way—denying a daughter breast milk, so that she dies of unavoidable (and unprosecutable) "natural" causes.[22]

How Much Say Do Mothers Have?

How could a mother, a woman herself, kill a daughter *because* her baby is female? To discriminate on these grounds would seem to validate her own

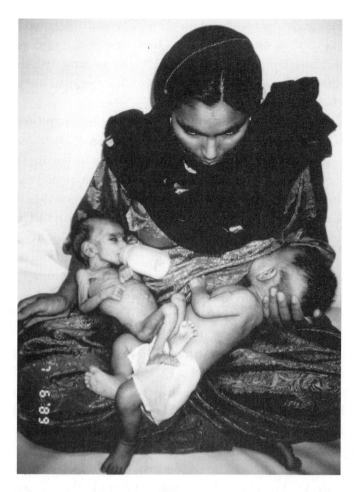

Fig. 13.2 At birth the female twin was taken by the mother-in-law and bottle-fed, while the son remained with his mother and was breast-fed. When they were reunited at five months of age at a clinic, the difference between them was all too apparent. Intervention was too late. The bottle-fed daughter, limp and marasmic, died shortly after this photograph was taken.[23] *(Photograph by Gul Nayyer Rehman, courtesy of Dr. Mushtaq A. Khan, Children's Hospital, Islamabad)*

inferiority. It is interesting to note that in places like China and Bangladesh daughters are most at risk in families that already have one or more daughters—in precisely those families where the mother has already nursed a daughter. She can remember what it was like to love a baby girl. It is hard to believe, yet maternal compliance with daughter infanticide cannot be understood without taking into account her situation.[24] She lives with her husband, among his relatives, dependent upon them. The well-being of the children she rears will rely on their good will. Quite simply, the men of the family wanted sons, therefore so did the women. From an early age, these women were conditioned to place their hopes on sons they would bear one day, and on the sons born to their sons.[25]

Even today, in many societies mothers without sons are pitied and looked down upon. Wives with sons are more highly prized, their male children favored. "Soon after I delivered my son, my parents-in-law moved us into a larger apartment," recalled one Korean woman who happened to be a winner in this chromosomal lottery. Such pervasive conditioning makes it impossible to view maternal preferences separately from her husband's family's interests.[26] One modern mother who disapproved of this prejudice nevertheless voluntarily opted for an (illegal) abortion when told that her second child would be another daughter. She knew girls were scarce, but after agonizing, still chose not to bear another daughter. Often, the matter is literally taken out of a mother's hands, as in the case of the Pakistani twins where the mother-in-law bottle-feeds the daughter, consigning the girl (but not breast-fed son) to die of dysentery and malnutrition (see figure 13.2).

Such cases led anthropologist Susan Scrimshaw to argue in a much cited passage that "the decline of infanticide may result in more suffering for older infants and children, and even adults, than when an infant's fate, be it life or death, was determined swiftly, early and irrevocably." Scrimshaw was not advocating infanticide. Rather, she was making a realistic and compassionate comparison between one fate and a "far crueler" alternative.[27] Similar logic leads many educated people in Asia—including medical personnel—to view sex-selective abortion not only as a family's right but as preferable to un-wanted births.[28]

Reasons for Preferring Sons

"Daughters are no better than crows" observes a Tibetan proverb. Variations on this theme can be heard throughout northern India. "Their parents feed them and when they get their wings, they fly away."[29] Daughters, people complain, leave at marriage; resources devoted to rearing them are lost to the patriline. With them depart substantial dowries, enriching their husbands' families while impoverishing their own. Parents dread the prospect of marrying off several daughters almost (but not quite) as much as they dread potential disgrace should a daughter fail to marry into a family of appropriate status, or be seduced and left pregnant but unmarried.

By itself, the "daughters depart" rationale begs the question of why the system is set up this way, with sons staying, daughters decamping. Nor does it explain *why* parents voluntarily fork over exorbitant dowries. Attention, then, gravitates to the traditional rationales for son preference, explanations of "pride and purse," sons' special labor value, the ritual role accorded to sons, and their symbolic value.

In one of the few studies of its kind, Mead Caine of the Population Council of New York quantified the value of labor provided by sons as compared with daughters in Bangladesh. By ten to thirteen years of age, a boy is a net producer. By age fifteen a son has repaid his parents for what it cost them to rear him, and by age twenty-one repaid them for one sister as well. Daughters, by contrast, though they work early and hard, leave home before they repay parental outlays.[30]

By themselves, neither family "pride" nor "purse" (economic interests) explains why sons earn more in the first place, why parents continue to favor them, or why parents send daughters away with large dowries.

The Reproductive Potential of Sons

A long history of male-male competition for mates has left a sexually selected legacy of men who are somewhat larger, and much more muscular, than women. This is one reason men make more effective allies than women. The other is that in patrilocal breeding systems, these allies will also be kin. Whether protecting access to females in the community or helping to maintain a patriline's rights over sources of production, males have greater "resource holding ability." This fact of life is not lost on parents in parts of the world where possession has long been ten-tenths of what law there is, and where resources have been inseparably linked to family survival through time. Where "sons are guns" (an old Rajasthani saying), the alternative to passing property to sons who can defend it against competing lineages is to lose control of a legacy.

In patriarchal social systems, a wealthy son finds himself in control of productive resources that women need. He will be in a position to attract multiple mates. In a stratified society such as Rajasthan's, families seeking social advancement compete among themselves to amass a dowry large enough to secure a place for their daughter in an elite household. This brings a prestigious alliance for parents along with the prospect of well-endowed grandsons. Should calamity strike, it is the only prospect for descendants surviving at all. Thus does son preference among elites lead to hypergamy, the custom by which women marry men of higher status. At the top of the hierarchy, however, hypergamy dooms daughters. There is no higher-ranking family for them to marry into.[31]

Selective elimination of daughters first attracted attention in the West during the years of the British Raj. Nineteenth-century travelers visiting Rajasthan and Uttar Pradesh in northern India remarked on the rarity of seeing girls among any of the elite clans. It was assumed that as part of purdah the daughters of these proud descendants of warrior-kings were kept in seclusion. "I have been nearly four years in India and never beheld any women but those in attendance as servants in European families, the low caste wives of petty shopkeepers and [dancing] women," wrote Fanny Parks in her 1850 travelogue through northern India. It did not occur to the observer that *there were no daughters.*[32]

Bit by bit, the light dawned. One British official stumbled on the phenomenon of missing daughters while engaged in negotiations with local landowners. He mistakenly referred to one of these mustachioed men as the son-in-law of the other, evoking sarcastic laughter. This was scarcely possible, they told him. The birth of a daughter would be such a calamity to families of their rank that she would never survive. It was unthinkable that *any* of *their* daughters would reach marriageable age. Among the most elite clans such as the Jhareja Rajputs and the Bedi Sikhs—known locally as the *Kuri Mar,* or "daughter destroyers"—censuses confirmed the near total absence of daughters; lesser elites killed only later born daughters. Overall, including lower-ranking clans who kept some or all daughters, sex ratios in the region were as high as 400 little boys surviving for every 100 girls.

Public outrage against infanticide among nineteenth-century Britons back home led to the anti-infanticide laws of the 1870s. British colonial legislation reduced infanticide but did little to alleviate the often lethal neglect of girls who survived. When a nineteenth-century British official asked a landholder from Uttar Pradesh why the majority of Rajput families continued to eliminate their daughters in spite of British laws against it, his reply was to the point: "The father who preserves a daughter will never live to see her suitably

married, or [else] . . . the family into which she does marry will perish and
be ruined." The man then went on to itemize specific cases confirming his
point that "those who preserve their daughters never prosper" and end by losing their land.[33]

In a world fraught with ecological peril, recurring droughts, famines, and
warfare, the best hope for long-term persistence of a lineage was concentration of resources in a strong, well-situated male heir with several wives or
concubines. If family circumstances make this tactic doubtful, a daughter or
two provide insurance against total extinction of the family line. If a family is
truly wretched, the best it can hope for is that daughters will be able, as
slaves, wives, or concubines, to move up the social scale into positions where
their children might possibly survive. Such systems did not originate because
men sought to sire as many offspring as possible, although many did. Rather,
the goal—both subliminal and consciously stated—was to ensure that at least
some of their own lineage, "honor" and advantages intact, were represented
in subsequent generations. Ultimately, this conservative course tended to
prevent local extinction of the family, and in that way was correlated long-
term with lineage survival.

From turbaned warriors on the dusty plains of Rajasthan to modern
urbanites, we are endlessly fascinated by how families fare over time. Witness
the worldwide popularity of such TV programs as *Dallas, Falcon Crest,* and
Dynasty. Whether it is a television family or their own, people are easily
drawn in. They want to know how different characters will fare in the high-
stakes game of marriage, reproduction, and maintaining access to resources.
Who will survive and prevail? Who succumb? People discuss such matters ad
nauseam. Voyeurs and gossips weigh the merits of alternative solutions to
each family's posterity problems. We are a species obsessed by strategies of
heirship, and superbly equipped to devise them.[34]

In nineteenth-century Rajasthan, where periodic droughts and famines
were a certainty, survival of family lines required extreme measures. Heart-
less? Definitely. And ruthless. But prevailing rules for deciding which sex offspring will contribute most to family ends were devised over generations.
Outcomes of successive trial and error, observation of the trials of others,
imitation of those who succeed—these became codified as preferences
for particular family systems. Adaptive solutions were retained as custom
because families that followed these rules survived and prospered.[35]

Ideology Alone Cannot Explain Sex Preferences

Sex preferences obviously have a lot to do with ideologies. Yet if genetic survival of family lines is at stake, evolutionists would expect a biological basis for the underlying emotions. They would also expect parents in other species to bias investment by sex as well.

Animals have no traffic with symbolism, gender constructs, or concepts like "old age insurance." Hence it is sobering to discover that humans are *not* the only creatures shaping offspring sets to achieve particular compositions. When they can, many animal mothers bias sex ratios prior to conception, selectively abort fetuses, and differentially nurture sons and daughters. Humans are merely the only animals to do so consciously and to articulate reasons for their biases. Only the mechanisms differ. As Aldous Huxley put it: "Ends are ape-chosen; only the means are man's."

One difficulty with research on sex ratios is that unless the sample sizes are very large, it can be devilishly difficult to be sure that small fluctuations in the proportion of sons and daughters are not due to chance. Under a range of circumstances, birds, fish, reptiles, and mammals invest differentially in daughters and sons. It is instructive to take a closer look at the pattern of sex-biased parental investment in these other animals before returning to the question of why humans bias sex ratios as they do, using techniques as crude, cruel, and wasteful as many routinely do. Far more efficient mechanisms for biasing sex ratios prior to birth are evolutionarily feasible. Fig wasp mothers, for example, evolved the capacity to custom-configure the sex ratios of their clutches. (See p. 65.) Somehow assessing which sex offspring will be reproductively most advantageous, the mother adds or withholds Y-bearing sperm as she lays each egg. When William Hamilton published his 1967 paper "Extraordinary Sex Ratios," he launched one of the wildest and woolliest pursuits within evolutionary biology, known as "sex ratio theory."

More "Extraordinary Sex Ratios"

In turtles, alligators, crocodiles, and many fish, a mother's task is simple. Sex is not predetermined when the egg is laid but gradually crystallizes during embryonic development, determined by temperature or other environmental conditions. A mother American alligator, for example, ensures that most of her eggs hatch female simply by locating her nest in a sunny spot. If, on the other hand, she clambers ashore and lays her eggs in a shady part of the beach, her eggs develop into males. In the case of some fish, like Atlantic silversides, the adaptive rationale for environmentally determined sex seems clear-cut. Fry released into the cool waters at the outset of each breeding season are always female, while those born later, after the water has warmed, are mostly male. In a world where big mothers will be more fecund ones, "his" and "her" time-sharing of the birth season means daughters born early have more time to grow big before they lay eggs. Biasing sex ratios in mammals is more complicated, and less well understood.

Skews in secondary sex ratios have been documented in mammals, but deviations from 50-50 are rarely so pronounced as in fish or wasps—with one notable exception, wood lemmings. These denizens of fir forests in Northern Europe have the most skewed sex ratios of any mammal known. Wood lemming mothers produce three to four times as many daughters as sons. Their secret is a curious alteration on the sex chromosomes that causes genes carried on the Y chromosome to remain unexpressed. In humans and other mammals, a female with just one X chromosome (denoted "XO") would not be fertile, but for some reason, these "XY" lemmings exhibit female phenotypes and are fertile.

Just why such a capacity evolved is not known. Zoologist Nils Stenseth suggests that manipulative lemming mothers have adapted to reproductive cycles characterized by an inbreeding phase. Wood lemming sons in the past confronted the same local competition for mates that wasp sons confront when circumstances force them to breed with their sisters inside a fig. This very "dominant" X chromosome allows XY sons to be transformed into daughters.

Like other small arctic mammals, lemmings are prone to excesses, population explosions followed by population busts. In bad years, the population may crash. A pregnant female lucky enough to survive would find herself alone in a lemmingless land, with no females for her sons to mate with. What better tactic at that point than producing only as many sons as needed to fertilize daughters—like Hamilton's fig wasps. Her grandchildren will move out to colonize a new wide-open niche.[36]

Wood lemmings are the only mammals with chromosomal sex determination known to bias sex ratios to such hymenopteran extremes. But more modest sex-ratio biases are widely documented, including spontaneous abortions that are sex selective—a surprise, perhaps, to anyone who assumes that abortions are unnatural.

Sex-Selective Abortion in Animals

Few funding agencies are interested in spending money to find out for sure if other animals bias their sex ratios. Fewer still feel compelled to study spontaneous abortion of daughters in aquatic rodents. Fortunately, though, governments are very interested in eliminating introduced pests. This is why Britain's Ministry of Agriculture set up a massive program to trap coypu, large (10-kg) guinea pig–like animals brought to Europe from South America for fur breeding. When some of the coypu (also known as nutria) escaped, they proved as footloose as they were furry and prolific, spreading like kudzu weed across the marshlands of eastern England.

The coypu trapped in this pest-control program provided the first opportunity to test sex-ratio theory in free-ranging mammals. Hired by the government to eliminate his quarry, biologist Morris Gosling decided to inspect their innards in the process. He dissected 5,853 coypu. Of these, 1,485 had embryos old enough to count and to sex. Examining them, Gosling made the first of several startling discoveries.

Prior to fourteen weeks of pregnancy, a coypu uterus was as likely to contain a mostly male as a mostly female litter. Later in gestation, however, it was hard to find a mother pregnant with a small litter (of four embryos or less) that was anything other than mostly male. The only plausible explanation was that females carrying small, mostly female litters were spontaneously aborting them. Surprisingly, the fattest females in best condition were the most prone to do so. Were these abortions really reproductive failures, Gosling wondered, or were they adaptive maternal management?

Et Tu, Coypu?

Late in her pregnancy, after week fourteen of her nineteen-week gestation, a particularly fat coypu carrying the "wrong" type of litter for her circumstances spontaneously aborts. By this point, she would have laid down fat stores needed to tide her through lactation. So what can a fat female at this stage gain by bailing out so late? What she gains is the opportunity to do even better reproductively in her next pregnancy. Instead of squandering a somatic windfall on a handful of daughters, she aborts and quickly conceives again— possibly conceiving a mostly male litter, or, failing that, at least a litter large enough to take advantage of being in such fine fettle.

"Abortion might be advantageous," Gosling reasoned, if "it allows the female to transfer resources to a litter that is likely to achieve higher RS [reproductive success]." Females in good condition who find themselves pregnant with small, mostly male litters, can count on producing especially large-bodied, competitive sons. But fat females pregnant with mostly daughters reap no special reward. Pregnant coypu are somehow assessing their own condition, and aborting or continuing with pregnancies accordingly.[37]

It was an astonishing observation, but not totally unforeseen. Just over a decade earlier, in 1973, two graduate students, biologist Robert Trivers and mathematician Dan Willard, had published a paper in *Science* predicting Gosling's result.

Custom-Made Families

The Trivers-Willard hypothesis states that wherever variation in reproductive success is greater for one sex than for the other, and where the reproductive success of individuals of that sex depends on maternal effects, then mothers in good condition should favor the sex with the greatest variance in reproductive success. Mothers in poor condition should favor the sex with the least. Under most circumstances, the sex with the greatest variance in reproductive success, and the one that benefits most from maternal advantages, is sons. This is why in a species like coypu, mothers in good condition should theoretically prefer sons (or else a very large litter), while those in poor condition should prefer daughters. Just *how* mothers might do this is a mystery. Some have speculated that sex ratios are biased prior to conception by different hormonal conditions in the mother and differential survival of X- and Y-bearing sperm inside the mother en route to the egg, or else through differential survival of male and female embryos.[38]

In devising the theory, Trivers and Willard actually had large mammals, like deer or caribou, in mind. A male deer whose mother was healthy and well fed would grow into a particularly large and competitive stag, able to out-compete and exclude rivals born to mothers in poor condition. The mother of a noncompetitive son would be better off producing a daughter: even a hind in poor condition should be able to conceive and pull through at least some offspring.

Today, the logic of Trivers-Willard has been found to predict sex ratios at birth among animals ranging from the noble red deer of Scotland to pudgy possums ambling about on the forest floor of Central America, not to mention footloose coypu everywhere. The hypothesis even explains the near complete specialization in daughters by low-ranking spider monkey mothers in

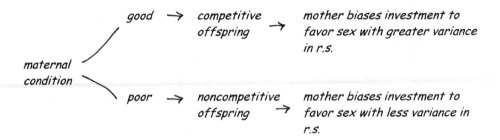

Fig. 13.3 The Trivers-Willard hypothesis seemed to explain sex-biased termination of pregnancy in coypus.

Peruvian rain forests. It applies when the main determinant of reproductive success is access to females.[39]

But what happens when the critical factor is not access to mates but resources? What if offspring of one sex are better than the other at protecting a territory, or converting its resources into reproductive success? What if offspring of one sex do more than the other sex to enhance or protect the value of local resources at the parents' disposal—known as "local resource enhancement"?[40]

Today the classic demonstration of biased production of daughters or sons depending on which sex most enhances the value of parental resources derives from a remarkable study of a rare species of bird known as the Seychelles warbler.

The Seychelles are a motley group of islands in the western Indian ocean, some rocky and waterless, others lushly tropical. These islands provide the natural "laboratory" that permitted Dutch ornithologist Jan Komdeur to prove conclusively that bird parents adjust sex ratios, producing offspring of the sex most likely to enhance the family's situation, depending on the circumstances prevailing when they hatch. These birds clinch the case that animals can custom-tailor their families.[41]

Up until 1988, the entire world population of Seychelles warblers was confined to a single island. Three hundred and twenty perky, white-chested little birds the color of cinnamon toast had saturated locally available habitats on that island. Breeding pairs were spread out in territories, where they remained for up to nine years, producing a clutch once a year, usually just one egg per clutch. Although warblers can breed in their first year, daughters remained where they were born, helping parents catch insects to feed younger siblings. When this allomother was removed, reproductive success of the parents went down. But there is a catch.

If insects are scarce, having helpers around who compete with their parents for sustenance is more a liability than an asset. In line with this calculus, parents on poor territories who do not benefit from having helpers produce mostly sons (who are not inclined to stick around). Noticing this, the researchers decided to experiment. Parents were transplanted to new territories under controlled conditions.

Warbler pairs placed on food-rich, wide-open territories could presumably afford "au pairs." As predicted, 87 percent of these privileged parents produced daughters, the sex most inclined to stay and help out. Of parents placed on poor territories, only 23 percent had daughters fledge. How? Not known. One possibility might be that the birds use some sort of "starting rule"—incubating eggs of the "right sex" but abandoning nests containing the wrong sex and starting over. This much is certain: Seychelles warblers are adaptively configuring offspring sets in response to family history and local conditions just as surely as some human parents are. It is unlikely, however, that the mechanisms in humans are the same. Rather, there appears to have been selection on the human psyche for general decision rules that produce outcomes similar to those physiologically produced in other animals. Far from locking parents into some preordained response, however, a biological basis for these preferences should make parental attitudes toward sons versus daughters imminently changeable. By now, this claim will seem curious only to readers who still assume that evolved traits are necessarily immutable— which they are not.

When the "Rules" Themselves Are Contingent . . .

The perennial question "which sex to produce" can be mind-boggling, especially in such flexible primates as baboons and macaques, "weedy" species like humans are, readily adapting to diverse habitats. As in all the well-studied Old World cercopithecine monkeys, baboon and macaque daughters inherit rank from their mothers. Because daughters remain nearby, it behooves a high-ranking mother to produce the sex that will benefit most from her own status, as well as bolster matrilineal interests by supporting kin (another form of local resource enhancement). In habitats like Amboseli, where food is scarce, high-ranking mothers do just this—they overproduce daughters. The same pattern can also be documented for some populations of macaques.

Year after year, mothers in the highest-ranking matrilines consistently produce significantly more daughters than sons, while low-ranking females produce few daughters and more sons. Low-ranking females not only pro-duce few daughters, but such daughters as they do produce are more likely to die than are sons born to mothers of equivalently low rank. Based on captive studies of bonnet macaques, Joan Silk showed that whereas sons who depart their natal group can leave the disadvantages of their mother's low rank behind, daughters cannot. In her study, no daughter born to a low-ranking mother managed to produce a single surviving offspring. When competition for local resources is intense, a daughter born to a high-status mother is the right sex in the right place at that time.[42]

Recall that among the baboons Jeanne Altmann studied at Amboseli, infants have a 25-percent chance of dying during each of the first two years of life. But if that baby is a daughter born to a high-ranking mother—the "right sex"—the baby's survival chances go up twofold, and are higher than survival chances for a son born to a mother of the same status. Such daughters also breed sooner. On average, mothers who get the sex right contribute an extra half-grand-offspring to the next generation. Mothers at Amboseli produce no more than seven offspring in their lives, of which on average only two sur-vive. Given how little these baboon mothers have to show for a lifetime spent producing and carrying babies, such bonuses add up.

Generation after generation, cumulative reproductive advantages mean that mothers in these matrilineal systems compete for more enduring stakes than the isolated copulations males fight over. A male who hitches his repro-ductive star to a successful matriline by siring a daughter in one, secures his ticket to posterity. Similarly, if a male's mate is a subordinate female, both parents benefit from son production. Lowborn sons, like poor country boys, strike out for distant opportunities, leaving natal disadvantages behind. But in some cercopithecine monkeys like macaques, there is another reason for sub-ordinate mothers to bias toward sons. Females from dominant matrilines maliciously harass daughters born to competing mothers, sending a not so subtle message: "We may tolerate your sons for a time, but your daughters— who will be permanent residents—are not welcome." These bullies inflict much wear and tear on low-ranking mothers, especially those carrying daughters. Silk hypothesized that such penalties imposed upon low-ranking mothers who produce daughters has led to selection on subordinate mothers to either avoid conceiving, or avoid gestating, daughters.[43]

Yet even this sophisticated calculus is not the whole story. When environmental conditions change, the mother macaque or baboon pulls out a new rule book.

. . . and the "Wrong" Sex Shall Become the "Right" One

Year after year the evidence grows stronger that in habitats like Amboseli, daughters are a liability to low-ranking mothers. Sons offer the best prospects. Yet researchers elsewhere document different patterns. In some baboon and macaque populations, no effect of maternal rank on sex ratios is found at all. Others exhibit the mirror-image of the Amboseli pattern, with high-ranking mothers overproducing sons, low-ranking ones daughters, just as Trivers and Willard predicted.[44]

Different teams of researchers were reporting different patterns, each group suspecting the others must be getting it wrong. Those who found no statistically significant differences assumed that the other two groups were infected by "sex-ratio fever" and in their theoretical delirium were imagining patterns in what was only random variation.

In 1991, Carel van Schaik and I were among the primatologists swept up in what we jokingly referred to as the "wild, wild world of sex-ratio research." What if, we wondered, the researchers weren't wrong. What if the monkeys were changing the rules? We noticed, for example, that it was the macaque and baboon populations from wide-open habitats with plenty of food and room for expansion that were least likely to conform to the "Amboseli pattern." Outright reversals of that pattern (with high-ranking mothers overproducing sons, low-ranking ones daughters) were most often reported in large outdoor breeding colonies where the combination of ample food and space contributed to very high birth rates. (Producing babies, after all, is what breeding colonies are for.) This is when it occurred to us that under ecological conditions conducive to rapid population growth, the differences in male and female reproductive potential so critical to the logic of the Trivers-Willard hypothesis become relevant. At this point, a mother's determination of the optimal sex of offspring for her circumstances does a flip-flop.

We reasoned that in rapidly expanding populations, where both high- and low-ranking females can successfully breed, the greater reproductive potential of sons born to mothers in good condition takes priority over the enduring value of advantageous maternal rank. Under the arduous conditions at

Amboseli, matrilineal access to scarce resources is the mother's top priority. But in high-growth populations, monkey mothers march to a different drummer, depending on whether the most important factor limiting the breeding success of their offspring will be access to resources or access to mates.

What Keeps Human Sex Ratios Nearly Equal?

The existence in animals, especially other primates, of heretofore undreamed-of capacities to adjust their production of sons versus daughters in adaptive ways raises an awkward question. Given long-standing biases in favor of a particular sex, why hasn't natural selection led to subsets of human mothers who adjust to variable local conditions by automatically producing the desired sex? The existing system is not only cruel (which is not relevant to Mother Nature), it is wasteful (which *is*).

If it is possible for selection to act on mothers to bias sex ratios at birth, why stop with a paltry six extra sons per hundred daughters? In populations where from time immemorial parents have discriminated against daughters, why don't we see sex ratios at birth in the vicinity of 200:100? This would save parents much wasted effort: all the energy, the opportunity costs, the time and risk of a pregnancy to produce a baby her parents won't even keep. Why, then, are human sex ratios at birth so nearly equal, roughly 51 percent male, 49 percent female?

When biologists are asked why sex ratios consistently hover conservatively close to parity, more likely than not they will invoke "Fisher's principle of the sex ratio." This time-honored axiom of population genetics explains why roughly equal numbers of the two sexes are produced among so many species of birds and mammals. In the 1930s, British biostatistician Sir Ronald Fisher reasoned that so long as producing sons costs the same as producing daughters; and so long as outbreeding prevails (that is, brothers don't breed with sisters, as fig wasps do); and so long as all individuals have roughly the same opportunity to breed (a big *if,* as it turns out); then parents should allocate equal investment in sons and daughters.

Imagine a population in which certain parents specialize in one or the other sex. Let's say most mothers produced sons. As offspring mature, they will breed in a lopsided world, top-heavy with males. Too bad for the sex in excess. Although every scarce female will get to breed, only a random subset of males will manage to. Too bad also for the parents that overproduced sons, because, on average, son-producers will be penalized by having fewer grand-children. The mother lucky enough to produce daughters, on the other hand, will be rewarded by disproportionately more grandchildren—at least tem-porarily.

Over time, natural selection should favor parents that produce the rare sex, with the predictable outcome—a glut of daughters. Once again, the sex ratio should gravitate back to favor son-producers. And so it goes, the pendulum swinging first one direction, then the other, favoring first daughter-producers, then son-specialists. The outcome, according to Fisher, is a population with more or less equal numbers of sons and daughters.[45]

Fisher's principle is the conventional explanation for why wildly skewed sex ratios evolve only under special conditions. But such special conditions turn out to be not so unusual. Supposedly, Fisher's principle explains why most human sex ratios are only as mildly skewed as they are. The reasons slightly more sons are born on average is that males are more vulnerable (both in utero and in infancy) to dying before the end of parental investment; thus, by producing slightly more of them, parents are merely equalizing investment in sons and daughters.

Yet other animals—baboons and macaques, for example—deviate from Fisherian equality when one or the other sex costs less or provides a bigger reproductive payoff. Why don't humans?

It is possible, of course, that the phenomenon occurs but has somehow gone undetected. For example, if parents biasing toward sons were lumped with those biasing toward daughters, the average sex ratio would come out 50-50. Indeed, deviations from the expected, approximately equal, human sex ratios at birth are sometimes noted.[46] Occasionally groups surface with spectacularly high sex ratios that cannot be attributed to differential neglect or infanticide. These may (or may not) have to do with customs that affect the timing of conception.[47] Furthermore, every so often geneticists stumble on a rare pedigree, such as the English family that for ten generations produced daughters in 32 of 35 births, or the French family that produced exclusively daughters (72 of them) over three generations. Yet these could be explained as chance occurrences.

Massive screening has unearthed only a handful of deviant cases, and none so extreme, nor so precisely calibrated to reproductive possibilities, as the wildly biased sex ratios readily located among wasps, wood lemmings, warblers, and spider monkeys. Overall, deviations from the standard human sex ratio of 102 to 106 boys per 100 girls are rare, prompting a puzzled commentary from George Williams, who wrote in a now famous passage:

> I find it rather mysterious that adaptive control of progeny seems not to have evolved. [That this matter is left to chance] seems to contradict evolutionary theory. . . . Instead, deviations from random sex determination are trivial at best.[48]

There could be an error in sampling, or else the theory could be wrong. Or human sex ratios might be adaptive all right, but parents postpone their adjustments in parental investment until after birth—as many Asian families have long done. Much evidence points to this third possibility. Humans confront the same posterity problems other animals do, but resolve them differently. Instead of innate mechanisms that bias production of sons or daughters at conception (as in wasps), or differential retention of mostly female litters (as in the coypu), human mothers consciously choose sons and daughters *after birth,* in line with parental evaluation of what the repercussions will be for long-term family goals. The underlying psychology—although not the outcomes—are probably similar when modern American parents make choices about how much money to spend on toys for their children, or certain medications, like growth hormones.

Each year in the United States, parents spend 60 percent more on toys for boys (Legos and G.I. Joes) than on toys for girls (disproportionately dolls). Parents are twice as likely to treat a growth hormone inadequacy in a son as in a daughter. Part of their calculation is surely not just whether they wish to invest more in sons than daughters, but which sex they feel will benefit more from the intervention. Height, to take one example, is a far more important predictor of success (including salaries and marriage options) for sons than it is for daughters.[49]

Reassessing the Rajput Case

No research on biased sex ratios in birds or mammals had been done when anthropologist Mildred Dickemann first encountered the logic laid out by Trivers and Willard in their 1973 paper. Social scientists at that time paid scant attention to the idea that there might be innate human predispositions that enhanced inclusive fitness and the long-term survival of family lines.

Devaluation of daughters was viewed as a purely cultural construct. It was assumed to be the outcome of free-floating minds spinning infinitely variable webs of meaning out of locally received traditions.[50]

As far as cultural anthropologists were concerned, the ideology of son preference along with the custom of paying dowries to marry off daughters sufficed to explain female infanticide. What other reasons could there be? Yet Dickemann was struck by how well the patterning of son preference in the north Indian case conformed to predictions of an evolutionary model that applied to animals generally.

Trivers and Willard proposed that parents in good condition should prefer sons, those that were disadvantaged, daughters. They even specified that this logic would be found in socially stratified human societies, where women marry up the social scale, whenever the "reproductive success of a male at the upper end of the scale exceeds his sister's, while that of a female at the lower end of the scale exceeds her brother's. A tendency for the female to marry a male whose socioeconomic status is higher than hers will, other things being equal, tend to bring about such a correlation." Trivers and Willard's logic even explained the most puzzling feature of daughter slaying in the Rajput case—why the most elite families were the most likely to kill half of their off-spring. By contrast, sub-elites were left paying exorbitant dowries to place daughters in one of these elite households, impoverishing their sons in the process. The poorest subcastes, who really did not have enough resources to feed their children, were the ones who welcomed daughters and did not kill them.[51] None of this made sense unless one accepted the assumption that parents were not counting offspring but looking further down the line, toward grandchildren and beyond, toward the survival of a family line.

Reversals of Fortune Leaving Daughters Preferred

Eliminating daughters at the top of the hierarchy produces a vacuum sucking up marriageable girls from below, and creating a shortage at the bottom. Families don't pay dowries to place daughters in families with the same or lower status than their own. They demand payment for them instead. At the bottom of the heap, sons whose families cannot cough up the required *bride-price* remain celibate. Far from calamities, daughters are the most valuable commodity low-status families possess.[52]

Referring to a daughter as a commodity will strike many as extraordinarily callous. But we are not talking about postindustrial Western populations that for generations have lived in an unprecedented state of ecological release, freed from concern about famines. Continued survival of such parents and their children rarely depends on choices mothers make about how much food to allocate to one child versus another. But not all mothers are so fortunate. Daughters not only offered the only prospect for upward mobility, in many cases they provided the only possibility at all of continued survival of a family line.

In parts of the world where drought and famine are recurring hazards, the landless and dispossessed invariably have the worst chance of making it through. Under such harsh circumstances the likeliest survivors will be offspring of mothers who marry into families with access to resources, like arable land.[53] Hypergamy (girls marrying up) is not a fluke. It was a long-standing necessity for lineage survival. Nor can it be denied that decisions leading to it have genetic outcomes.

Centuries of hypergamous mating have left a trail of genetic markers, like breadcrumbs through the forest of the Indian caste system, documenting the different paths followed by the two sexes as they married and produced offspring. An examination of genetic traits carried in mitochondrial DNA (DNA found in somatic and egg cells but not in sperm), which is transmitted only from mother to offspring, showed that these mother-transmitted traits are spread widely beyond traditional caste boundaries. For centuries, they have been carried by brides and concubines moving up in the world by marrying into higher-caste families. By contrast, paternally transmitted markers, traits passed from father to son on the Y chromosome, are less mobile. Father-transmitted traits remain localized, rarely spreading beyond the caste where they originated.[54] This may be one reason why male traits are more vulnerable to extinction than those carried by mothers. Thus do customs previously viewed as purely cultural have profound demographic and genetic consequences, as well as deep roots in human motivations and their decision rules regarding children.

Human Nature and Human History

The earliest evidence for sex-biased infanticide derives from the DNA of baby skeletons—all less than two days old and without apparent defect—excavated from the sewer of an ancient brothel in Roman Ashkelon on the southern coast of modern Israel. Fourteen of the nineteen victims of what archaeologists suspected was infanticide were male. If their mothers were indeed prostitutes, one assumes they came from the lowest rank of society: daughters but not sons of these women would have value. A preference for sons among elites mirrored by a preference for daughters among the dispossessed is a pattern that still persists. Daughter preference can still be documented today among Hungarian Gypsies and other disadvantaged groups. Consider what happened with the late 1980s fall of communism. Across eastern Europe, economies and social services were disrupted, leading to an increase in both misery and unwanted pregnancies. Not surprisingly, the incidence of neonaticide has increased, but with an unusual twist. Prior to 1990, sons and daughters were about equally likely to be killed. After 1990, Slovakian researcher Peter Sykora documents that the victims are disproportionately male—21 of 27 in the neonaticides in his sample.[55]

Large chunks of Western history can be understood only by paying attention to such patterns. Human fates can be read as artifacts of differential treatment of offspring by their parents. Which sons inherited land and continued dynasties, which departed instead to colonize new worlds. Which offspring were predestined to live out their lives in monasteries (or in convents), which daughters were dowered and sent off to distant kingdoms. Nowhere is this point better made than in the writing of the archaeologist and social historian James Boone.

Using medieval Portuguese genealogies, Boone traced the fates of sons and daughters among both the elites—royalty and landed gentry—and those who served them, bureaucrats and soldiers, over a two-hundred-year period (from 1380 to 1580). Dukes and counts at the highest social ranks left more surviving legitimate offspring (4.7 offspring on average, with no reliable counts for illegitimates) than did cavaleiros and military men below them (2.3 legitimate children on average). For both sexes, firstborn offspring fared better. Later-born sons fought in the Crusades farther from home, stayed away longer, and were more likely to die in far-off places like India than firstborn sons, who often went no farther than Morocco and soon returned to marry and take over family holdings.

Redundant daughters were similarly banished, not to distant lands, but to convents. Italian novelist Alessandro Manzoni provided an apt description of this predestined claustration in his description of the proud Milanese patriarch who "destined all the younger children of either sex to the cloister, so as to leave the family fortune intact for the eldest son, whose function it was to perpetuate the family." This practice brought great unhappiness to younger offspring of both sexes.

In his 1827 epic *I promessi sposi* (The Betrothed), Manzoni sums up the plight of a later-born daughter:

> Still hidden in her mother's womb . . . her state in life had already been irrevocably settled. All that remained to be decided was whether it was to be that of a monk or a nun, a decision for which her presence but not her consent was required.[56]

Among Boone's medieval Portuguese, between 10 and 40 percent of daughters at any given time were cloistered in convents. Elite daughters who married produced an average of 3.7 children, about the same as the number (3.3) of surviving children left by sub-elite women, many of whom had moved up the social scale when they married. Overall, the reproductive success of daughters born to lower-ranking families was higher than that of their brothers, while at the top of the hierarchy—as among the north Indian Rajputs—the reverse was true. When Boone fed these data into his computer to simulate how this situation would play out through time, elites produced significantly more grandchildren in the third generation through sons than daughters, while lower ranks did better with daughters than sons.[57]

Among "The Despised Ones"

It was reversals of fortune such as these that attracted the notice of anthropologist Lee Cronk when he went to Kenya to study the Mukogodo. Cronk's study is unusual, because he specifically focused on those on the lowest rungs of the local ladder.

The Mukogodo are former foragers pressured by economic necessity to attach themselves as a disadvantaged "subcaste" to Masai pastoralists, adopting Masai language and values but never achieving equal status. Locally, the name Mukogodo means "the despised ones" or, more literally, "poor scum."

As is typical among pastoralists, the Masai prefer sons. The Mukogodo, who emulate them, *claim* to as well. But the actual behavior of Mukogodo mothers and the sex ratio of their offspring (there are about 67 little boys for every 100 girls) tell a different story. Mukogodo mothers breast-feed daughters longer than sons, and are more inclined to pay to take a sick daughter than a sick son to the medical clinic. Partly for this reason, daughters are healthier and more likely to survive than sons.

Out of this strange union of two cultures has emerged a hypergamous marriage system structured along the lines Dickemann identified in the rigidly stratified clans of precolonial Rajasthan: women flow up the hierarchy, with daughters preferred over sons at the bottom. Because so many Mukogodo women become primary or secondary wives to Masai up the social scale, many Mukogodo men, with smaller herds of livestock to draw on for bride-price, have difficulty obtaining wives at all. With so many Mukogodo sons growing old wifeless, their average completed fertility is below that of the average Mukogodo daughter.[58] It is not possible to know for sure which mothers value more, the material benefits daughters bring, counted in livestock, or the grandchildren; but my guess is that over evolutionary time the two were so intertwined as to make them inseparable so far as a mother's internalized preferences for different offspring are concerned.

Economics of Daughter Preference

Outright daughter preference is unusual, but not necessarily confined to the disadvantaged. Among the matrilineal Tonga people of southern Zaire, daughters are essential for perpetuating the *basimukoa,* or matrilineage. The more prosperous the matrilineage, the more pressure to bear daughters. Not surprisingly, there are *two* cries of joy at the birth of a baby girl, only one for a boy. Too many sons, and the mother comes in for criticism from kin. More to the point, males die in childhood at far higher rates than females. Of recorded births, only 92 boys are reported for every 100 girls. When twins of mixed sexes are born, the boy is neglected, and more likely to die. Same-sex male twins die at five times the rate of singletons, suggesting that parents do not go out of their way to keep them alive.[59]

Sometimes parents, even those living in areas with a long tradition of son preference, come to prefer daughters because women have found a special economic niche for themselves. This is the case with daughters born on the island of Cheju Do, off the coast of South Korea. Cheju Do is renowned for its women abalone divers, called *haeyno.* Because this occupation is relatively well paid, daughters provide more security than sons. When a woman on Cheju Do learns she is pregnant, she prays for a girl.[60] Financial independence of these women has also led to the highest divorce rate in Korea. In this respect, Cheju Do has come to resemble some Western countries where families are in transition between long-standing patriarchal traditions and brave new worlds where legal protections along with economic opportunities for women mean they can afford to survive and rear a family with or without a male provider.

Fine-Tuning Family Configurations

Parents can be remarkably specific in their requirements for certain offspring sets. There are time-honored traditions specifying which sex stays or inherits, which leaves empty-handed or marries out with a dowry instead, which child lingers on as a celibate spinster baby-sitting for the designated heir.

In many areas of Asia, the ideal family is composed of two to four sons and one or two daughters. Thus, it should not surprise us to occasionally encounter "missing boys" along with all the missing girls—albeit not in such vast numbers. Anthropologist G. William Skinner was among the first to predict and document just such a pattern to missing children. In his most recent study of census data from China's lower Yangtze region, Skinner and co-worker Yuan Jianhua documented 1.2 million missing girls, mostly higher-birth-order daughters, but also some 60,000 missing boys, mostly from families that already had several sons.

Culturally mediated parental preferences can play out with chilling predictability. Studies of child survival among villagers in the Punjab and in Bangladesh make it clear that it is not just daughters in these families that are at risk, but daughters with one or more older sisters. In one village in Bangladesh, such daughters have a 90-percent higher chance of dying before the end of childhood than do girls without any older sister. A boy with the bad luck of being born after two or more older brothers has a 40-percent greater chance of dying than an only son does.[61]

———

Parental commitment to offspring can depend on how nearly the child's sex and birth order conform to a desired norm. Among the first to empirically demonstrate this was sociobiologist Paul Turke, in fieldwork among Pacific islanders on Ifaluk Atoll. Daughters among these fisherfolk are more productive than sons. They also help parents to rear younger siblings more than sons do. No wonder daughters are preferred. Parents who achieve the ideal configuration, producing a daughter first and then a son, were better off and reared more surviving offspring than those whose firstborn child was male. Overall, mothers who bore a daughter early in their reproductive career had higher lifetime reproductive success than women who bore a son first.[62]

In patriarchal societies in saturated habitats, such mild preference for an initial daughter may be taken to extremes. Among eighteenth- and nineteenth-century farmers on the Nobi plain of Japan, the ideal pattern of "first a girl then a boy" has a name: *ichihime nitarô*. Sons are the preferred sex, but if they can, parents arrange things so as to have a little allomother on hand to help rear the primary heir, to make sure he is as healthy and good as he can be.[63] Parents were not above loading the demographic dice in an astounding gamble. Those young enough to be confident of plenty more chances to try for a "jackpot" configuration might eliminate even the much desired son if he happened to be born first, thus enhancing the odds of achieving the ideal *ichihime nitarô*. Thereafter, parents in Tokugawa Japan used infanticide to space births and—if conditions were sufficiently auspicious—to achieve as nearly as possible an ideal configuration of well-spaced, gender-balanced offspring with a fully qualified firstborn son coming of age just as his father was ready to retire.[64] Clearly, the "mania for sons" was never so simple as an across-the-board preference for male children, solvable by an across-the-board biasing of the sex ratio.

Why Humans Bias Investment *After* Birth

Humans, like other animals, use flexible "decision rules" to bias investment toward daughters under some conditions, toward sons under others. But unlike a mother wasp, who sizes up demographic prospects and then commits herself to producing mostly daughters or mostly sons, humans with very few exceptions leave the matter open until after birth. Then they evaluate contingencies like birth order, offspring quality, available assistance, even inheritance prospects. Given the importance of history and how extraordinarily flexible human breeding systems are, and how variable the environments in which they live can be, parents with innate propensities to produce one or the other sex would have been wrong as often as they were right.[65]

Where environmental conditions, marriage and residence patterns, or laws can change on short notice, the better part of evolutionary valor is to postpone irrevocable decisions till the last feasible moment. Conscious strategists constantly update information about local prospects for sons versus daughters. Chronic tensions between maternal and patrilineal interests are resolved quite differently as new subsistence opportunities open up while others close, as daughters once of no use suddenly become net assets, and so on. Unlike other creatures with pressing reasons to bias parental investment prior to birth, the sheer variability of the human condition makes that degree of precommitment ill-advised. Furthermore, unlike other mammals, the sheer duration of parental investment in the human case, and the myriad forms it can take—food, educational costs, marriage payments, inheritances—means that parents have many ways and myriad opportunities to bias investment in different offspring.

Imagine a mother in unusually good condition who somehow biased production toward a daughter in preparation for a nice matrilocal life, and then found herself captured by some warlike, patriarchal tribe where only sons were valued. Her physiologically based "decision" would have been a mistake. Far better to pursue the Fisherian course of equal investment prior to birth, and then fine-tune investment in sons and daughters after birth, responding to local cues and customs.

Biologically Based Behaviors Are Changeable

Faced with constraints, parents readily value some offspring over others. This is the bad news. The good news is that nowhere in the human psyche are specific sex preferences—such as a mania for sons—engraved in DNA. As widespread as son preference happens to be, there is nothing to indicate that it represents an innate or universal preference on the part of mothers or fathers. There is no all-purpose psychoemotional straitjacket where daughters and sons are concerned. In societies with strong patriarchal traditions, however, it may take special circumstances for daughters to become as desirable as sons—especially if parents expect to have only one child.

Sex of offspring has been a long-standing concern for Westerners, too. Even those who claim they "don't have a preference" find that they do when pressed to imagine a situation in which they will have only one child. Instead of infanticide, however, Western parents have adjusted parental investment through time by designating some sons for the church, some later-born daughters to become "spinster aunts" (the fate George Eliot's family had in mind for her). In the United States, it has been only in the last century or so that married women had rights to own property in their own name, and only since the Married Women's Property Acts passed in England and the United States in the latter part of the nineteenth century that daughters began to

Fig. 13.4 Traditionally daughters from poor and low-status families in northern India—like this woman road worker—were buffered from discrimination both by the "bride-price" they commanded at marriage and by the wages they were able to earn. In north India today, however, men increasingly fill even the meanest and most poorly paid jobs, such that discrimination against daughters is creeping down the social scale, erasing much of the protection daughters of the dispossessed once enjoyed. *(Sarah Blaffer Hrdy / Anthro-Photo)*

inherit on an equal footing with their brothers. Protected by law, daughters today are actually somewhat more likely than sons to finish college, and are beginning to have athletic and career opportunities equivalent to those long open to sons. For many one-child couples, daughters are actually the sex of choice. But these are very recent transformations, still virtually experiments—fragile ones at that—following as they do on the heels of long-standing biases favoring sons.

Fig. 13.5 Only in the past few decades have Western women in countries like the United States had educational and athletic opportunities equivalent to those available to men. These young women are collegiate varsity rowers, evidently as proud of their strength and competitiveness as of their femininity. No one knows yet how such novel social experiments will pan out. *(Courtesy of David Joffe)*

Western folklore about sex determination could fill volumes. The Greek philosopher Anaxagoras believed that the left and the right testes differed, so that by tying off the weaker (left one) a man increased the odds of a son. Aristotle recommended facing north during sexual intercourse, because he believed a cold southern wind would induce conception of daughters. For the more literal-minded, homespun recommendations for siring sons prescribed wearing boots to bed.

Not all of this is ancient history. Eschewing such folklore, New Yorkers in the 1960s turned to Dr. Landrum Shettles, who prescribed a regimen of vinegar douches to privilege X-bearing sperm, a douche of baking powder to promote the fortunes of Y-bearers. Shettles was followed in the 1980s by his West Coast counterpart, physiologist Ronald Ericsson, founder of Gametrics Ltd. of Sausalito, California. Ericsson promised parents sex selection using a special technique to separate faster Y-bearing sperm from the more sluggish X-bearing ones. He advertised his central premise with vanity license plates on his car that read "X or Y." There was even a brief period when North Americans could go to a drugstore and pick up a "Gender choice child selection kit" for $49.95, complete with thermometer and paraphernalia for monitoring vaginal mucus, to determine precisely the moment for conceiving a son or daughter. When the U.S. Food and Drug Administration decided that claims implied by pink and blue advertising on the box were not substantiated, the kits were pulled off store shelves. Today, prenatal sex testing is widely available in the West. Anyone determined to use it for sex selection can manage to do so without breaking any laws.

————

This chapter began with sex-selective infanticide in China. I am not immune to the distress this stark topic generates. All the more reason for dispassionate analysis. For humans are, above all, resourceful creatures. They do not readily abandon self-interest for the common good, or for someone else's good. Humans do not easily, and without good cause, abandon the nepotistic urges that brought us from a paltry ten thousand souls a few hundred thousand years ago to the six billion on Earth today. Philosophizing about topics like whether humans have free will (and if so, how to employ it) is far removed from the daily concerns of most humans.

It is the common humanity of the parents that is at stake here, not ethnic or cultural differences. Those who would rush to Beijing to deliver passionate diatribes would do well to maintain some historical perspective. While Chinese infanticide rates have declined dramatically since the nineteenth century, during that same period rates of child abuse, neglect, and infanticide have skyrocketed in countries like the United States, although sex of the offspring has relatively little to do with it.[66]

Infanticide in China is already illegal. Since 1987, laws against disclosing the sex of a fetus to parents who might subsequently practice sex-selective abortion make Chinese laws related to sex-selective abortion tougher than such laws in the West. It is hard to see, therefore, what sense there would be to additional legislation making preferential female infanticide or sex-selective abortion *more* illegal in China than it already is. Incentives are liable to be more effective than prohibitions. The most effective remedy may be widely available contraception for birth spacing combined with educational and employment opportunities that create attractive futures for daughters, including scholarships and job opportunities that will benefit their families. Countries convinced that mandatory birth control is essential for the long-term welfare of their people might want to consider special vouchers for daughters-only families—good for extra grandchildren.

Old Tradeoffs, New Contexts

What fury hostile to humankind
First led from Nature's path the female mind,
Her innocent sense by . . . fashion's law repressed,
And so a babe denied its mother's breast?

—Luigi Tansillo, from "La Balia," translated by William Roscoe in 1798

Throughout human history, and long before, mothers have been making tradeoffs between quality and quantity, managing reproductive effort in line with their own life stage, condition, and current circumstances. As a result, infancy has not always been the warm, safe-in-the-arms-of-love tableau many of us imagine. It was, instead, a perilous bottleneck that each individual contributor to the human gene pool had to pass through. Historical records provide ample documentation as to how tight a squeeze that sometimes was.

Of 21,000 births registered in Paris in 1780, only 5 percent of them were nursed by their own mothers. It is a riveting statistic that has come to characterize an era, France's "heyday of wet-nursing."[1] The numbers provide evidence of maternal indifference on a massive scale and today are often held up as the prime exhibit in the case against the existence of maternal instincts in the human species. But I don't think that's what they actually prove.

These much-cited numbers derive from Lieutenant-General Charles-Pierre LeNoir, a police official whose job it was to monitor the referral bureaus used by working parents to locate wet nurses. LeNoir was also responsible for investigating complaints about wet nurses who failed to live up to the terms of their contracts, as well as registering the disappearance of infants lost in the shuffle.

Of the 20,000 babies nursed by women other than their mothers, the luckiest 25 percent were born to propertied parents who placed their children directly with wet nurses. Often such elites would rely on rural tenants or other contacts to find acceptable candidates. Some of the wet nurses, nannies hired to lactate as well as caretake, would live with the family under maternal supervision. The unluckiest 25 percent of babies were delegated to foundling homes as described in chapter 12. It was up to these institutions to locate someone to feed them, if they could.

The remaining wet-nursed babies were mostly born to the middle class— artisans, shopkeepers, or traders. These were the "Bourgeoisie de Paris," but often only barely. Within this social class, a mother's salary or her unpaid labor was critical for the family's economic well-being.[2] Typically, such mothers were neither unmarried nor destitute. They relied on professional intermediaries to find wet nurses for their babies. Hence the edge to the feminist query raised by philosopher Elisabeth Badinter in *Mother Love: Myth and Reality.* If such a thing as maternal instinct exists, how could so many thousands of mothers be so unfeeling as to ship their newborns off to be suckled by an unknown woman?

"Discretionary" Distancing

Twentieth-century debates over the existence of maternal instinct focused on such "discretionary" delegation of care. It was not the desperate mothers, who arguably had no choice, that attracted notice, but the bourgeois mothers who presumably could afford to keep their babies near them—and yet did not. Greuze's painting of the farewell kiss (figure 14.1) shows what went on outside the house. In large French cities, a middleman, called a *meneur,* would pick up the newborn. What happened inside the home, or inside people's heads, was less clear. An account by an eighteenth-century Frenchwoman—a disciple of Rousseau—offers a glimpse. The writer, Jeanne-Marie Phlipon de Roland, has just visited an acquaintance who, though hopeful of a male heir, had given birth to another daughter. "Mme. D'Eu gave birth yesterday at noon to a girl," Madame Roland wrote.

> Her husband is completely ashamed of it; she is in a foul mood over it. . . . The poor baby was sucking its fingers and drinking cow's milk in a room far removed from its mother, waiting for the hired woman who was to nurse it. The father was in a great rush to have the ceremony of the baptism over, so the little creature could be sent to the village. . . .[3]

The husband seems to be deliberately structuring this situation to minimize the mother's contact with her baby—in a "room far removed." It was a

Fig. 14.1 *La privation sensible* (The Painful Deprivation) by Jean-Baptiste Greuze (1725–1805) depicts the pickup of a newborn by an itinerant entrepreneur who will transport the baby to a wet nurse in the country. Little of the vast literature on this topic deals with the psychological effects on children, yet their distress must have been on the artist's mind. The painting has two focal points: the mother kissing her baby goodbye; and, below, two saucer-eyed, fearful children.
(Courtesy of Bibliothèque Nationale, Paris)

procedure that allowed few opportunities for infant cues to elicit nurturing emotions, thus inhibiting formation of any bonds between mother and infant. Absence of maternal responses under these conditions tells us little about innate potentials of the mammal in question.

Once out the door, the baby might find her wet nurse waiting in the *meneur*'s cart, ready to hold her and feed her during the long, rough trip back to the wet nurse's rural home. Otherwise only the *meneur* would show up, leading a horse with baskets strapped to its back. Instances of babies lost along the way occasionally surfaced in police reports for Lyons and Paris. For babies who reached their destination, it was still less than certain that the woman waiting there would have sufficient milk. No wonder peasants who heard a church bell ring simply shrugged, "It's nothing, a little Parisian died!"[4]

Propaganda About Hired "Killers"

Eighteenth- and nineteenth-century authorities became increasingly concerned. They were worried about the high levels of infant mortality and population decline, as well as "public morality" (that is, they were distressed by the sight of women working outside the home). References to "natural law" and the "sacred duty" of mothers abound in testimony before committees drafting legislation on wet-nursing and infant abandonment.[5]

Reformers, who had a stake in romanticizing instinctive maternal devotion, likewise had a vested interest in identifying the use of a wet nurse with the worst possible motivations. It became convenient to lump a range of parental choices into one category—wet-nursing—and to identify a wide range of intentions under one motive: infanticide. Such propaganda was especially rife by the time France belatedly passed the Roussel Law of 1874, which was designed to protect infants from the worst excesses of wet-nursing.

Medical doctors called in to testify as expert witnesses stressed the murderous intentions of mothers who hired other women to nurse their babies. French reformer Dr. Alexander Mayer described the practice of "abandoning, a few hours after its birth, a cherished being, whose coming was ardently desired, to a coarse peasant woman whom one has never seen, whose character and morality one does not know," condemning the practice as "barbaric."[6] Parisian mothers, he contended, were sending babies off to distant wet nurses "with the desire of not seeing them again."[7]

The notion that wet-nursing must be a disguised, nonprosecutable form

Fig. 14.2 Father brings his infant to consult a *recommandaresse,* a woman who for a fee procures a wet nurse. *Le Bureau de Nourrices* (The Wet Nurse Office), Paris, 1816. *(Courtesy of Wellcome Institute Library, London)*

of infanticide, with wet nurses serving as contract killers, made effective propaganda and was quickly absorbed into common parlance. In England, *angelmaker* was common slang for wet nurse; the German equivalent was *Engelmacherin.* In France, *faiseuse d'ange* was extended to include abortionists. The underlying logic appears to be that any woman who gets pregnant and then does not carry the fetus to term, or who after birth does not care for the infant at any cost, is worse than just unnatural; she is murderous.[8] (This same attitude persists today among many who oppose reproductive choice.)

In 1865, Dr. Mayer correctly prophesized that "The whole thing is so revolting to good sense and morality that in twenty years people will refuse to believe [wet-nursing] ever happened."[9] Today, scholars who recall this era tend to follow Dr. Mayer's lead. "It must have been common knowledge," writes twentieth-century psychoanalyst Maria Piers in her book *Infanticide,* that the wet nurse the parents hired was "a professional feeder and a professional killer. . . ."[10] Wet nurses, proclaims another modern commentator, "were surrogates upon whom parents could depend for a swift demise for unwanted children."[11]

In the absence of other forms of birth control, women's maternal responses were heavily influenced by an amalgam of old and new rules. Old mammalian decision rules for dealing with tradeoffs between subsistence and reproduction were reinforced by a conscious pragmatism on the part of mothers. For example, if she continued to care for a particular infant, would she lose her job? If she lost her job, how would she and her family survive? On the other hand, could she improve her lot (the nest egg she might accrue, the better home she might provide), if only she were free from current encumbrance by the infant? In fact, few mothers were seeking to kill their babies. Many, however, were trying to reduce the toll infants born inopportunely would take on their well-being and future prospects. Add to that equation the heavy hand of fathers who were, among other things, eager to resume conjugal relations.

Propaganda about maternal intentions notwithstanding, allomaternal sharing of milk must have first occurred among foragers where women cooperated to keep each other's babies from fretting. Wet-nursing in this much earlier context provided a means for individuals to keep infants alive and contented, not kill them. How might this first voluntary sharing of milk have been transformed into the commercialized networks that we know about from more complex, stratified societies? We cannot hope to understand how tens of thousands of mothers became enmeshed in an intricate traffic in mother's milk, or evaluate what the wet-nursing era does or does not tell us about human "maternal instinct," unless we start at the beginning.

Mother's milk, with its special immunological and nutritional properties, has always been too valuable to share indiscriminately. Among other primates, it is rare for a mother to let another female's offspring nurse from her.[12] When milk is provided by allomothers, it is volunteered by kin on a short-term, opportunistic basis. Alternatively, an older infant, the monkey equivalent of a toddler, might take the initiative, latching on to a related female's nipples and being tolerated.[13] Such suckling is more nearly a quick pick-me-up, tiding a youngster over, than a primary source of nutrition.[14]

A look at ethnographic accounts of mothers and their daughters, sisters, or cowives who proffer breast milk to one another's offspring reveals a similar pattern of casual reciprocity, opportunistically offered and received. From Efé net-hunters in the Ituri Forest to the fisherfolk of the Andaman Islands, allomaternal suckling was a mutually beneficial courtesy extended by coresident women—affines, neighbors, and blood kin.[15]

How Flexible Lactation Is

Evidence that such casual wet-nursing was ever an important part of Pleistocene lifestyles is purely circumstantial. Nevertheless, several features of woman's biology improved the odds that lactating allomothers would have been available. So far, the only pheromone identified for humans is the mysterious substance that causes one woman to synchronize ovulation with another. If synchrony of ovulation meant that women living together gave birth at the same time, this would facilitate reciprocal suckling. But lactation in women, as in most primates, is extraordinarily flexible anyway. This is why mothers who stop lactating for a time (as during illness) can resume and begin rebuilding their milk supply as soon as they recover. Milk supply builds up in response to infant demand, and lactation can be sustained almost indefinitely until either mother or infant shuts down production through weaning. (This is how novelist Jane Austen came to be the seventh child in her family to suckle from the same wet nurse.)[16]

In a pinch, lactation can be induced without an allomother ever becoming pregnant. Adoptive mothers—girls as young as eight, grandmothers as old as eighty[17]—have lactated. But this took more than a miracle. Breasts have to be kneaded and massaged past many women's endurance, and nipples sucked (some women use baby animals) long enough to trigger endogenous production of prolactin and oxytocin.[18] In allomothers able to produce milk, there is no colostrum, but otherwise the composition of induced milk is adequate to sustain infant growth.[19]

Anthropologists have not paid much attention to induced lactation. Yet there is a telling pattern in the dozen or so accounts that exist. Whether from India, Africa, Indonesia, North or South America, when induced lactation *is* mentioned, the milk provider is most often an old woman, usually one nursing orphaned or fostered grandchildren.[20] Besides being more willing and not otherwise engaged in reproductive pursuits, grandmothers are ideal in another respect. For physiological reasons, a woman who has already lactated is three times more likely to induce lactation successfully than a woman who has never borne a child.[21]

Coerced Wet-Nursing

Among those foragers and horticulturalists who practice wet-nursing, women voluntarily offer their breasts as a favor to another woman's child. Even when disputes erupt over who stays in camp and who goes off to forage,[22] the benefits are so obviously reciprocal that matters resolve themselves. More exploitative forms of nonreciprocal wet-nursing could not arise until there was one class of mothers able to compel lower-ranking mothers to make their breasts available.

There are multiple precedents for coerced wet-nursing in other mammals, especially those with cooperative breeding systems (discussed in chapter 4). The behaviors involved are neither very specialized nor unusual. Take the pack of wild dogs in which the dominant female killed all but one of the pups in a subordinate female's litter. As the subordinate continued to suckle her lone surviving pup, the dominant mother's ten pups, already larger than the lone survivor, took over her teats. The wet nurse's last pup remained stunted and, when the pack moved, fell behind and would have died had the observers not rescued him.[23]

But as to when in *human* prehistory one mother first appropriated the milk of another, no one has offered even a guess. By the third millennium B.C. it occurred to a Sumerian mother (the wife of Shulgi, ruler of Ur) while singing her son to sleep to promise the child a wife when he grows up, and then a son—complete with a wet nurse:

> *The nursemaid joyous of heart will sing to him;*
> *The nursemaid joyous of heart will suckle him. . . .*[24]

In the time of Homer, in the eighth century B.C., some wellborn sons (like prince Odysseus) were suckled by servants, while others in the same population were nursed by their own mothers.

Some wet nurses were themselves from privileged backgrounds, their status further elevated by contact with small scions. In ancient Egypt, wet nurses were recruited from the harems of the pharaoh's senior officials (an ingenious way to elicit loyalty), and these allomothers subsequently appeared on the guest lists for royal funeral feasts. Around 1330 B.C., King Tut built a tomb to honor his wet nurse.[25] The child of one royal wet nurse from ancient Egypt was permitted to use the title "milk-sister to the king." Similar respect was accorded wet nurses in India, China, Japan, and the Near East.[26] In Arab cultures, Islamic law provides for three kinds of kinship: by blood, by marriage, and by the happenstance of two individuals having sucked milk from the same woman.

Less fortunate wet nurses were effectively slaves with wretched options. Dozens of texts and manuals survive telling parents what to look for in a good wet nurse. Virtually all advise against selecting a woman who is pregnant or still nursing her own infant. Given that wet nurses are often not well nourished, there was a legitimate concern that the nurse might not be able to make enough milk for two infants. Without so much as a comment on the implications, parents were advised to find a wet nurse who had recently given birth and whose milk was still "new." Thus do the manuals display a stark disregard for the well-being of the wet nurse's own infant. Her seemingly dispensable baby is assumed to have died, been weaned *very* early, or been farmed out to another woman, possibly to be fed something other than mother's milk. "Pap," a gruel mixture of water and ground meal used in "dry-nursing," was usually lethal for newborns.

The fifteenth-century correspondence between an Italian merchant and his wife chronicles that enterprising woman's efforts to find a suitable wet nurse for one of her husband's clients.[27] She has her eye on a particular slave whose own infant seems likely to die. The merchant's wife makes no secret of her disappointment when the slave's baby survives. Historian Richard Trexler notes that about 30 percent of infants sent to foundling homes during this period of the Renaissance were the offspring of slaves, whose owners had other uses for their milk.[28]

The Wet Nurses

Of all the protagonists in these transactions, we know least about the wet nurses themselves. Whether slaves or just destitute peasants, the price of continued survival was providing milk to unrelated offspring at the expense of their own. Some wet nurses may have been country girls hoping to earn a dowry and then marry and reproduce in earnest. Many no doubt became quite attached to their charges. Yet few would have been permitted contact with their charges after weaning. We know almost nothing about the psychological trauma these ruptured attachments caused infants and their caretakers.

The demographic consequences, however, are known. Lactational suppression of ovulation delayed the wet nurse's next conception. Yet this long interval between births was not offset by increased survival of her own infants. Only rarely did her circumstances improve, permitting her to offset early losses by producing healthier offspring later. For her, wet-nursing was a losing proposition all around.

There are hints that in spite of the hopelessness of their position, wet nurses sometimes tried, with varying success, to subvert a system heavily biased against mothers nursing their own children. From Moses' mother to the Russian women who bribed foundling home personnel, some mothers managed to get themselves paid to nurse their own babies. Wherever they could, mothers strategized to improve their lot. Rarely could they succeed. Nevertheless, substitution of one baby for another—the source of much topsy-turvy merriment in Gilbert and Sullivan—was taken seriously enough in ancient Mesopotamia to merit dire punishment. Switching babies was specifically prohibited in the Code of Hammurabi (1700 B.C.). If a wet nurse was caught doing so, "they shall cut off her breast."[29]

High Fertility Plus High Survival

From medieval times onward, wet nurses—paid, indentured, enslaved—could be found in elite households in Europe, Asia, and the Near East. Care was taken to select a nonpregnant nurse with a healthy supply of milk. Living in aristocratic households, closely supervised, infants nursed by such wet nurses had about the same survival rates—sometimes better—as infants nursed by their own mothers. For a baby born in Lieutenant LeNoir's sample from eighteenth-century France, survival chances were around 80 percent both for the tiny fraction nursed by their own mothers and those with the good fortune to be wet-nursed in their parents' home.[30]

Far from increasing infant mortality, wet nurses situated in privileged homes permitted elites to bypass a normal mammalian constraint. By com-

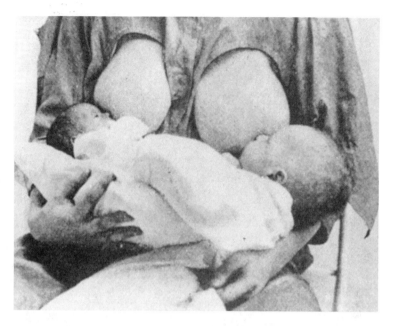

Fig. 14.3 Prior to World War II, hospitals in the United States still hired wet nurses to feed premature babies. The wet nurse was allowed to continue nursing her own infant, both for her peace of mind and because stimulation of one nipple by the sucking of the stronger baby produced a let-down reflex that made it easier for the weaker "preemie" to obtain milk. Hospital administrators calculated that wet nurses provided two to three hundred ounces of milk in exchange for a salary of eight dollars a week.[31] *(Courtesy of Syndics of Cambridge University Library)*

mandeering the milk of other women, elite wives became pregnant again much sooner without subjecting their infants to higher mortality. They circumvented the tradeoff between "quantity" and "quality" of care. In fact, some infants (especially if they were daughters, who might otherwise have been weaned early so their mothers could get pregnant again hopefully to produce a son) were wet-nursed *longer* than they would have been breast-fed by their own mothers.

For elites, wet-nursing meant high fertility *plus* high probability of infant survival. In a not atypical case, one eighteenth-century British duchess who gave birth to her first child at age sixteen, a year after her marriage, continued reproducing for thirty more years, until her twenty-first child was born when she was forty-six.[32] Eight surviving offspring—which would be a record-breaking level of reproductive success for a hunter-gatherer—was merely *average* for women in her circle. Typically, wives gave birth almost

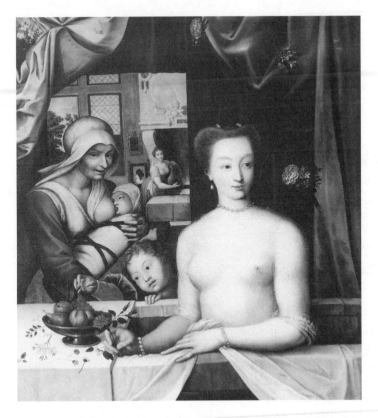

Fig. 14.4 Once the use of wet nurses became an established custom, mothers had various incentives for hiring them. Gabrielle d'Estrées bore Henri IV of France three children out of wedlock. Renowned for her beauty, detested for the wealth and influence it brought her, she might have achieved her ambition to see one of her sons succeed to the throne of France had she not died at the age of twenty-six. Her decision to use a wet nurse may or may not have been related to the production of plentiful heirs. More likely, her choice was dictated by convenience, her ambitious preoccupation with machinations at court, and her desire to preserve compact, symmetrical, youthful-looking breasts. Even more than for most women, vanity was relevant to a courtesan's self-interest. *(Photographie Giraudon; courtesy of Musée Condé, Chantilly)*[33]

annually for the first decade of marriage, slowing to a more gradual pace for the second. In isolated pockets of Europe, rapid production of many children at short intervals continued to be the norm into the nineteenth century.[34]

Fine-Tuning Parental Investment

The longer a mother nursed her baby, the more likely it was to survive. Mothers could safely wean early only when water did not cause dysentery, and where alternatives to breast milk were nutritious and palatable. But unless an observer is right there counting "time on the nipple" and "time off" (the way primatologists do), it's rarely possible to know when weaning actually occurs. With wet nurses it is. When payment stopped so did the likelihood that the infant still had access to breast milk.

Consider a remarkable study of the intimate lives of Renaissance Florentine families. From household diaries, called *ricordanze,* historian Christiane Klapisch-Zuber determined that one family out of three was more likely to have a son nursed *in casa*—the costlier, safer arrangement preferred by fifteenth-century elites. This in-house wet nurse was supervised by the mother. When infants were sent to wet nurses away from home, they were more likely to be daughters—69 percent of daughters born, compared with 55 percent of sons, and mostly these were later-born, younger sons, "heirs to spare." Even then, parents paid more so their sons would be nursed a month and a half longer on average than their daughters.[35]

How Could Love, If "Natural," Be Discriminatory?

"How could it be that love, if it were indeed natural and spontaneous, would be directed toward one child more than another?" asked Elisabeth Badinter with her usual crisp logic. How could a mother care assiduously for a first-born son and then "send the younger children away for many years?"[36] Yet, unequal treatment of progeny is only a problem for those who equate biology with genetic determinism, who assume that irrespective of maternal age or condition, or of the viability or even sex of her progeny, all mothers are the same, an invariant phenotype MOTHER.

This, of course, is true if the invariant constant (the 50-percent chance of sharing genes by common descent) is all that matters. But in the pragmatic and not-at-all-nice domain of Mother Nature, mothers evolved to factor in costs (which, in the human case, can range from mother's age or physical condition to a conscious awareness of future costs) as well as to factor in benefits (for example, a social milieu that offers sons better opportunities than daughters).

Evolutionarily, the simplest way to explain maternal behavior is as a special case of Hamilton's rule (see above, page 63) to explain altruistic acts between related individuals. Applied in this context, Hamilton's rule is not so much about genes (after all, no one has any idea what is going on at the level of genes, or what mechanisms are involved) as it is about predicting when one individual should incur a cost on behalf of another. At this level, Hamilton's rule is a formally organized metaphor for how natural selection shaped the economy of maternal emotions, with C being the cost to the donor, B being the benefit to the recipient, and r being the degree of relatedness:

$$C < Br$$

Far from disproving the relevance of biology, models based jointly on Hamilton's rule and life-history theory predict that indeed, yes, parents would invest differentially in offspring according to their circumstances. A mother would alter her commitment according to how likely an infant of a particular sex or condition was to enhance the well-being of the family or translate parental investment into long-term reproductive success. Like the cases of sex-selective infanticide in the previous chapter, decisions about wet-nursing are another empirically measurable index that reveals biases in parental decision-making. The life story of the French diplomat and statesman Charles-Maurice de Talleyrand-Perigord is a case in point, one that probably had lasting repercussions for the history of nations as well as individuals.

Talleyrand was the second-born son of an ancient and powerful family that had—like many others in the eighteenth century—fallen on financial hard times. After their first son and heir was born and wet-nursed at home, the family opted to economize on wet-nursing costs by sending their second son to be nursed on the outskirts of Paris. When the older son died, his parents immediately retrieved Talleyrand, the son now designated their heir. To their dismay, the toddler had been allowed to tumble off a chest while with the wet nurse, had injured his foot, and was crippled for life. Hence when a third son was born, a family council was convened and it was decided that Talleyrand would not be a credit to the family and should forfeit his right of primogeniture and enter the church, a vocation he eventually abandoned. The rest is history, History writ large, as this clever, calculating, and utterly cynical man became a key advisor to Napoleon.

Costs of Rapid Reproduction Borne by Wives

When wives not using other forms of birth control delegated breast-feeding to wet nurses, one consequence was shorter birth intervals. Because their wet-nursed infants (especially those nursed in-house) generally survived, the outcome was larger completed family sizes for well-off families. Some of the unintended consequences for the mother, however, included obstetrical diffi-

Fig. 14.5 The French statesman Charles-Maurice de Talleyrand-Perigord (1754–1838) recalled that in his entire life, he had not spent more than a week under the same roof as his parents. Little is known about the psychological consequences of his strange childhood. However, his biographer tells us that the steely-eyed Talleyrand "became a byword for lack of principle in an unprincipled age."[37] *(Painting by Z. Belliard, Courtesy of Archives Départementales de l'Indre and the Château de Valençay)*

culties from rapid births—ranging from headaches and anemia to the cervical lacerations, pelvic infections, and prolapsed uteruses itemized in lurid detail by historian Edward Shorter in his *History of Women's Bodies*. While elite fathers, especially those married to successive wives, enjoyed unprecedented reproductive success, their wives more nearly suffered it; many died young and most (like Emma Darwin, who bore 10 infants) dreaded frequent confinements.

From such glimpses as we are permitted, it is clear that the pronatalist push was mostly coming from husbands and from their families rather than from mothers themselves. Mothers wanted children—just not quite so many. Having produced the requisite male heir, some women *insisted* on nursing so as to step off the reproductive treadmill. Well-spaced, healthy children (quality over quantity) suited *her* better.[38]

But more than conscious pronatalism was at stake. In France, husbands had an extra incentive to get their babies to wet nurses. They wanted their conjugal privileges, and the Catholic church discouraged sex between husbands and nursing mothers, perhaps for the same reasons so many cultures do. Postpartum sex taboos are found across traditional societies from North and South America to New Guinea and Africa, presumably as an extra precaution, a failsafe to guard against a new sibling being born too soon.[39] Ironically, a custom that originated because it increased infant survival by guaranteeing longer birth intervals more often had the opposite effect in Catholic countries. Sending infants out to wet nurses probably led to far more infant deaths than abbreviated birth intervals due to the occasional impregnation of a nursing mother.

Maternal resistance to rapid births may have contributed to the early adoption of birth control in France. For reasons not yet well understood, birth rates among French elites dropped faster and sooner than elsewhere in Europe, where the demographic transition did not get under way until the end of the nineteenth century. Rousseau's campaign to convince parents to treat all children equally may have been partly responsible. But pressure to limit family size as a way of maintaining family status was already building.

Since, formerly, status and well-being tended to be correlated with reproductive success, it is not surprising that mothers, especially those in higher social ranks, put the basics first. When confronted with a choice between striving for status and striving for children, mothers gave priority to status and "cultural success" ahead of a desire for many children.[40]

If instead of channeling wealth to one son (through primogeniture) parents had to treat all offspring equally (which became the law of the land in France in 1804 with the adoption of the Napoleonic Code), then producing fewer heirs was essential if a propertied family was to maintain its social and economic standing. For this and other reasons (again, poorly understood), mothers anxious about maintaining their descendants' privileged status took steps to reduce their family sizes. These smaller family sizes (closer to five than ten) more nearly resembled those of their foraging ancestors than the large offspring sets characteristic of the eighteenth century. Smaller families reduced pressure to economize on the amount of investment parents committed to each offspring and permitted mothers the luxury of an Enlightenment ethic regarding child-rearing. Many people think of the smaller, post–demographic transition, family sizes as unusual. But they are not. They are more nearly a *reversion* to an earlier species norm.

Spread of Commercial Wet-Nursing

Long before the demographic transition, wet-nursing had seeped down the social ladder, leading to a wide-scale traffic in mother's milk. Commercial wet-nursing peaked in eighteenth-century Europe, but precursors can be traced far earlier. Documents written on papyrus in 300 B.C. indicate that free women in Hellenistic Egypt, not just slaves, were contracting to provide breast milk.[41] By the second century A.D., selling access to a woman's breast was a routine commercial transaction, and in the vegetable market at Rome's Forum Holitorium, buyers met with sellers in a place called the *lactaria* (designated by a column).

Commercial wet-nursing was already far removed from the casual, voluntary nursing found among foragers and horticulturalists. Eventually, having a wet nurse became a symbol of high status. Most likely, the practice of enforced or contractual (rather than reciprocal) wet-nursing first became established among elites. Thereafter it became an enviable option that sub-elites sought to emulate. Only later did the practice of using a wet-nurse spread among working mothers. As the custom spread from elites to sub-elites,[42] wet-nursing offered a novel solution to a perennial dilemma: How is a mother to engage in status- or subsistence-related activities without being burdened by an infant? Once again, mother's immediate well-being and perhaps the needs of older children past the perilous first years of life took precedence over caring for a particular infant.

———

Down the economic scale in the pre-bottle world, the use of wet nurses was a late addition to the human repertoire. It had little to do with sparing a wife from the drudgery of nursing her baby, or helping her keep shapely breasts, or causing her to become pregnant again sooner—though these were outcomes. Rather, the mother's labor was essential either to her own subsistence or (as in the case of butchers and artisans) to the family economy. Nursing an infant interfered with her efficiency at a time when the line between bourgeois status and destitution was perilously narrow. Given the tradeoffs, she opted for the wet nurse. Nor would it surprise me to find modern mothers once again seeking to hire others to nurse their babies as they read articles touting the benefits of breast-feeding for their children's security, intelligence, and immunological systems. Many more will take advantage of breast pumps to provide a personal stock to daycare providers.

Quest for Affordable Care

During the eighteenth century, the population in France grew from 20 million to 27 million.[43] In the countryside, poor harvests plus the fractioning of small landholdings among several sons swelled the numbers of the dispossessed. Providing for oneself was difficult, providing for a family even tougher. A man whose wife did not work could not expect to earn enough to support a family. Hence, the arrival of additional children meant disaster if the mother had to quit working.

Desperate peasants migrated to cities, but their subsistence remained marginal. Rapid urbanization combined with slow industrialization translated into few opportunities, not only for parents but for such children as survived. Incomes were low, rents high, the price of bread rising faster than wages. Many of the French mothers who sent their infants to wet nurses were working women barely eking out a precarious existence—"bourgeois" but only by the skin of their teeth. The more involved the wife was in helping her husband—whether in a store or some craft like silk-making—the more likely the family was to use a wet nurse.[44] At any given point in time, need for the mother's labor was a better predictor of what would be done with the infant (kept at home, sent to a wet nurse, left at a foundling home) than were infant mortality rates. How could this be?

Historian George Sussman calculates a typical budget for a family of artisans. Nearly half of the family's earned monthly income would go for food, 15 percent for clothing, 6 percent for light and heat, and another 13 percent for rent. In addition, each wet-nursed child could cost eight livres a month—20 percent of the family budget, so long as the family could afford to keep up the payments.

In short, working mothers sent their babies farther and farther away, not to get them legally killed but primarily in quest of affordable care. It was not maternal nature (always contingent on circumstances) that changed through time, but maternal options. Rubber nipples and pasteurized milk would not be available until the end of the nineteenth century. Practically speaking, wet nurses were the only safe alternative to a mother breast-feeding her own baby. Trouble was, working mothers were in competition with elites who paid as much as several hundred livres a year to have wet nurses live in their homes. (Wet nurses themselves were often hard-pressed to find low-cost care for their own displaced infants.) Parents in search of affordable wet nurses would also have been in competition with foundling homes.

Yet somehow, bourgeois parents did locate wet nurses, and in a world without other forms of birth control, wet-nursing produced the same effect among those mothers as it had among the elites: a staggering hyperfertility, but with a cruel difference. Among sub-elites, high fertility came coupled with high rates of infant mortality.

Working women routinely produced twelve to sixteen children. In his remarkable social history of bourgeois families in Lyons during this period,

French demographic historian Maurice Garden describes one butcher's wife who produced twenty-one children in twenty-four years.[45] Survival chances of these children were directly correlated with how much their parents paid for wet-nursing. Abandonment was free, but about the time Lieutenant-General LeNoir was tallying his statistics, mortality rates in the foundling homes around Paris hit 85 percent. Also about that time, one year of wet-nursing in a rural location went for 100 livres and reduced mortality rates by half—to 40 percent or lower.

Even six months of wet-nursing could make the difference between life and death. Around 10 percent of working parents eventually defaulted on payments to the wet nurse, with the result that their babies ended up in a foundling home. Yet the head start provided by just six months at the breast of a paid wet nurse boosted survival chances of an infant subsequently consigned to a foundling home. In-house wet-nursing, while more than twice the cost of a rural wet nurse, cut the mortality rate by another half, down to 20 percent—the same as if a mother nursed the baby herself.[46]

Daycare as Modified Wet-Nursing

Perhaps the closest analogy to what went on with wet-nursing is what we modern mothers go through when we scrounge and compromise to locate affordable infant daycare, mercifully nine-to-five (not including weekends), and not nearly so wretched nor so deadly as wet-nursing.

Of 21 million children under the age of six in the United States in 1995, 12 million were in daycare. Of infants less than one year old, 45 percent were in some kind of daycare.[47] Mothers seek this care in a market where wealthy, nonworking mothers, highly paid women professionals, ordinary working women, and mothers pushed off welfare into the labor market at minimum wages, not to mention government agencies seeking to place foster children, are all competing for alloparental care, a commodity not in abundant supply to begin with. A 1998 New York Times article on the "acute lack of daycare" faced by mothers being pressured from "welfare" into "workfare" notes that

> three quarters of the mothers in workfare rely on unlicensed baby sitters who are also paid by the city. The lucky ones enlist trusted relatives or close friends. The less fortunate leave their sons and daughters in crowded, dirty apartments with caretakers they barely know. . . .[48]

Just like working mothers today, eighteenth-century European mothers weighed infant care from wholesome nurses close to home against affordability. A critical difference was that the severely sub-par eighteenth-century infant daycare/nightcare entailed far less parental supervision, far greater risks to infant health, and far worse psychological risks from ruptured emotional attachments.

Alternative Ways to Mitigate Costs of Care

Dual-career mothers, whether they forage or go to work, have always sought ways to mitigate the costs of infant care. Today, mothers hire nannies, leave children in government-run crèches, *maternales,* or daycare centers; they delegate childcare to kin; or else they continue caring for infants themselves but reduce the amount of care given to each infant. Reductions range from leaving an infant in a car-seat for fifteen minutes while running an errand, to neglect so pronounced it results in an infant's failure to thrive. In the more extreme cases, consequences of these tactics are measurable. More often they are not. Even when maternal retrenchment affects morbidity or infant mortality, the effects of maternal decisions do not leave a measurable trace.

Societies where infanticide is unthinkable, where infants are never sent to suckle at some hired woman's breasts, nor left in bundles along roadsides, nor swaddled and hung from trees, tend to be societies in which women have some degree of reproductive autonomy and access to some fairly reliable form of birth control. Or else, they are societies where mothers have at their disposal social customs or institutions for delegating some care to allomothers.

Just as European infants were once left in foundling homes or sent to wet nurses at near epidemic levels, today we witness poor mothers from the Philippines, Central and South America, South Africa, and Asia leaving children behind to be bottle-fed and somehow cared for by kin while they themselves go far away to work as housekeepers and caretakers for other people's children. Solutions differ, but the tradeoffs mothers make, and the underlying emotions and mental calculations, remain the same.

Fostering Out, and Other Ways
to Delegate Care to Alloparents

"Where every child is a wanted child" is, appropriately enough, the slogan for Planned Parenthood, an admirable organization that seeks to help mothers

Fig. 14.6 A South African girl and her nurse. Mothers who hire nannies often prefer not to think about where the nanny's own infants are. *(© Ian Berry / Magnum)*

time births and space offspring in line with maternal health and family needs. In this ideal world, every child is wanted because maternal conditions and motivations are taken into account. In terms of Hamilton's rule, the same outcome could be produced other ways—for example, by reducing the cost of each infant and/or by spreading that cost among alloparents.

Some years ago David Kertzer, the anthropologist studying Italian foundlings, pointed out that poverty, by itself, was a poor predictor of infant abandonment. Sardinia was a case in point. During the same period when large numbers of infants were being abandoned in Tuscany, Sicily, and elsewhere in Italy, virtually no babies were abandoned in this poorest of Italy's regions. Between 1879 and 1881, when 69,000 babies were left at foundling homes in Sicily, only *fifteen* were abandoned in Sardinia. Kertzer attributes this near absence of infant abandonment to Sardinia's mother-centered family organization. Daughters remained near kin, so that even unmarried girls would have a "supportive network of female kin."[49]

As has always been true, availability of matrilineal kin—sisters, mothers, and grandmothers—makes for an especially reliable source of allomaternal assistance. Not quite a beehive, but far more valuable than a village, an extended family of matrilineal kin turns out to be a wonderful resource for rearing human infants.

————

Imagine a place where household incomes are low and fertility is as high as anywhere in the world. Many fathers are only sporadically in residence with the mothers of their children; and fathers, when they *are* on the scene, may be unpredictable regarding which children they invest in, and how much. A substantial number of women conceive at a young age, often prior to marriage or formation of any stable relationship. Even with the assistance of the father, parents may not have sufficient resources to provide for all their children. There is no government assistance of any kind for mothers with dependent children. Nor are there government-sponsored institutions to take in foundlings, nor churches to accept oblates—children given as gifts to the Church, a common practice in medieval Europe. Yet abortion is uncommon and infanticide rarer still. Wartime catastrophes aside, infants are never abandoned outright. And despite the rarity of baby bottles, the practice of one mother hiring another to provide breast milk to her baby is unknown.

This description applies to large areas of sub-Saharan Africa, especially those areas that the economist Ester Boserup characterized years ago as having "female systems"—by which Boserup meant relying on hoe agriculture, in which women are highly valued for their labor. It is a traditional way of life associated with maternal inheritance of garden plots and the existence of strong networks of female kin.[50] Many Africans would regard the thousands of European mothers who abandoned their infants at foundling homes as extraordinary creatures, acting outside the range of human behavior. Instead, among them one hears traditional proverbs like "If you have a child, you have a life."[51] Even in the face of rapid change in some areas—including the devastating effects of AIDS, which has begun to undermine and overwhelm these time-honored childcare networks—key aspects of this system remain intact.

Throughout most of Africa, children are passionately desired by both sexes, although fathers want more of them than mothers do. Mothers breastfeed infants for a long time, and, with varying success, maintain postpartum bans against sex while the mother's current child is deemed too young to make way for a new sibling. It is part of the usual maternal campaign for raising viable children in the face of the male desire for quantity. Only as fathers begin to invest more in each child—to "do more of their share," as is happening with modernization—do husbands begin to appreciate their wives' concern for spacing or limiting births.

A number of anthropologists have commented on the resistance of traditional African men to any form of birth control, especially condoms. Many women who take the pill must do so in secret. The primary interest of these women does not seem to be to quit childbearing, but rather to space births more widely apart (which, of course, does ultimately limit completed family size).

Adults genuinely enjoy indulging little children. And why not? Children are viewed as ties to the ancestors as well as links to the future in a world of mystifyingly rapid change whose idiosyncrasies children grasp more readily than their elders. Grown children, especially those who find a niche for themselves in modern Africa, offer the only old-age insurance there is. Even a remote father who rarely sees his children—and relatively few fathers provide a great deal of direct care—derives prestige and political clout from having them. Grown children acknowledge the "very great debt" owed to those who cared for them when young. It is not just gifts of food and cash from the parents, but this prospect that their former charges will feel obliged to care for them one day, that makes foster mothers eager to take youngsters in.

Such foster mothers—who may be real or fictive kin—are often postreproductive and are generally given the generic title "granny." Across large areas of west, east and southern Africa grannies provide the closest real-life example of Boswell's vision of "kind strangers" and recycled children. Mothers routinely send one or more children to live in a foster home, either with one of these "grannies" or among prosperous connections who can offer advantages in schooling, better nutrition, or other opportunities. The genetic mother and the man socially recognized as the father stay in touch and send gifts of food and money.

But kinship and the gifts are not the only reason grannies take in charges. As one of them put it, "You never know whether the children will love you more than their own mothers and thus bring you sufficient benefit." A commentator observed more philosophically that "Children are like young bamboo trees; you don't know which of the shoots will be cut away and which will remain"—nor how any given child will turn out.

Caroline Bledsoe, an anthropologist who has studied the fosterage system among the Mende people of West Africa, describes what happens when a particularly enterprising or lucky child makes good. Numbers of relatives, including long-absent fathers, show up to stake their claim. The logic of the lottery is explicit in the mind of the West African man who told her: "It is good to bring up or mind a lot of children because you don't know which one will be successful. However it may be, one *must* be successful, and you will get your reward out of your expenditure."[52]

People understand that mothers might not want children born too closely together. But not wanting a child is incomprehensible. Serious misunderstandings have occurred when West African parents living abroad took advantages of what seemed wonderful opportunities to foster their children with Western families who could offer them a better education, higher standard of living, and valuable contacts. The Westerners mistakenly believed that they were adopting unwanted children. The parents, however, assumed the children were only being temporarily fostered and expected them to be returned.

African villagers are genuinely shocked when ethnographers inquire about infanticide. They are mystified by the mindsets of foragers in parts of the world like lowland Amazonia, where the proverbs have quite a different ring to them. Africans cannot imagine anyone saying, as some Amazonian tribespeople do, that "infants are not precious to us."[53]

A story told by the demographer Nancy Howell illustrates how attitudes of sedentary African horticulturalists with readily available caretakers differ from other Africans who still live as nomadic foragers. Among the nomadic !Kung San, the number of babies a mother can care for is limited by the fact that she must be able to carry them wherever she goes. A group of Bantu women came across a !Kung woman who had gone into the bush to give birth. The poor woman, still somewhat dazed, had just realized that her baby had been born with a birth defect, and she felt it was her duty to abandon the

Fig. 14.7 In the West African system of fosterage, most children remain with their mothers until after they are weaned. However, when—as in the case of this Mende "granny"—the foster mother provides breast milk as well as care, the distinction between European wet-nursing and West African "fostering out" is blurred. *(Courtesy of Caroline Bledsoe)*

baby as soon as possible. Yet the Bantu, with their very different ethic, persuaded her to keep it.[54]

Among the Bantu, a woman's experience is quite unlike that of a !Kung mother. A number of people (primarily mother's kin) routinely care for each child. At any given point in time in a broad area of West and South Africa, up to 40 percent of mothers of childbearing age will have sent infants already weaned, or in the process of being weaned, to live with a "granny" or to the home of a well-to-do relation. In some areas, availability of a maternal grandmother is significantly correlated with the child's survival.[55]

Even a very inopportune child is still wanted by parents, so long as others are available to help rear him. Young mothers, or those who are unmarried or anticipate inadequate support, are most likely to rely on a granny.[56] These belong to the same class of mother who among the Ayoreo would be most prone to commit infanticide, who in eighteenth-century Europe would abandon their babies at a foundling home, or who in impoverished parts of urban Brazil would opt not to breast-feed a baby even though this means a much greater risk that it will die.[57] Mothers also send their children to grannies if they remarry, or if they fear cowives may maliciously target their children.[58]

Yet the generosity of "grannies" has its limits. When resources are short, foster children may get short shrift—particularly those whose parents send few presents or who are not really relatives. Children who show little aptitude for becoming one of those "little bamboo shoots" that will prosper and reward their caretaker may be overlooked, or left to scavenge food, or denied medical care by their foster parents. Odd remarks, inconsistent with the idyllic (and generally correct) impression that every child is a wanted child, are sometimes overheard. Anthropologists Robert and Sarah LeVine, who worked among the Gusii in Kenya, cite a saying worthy of a Dogon cowife: "Another woman's child is like cold mucus. . . ." As the LeVines put it, "In a society that values children above all else, we found nevertheless that some children were valued more highly than others."[59]

Even in traditional Africa, marginalized children are likely to be neglected. Population growth has aggravated this situation, as there are so many more children today than adults. The AIDS epidemic has further worsened this imbalance. Not only does this dread disease produce orphans, but the mode of transmission makes it likely that whole families—fathers, wives, cowives—will be affected. There are no longer enough "grannies" to go around. In some areas of modern Africa, particularly in those cities where up to one-quarter of pregnant women are HIV-positive, child-rearing networks are breaking down, leading to wholesale abandonment of orphans.[60]

Continuum of Maternal Commitment

Mothers have always had to make the most of resources at hand while coping with the sliding scale of paternal and alloparental help available. Mothers make tradeoffs compatible with their own subsistence, the needs of different children, and their own future reproductive prospects. These tradeoffs are made in a world of constantly shifting constraints and options. In foraging societies, for example, suckling infants are far more costly than older children, who are at least mobile. Not so in our own increasingly technological society, where costs of child-rearing (for example, college tuition) go up—not down—with the child's age.[61]

Some mothers encounter completely novel options—like breast milk for hire in the eighteenth century, or new birth technologies in the twentieth—that allow women forty-five and older to give birth. A woman who postponed reproduction in the Pleistocene was probably waiting out a famine, or looking forward to a situation with more stable allomaternal assistance. In the twenty-first century, career women will count on amniocentesis, in vitro fertilization, and procedures combining the DNA in their own eggs with cytoplasmic material from a younger woman's eggs to keep reproducing beyond their prime reproductive years. Such techniques will reduce risks to a woman from delaying reproduction (allowing her to achieve a desired professional or social status prior to bearing children), but are likely to introduce other risks or tradeoffs yet unknown.

No social creature, even the most independent woman, makes such decisions in a vacuum. In addition to laws, technologies, and protection from environmental hazards, there are today, as in the past, people both more and less powerful than the mother herself who shape the reproductive options available to her. Now, as in the past, mothers do not live in any one type of family arrangement. Nor is there any one species-typical level of maternal commitment to infants. Without question, historical context matters a great deal. But to interpret variation in the way mothers respond to infants as meaning that somehow a woman's biology is irrelevant to her emotions, or that there are no evolved maternal responses, is to misread both the human record and a vast amount of evidence for other animals.

No one suggests that the hundreds of thousands of mothers in eighteenth-century Europe who sent babies to wet nurses, or the mothers who abandoned infants outright at foundling homes, were typical of all mothers at all times. In a state of ecological release, in which the costs of caring impinge far less on the mother's health and well-being, mothers can afford the luxury of loving each baby born. This is especially likely to be so when women have the inestimable privilege of consciously planning when births will occur. Nevertheless, the "unnatural" mothers chronicled here can only be the visible tip of an iceberg of maternal ambivalence that left no record.

These nuances of maternal emotions and the many "little decisions" mothers make are rarely measurable. For every mother who abandons her infant outright, there have to be thousands of other mothers who abjured such draconian remedies yet nevertheless fell short of commitment in ways that lowered infant viability.

Legacy of Ambivalence

And what does this degree of ambivalence mean for the notion of "maternal instinct"? So long as we are clear about what we mean, there is no reason not to use the shorthand *instinctive* to describe the adoration of their babies that mothers feel. Like all primate females, women and girls find babies utterly enticing and attractive, and most are eager to hold and care for them. This is especially likely to be true of a recent mother because of the hormonal changes during pregnancy and birth that lower her threshold for forming an affiliative bond with an especially attractive (in terms of its smell and odd appearance) little stranger, and such bonds intensify during lactation. Virtually all female primates, if they remain in close proximity to a small baby long enough, learn to recognize and form an attachment to that particular baby.

Every human mother's response to her infant is influenced by a composite of biological responses of mammalian, primate, and human origin. These include endocrinal priming during pregnancy; physical changes (including changes in the brain) during and after birth; the complex feedback loops of lactation; and the cognitive mechanisms that enhance the likelihood of recognizing and learning to prefer kin. *But almost none of these biological responses are automatic.* To survive over evolutionary time, all of these systems had to pass through the evolutionary crucible so well summarized by Hamilton's rule. One way or another, whether the cue is fat deposits that influence activation of the ovaries, or signals indicating that social support is forthcoming, probable costs and potential benefits are factored in. In humans, whose infants are so costly, and for whom conscious planning (thanks to the neocortex) is a factor, maternal investment in offspring is complicated by a range of utterly new considerations: cultural expectations, gender roles, sentiments like honor or shame, sex preferences, and the mother's awareness of the future. Such complexities do not erase more ancient predispositions to nurture. All the systems in this messy composite are vetted according to costs, benefits, and genetic relatedness to the infant recipient of altruistic maternal acts. But none of this guarantees perfect synchrony between systems. We should not be surprised that conflicting motivations emerge at both conscious and unconscious levels, in ambivalent maternal emotions psychoanalyst Rozsika Parker sums up as feeling "torn apart."[62]

We are still far from understanding how genetically influenced receptors in the brain, thresholds for responding to different chemical signals, hormone levels, and feelings of anxiety or contentment interact to produce the myriad "decisions" that continuously affect maternal commitment. Yet this fact remains: human infants are so vulnerable and dependent for so long a time, that the level of commitment to them by the close relative on the spot at birth, primed to care, and lactating, is the single most important component of infant well-being.

Throughout human evolution, the mother has been her infant's niche. Physical and social circumstances affect the baby as they affect her. Whether or not an infant ingests colostrum, nurses for five months or five years, whether a mother keeps her infant nearby or turns him over to an allo-mother, each represents a maternal decision with implications for infant survival. Demographically and statistically, multiple, small maternal decisions about how much to invest, and for how long, in any given infant add up to life-or-death outcomes for human offspring.

Other ape neonates cling for dear life to the mother's hair. For human babies survival is more complicated. There is no environmental hazard more omnipresent or immediate in its impact than a retrenchment in maternal care. Consequences of "little maternal decisions" are far more perilous in some habitats than others. Maintaining maternal commitment was once as important for an infant's survival as oxygen, and often it still is. Yet there is little in the routine ethnographic or historical descriptions of mothers to suggest that maintaining adequate maternal commitment was a problem. Is this because it was not? Or is it, as I believe, because an idealized view of maternal commitment has been taken for granted for so long, and because decisions leading to small retrenchments are so unremarkable?

As birth intervals grew shorter through the course of human evolution and recent human history, pressure on mothers to delegate caretaking to others became even more intense. Whenever they safely could, or when they had little choice, mothers handed babies over to fathers or alloparents, weaned them early, or swaddled them and hung them from doors. At the psychological level, these decisions differ little from those a contemporary mother makes everyday when she asks her neighbor to baby-sit or contracts for more or less adequate daycare. She is playing the odds, and evaluating her priorities. Hence I am dumbfounded when I hear contemporary politicians lament "the breakdown of the family" in the modern world. A recent editorial in the *Wall Street Journal* complained that "the denigration of the marriage—the famous 'nuclear family'—was feminism's greatest failure" [leaving in its wake] "a legacy of children without fathers and women without husbands. . . ."[63] Although it is heartening to know that the editors of the *Wall Street Journal* are concerned about teen pregnancy and children growing up in fatherless homes, it is faulty logic to indict feminism for social problems that are as old as the conflict of interests between men and women, and probably far older.

Hominid fathers have been choosing between investing in the children they already have and finding new mates with whom to sire more for as long as there has been a division of labor between hunters and gatherers and a practice of hominid males sharing food with immatures. Worldwide, such tensions have far more to do with the prevalence of female-headed households than feminism possibly could. Approximately one-third to one-half of all households in the world are female-headed, most of these in countries too poor to have been touched by a social movement that is less than two centuries old and so far primarily a luxury that only educated Western women have been able to afford.[64]

Over the millennia, mothers have factored into their decisions information about the effects a particular birth would have on older children; the probable response of father or stepfather; the infant's own prospects for survival; and the prospect of translating her efforts into subsequent reproductive success. Unlike other primates, women possess the capacity to foresee outcomes. Being born to a wise and foresightful mother would seem like good fortune to most of us. But it would be a blessing that brought with it peculiar hazards.

This is why the human infant, though born especially helpless, has had to become psychologically sophisticated in specialized ways, attuning himself to the task of assessing and extracting commitment from those closest to him, especially his mother. Infancy and childhood comprise the first perilous bottleneck every contributor to the human gene pool must pass through. Small downward adjustments of maternal priorities regarding a given infant, which cumulatively amount to life-or-death decisions, have had an enormous impact on the direction of human evolution. Degree of maternal commitment was itself a selection pressure impinging on each newborn. What, then, were the evolutionary consequences on the bodies, minds, and temperaments of human infants?

PART THREE

An
Infant's-Eye View

Why is our need to belong to a family—any family—as vital as air?
—Amy Tan, 1998

Mother Nursing Her Baby by Mary Cassatt
(Art Institute of Chicago)

Born to Attach

*The movements of expression in the face and body . . . serve as the first means
of communication of the mother and her infant. . . . [They] give vividness and
energy to our spoken words. They reveal the thoughts and intentions of others
more truly than do words, which may be falsified.*

—Charles Darwin, 1872

W hen people imagine Charles Darwin, they think of the fresh-
faced young naturalist who set out on the *Beagle,* or the balding,
sideburned gentleman with bushy eyebrows who settled at
Down House to write *The Origin of Species, The Descent of Man and Selection in
Relation to Sex,* and *The Expression of the Emotions in Man and Animals,* the works
that would make Charles Darwin, the bearded old man in a black cape, the
most influential biologist of all time. Few think of Darwin as a congenitally
sensitive little boy who lost his mother at an early age. But that's how John
Bowlby saw him. For Bowlby had this maternal habit of looking at the world
from a child's or an infant's point of view.

Bowlby's last work, published posthumously in 1990 (the same year he
died), was a psychiatric case study entitled *Charles Darwin: A New Life.* He
toyed briefly with the notion of titling it *The Origin of Charles Darwin.*[1] For the
mother in me, the image of Darwin as vulnerable child is perhaps even more
meaningful than the superb intellectual biographies that also exist.[2] Bowlby
traced Darwin's lifelong illness and his peculiarly humble and dogged style of
doing science to early maternal deprivation.[3]

Darwin on John Bowlby's Couch

As Bowlby saw it, the nineteenth-century father of human ethology as well as
evolutionary psychology, and the first naturalist in the world to apply ethological methods to a detailed study of infant development, was startlingly,
poignantly, out of touch with the early development of his own emotions.
"Throughout his scientific career, unbelievably fruitful and distinguished
though it would be," wrote Bowlby, "Charles's ever-present fear of criticism,
both from himself and from others, and never satisfied craving for reassur-

ance, seep through. . . . Unflagging industry and a horror of idleness were to dominate his life."[4] Bowlby traced a mystifying array of psychological and bodily symptoms to Darwin's ruptured attachment and insecurities in childhood.[5]

This was Bowlby's take on Darwin's lifelong insecurities, anxiety, depression, chronic headaches, fainting sensations, dizziness, "swimming head," "black dots," hysterical crying fits, ringing in the ears, itching, eczema, nausea, and vomiting—all the ills that plagued Darwin for much of his adult life, what Darwin himself would sum up as "One long struggle against the weariness and strain of sickness."[6]

There is no dearth of alternative hypotheses to explain the hodgepodge of symptoms famously known as "Darwin's illness." Some years before Bowlby's diagnosis, an Israeli parasitologist had suggested that Darwin contracted Chagas' disease, caused by a blood parasite, while traveling in South America as a young man. A historian suggested that a weak immune system left Darwin susceptible to allergic reactions to the preservatives in which he stored his specimens. Psychiatrist Ralph Colp, in a very detailed examination of the timing of the onset of Darwin's symptoms, argued that they resulted from Darwin's anxiety at the prospect of challenging his society's most deeply cherished convictions. Some have even suggested Darwin's problems might have been psychosomatic symptoms exploited by a confessed workaholic to avoid tiresome social obligations.[7] Enviously, I have on occasion wondered just how much this shy, serious country gentleman actually minded being "forced to live . . . very quietly . . . able to see scarcely anybody and cannot talk long with my nearest relations"?[8]

Bowlby Examines Darwin's Childhood

Darwin was an impressionable, innately sensitive eight-year-old when his mother died. Not permitted to grieve in ways that would help him cope with his feelings of abandonment and despair at the disappearance of this beloved figure, Darwin became susceptible to what Bowlby termed "hyperventilation syndrome," an old-fashioned diagnosis doctors now subsume under the more general diagnosis of "panic disorder."[9]

> My mother died in July 1817, when I was a little over eight years old, and it is odd that I can remember hardly anything about her except her death-bed, her black velvet gown, and her curiously constructed work-

Fig. 15.1 Catherine Darwin shown with Charles. He recalled that "When my mother died, I was 8½ years old, and [my sister Catherine] one year less, *yet she remembers all particulars and events of each day whilst I scarcely recollect anything* . . . except being sent for, the memory of going into her room, my father meeting me—crying afterwards.[10] *(Courtesy of the Darwin Museum, Down House, and English Heritage Photographic Library)*

table. I believe that my forgetfulness is partly due to my sisters, owing to their great grief, never being able to speak about her or to mention her name.[11]

As far as is known, neither Charles nor anyone else in the Darwin family, not father or sisters, ever mentioned his mother again. In his autobiography, Darwin simply recounts that shortly afterward, he was sent away to school. Thereafter, Darwin's relationship with his mother was replaced by less satisfying substitutes. He was lorded over by an intimidating father and scolded by moralistic sisters bent on making him a "better" person at the expense of his own inclinations.

Years later, in what Bowlby seized upon as a crucial clue for his diagnosis, Darwin wrote to a cousin who had just lost his young wife: "Never in my life having lost one near relation, I daresay I cannot imagine how severe grief such as yours must be."[12] For a man of "enlarged curiosity," arguably a genius, not to be able to "imagine" a loss such as he had actually suffered—well, frankly, Bowlby has a point. This does seem odd.

From an early age Darwin felt browbeaten by his overly allomaternal sisters, and when their admonitions distressed him, he "made himself dogged so as not to care."[13] To his anxiety, at the prospect of lost affection, and other internalized self-doubts (which spilled over into Darwin's science in marvelous ways, eliciting in him a healthy respect for facts that did not conform to his theories), was added this extraordinary perseverance and propensity to rescue his sense of self-worth through fastidious labor. "It's dogged as does it"—a common phrase of the time (used by Trollope as a chapter title in his novel *Barchester Towers*) became Darwin's favorite motto.

Avoiding what distressed him most, Darwin equipped himself to explore a different set of dangerous ideas. Yet he continued to suffer odd lapses. Bowlby was particularly struck by an anecdote recalled by Darwin's granddaughter, Gwen Darwin Raverat, who used to play word games with her grandfather at Down House.

"The Letter Game: Word making and word taking" was an old Darwin family staple, not unlike Scrabble. The object, recalls Raverat in her spirited memoir, *Period Piece*, was to arrange separate letters into any word in the dictionary by stealing a letter and adding it to another word already in play. Raverat recounts her grandfather's consternation one day when a fellow player added an "M" to the word "OTHER": Darwin looked at it for a long time, unable to recognize it as a word:

> There was the story of my grandfather (C. D.) who, on seeing the word MOTHER on the board, looked at it for a long time, and then said "MOETHER; there's no word MOETHER." I feel that the Psychologists might get a great deal of fun out of this anecdote—I beg their pardons, I don't mean fun, but Important Information; clues to the

conception of *The Origin of Species* on the one hand, or to his ill health [which] could doubtless be proved by this story to be the direct consequences of the early death of his own MOETHER."[14]

His granddaughter's lighthearted interpretation of Darwin's lapse apparently stimulated Bowlby's more serious diagnosis of maternal deprivation.

From his own childhood, Bowlby was all too familiar with the distant relationships typical in well-off British families, and focused on this aspect of Darwin's childhood. Bowlby hypothesized that Darwin's early loss and suppressed recollections of it contributed to his lifelong vulnerability to anxiety and to his fear of losing anyone's affection or of physical loss of anyone he cared for.

Had Darwin himself set out to analyze this painful terrain, to try to understand the tactics that immature humans use to avoid being abandoned and to cope with varying degrees of maternal deprivation, even so creative a mind as his own would not have been intellectually prepared for the journey. For one thing, Darwin was not prepared to imagine the mother as a multifaceted strategist. This constrained his ability to think about the full range of selection pressures that would have acted upon infants in the evolutionary past. The first step toward a broader understanding of the psychology of babies was taken by Bowlby, who set himself the task of reconstructing what it would take for a baby in the Pleistocene to be among those that survived long enough to grow up. Bowlby was the first to really use Darwin's theory of natural selection to explain infant emotions.

Connoisseurs of Mothering

Bowlby proposed that all primates are born preprogrammed to form a powerful emotional attachment to their mother or other primary attachment figure to whom the infant strove to stay close at all times. Early in life, he believed, infants form an "internal working model" about what to seek and expect from relationships, based largely on the extent to which their own early feelings are reciprocated.

Darwin and other early pioneers of human ethology had hinted at infant motivations, but Bowlby was the first to characterize infants as *both* highly vulnerable *and* precociously social actors with their own agendas for ensuring continued protection and access to the nurture necessary for survival. This way of conceptualizing infant development became known as "attachment theory." By the end of the twentieth century, Bowlby's insights would be revised, expanded, and incorporated into the new view of mothers then emerging as scientists explored the full range of selection pressures acting on these little monitors of maternal commitment.

For in Mother Nature's great gallery of creations, human infants are both masterpieces and connoisseurs—connoisseurs of mothering. Embryos evaluate a mother's chemistry and have already begun to register her utterances. Infants memorize her scent and assess her glances, her warmth, her tone of voice. Above all, infants are exquisitely sensitive to signals of maternal commitment. Will she stay close or (most dreaded prospect!) disappear?

If human mothers were automatically nurturing, their infants would not need to be so attuned and keenly discriminating. Although this is not the part of the story that Bowlby stressed, the connoisseurship of infants is the strongest evidence there is for how contingent maternal commitment has been through the environments of greatest evolutionary relevance in respect to such traits. For in the course of human evolution their increasingly slippery and hairless mothers had become more fertile than any ape ever before, with consequences for how unconditionally loving a human mother could afford to be.

If a baby finds a rubber pacifier soothing it is because for at least fifty million years, primate infants so engaged could feel secure, because a baby sucking on a nipple is a baby likely to have a mother close at hand. A taste of sugar-water in its mouth produces even more sustained calm,[15] because for a mammal such sweetness is associated with mother's milk. More subtly, as Darwin remarked, each touch, intonation, and expression, as well as how long it takes the mother to respond to signals of distress, contribute to an accumulating internal dossier on her "thoughts and intentions."

Even Bowlby, however, continued to think of a mother and a securely attached infant as one harmonious unit. Attachment theory developed independently of revolutionary changes in Darwinian thinking introduced by Hamilton (on kin selection) and Trivers (on the potential for conflict between maternal and infant interests). In the past decade, these ideas have forced geneticists and developmental psychologists alike to reassess the tensions inherent in this close, interdependent relationship involving two genetically nonidentical, divergently self-interested partners: mother and offspring.

Harmonious But Also Discordant Interests

By the 1990s, evolutionary biologist David Haig was projecting Trivers's counterintuitive analysis of "mother-offspring conflict" back in developmental time. Haig called attention to the invasion of maternal tissue by the fetal placenta, whose secretions begin to mimic the mother's own hormonal messages to herself, usurping control over certain features of her physiology. No longer necessarily assuming a harmonious intimacy between mother and developing embryo, Haig's view transformed gestation into a complicated "tug-of-war" over maternal resources.

Gestation, a host of complex transactions across the placenta, would come to be viewed as a high-stakes physiological game as mother and embryo maneuver for advantageous shares of her bodily resources. But this is a game with specific boundaries: if her potential losses rise too high, the mother bails out; the infant wins only if the mother is still playing at birth and through the period of infant dependency. These newborns emerge immobile, unable to forage or regulate their own temperatures, defenseless and exposed to diverse dangers. Nonhuman primate infants can at least cling; human babies—born developmentally so early that some anthropologists refer to them as "exterogestate fetuses"—can't even do that.

Once evolutionary theorists began to elucidate the innate conflicts of interest between a mother and the rapidly growing fetus within her, between an infant who desires to nurse longer and a mother who is ready to wean it, all the physiologically inefficient and sometimes noisy *Sturm und Drang* of Mother Nature's nursery takes on new dimensions. "Of course!" we exclaim in hindsight. Of course human infants are born connoisseurs: over generations, those that were not so savvy were more likely to miss cues critical for survival in an already hostile world.

In the human case, the costs of rearing infants (and then children) go on and on, long past weaning, even though members of the genus Homo pursuing novel lifestyles had become more fertile than any ape ever before. Meanwhile women were giving birth after shorter-than-ever-before intervals, fast-arriving primate newborns had not only to fear unwanted attentions from unrelated intruders (as other primates do) but also to become sensitized and evolve so as to dread (and if possible counteract) any hint of reluctance to commit on the part of a mother whose ministrations right after birth are all too desperately needed.

From an infant's perspective, what are the evolutionary consequences of contingent maternal love? Should it surprise us that across all 175 species of primates, none rivals the psychological sophistication and plump pulchritude of the human neonate? Parental preferences have produced runaway selection for their allure.

An infant's basic survival instructions include not just attaching to mother but appealing to her. At the slightest hint of ambivalence, baby hominids go into overtime, bent on keeping the engine of maternal commitment chugging away at full throttle. Yet even if she fails them—and this is the truly amazing part—human infants appear to be equipped with a few contingency plans of their own.

Over tens of thousands of years, in worlds where infant survival often depends upon maternal calculations, tradeoffs, choices, and prioritizing, an infant had to be appealing in order to extract more rather than less care from his mother, or in extreme cases to be cared for at all.[16] Maternal ambivalence and infantile allure are scarcely areas where many of us expect (and even fewer of us ever want) to find natural selection callously about her work. But diverse strands of evidence lead to the unavoidable, if disturbing, conclusion that natural selection has indeed operated in this realm. At no time has Mother Nature been more persistently about her "bad habits" than in the recent past of *Homo sapiens*.

There are many reasons why human immatures are vulnerable. But I have become convinced that in worlds before birth control, a mother's assessment of her infant's prospects to survive, grow up, and find a niche first in the family and then in the world has to be counted among the primary factors affecting infant survival.

If, from a mother's point of view, deciding to suckle any one child means sacrificing some other option (usually the option to conceive again in the near future), if being pro-life always necessitates a choice, then survival from a human infant's point of view meant being born sufficiently attractive to its mother to lure her into the first, largest, step toward lifelong commitment: the establishment of lactation. From ovulation to conception to birth, the mother's germ line (or hereditary material) embarks on a perilous odyssey, fraught with staggering wastage. By the end of this section on why infants are the way they are, it should be clear why nature's real heroes and heroines are the eggs, the embryos, the infants who against staggering odds survived at all.

———

Maternal ambivalence is treated today as if it were a deep secret only just being unveiled, with a rush of brave new publications in both fiction and psychoanalysis revealing "the maternal heart of darkness, a territory all mothers and clinicians know, but few discuss."[17] But this new post-Bowlby way of thinking about mothers and offspring, applying an evolutionary perspective to both their discordant and common interests, alters our initial expectations. It leaves many mothers—myself included—wondering how any one of us could have been so naïve as to ever envision the mother-infant dyad in the idealized glow of harmonious unity.

Far from being surprised or shocked, we should be asking ourselves how we failed to expect these ambivalent emotions in their every nuance. There are good reasons *why* infant demands sometimes seem so insatiable, and there are equally good reasons *why* mothers sometimes find such servitude overwhelming and resist them. There are also sound evolutionary reasons why such tensions would have an important impact on the developing child's view of the other people in his or her world. For what Bowlby termed the baby's "internal working model" of relationships would in fact constitute the best predictor any developing human could have about what to expect. The internal working model is more nearly a working hypothesis about how much support from mother and other kin will be forthcoming, and to what degree.

When a Human's Life Begins

If, as David Haig argues, this tug-of-war begins at the moment of conception, then from the perspective of a precivilized mother, abortion and infanticide are equivalent, except that in earlier times the former (abortion) was riskier for the mother, whereas today birth followed by infanticide is. There are certainly many important legal and ethical distinctions to be observed here, but none that Mother Nature would take note of. This haunting realization raises unescapable ethical dilemmas for those committed to a woman's (legal) right to choose how, when, and whether she'll sign over part of her existence to another being, as well as for those who take seriously a child's need to be born wanted by someone prepared to care for it.

If we are to rescue from mysticism the debate over reproductive rights, we cannot ignore these issues. If we accept that the fetus is a dependent organism that happens also to have its own agenda, at what point in the process of gestation can we honestly declare that the life of an individual has still not begun? How can it be legal to buy, sell, or inherit embryos (as it is in many states in the United States) but not babies? On what grounds do humans accord the mother's interests priority over those of her fetus? "Equal Rights for Unborn Women" reads the slogan on an anti-choice T-shirt. What to do about seemingly irreconcilable moralities: the rights of the unborn versus the bondage of the born? If human DNA is 98 percent identical to that of a chimp or a bonobo, what is it that makes the information encoded in this DNA human rather than ape? What is responsible for the transformation of the potential encoded in human DNA into a being with the unique cognitive and emotional capacities that make us "human" and distinguish us from all other animals? Scientists estimate that a mere 50 or so genes—out of the vast number of genes that chimpanzees and humans share in common—account for the cognitive differences between the two species; that fractional genomic disparity combined with differences in several regulatory genes that control the timing of gene expression make all the difference.[18]

In particular it is attributes like language—as George Eliot and her common-law husband, the nineteenth-century evolutionary psychologist George Henry Lewes, pointed out long ago—that allow "the mother to transform a maternal instinct into a maternal sentiment." Language is integral to the symbolic capacity that allows humans to understand cognitively what others are expressing at the same time as we understand at an emotional level what others are feeling. It is because "a human mother can appreciate the claims of offspring in general, not just her own" that a woman "can have an emotional appreciation of the claims of the helpless which is denied a baboon," Lewes wrote.[19]

What makes us humans rather than just apes is this capacity to combine intelligence with articulate empathy. But all humans develop this empathetic component in the first months and years of life as part of a unit that involves at least one other person. This is what the psychoanalyst and pediatrician D. W. Winnicott meant when he said, "There is no such thing as a [human] baby; there is a baby and someone." No matter how sophisticated the in vitro technology, or even the capacity to clone one human organism from another, the DNA of *Homo sapiens* does not develop these uniquely human capacities without the intervention of other humans, sustained interactions between a genetically engineered baby and its interacting caretakers.

To be distinctively human—different from, say, a genetically very similar chimpanzee—is to develop this unique empathetic component that is the foundation of all morality. Such unique capacities do not, cannot, develop in a social vacuum. In chapter 23 ("Alternate Paths of Development") I will speculate on just how and why this quota of compassion for others unfolds as it does, and why I believe so strongly that to be pro-life, meaning in favor of *human* life, means being pro-choice.

But I am getting ahead of my story, which begins with Darwin and his biographer-analyst John Bowlby, who set out to investigate exactly why it is so important for human immatures to feel securely attached, and ends with my speculations as to why receiving signs of commitment from others is essential for developing uniquely human capacities.

Meeting
the Eyes of Love

A child forsaken, waking suddenly,
Whose gaze afeard on all things round doth rove,
And seeth only that it cannot see
The meeting eyes of love. *

—George Eliot, 1871

D
arwin embraced the entire natural world and every species in it as his subject. Bowlby applied Darwinian concepts on a more intimate scale. Bowlby's focus was the infant's perception of his or her interpersonal world.

Bowlby's classic trilogy on attachment and loss—*Attachment* (1969); *Separation: Anxiety and Anger* (1973); and *Loss* (1980)—sought to explain the evolutionary origins of the emotional attachment that humans forge with mothers, and allomothers. One goal was to explain the desperation, rage, and despair experienced when these attachments are ruptured.

Producing "Attachment Theory"

If the field of evolutionary psychology is in any sense the revolutionary "new science" touted on the cover of magazines,[1] and not just sociobiology-by-another-politically-more-palatable-name (as much of it is), that new science began with Bowlby, four decades earlier. Through Bowlby, significant portions of psychoanalysis, and eventually psychology, were, in his words, "recast . . . in terms of evolutionary theory" and brought in from the cold, where they had previously been left out "beyond the fringe of the scientific world."[2]

---

* These are the verses from Eliot that Bowlby used as an epigraph for chapter 2 of *Attachment* (1969).

Just before the outbreak of World War II, Bowlby went to work as a psychiatrist at a center for troubled youngsters in London. He was struck by the blank emotional responses of some of the delinquents assigned to him for treatment. Many of them shared a common developmental history. Either they had been physically separated from their mothers while young, or something prevented them from forming a close attachment to them. Bowlby's admittedly subjective observations became the centerpiece for his early paper "Forty-four Juvenile Thieves: Their Characters and Home-life," which was published just after war's end.

Bowlby was already thinking about mother-infant separation as a source of developmental disturbance even before he had an opportunity to view fellow psychoanalyst Rene Spitz's documentary film chronicling the lives of children in a Mexican foundling home titled "Grief: A Peril in Infancy." Spitz's aim was to capture the emotional distress, anger, and resentment, followed by self-protective emotional disengagement, of infants coping with abandonment. As an all-pervasive gloom settled over the little form in the crib, a formerly cheerful (preseparation) toddler was transformed into a "frozen," "passive," and "apathetic" automaton, often leaving war-sensitized viewers with the mistaken impression that they had just seen a film about concentration camp victims rather than children in an ostensibly benevolent institution.

Spitz's goal was emotional impact. This aim took precedence over scientific methodology. The main message of his black-and-white silent film could be summarized by one of the title cards flashed on the screen: "The cure: Give mother back to baby."[3] As a result, both the film and Spitz's subsequent papers (published in 1945 and 1946) were savaged by critics who pointed to inexplicably fluctuating sample sizes and to sources of bias, especially the fact that Spitz's study subjects had been available throughout for adoption so that the healthiest among them might have been skimmed off, leaving only the most damaged children for Spitz to study. The then president of the New York State Psychological Association described these attacks on Spitz's reputation as "a kind of hydrogen bomb . . . of destructive criticism; not a paragraph is left standing for miles around."[4] But this firestorm did not alter the impact of his film: it touched the consciences of institutional administrators throughout the Western world, and it solidified young Bowlby's conviction that his hunches were well founded.

———

By associating himself with Melanie Klein, a pioneer in the psychoanalysis of children, Bowlby entered "the only discipline [then] committed to the systematic study of emotions and relationships."[5] At that time it was essentially either Freudian psychoanalysis or nothing. Early on Bowlby, like Klein, accepted Freud's notion that the child's early relationship with the mother becomes a template for future relationships. But he grew increasingly frustrated with this focus on interior, imagined worlds.

When Bowlby saw a disturbed and hyperactive boy and an even more emotionally disturbed mother, he felt certain that the boy's difficulties had something to do with their relationship; but talking to (let alone attempting to treat) the mother was off-limits. For Bowlby, ambivalent, remote, negligent, or abusive parents were not necessarily imagined; some were all too real. Yet as a psychoanalyst, his job was to work through the child's fantasies. "I held the view that real-life events—the way parents treat a child—is of key importance in determining development, and Melanie Klein would have none of it. The object relations she was talking about were entirely internal relationships." The disapprobation was mutual. With regard to Bowlby, Anna Freud simply shook her head and said: "As analysts . . . we do not deal with happenings in the external world as such but with their repercussions in the mind."[6]

Bowlby Looks to the Animals

From the outset, Bowlby was attracted to the fledgling field of ethology then emerging in Europe, where researchers Konrad Lorenz in Austria and Niko Tinbergen in Holland were determined to observe their animal subjects in natural environments—the real world. At the time, however, there were no bridges between psychologists studying humans and animal behaviorists. "I badly needed a guide," Bowlby recalled.[7]

In 1954, at a meeting in London of a Study Group on the Psychobiological Development of the Child, convened by the World Health Organization, Bowlby and a young ethologist from Cambridge University named Robert Hinde found themselves as the "second string" stand-ins for Lorenz and Tin-

Fig. 16.1 Goslings are born programmed to attach to the first figure that looms before them, usually their mother, but in this instance the phylogenetically distant biped Konrad Lorenz. Following mother is the goslings' best prospect for keeping safe from predators. The astonishing thing Hinde told Bowlby was not that the hatchlings became "imprinted" and subsequently followed the first identifiable Pied Piper they encountered, but that this slavish devotion had *nothing* to do with food. *(Photo by Thomas McAvoy, © Life Magazine)*

bergen, who had declined to participate. Hinde, who studied the behavior of chaffinches, coots, and great tits, had been trained to search for both the immediate physiological causes as well as the ultimate, evolutionary causes of animal behavior. But his connections with psychology predisposed him to consider humans as well. In this bird man, Bowlby found his "guide"—a guide not only to thinking about mothers and infants in terms of what was going on in their real world, but also with consideration to what went on in past worlds, where the ancestors of these infants had also needed to survive.

Bowlby returned the favor by drawing Hinde into the study of primate, and eventually human, relationships.[8]

Where Does the Need to Attach Come From?

At the time, Freud and his followers, as well as the behaviorist school in America, all assumed that "The reason why the infant in arms wants to perceive the presence of its mother is only because it already knows by experience that she satisfies all its needs without delay . . ." They assumed the infant was conditioned by the mother's satisfaction of its "need for nourishment." But now it appeared that the infant's need for contact with a familiar figure was something quite separate from hunger. The "cupboard love" theory left a great deal about an infant's love for his mother unexplained.[9]

Hinde alerted Bowlby to the bizarre experiments on "mother love" that Harry Harlow was doing at the University of Wisconsin. Harlow would provide infant rhesus macaques a choice between terrycloth-covered but nippleless "mother-surrogates" that permitted infants to cling to a warm, soft surface, or similarly configured dolls made of bare wire but equipped with milk bottles. The baby monkeys spent only as much time on the wire mothers as was necessary to feed before scrambling back to the comfort of the soft surrogates, where they felt more secure. Clinging to a surrogate mother had everything to do with this notion of "security" and little to do with satisfying hunger.

So powerful was the baby's urge to cling to a familiar figure, even an "abusive" surrogate mother was preferable to being detached. When Harlow presented baby macaques with mechanically devised spike-sprouting surrogates or mothers that suddenly puffed blasts of air, the infants gripped all the tighter to the only security they knew. About the only mother perceived by infants as worse than no mother was the refrigerated model Harlow engineered to be literally cold as ice; the tortured infant assigned to her huddled despairingly in the corner.[10]

Harlow's worst-case scenarios revealed baby monkeys desperately seeking security from the closest thing to a mother they could locate. Bowlby's mind was now alive to the possibilities of experiments with more normally reared infant monkeys. Would infants separated from their mothers respond with the same sequence of protest and despair that had been observed in hospitalized human infants? From this followed the next, bigger, question: Would maternal deprivation in infancy have effects detectable later in the monkey's life?

Fig. 16.2 One of Harlow's surrogate mothers.
(Harlow Primate Lab, University of Wisconsin)

Effects of Separation in Other Primates

The goal was to determine under controlled experimental conditions whether periods of separation from the mother during infancy had long-term effects on how secure or confident that individual felt. Like infant humans, all monkey and ape babies seek, in Bowlby's words, "secure attachment to a trusted figure" during the first months of life. Only in the most primitive-seeming surviving prosimians—such as bush babies (galagos), ruffed lemurs, or the saucer-eyed nocturnal aye-ayes, who give birth to multiple young and leave them in nests—are baby primates ever out of direct, tactile, body-to-body contact with a caretaker. In all other primates, infants are carried everywhere by mothers or allomothers.[11]

Experiments with rhesus monkeys undertaken by Hinde at Cambridge University showed that infant monkeys whose mothers were temporarily removed (so that the infants themselves were left in the same environment but deprived of this contact) sequentially exhibit "protest," "despair," and finally "detachment" as the infant begins to reorganize its behavior in light of its loss of security.[12] Monkeys in the wild act the same when separated from their mothers. An infant langur temporarily separated from its mother may climb high in a tree and make high-pitched birdlike calls. (Mothers, too, call piteously, and for hours—the unpublished side of the macaque studies.) Whatever was going on in the attachment of human babies to their mothers, the origin of such emotions went far back in time, as do the emotions of human observers who find these cries heartrending.

———

A number of experiments were done in which five- to six-month-old rhesus macaques were separated from their mothers for either one six-day period, two such periods, or one long thirteen-day period. In some experiments (code-named "Mother Goes to Hospital"), the mother was removed from the cage, while the infant remained with other group mates. In spite of the company, the baby monkey immediately missed the mother, looked everywhere, puckered its lips into plaintive whoo-calls, and ceased playing. The infant gradually sank into a motionless apathy that few would hesitate to call "despair." The longer the absence, the more depressed the infant became, and the longer recovery took when its mother was returned. Infants who suffered multiple separations had more pronounced, longer-term effects than those separated only once.[13]

When mothers were returned, infants clung yet more tightly. Many regressed in behavior, acting younger than they really were. Although "too old" for such behavior, the babies spent more time clinging to the mother than before the separation. Reluctant to leave the mother to explore, the reunited infants succumbed to violent temper tantrums if the mother moved away.

Just *how* distressed the infant was after separation depended in part on how secure the infant's relationship with mother had been before. The more the mother had been trying to separate herself from the infant (that is, if she had been at all rejecting before the experiment), or to the extent that the infant had any reason to feel insecure about her commitment beforehand, the removal of the mother was that much more distressing. Compared with secure infants, insecure ones given a "second chance" on the mother's return clung all the more desperately.

Twelve months later, infants separated from their mothers for either one or two six-day periods were compared with control infants that had remained continuously with their mothers. The separated infants explored less and played less. The most striking difference emerged when infants confronted anything novel. Two-year-old monkeys tend to be insatiably curious. Yet, the more reason the infant had to feel insecure, the more hesitant it was to explore even something as innocuous as a banana left in the adjoining cage. More than two years after the original experiment, infants who had been separated from their mothers remained more timid than infants whose mothers had not been removed.[14]

Separated infants also exhibited measurable physiological changes in sleep patterns and heart rates. Hormones like cortisol, produced by the body in response to physical and psychosocial stress, were secreted at higher levels as infants mobilized bodily resources to cope with the impending challenges of an "orphaned" situation. As less invasive procedures (analyzing samples of saliva rather than blood) became available, similar responses were recorded for humans; separated babies had elevated cortisol levels and their hearts beat faster.[15]

From these experiments it was obvious that the infant's relationship with its mother (whether the infant was securely attached to her before separation) affects how the infant responds to brief separations and the ensuing reunification. Such studies make it possible to observe how secure versus insecurely attached infants react. But something else was needed.

Ainsworth's "Strange Situation Test"

Like Darwin's theory of natural selection, Bowlby's theory of attachment was one of those grand explanatory schemes evoking a mystical allegiance from adherents. Decades before anyone had solid evidence that natural selection actually occurred in the wild, Darwin's hypothesis attracted ardent support as well as criticism. So with attachment theory. Bowlby, his coworkers, and their followers fervently believed in the importance of secure attachment for subsequent development, but scientifically they were hamstrung. To measure an emotion like love, to measure attachment, was like trying to apply a ruler to a wave.[16]

What finally brought attachment theory firmly into the realm of scientific study was a deceptively simple twenty-minute procedure called the "strange situation test." Invented by Bowlby's associate Mary Ainsworth, this ingenious procedure could be done under controlled conditions in a lab no more elaborate than a room with some toys, which made it possible to classify particular qualities of the relationship between infant and mother.

During research on infants growing up in rural villages in Uganda, Mary Ainsworth meticulously recorded the newborns' progression from undiscriminating acceptance of anyone who holds and comforts them to differential responsiveness, in which infants begin to prefer the familiar figures (not necessarily the mother) who comfort them most often. Since the mother was often the main caretaker, Ainsworth observed a highly differentiated preference for this most trusted person during a period when the infant typically cries if mother leaves and exults in her return. Infants not only protest when mother leaves, but actively take the initiative, purposefully crawling after her.

Near the end of this phase, around six to eight months, uneasiness at the appearance of strangers intensifies. This was the knowledge that Ainsworth employed to begin classifying attachment.[17]

After her return from Africa, while continuing her research at Johns Hopkins, Mary Ainsworth came up with the exact protocol of the now widely utilized "strange situation" procedure. According to prearranged plan, mother and infant arrive together in the lab; shortly thereafter the mother slips out, leaving her infant alone in an unfamiliar setting with a kindly disposed stranger. Then the mother returns. This process is repeated. Exactly what the infant does (or does not do) to reestablish physical and psychological contact with her mother after these absences is carefully observed, coded, and classified.

Virtually any normally reared primate infant, whether monkey or human, separated for a period from its mother, responds to her return by rushing up to her, jumping aboard, and holding tight. If the relationship with mother before the separation had been a bit tense (if, for example, a mother had disappeared on a previous occasion, or if the mother had begun to reject the infant as a part of weaning), the insecure infant would cling all the more tightly upon her return.[18] If a human infant feels very secure, he may scarcely notice his mother's absence, and continue to play and explore. Upon her return he merely looks up, a smile lighting his face, glad to see her—but not obviously relieved because he had not been worried. Whether easygoing or more stressed, six-month-old monkeys or two-year-old toddlers obviously derived comfort from their mothers. Mary Ainsworth assumed that all human infants were going to feel this way.

———

Probably no one was more surprised by the results of the first systematic applications of her "strange situation" procedure to humans than Ainsworth herself. As anticipated, the majority of human infants acted like baby macaques; they were distressed by the mother's absence, rushed to her on her return, and were comforted. But not all the toddlers acted this way. Some infants seemed wary and distressed even before separation from the mother. They were preoccupied by their mothers before as well as after, and failed to be comforted by her on her return. Others refused to look at her. Some actu-

Fig. 16.3 Mary Ainsworth (1913–1999) together with Bowlby in 1986. "He turned my whole career around in another developmental pathway," she said, recalling their first encounter thirty years earlier.[19] Ainsworth's "strange situation" procedure made it possible to scientifically test hypotheses related to mother-infant attachment. *(Courtesy of Erik Hesse)*

ally avoided the mother or seemed unnervingly nonchalant, as if they did not care.

These infants were classified as insecurely attached to their mothers. The insecure infants were divided into two categories: the first were "insecure/ambivalent" about their relationship to their mother. They focused on their mother but seemed hesitant to trust her, and were distressed when she left but not necessarily comforted by her return. In the second category were the "insecure/avoidant." "Avoidant" toddlers failed to show distress on separation and—against expectation—actually sought to elude their mother on her return. Researchers noticed that the mothers of these "insecure/ avoidant" infants seemed negative and even rejecting in their treatment of their infant.[20] When observed at home, avoidantly attached human infants actually exhibited the sort of anger that simmers quietly until suddenly, seemingly unprovoked, the infant strikes out at the mother.

When the Primary Attachment Figure Is Herself Frightening

A subset of infants could never be classified. This led Ainsworth's former student Mary Main, a professor of psychology at the University of California at

Berkeley, to reexamine these anomalous cases. A third category of "insecurely attached" infants that Main and coworkers termed "disorganized/dis-
oriented" was identified.[21] When the mother returned after separation, the
"disorganized" infants seemed confused, both seeking the mother and avoiding her. The attachment figure is "the primate infant's haven of safety in times
of alarm," wrote Mary Main. For this reason the infant who is frightened
by an attachment figure "should experience simultaneous tendencies to
approach and take flight. Conditions of this kind present an attached infant
with a paradox that cannot be resolved in behavioral terms." Together with
Erik Hesse, Main hypothesized that infants repeatedly exposed to frightening
behavior by their caretaker, or else to caretakers who themselves seemed
frightened, were presented with an irresolvable dilemma. As a result, the
infant was unable to marshal any coherent strategy.[22]

Ghosts in the Nursery

Ainsworth's "strange situation" procedure made it possible to classify and
measure security of attachment. Researchers could replicate each other's
findings and also document correlations between patterns of attachment and
different physiological responses. Longitudinal observations (studies made
over a long time) have now made it possible to compare early attachment
classifications with subsequent performance of the same individual in school,
and with assessments of particular personality features in the same individual
made later in life. Ainsworth's "strange situation" procedure has been repeated now in literally hundreds of studies around the world. Generally
speaking, securely attached infants tend to be more socially secure at the time
they enter preschool, and to respond better to instructions from teachers,
than children who have been classified as insecure and disorganized in their
attachment to a key caretaker.[23]

The little girl who avoided her mother's gaze at age one was more likely to
turn a deaf ear to her teacher at age six. Thus a child's treatment, even by
those who do not know her history, has a tendency to repeat and reinforce
past experiences with other people in new situations.[24]

From Spitz's films and Bowlby's first book, attachment theory, and along
with it characterizations of infants as either "securely" or "insecurely" attached
to their mothers, has flowed into the mainstream of thought about children's
emotional needs. Bowlby's ideas about attachment had practical applications

Fig. 16.4 Some infants responded to the mother's return with confusion or apprehension. Some would seem confused and put hands to mouth as shown here or bow their head and step backward or generally appear fearful. Such children could not be classified within Ainsworth's traditional categories. Subsequently, Mary Main proposed assigning them to a new "disorganized/disoriented" attachment status. *(Drawing by T. Rigney, courtesy of Mary Main)*

and entered the mainstream long before they were scientifically substantiated by Mary Ainsworth's "strange situation" procedure.

Longitudinal studies of the same infants make Bowlby's early speculations seem increasingly prophetic. In *Attachment* he had written: "the inheritance of mental health and of ill health through family microculture is no less important, and may well be far more important than is genetic inheritance."[25] And, more often than not, securely attached babies do grow into socially secure schoolchildren who mature into adults who form stable attachments and rear secure children, while insecure attachments breed more insecure attachments.

Transplanted back into psychoanalysis, Bowlby's scenario fell on newly fertile Freudian soil. It yielded the vivid metaphor of a "ghost in the nursery." By 1975, psychoanalyst Selma Fraiberg and her coworkers would declare:

> In every nursery there are ghosts. The intruders from [a family's] past have taken up residence in their nursery claiming rights of ownership. They have been present at the christening for two or more generations. While none has issued an invitation[,] the ghosts take up residence and conduct the rehearsal of the family tragedy from a tattered script.[26]

Empirical support for the "ghost" was not far behind. Mary Main devised an interview that permitted researchers to assess, decades later, how adults felt about their own attachment histories. She and her coworkers were finding that adults' recollections of their childhoods predicted how they would relate to their own infants, even how they would talk to them. It was the mother's manner that they particularly remembered, and subsequently reenacted with their own children.[27]

General patterns of attachment were proving consistent over time: insecure infants were more likely to become insecure children, and hence likely to become parents who produce insecure infants. These effects seemed to track separately from inherited differences in temperament and personality traits.[28]

Styles of Mothering Writ Large

In ways not yet understood, "ghosts in the nursery" interact with inherited temperament and local customs to generate regional differences in child-rearing. Studies done in northern Germany record surprisingly high proportions of avoidantly attached infants, while south German and Japanese babies are generally secure. Even when insecurely attached, they rarely avoid their mothers or exhibit indifference; rather, they resist by pushing their mothers away.[29] Similarly, children reared communally in Israeli kibbutzim are almost never classified as avoidantly attached, so long as they retain sustained contact with their mothers.

From the age of six weeks, kibbutz infants are cared for in small communal groups, nine hours a day, six days a week. They were visited frequently by their mothers, who came each day to feed and bathe their babies. Some of these infants were in communal care during the day but slept at home at night. Of those sleeping at home, 80 percent were classified according to the Ainsworth procedure as securely attached. But of those spending the night away from home, only 48 percent were.[30] Boarder babies were deposited in large communal dormitories, where they were mixed in with all children under twelve and attended by a couple of custodians. A child waking in the night would probably not encounter anyone familiar enough to qualify for what Eliot and Bowlby thought of as "the meeting eyes of love," the baby's primary reassurance that care will continue to be forthcoming. The message is fairly clear: effects of communal rearing depended on both the quality of the care itself and on the child's experiences within its own family, in his or her quest for commitment from caretakers.

Compared with Darwin's "dangerous idea," the evolutionary theory that philosopher Daniel Dennett has termed "universal acid" because it cuts so deeply into human conceits about our place in the universe, Bowlby's intellectual acid was less corrosive. Yet for psychoanalysts, for feminists, and especially for any woman with ambitions, it burns very deeply indeed.

Situating infant emotions in a tangible world trivializes psychoanalytical preoccupations with imagined, interior worlds. For infants, the world really is a dangerous place. By situating the mother (or other primary caretaker) at the center of each developing infant's universe, Bowlby's theory of attachment stings most smartly where it pricks the conscience of every mother who is aware of her infant's needs but who also aspires to a life beyond bondage to them.

To this day, the sting of Bowlby's acid evokes in me the haunting words of Anke Ehrhardt, a preeminent woman scientist, a woman every bit as impressive for her warmth and grace as for the accomplishments that made her the world's expert on children's development of gender identity. Over breakfast at a scientific conference in Prague this extraordinarily nurturing woman confided why she consciously decided never to have children. She said it was because she "knew too much" about what they need.

"Secure from What?"
or "Secure from Whom?"

You know——at least you ought *to know,*
For I have often told you so——
That Children never are allowed
To leave their nurses in a crowd;
Now this was Jim's especial Foible
He ran away when he was able,
And on this inauspicious day
He slipped his hand and ran away!
He hadn't gone a yard when——Bang!
With open jaws a lion sprang. . . .

—Hilaire Belloc, 1938

B owlby sought to explain infant desires and fears in terms of past environments of evolutionary relevance. He recognized that the endocrinological, sensory, and cognitive makeup of infants were composites of ancient dramas encompassing innumerable past lives. Any ancestral ape that survived long enough to grow up and reproduce can be assumed to have spent the first months of life in continuous ventro-ventral (stomach-to-stomach, chest-to-breast) contact with its mother, and to have been motivated to remain near to that mother for many months after. Separation from mother spelled disaster, a point Bowlby loved to make by citing Belloc's poem about ill-fated "Jim."[1]

A wild chimp baby orphaned before age three does not survive, even if adopted by a solicitous older sibling. Even after age five, losing a mother is a serious, life-threatening liability for a wild ape. Vigor is severely jeopardized, and loss of mother almost certainly produces a suite of costly coping responses like rocking, hair-pulling, self-clutching, reduction in time spent playing, and long-term timidity. Human apes in a foraging context fared little better. Of the Aché infants who lost their mothers prior to one year of age, not one survived.[2] For pre-Neolithic, still-nomadic humans the minimum age for close mother-infant contact must have been quite variable, depending on local conditions. It is extremely unlikely, however, that an infant could have been weaned earlier than two years without severely jeopardizing his or her life chances. Younger than four years, the mother was still gambling.[3]

Little we have learned since runs counter to Bowlby's principal assumption that during humankind's "Environment of Evolutionary Adaptedness," infants remained in close contact with mothers. Nevertheless, ethnographic evidence from foraging peoples such as the Aka and the Efé, plus new evidence from other primates, suggests that fathers, and especially alloparents, were more important alternatives to *continuous* one-on-one contact with the mother than Bowlby had realized.

When it was safe to do so, or when she had little or no choice, a mother delegated care to others. However, this does not mean that the infant calculates the tradeoffs involved the same way that the mother does. As psychoanalyst Rozsika Parker puts it, whenever "a mother and a child recollect their relationship, two separate narratives emerge."[4]

From the infant's point of view, being close to its mother would always be the infant's top priority, even if the mother might benefit from more distance. For one thing, the typically dilute, low-fat constitution of primate mother's milk means that breast-fed babies had to suckle more-or-less continuously to feel satisfied, especially when, as in the case of nomadic mothers with lean diets, mothers were just barely keeping up with infant demand. An infant constantly on and off the breast, in turn, ensured widely spaced births, a beneficial side effect from staying close. Beyond security and nutrition, continuous access to the mother's breast means that a mother's subsequent conception is held at bay. Among other things, then, staying close to mother protected the babe-at-breast from competing demands on the mother by siblings not yet conceived and the threat that a mother who finds herself pregnant will wean at once. One of the most serious challenges that could confront an infant in a world where the alternatives to mother's milk were neither digestible nor safe was for its mother's prolactin levels and other ovulation-inhibitors to slip so low that she ovulates and conceives before her current infant can be safely weaned. Yet under the newly settled living conditions of the Neolithic this must have been a common occurrence. Babies eating solid foods sooner would have suckled less, communicating to mothers

who were also eating more and walking less that her baby was no longer ben-
efiting from milk before this was really true. Premature conceptions pitted
the needs of an older infant against its younger sib's.

Anything But Blank

Far from being a blank slate, a human infant is equipped with a number of
behavioral systems ready to be activated. These systems range from the
simpler "fixed action patterns" (clutching, grasping, rooting for the nipple)
to more sophisticated cognitive capacities that require dynamic feedback
from the environment and practice to learn. Although not entirely preprogrammed (like the Moro reflex, discussed earlier), these learning systems are
biased so that the relevant processes are activated by stimuli falling within
one or more broad ranges, terminated by stimuli falling within other broad
ranges, and strengthened or weakened by stimuli of yet other kinds.[5] And
what provides the critical stimuli for infant development to unfold? The
more-or-less continuous presence of a sympathetic and responsive caretaker.

––––––

Like Bowlby's Environment of Evolutionary Adaptedness, his articulation
of "learning biases" has become integral to the vocabulary of evolutionary
psychologists. By now, Bowlby's ideas about the grooves Mother Nature
engraves on every baby's slate are themselves entrenched in the working
assumptions of cognitive psychologists. Infants, we now know, don't process
random stimuli but seek out and fixate on specific patterns, like the components of a human face. They prefer curves (like cheeks or eyebrows) to
straight lines, strong contrasts of light and dark (like pupils surrounded by
the whites of eyes), and acute angles (the corners of eyes) to obtuse angles.
Infants are captivated by movement within a frame (lips talking in a face).
"When you add up all these innate preferences," writes infant psychologist
Daniel Stern in his beguiling *Diary of a Baby,* "they almost spell FACE."[6]

Not just any old face, though. Newborns preferentially seek out faces with
humanlike configurations: two eyes, a nose, and a mouth. They distinguish
symmetrical "pretty" faces and look at such faces longer than at lopsided or
grotesque ones, and perhaps seek out faces that are feminine. Babies also look
longer at familiar faces, which most psychologists take to mean that infants
"prefer" them.[7]

Head propped up on a pillow, a newborn turns at will. She lingers longest facing in the direction of the pad her mother wore inside her bra, savoring that smell in preference to the smell wafting from the other side, from a pad worn by someone else. Infants are especially attuned to melodious, high-pitched voices that rhapsodize in "babyese." By the third trimester of pregnancy, fetuses can and do hear through the womb; their hearts beat faster when a tape of the mother's voice is played than if it is the voice of a stranger.[8] Not surprisingly, then, experimenters can demonstrate that within three days after birth the neonate prefers mother's voice to that of a stranger.[9] The sight of mother matters, but so does the mother's affect, how she seems to feel toward her baby. Infants confronted with a mother who is depressed (or a mother instructed by experimenters to wear an immobile, emotionless mask) find it unnerving.[10]

Nothing blank about this slate: human infants are predisposed to seek out a familiar, feminine, person: a person likely to be its mother. When they find her, even brand-new babies are prepared, during alert windows of opportunity, to open communication by responding to and imitating caretakers.

Animal and Human Predators

Identifying mother is a critical first step to attachment, but attachment for what? To Bowlby, the moral of Belloc's ditty was made clear in the succeeding lines: "And always keep a hold of Nurse / For fear of finding something worse." In the Environment of Evolutionary Adaptedness Bowlby had in mind, "worse" meant predators: possibly hyenas, but most likely big cats—tigers and leopards in the Old World, jaguars and mountain lions in the New. "Protection from predators is by far the most likely function of attachment behaviour," wrote Bowlby.[11] That attachment to mother keeps infant secure from predators has been more or less the party line ever since. But let's examine the idea further.

There can be little doubt that any Pleistocene cat worth its paws would indeed devour an unprotected immature. From South Asia (tigers) to South America (jaguars) big cats still occasionally kill and eat grown men. Owing to their small size and defenselessness, immatures would be especially vulnerable. During one particularly bad year (1878), British colonial officials in India reported 624 humans—mostly children—killed by wolves. Grown-ups are too big for wolves to take, and infants are normally kept close for safety. But the local practice of sleeping outside on stifling-hot summer nights permitted

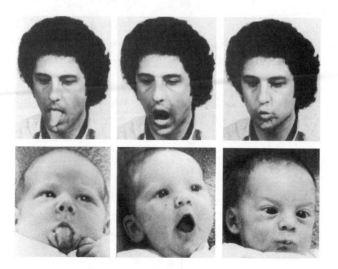

Fig. 17.1 Little *"Homo imitans,"* quipped psychologist Andrew Meltzoff, about the infant subjects, ages twelve to twenty-one days, that he induced to imitate different facial expressions.[12] *(Courtesy of Andrew N. Meltzoff)*

these wolves their chance: they would slip noiselessly into open-air bedrooms and make off with a young child.

Every forest culture has its own version of Belloc's rhyme to terrify children into staying close. All parties involved are either consciously aware of the dangers, so that children rarely wander from camp on their own, or innately predisposed (in the cases of toddlers) to stay near mothers.

But were wild animals the only dangers? Worried about an influential alternative hypothesis to explain an infant's attachment to his mother (namely, the idea that infants learn valuable social skills from her), Bowlby posed additional questions for himself. Why, he asked, "should attachment behaviour persist into adult life, long after learning is complete?" and "why . . . should it be especially persistent in females?"[13] Both questions could be satisfactorily explained by his anti-predator hypothesis. But so might they also have been explained by a third possibility: the threat to infant primates from unrelated males, predators of the same species. This possibility, if it occurred to Bowlby at all (evidence for this phenomenon was still skimpy during the years he was developing attachment theory), would have been dismissed as an aberration, of no relevance in the EEA.

———

As documentation improves over the twentieth century, records for nomadic foraging societies, as well as for some more settled pre-Neolithic societies in ancient Greece, show that children were intentionally targeted by stepparents, marauders from other groups, and even by competing mothers. Such practices did not necessarily end with the emergence of modern "civilization." Millions of children died at proportionately far higher rates than adults in episodes as disparate as the Stalin-induced famine in the Ukraine (1929–33) and Hitler's Holocaust. The point here is that whether in tribal raids or politically motivated genocide, children are the most defenseless and suffer the highest mortality.

Yet I believe it is true (wartime statistics aside) that in humans infanticide by invading males is less common, and a less important source of mortality, than in other primates, such as langurs, howler monkeys, gorillas, and chimpanzees. One possibility is that as the threat of infanticide became a selection pressure in its own right, greater paternal protection was selected for as well. Hominid families may have become especially effective at protecting otherwise defenseless immatures and mothers burdened by caring for them.

Males inclined to stay close to their families so as to protect their offspring had more infants survive. Not only did infanticide select for maternal counterstrategies to forestall male aggression against infants (leaving their group; mating with multiple males), but it may also have favored fathers inclined to defend their progeny. Human families had to protect young from many types of predator. Conspecifics, however, may have comprised the most worrisome class of predator.[14] Nor were kin and other group members, especially stepfathers and cowives, necessarily excluded from this roster of potential killers. The historical record provides its own dismal chronicle of pedicides, ranging from the Empress Livia's plotting against Octavius's heirs to little English princes strangled in the Tower.

Rethinking "Fear of Strangers"
Even though we cannot know how rare or common infanticide by individuals other than the mother might have been among our ancestors, the sheer weight of accumulating evidence for infanticide by unrelated males is causing some primatologists to forge a very different set of assumptions about the conditions under which hominid infants might have evolved. They accept as a given the proposal (which to many still seems bizarre) that members of their own species may be a threat to the survival of immatures.

Primatologists can argue endlessly—and no doubt we will—about infanticide's implications for the evolution of human families, and its implications for primate social structure generally. Yet one observation strikes me as indisputable: for a wide variety of primate species, including humans, an incursion of unrelated males who never mated with the mother is potentially bad news for immatures—especially those not yet weaned. Hence we need to reconsider some aspects of child development, such as the emergence of an acute panic response, specifically when infants are approached by strangers. It is a phobia that appears almost universally in older infants, just about the time that infants, although likely to still be nursing, become physically able to crawl away from their mothers.

––––––

Prior to five or six months of age, infants smile indiscriminately at almost anyone. It is as if this is a honeymoon period during which an infant becomes familiar with its local community, mostly kin. Around six months of age, however, infants begin to respond quite differently to unfamiliar people. Novel objects continue to be interesting, but the "visual looming" of a strange human jolts the infant into vigilance. His pulse rate rises; he may begin to cry. In the words of developmental psychologist Daniel Freedman: "Animate strangers are the major source of fear for infants in the second half of the first year." Strangers strike toddlers as especially fearsome if they are encountered in a strange place, if they are tall, male, bearded,[15] or if the infant is used to living among familiar people (rather than, say, in an orphanage), and is not accustomed to new faces.[16]

At one level, a fear of ostensibly harmless strangers seems an irrational phobia. But return with me to the forests of Paraguay where the Aché live, and where anthropologists Kim Hill and Magdalena Hurtado decided not merely to ask people what they said they feared but to collect data on what actually killed them. Hill and Hurtado painstakingly verified the deaths of 881 of the 1,493 Aché born since 1890, all of them nomads who died prior to 1971, before the Aché settled near the mission stations where they mostly live today. They were able to determine cause of death in 843 cases. Fevers and intestinal disorders (especially among children in the process of being weaned) were common. Of those dying from lethal accidents, snakebite was

Fig. 17.2 Fear of strangers tends to emerge at around six to eight months of age and appears to be a universal developmental stage. This filmstrip was made among New Guinea highlanders by Irenäus Eibl-Eibesfeldt, of the Max Planck Institute in Germany. As a visitor seeks to pick up the infant, he protests and retreats to his father. *(Courtesy of Irenäus Eibl-Eibesfeldt)*

the likeliest cause—eighteen adults ages fifteen and over, and eight children between four and fourteen, died after being bitten by a snake. During the same period, nine adults (but no children, probably because of the sorts of precautions just discussed) were killed by jaguars. Such data for other forest-living hunter-gatherers hint at the same pattern: snakes and predators are not hazards to be taken lightly.[17]

Oddly, though, infants don't have an innate fear of cats—big or small. Snakes, it is true, do have a peculiar salience. All primates fixate on them, take special notice, and once frightened, never forget. Yet typically, infants first must learn to be frightened of snakes by watching other group members react to them.[18] Experiments demonstrate that young monkeys learn to be afraid of snakes more readily than they learn to be afraid of, say, flowers, and that young primates pay very close attention to snakes or any long ropelike object on the ground that even looks like a snake; but all the same, they do have to learn from others that snakes are scary. (Once learned though, snake phobias prove particularly persistent.)[19]

But strange humans (especially adult males) are another story. No one teaches babies to fear strangers. Their panic derives from a built-in prejudice so deep it persists in spite of every reassurance the parent offers. The most benign and mellow of men, a friend from work encountered in the super-market, elicits a panic not alleviated by anything short of asking the mortified well-wisher to back off (or insisting Granddad shave his beard).

Across primates, very little is known about spontaneous "fear of strangers" in developing infants. It does not appear that monkey infants pass through such a phase. Indeed, Ryne Palombit, who studied infant-killing by strange male baboons at Moremi in Botswana, recalls youngsters fascinated by novel males, attempting to approach and socialize with them. In several instances, the males they "befriended" were the same ones who later killed them. Unlike humans, monkey infants apparently learn to fear strange males through frightening experiences. Chimpanzees, however, are reported to develop spontaneous fear of strangers on roughly the same schedule as human babies.[20]

The dawning realization that infanticide may have been a chronic threat during hominid evolution provides another possible reason why strangers would be a useful addition to a little hominid's repertoire of fears. During the period when the Aché described above were still leading a nomadic life in the forest, 55 percent of mortality among children between birth and five years could be attributed to something an adult member of the same species did. Of children under the age of three, Hill and Hurtado documented three daughters and one son killed by their mother, and three daughters killed by their father (granted, these were hardly strangers). Two other children died from neglect; one was buried alive; five were left behind when the group moved; eleven children were sacrificed along with an adult; and two died

Fig. 17.3 Infanticide has long weighed on human imaginations. According to Greek myths about *Cronos Devouring His Children* (the title of this 1820 painting by Goya), the Titaness Rhea was a mother with a problem. Every time she gave birth, her husband devoured her newborns. At last, pregnant with Zeus, Rhea decided upon a ruse suggested to her by Mother Earth. She hid the new baby deep in a cave (some say it was the Cave of the Dictean Zeus on the island of Crete, where bells, still audible today, chimed every time the baby cried to disguise his whereabouts). This time, when Cronos accosted Rhea, she handed him an enormous boulder wrapped in swaddling clothes. "Wretch! He knew not in his heart," exclaimed Hesiod (ca. 720 B.C.) that the baby Zeus still lived and would soon overcome his father by force. Psychoanalysts have focused on the "oedipal" tensions between father and son. Sociobiologists paid more attention to the timeless maternal counterstrategy of deceiving males. Children hearing the story just noticed how scary it all seemed. *(Courtesy of Museo del Prado, Madrid)*

when their mother died. But there were twenty-four cases of homicide involving nine girls and fifteen boys ages birth to fourteen. One was shot, and twenty other children were captured by hostile Paraguayans. These threats to child well-being would indeed have been posed by relative strangers.

———

I would not argue with those who complain that contemporary British or American society is "anti-child." Still, one has to ask, compared with when? Compared with where? Truly special is the society in which children never need to be afraid, and frankly I think such a society is more feasible today than in the past.

Admittedly, this is an odd whodunit, a less-than-scientific form of reverse engineering: asking what an infant's separation anxiety tells us about past threats in our evolutionary past. But it leads us to posit novel, previously undreamed-of explanations for a universal phenomenon like "fear of strangers." In many traditional settings, for a toddler just able to move away from its mother *not to fear* strange adults would seem a dangerous fantasy, perilously out of touch with realities in the worlds where our ancestors evolved. In some contexts, fear of strangers would prove itself a phobia worth suffering from.

Empowering the Embryo

How is the offspring to compete effectively with its parent? An offspring cannot fling its mother to the ground and nurse at will . . . [Rather it] should attempt to induce more investment than the parent is selected to give.

—Robert Trivers, 1985

A flock of finches scrounges for scarce seeds during a drought year. Only the toughest seeds are left. A handful of individuals with the strongest beaks manage to crack them open, to survive till the rains return. This ruthless weeding is natural selection. Darwin summed it up in one sentence: "What a trifling difference must often determine which shall survive, and which perish."[1]

More often than not, however, this drama of "trifling differences" is played out among trifling organisms, among mere cells in the form of sperm and ova, among conceptuses, neonates, hatchlings, fledglings, and weanlings, that flit only briefly across the evolutionary stage. As Darwin put it, "Eggs or very young animals seem generally to suffer most."[2] The curtain falls before what most of us think of as the play has even begun, the actors gone almost as soon as they are conceived, leaving no trace.

Surviving to Infancy

Each individual that survives to breed has first to hatch (if oviparous) or (if viviparous) to set up shop and extract resources from the mother efficiently enough and long enough to be born viable. Selection may begin even before any of these creatures leaves the protection of a follicle, an egg, their mother's womb. For some creatures, only a fraction of a female's eggs are even ovulated to begin with.

Consider the human case. By six months' gestation, even while still an embryo, each future mother has within her fresh-minted ovaries her full complement of oocytes, some seven million—fourteen thousand times more than will ever ripen to ovulation. For a woman who successfully bears five infants, this is on the order of 1.4 million times more eggs than she will ever convert into progeny.

Why so many eggs remains a mystery. Even before she is born, a female still in utero herself is losing eggs at a rate of approximately one every four minutes. From then until menopause, when, down to her last 200 or so, a woman ceases to reproduce, attrition is tremendous. Even so, the redundancy is staggering. What is it for?

One explanation is that an extravagant number of eggs guards against duds. The worst-case scenario would be that none were viable when needed. Prior to ovulation all these egg-precursors are essentially identical to cells of the mother's body. Yet flukes happen. In the course of their development, subtle differences between eggs emerge, making one a better candidate for ovulation than another.

Since the follicle matures but only proceeds to ovulation and conception if all systems are go, a wealth of eggs (though perhaps not seven million?) accommodates quite a few false starts. Perhaps overproduction of oocytes permits the luxury of having only the best of several available follicles selected from those ready each cycle. No one really knows. My own guess is that the answer is related to the benefits for mothers of being able to start, abort, and start again, the better to control her reproductive schedule and with it the viability of the end product. Here is quantity in the service of quality.

Behavior of Oocytes

Scientists are just beginning to fathom some of the ways oocytes react to the world beyond their follicular walls. Improvement in maternal energy balance (a 5-to-10-percent increase in body weight) enhances the likelihood of higher maternal estrogen levels; these in turn enhance the probability that a woman will ovulate and produce a fertile egg that month. Laboratories producing in vitro fertilizations have noticed that eggs from larger follicles also have an advantage after fertilization, since they are more likely to implant successfully.[3]

Further complicating a complex situation, the follicle surrounding the egg is also receiving instructions from the oocyte within. A growth factor secreted by the oocyte alters the outer layer's responsiveness to signals from reproductive hormones in nearby cells. It is possible that, along with maternal condition, some attribute of the egg itself shapes the destiny of its follicle.

Fig. 18.1 This embryo never managed to convert itself into viable descendants. Still, it is hard to imagine a more durable embryo than this fossil found in deposits from China dating to 570 million years ago. It was on its way to becoming some kind of multicellular animal with a complex life cycle when it was frozen in time.
(Courtesy of S. Xiao and A. H. Knoll)

Follicles communicate with other cells in the vicinity, and respond to slight changes in the mother month to month. Moreover, inherent qualities of each oocyte and its surrounding follicle affect the probability of ovulation versus decay. "Many are culled, but few are chosen," human reproductive ecologist James Wood wrote before these latest discoveries. An updated epitaph for an egg might read: "Many are culled but few succeed in volunteering."[4]

Only a fraction of oocytes ever rouse from hibernation. Most remain cloistered in the ovary. Gradually they degenerate and are reabsorbed locally. Nunlike, they age celibate and in situ through a gentle process known as *atresia*. Only rarely is an egg just an egg's way of making another egg. Atresia, not ovulation, is the fate of most. Of the lucky ovulates, even fewer are fertilized, elevated to the status of conceptus. Every conception, then, occurs at the cost of alternative ovulations, essentially alternative lives. The chances of an egg becoming a daughter are in the vicinity of one in ten million. The odds that that daughter will survive to breed, that that egg's egg will perpetuate her line, go down from there.

Why So Many Eggs?

A great deal of evolutionary theorizing about sex differences rests on the central dogma of *anisogamy*—gametes differing in size, with tiny, agile, and abundant sperm actively vying for access to one large and passive resource-rich egg.[5] In many species of birds and insects female reproductive rates are limited by the energetic cost of producing eggs. However, this generalization gets universalized, extrapolated to humans from breeding experiments with creatures like fruitflies.[6] Not long ago, I overheard one graduate student in psychology tell another: "Men are more competitive than women are because

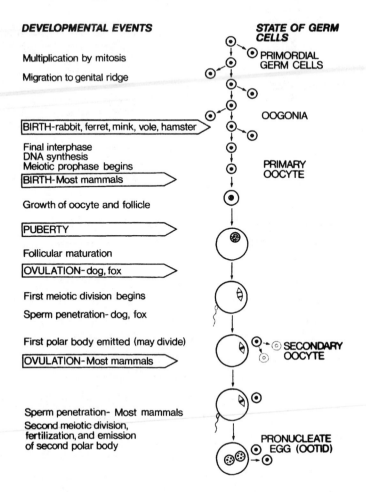

DEVELOPMENTAL EVENTS

Multiplication by mitosis

Migration to genital ridge

BIRTH-rabbit, ferret, mink, vole, hamster

Final interphase
DNA synthesis
Meiotic prophase begins
BIRTH-Most mammals

Growth of oocyte and follicle

PUBERTY

Follicular maturation
OVULATION-dog, fox

First meiotic division begins

Sperm penetration- dog, fox

First polar body emitted (may divide)
OVULATION-Most mammals

Sperm penetration- Most mammals
Second meiotic division,
fertilization, and emission
of second polar body

STATE OF GERM CELLS

PRIMORDIAL GERM CELLS

OOGONIA

PRIMARY OOCYTE

SECONDARY OOCYTE

PRONUCLEATE EGG (OOTID)

Fig. 18.2 Through oogenesis, the female germ line emerges as the tiniest of dots, gradually becoming detectable as a distinct cell line less than one month after conception. Clustering near the conjunction of the yolk sac and the sac developing into the esophagus and gut, these eggs-to-be slowly maneuver their way, by about day 35 of gestation, to the innards of the embryo. There they lodge in the genital ridges that will develop into the gonads. From that point on, the eggs change little until they multiply through mitosis, the process of cell division that preserves the full complement of 46 chromosomes in each of the daughter cells. These cells stop dividing around the sixth month, when the oogonia begin active DNA synthesis in preparation for the first stage of meiosis. This halving of the chromosomes makes a productive union possible one day when an appropriate sperm arrives at the *zona pellucida,* signals its eligibility (no one knows how), and is embraced by the egg and drawn past her barricades. In the meantime these gametic Sleeping Beauties slumber on unchanged for the next fifteen to twenty years in suspended animation, waiting for Sperm Charming. *(From Austin and Short 1982, courtesy of Cambridge University Press)*

a male's sperm cell is smaller than an egg." Obviously, though, life is more complicated. Reality must intrude.

At the level of wastage we are talking about here, relative cheapness is like talking about the price of hydrogen versus the price of oxygen, so abundant are they both as to make the relative cost of either sex's germ line incidental to the big picture. Even given the fading viability of her eggs between genesis and menopause, each woman is born with more ova than she could possibly convert into offspring. What is costly, as well as what makes human life so valuable, is not the egg, or even the conceptus, but the *nurturing* it takes to rear a human being. My own guess is that *it is because infants are so costly to rear that in species like humans women have so many eggs.*

How do I arrive at this idea? Estimates of just how many conceptions (once ovulated and fertilized) make it past the first cut to implantation are problematic since very early pregnancies are rarely detected. In the case of primates, we can only estimate the risks of fetal loss between conception and implantation. Information about losses is more accurate after implantation. For well-nourished monkeys, safe from predators and monitored for diseases in breeding colonies, as well as for humans living in modern cities, roughly 20 percent of pregnancies detected at implantation end in fetal death. Rates vary tremendously with maternal nutrition, health, and stress levels. When conditions deteriorate, loss rates skyrocket.[7]

Much early mortality is attributable to nothing more than bad timing. Among humans, for example, mother's age is an important predictor of how likely she is to spontaneously abort. Teenage mothers, and especially women over age thirty-five, are at higher risk. A fetus conceived by a woman in her mid-twenties has the highest probability of being carried all the way to term.[8]

As always, there is the luck of the genetic draw. Most chromosomally abnormal conceptuses will be aborted even before they implant. Chance plays a big role, yet surprising opportunities for selection to act on embryos persist. Embryos might secrete hormones that forestall spontaneous abortion or jack up nutrients their mothers supply them.

Embryonic Decisions

Few people think about differences between one embryo and another, much less imagine embryos "making decisions," pursuing one developmental course rather than another. Yet it would be a mistake to assume that an embryo is an embryo is an embryo. To assume that they are alike simply attests to how rarely biologists find opportunities to examine the behavior of embryos. Recently, a few have begun to.

When Karen Warkentin, a graduate student at the University of Texas, noticed that eggs laid by red-eyed tree frogs at her study site in Costa Rica took anywhere from five to thirteen days to hatch, she began to wonder if (looking at it from the embryo's point of view) there might be a reason.

It is assumed that adult survival among these colorful green frogs, with their red eyes, orange hands, and bright blue-and-yellow-striped sides, depends on such "decisions" as remaining hidden from predators during daytime and waiting to go out in the open until night, when they can feed under cover of darkness. But what kind of decisions could embryonic tadpoles, still dependent on the yolk of their eggs, profitably make?

Warkentin knew that parents deposit gelatinous masses of eggs on leaves hanging above their future home in the pond below. That way, embryos have a chance to grow bigger before they confront aquatic predators, mostly shrimp and fish. But not quite all of their predators are aquatic. Should a snake happen upon their leaf, the eggs find themselves caught between a snake and a shrimp-ridden place. Warkentin knew that, at her study site, more than 50 percent of clutches are attacked by snakes. Hence she experimentally introduced a snake onto the leafy refuge of five-day-old eggs while leaving control eggs of the same age to develop undisturbed. Confronted with *certain* consumption if they stay put, but only a *possibility* of being eaten if they drop, the nearly mature eggs respond to snake attack by jolting themselves into hatching prematurely, rupturing surrounding membranes and sliding off the leaf into the pond below.[9]

As Warkentin watched a snake gulp a mouthful of *eggs,* which she assumed must all be doomed, one wriggling *tadpole* popped head-first back out of the corner of the snake's mouth! Eggs oblivious to wind and rain, even earthquakes, have evolved to react immediately to the vibrations of a snake's attack. Embryos with the developmental flexibility to fast-forward on a split second's notice are more likely to survive. Of those some will one day metamorphose into serenading males whose clucks and other breeding antics get all the notice, completely eclipsing the key role played by the embryonic precursor to that actor in the spotlight.

Elsewhere among the amphibians, predator-induced hatching is played out in reverse. For example, some salamander larvae *slow down* hatching if they sense the presence of predatory gangs of flatworms. The larger the sala-

Fig. 18.3 Eggs of red-eyed tree frogs are jolted into early hatching by a snake's attack. In some oviparous animals, ranging from robins to crocodiles, hatchlings continue to benefit from parental care and protection. In others, however, mothers endow eggs at laying with all the investment they will ever receive. In such cases embryos are on their own. They make their own "decisions" about when to be born in response to surrounding circumstances without so much as a knock or a consultation with their mom. *(Photo courtesy of Karen Warkentin)*

manders are when they hatch, the better endowed they are to escape being eaten. In this case predator-*delayed* hatching is the ticket to survival.[10] Only by being able to imagine embryos as active agents did researchers finally observe "predator-delayed" and "predator-induced" hatching, though the phenomena obviously had been occurring all along.

Parent-Offspring Conflict

Even when they are essentially able to reach across the driver's seat to adjust the accelerator determining exactly when they hatch, embryos can't do much about the size and make of their vehicle. They must confine themselves to the resources present when the mother produced the egg. They must, as it were, play it as they are laid.

However, among viviparous and suckling creatures like mammals—developing embryos whose mothers bear and then nurse live young—immatures can aspire to more control over when they are born and how long they are nurtured. They may also negotiate how much of their mother's resources they will be allocated. To imagine embryos and sucklings as active agents this way is a new way of thinking about gestation and lactation. Instead of a model

Fig. 18.4 Trivers's theory of "parent-offspring conflict" drew evolutionary biologists into an out-landish domain where immatures are elevated to the status of main characters on the evolutionary stage. One is reminded of Grancel Fitz's pre–World War II photographic series depicting "Big Baby" lording it over his mother. *(Courtesy of Keith de Lellis Gallery, New York)*

comprising all-giving mothers and passive babies, this is a very different world inhabited by empowered embryos and manipulative weanlings.

It was Robert Trivers, the same young biologist who had shown how relative parental investment by females and males shaped Darwinian sexual selection, who elevated previously "disenfranchised" life phases in immature organisms to the status of active players with agendas of their own.

In the early 1970s, Trivers began to empower immature life phases by talking about infants—formerly viewed as agendaless grubs, passive tickets to posterity purchased by parents—as active agents whose behaviors could affect their own fates. "In classical evolutionary theory parent-offspring relations are viewed from the standpoint of the parent," he wrote in his classic 1974 paper outlining what he liked to call the underlying "logic" of parent-offspring conflict. "Parents are classically assumed to allocate investment in their young in such a way as to maximize the number surviving, while off-spring are implicitly assumed to be passive vessels into which parents pour the appropriate care."[11] His was a revolutionary perspective.

Thinking like a revolutionary was something that came naturally to Trivers. On paper, another upper-middle-class, prep-school-educated white male graduate of Harvard, during his graduate school years doing fieldwork in Jamaica (studying lizards) he transformed himself in dress, speech, *ganja* habits—even a propensity for bar brawls—into a convert, body and soul, of the Caribbean culture he adopted. Thinking back to those days, and to a time when there were anti-sociobiology demonstrations at Trivers's lectures on the grounds that he was seeking biological justification for an oppressive *status quo,* I can only smile and shake my head. In fact, I doubt that Robert Trivers ever met an institutionalized status quo that he didn't feel driven to destabilize.[12]

———

Previous generations of psychologists had not failed to notice the existence of "conflicting opinions" between parents and offspring. But by and large these disagreements were viewed in terms of the conflict between biology and culture. As Trivers put it, the baby was viewed as innately selfish, animalistic and greedy, in need of socialization. "This theory sees conflict as arising from the innate barbarity of the child."[13] Predictably, in seeking an evolutionary explanation for the conflict, Trivers identified with oppressed immatures.

At the time, he was already serving as British biologist William D. Hamilton's self-appointed bulldog, promoting his then little appreciated ideas on how degree of genetic relatedness among relatives influences their behavior toward one another. Hamilton's theory of inclusive fitness preadapted Trivers to understand parent-offspring conflict. An infant is related to itself 100 percent but shares only half of its genes by common descent with its mother, and 50 percent with her subsequent offspring, if full siblings; more like 25 percent, however, if they have different fathers. The mother will be equally related to all her offspring. Taking all this into account, conflict as well as a large measure of cooperation is an expected feature of their relations. "In particular," Trivers noted, "parent and offspring are expected to disagree over

Fig. 18.5 Robert Trivers, the sociobiologist who proposed new ways of conceptualizing cooperation and competition within the family, was strongly attracted to the warmth and inclusiveness of
the extended mother-centered Jamaican family that he became part of after marriage to his first
wife, Lorna Stapleton. He is shown here carrying their son, together with Lorna (laughing), at
the Stapleton home in rural Jamaica. *(Courtesy of Robert L. Trivers)*

how long the period of parental investment should last, over the amount of
parental investment that should be given, and over the altruistic and egoistic
tendencies of the offspring as these tendencies affect other relatives."[14]

    To the extent that investment provided by the mother to one offspring
detracts from what is available to her other offspring (current or future),
mothers and offspring are likely to disagree over the allocation. Trivers
phrased the genetic logic underlying this conflict in simple cost-benefit equations. For example, a mother is expected to encourage altruistic acts among
her offspring (to whom she is equally related) when *the benefit to one of her offspring is greater than the cost to another.* The offspring forced to give up some
benefit, however, only agrees (without sulking or dragging its heels), if the
benefit is twice (or even four times) his own costs. Offspring are only

expected to stop begging or attempting to extract more from the mother when the cost to her (in terms of her ability to survive and invest in other off-spring) is more than twice the benefit that the offspring receives.

Many mothers exhibit an intuitive sense of what they are up against. In her biography Nisa recalls: "When mother was pregnant with Kumsa [her younger brother], I was always crying. I wanted to nurse!" She also wanted her mother to carry her. When the baby was finally born, Nisa's mother told her to get a stick, as they would have to bury the baby alive since it had arrived after too short an interval. Whereupon Nisa herself announced: "My baby brother? My *little* brother? Mommy, he's my brother! Pick him up and carry him back to the village. I don't want to nurse!" Feeling personally responsible for her younger brother's fate in this way, Nisa accepted being weaned, although occasionally when her mother was asleep she would remove her little brother from the breast and suckle herself.[15]

Tantrums

Weaning conflicts epitomized for Trivers the disagreement over how much and for how long a mother should provision her offspring. Because the off-spring is selected to devalue the cost of parental investment, it will favor a longer period of investment than the parent is selected to give. Just after birth, when suckling is essential for her infant's survival, a mother and baby are likely to be of one opinion. Indeed, it is often the mother herself who ini-tiates nipple contact. As the infant becomes increasingly independent of the mother, however, the burden for initiating contact with the nipple shifts: the infant is more motivated to suckle than the mother is to provide additional nourishment. What is an infant to do? As Trivers put it, "An offspring cannot fling its mother to the ground and nurse at will." Because the offspring is smaller and less experienced than its parent, and its parent is in control of the resources at issue, this competitive disadvantage forces the offspring to resort to psychological tactics.[16]

Tantrums are the most dramatic, outward, and visible manifestation of these tactics. Though they appear to be spontaneous outbursts, Trivers be-lieved tantrums could be quite calculated. Infants engage in manipulative exaggeration of emotional and physiological states. In various species, includ-ing chimpanzees and humans, older infants frequently employ them, and continue to use such tactics throughout their lives.[17] Yes, offspring are defenseless, but they can nevertheless take advantage of existing means of communication to exaggerate their need, or vulnerability, perhaps acting younger or pretending to be in greater physiological distress than is actually the case. Immatures may play on their mothers' concern for their well-being. If temper tantrums seem crazy, Trivers believed they were "crazy like a fox."[18]

When Hinde told Trivers about the rhesus mother-infant separation experiments that had just been completed at the Madingley facility (at Cambridge University), Trivers barely heard the parts related to attachment theory. His imagination was inflamed by what the Madingley observations seemed to reveal about the underlying logic of the infant's attitude toward its mother, and for a time, as he described it, he could think of nothing else.

Rhesus infants whose mothers had already begun weaning or who were rejecting prior to the separation, and infants whose mothers were taken away rather than vice versa, tried harder to stay close when the mother returned than did an infant whose mother had not already been trying to distance herself; such infants also strove harder for proximity than did the infants who had been removed from the cage. For Trivers, sensitive to the underlying tensions of the relationship, the implications were enormous. The infant monkey appeared to be acting on "the logical assumption that a *rejecting* mother who temporarily disappears needs more offspring surveillance than does a *non-rejecting* mother who temporarily disappears." It was an observation Bowlby had already made. But Trivers interpreted it in an even broader evolutionary framework.[19]

What struck him was "seeing that all the machinations Freud imagined going on early in life had reality (which I had formerly disbelieved), but that [Freud] had misinterpreted it. For two months I had what you might call a brainstorm. Night and day I thought about nothing else. . . ."[20] Throughout his career, Trivers, like some tribal shaman, would delve into his own subconscious, then return with novel insights about his own social situation, which he then translated back into the language of evolutionary biology. Of the experimental results from Madingley that had stirred him so deeply, he simply noted:

> These data are consistent with the expectation that the offspring should be sensitive to the *meaning* of events affecting its relationship to its mother. The offspring can differentiate between a separation from its mother caused by its own behavior or some accident (offspring removed from group) and a separation which may have been caused by

maternal negligence (mother removed from group). In the former kind of separation, the infant shows less effect when reunited, because from its point of view, such a separation does not reflect on its mother and no remedial action is indicated.[21]

Maternal-Fetal Contracts

In 1978, Trivers left Harvard for the University of California, and eventually Rutgers. He was succeeded at Harvard by an Australian geneticist named David Haig. Trivers jokingly referred to Haig as occupying "the Trivers Chair in Evolutionary Logic."[22] Given the role Haig would play in extending Trivers's ideas, this was a remarkably apt title, even though it might have come as a shock to the Harvard administrators who no doubt breathed a sigh of relief when one of their most iconoclastic faculty members packed his *ganja* and rolled west.

During a post-doc at Oxford, Haig had familiarized himself with the vistas from what Hamilton (who had by then joined the zoology department there) calls the "Narrow Roads of Gene Land." Haig was fluent in Dawkins-style theorizing about the machinations of "selfish genes."[23] But over time, he also began to invent a new set of metaphors to explain the behavior of genes. Today, Haig is more likely to describe genes as "social"—members of a team, participants in what Egbert Leigh famously termed a "parliament of genes" rather than as purely selfish. Or, to quote Haig: "The complex behaviors and structures that have evolved by the process of natural selection can be viewed as adaptations for the good of the relevant genes ('replicators') rather than for the good of individual organisms ('vehicles'). . . . But organisms can also be viewed as collective entities (like firms, communes, unions, charities, teams) and the behaviors and decisions of collective bodies need not mirror those of their individual members." Although still interested in "the internal conflicts that can disrupt genetic societies" (intra-genomic conflict), he is also intrigued by the "social contracts that have evolved to mitigate these conflicts," especially conflicts between maternal and fetal genes.[24]

It fell to Haig to trace implications of "parent-offspring conflict" back through ontogenetic time from a tantrum-throwing toddler to a placenta-building fetus. Like Trivers's, Haig's passion in science was not so much for experiments or new discoveries, but rather for making sense of puzzles in the world around him. Instead of apologizing for metaphors that ascribe agency to genes, he simply points out that "Natural selection is a purposeless process with purposeful products." Thinking of genetic outcomes in terms of "the language of purpose" is a useful shorthand.[25]

Haig was one of a handful of pioneers in the arcane new field of "genetic imprinting," the still poorly understood process by which the same gene is expressed differently depending upon which parent it comes from. Haig theorized about a special class of genes that could "remember" whether they came from the mother or from the father, and act accordingly to promote either maternal or paternal interests.

Apparently, some genes inherit molecular "imprints" from respective maternal and paternal germ lines, so that it is possible for the expression of genes to be turned either on or off—that is, silenced in the course of development or turned up—depending on parent of origin. For most genes, this would make no difference. However, a minority of genes—possibly no more than fifty or so genes out of some 75,000 genes found in a human—are either activated or silenced depending on whether they are maternally or paternally imprinted.

What first struck Haig was how many of the imprinted genes discovered so far involve growth, perhaps especially growth in utero. Given that growth during gestation and lactation occurs at the expense of the mother's bodily resources, Haig began to see self-serving method in the madness of parent-specific genes.

In animals like mice where females are inseminated by successive males, paternally imprinted genes would have no "compunction" against extracting as many resources as possible from the mother, even if the pregnancy depleted her capacity to reproduce in the future. They must be countered by maternally imprinted genes coding for thriftier development—for males would have a stake in the future welfare of a mate (past her period of maternal investment in their single shared litter) only if they had evolved in monogamous mating systems, which these mice had not.[26]

By 1996, Shirley Tilghman and other molecular geneticists at Princeton undertook a series of experiments with specially genetically engineered strains of mice to test Haig's predictions. Ideas that had once seemed little more than wildly speculative gleams in Haig's eye were correctly predicting fetal growth in mice depending on which parent's genes were in charge.[27] At a given genetic locus, experimentally manipulated mouse embryos received either all their genetic instructions from the father or all from the mother.

Untrammeled expression of paternally imprinted genes produced giant babies, 130 percent of normal size at birth. Fetuses whose growth factors were monopolized by maternal instructions were dwarfed down to 60 percent of normal size.[28] In 1998, the strangest imprinted gene of all was described—the *mest* gene, which is expressed only when it is inherited from the father, never expressed if inherited from the mother. Mother mice whose only copy of the mest comes from their mother exhibit deficient maternal responses and, in particular, fail to eat the placenta after birth, missing out on that last dose of fetal propaganda.[29]

It occurred to Haig that if he was right about the long history of competing agendas carried forward in each fetus bearing imprinted genes from mother and progenitor, then "the interplay of opposing forces may be essential for a successful outcome to the pregnancy."[30] It was like a tug-of-war, where players adjust themselves to their tugging opponents by pulling too—in the opposite direction. Haig speculated that some of the more puzzling inefficiencies of pregnancy, as well as some of its occasional pathologies, were actually by-products of this long history of conflicting agendas. No longer upright, the participants lean markedly in one direction or the other, so that if the tugging ever stops, they topple over.

The Fetus's Supply Line

The idea of competing interests led Haig straight to the primary organ that brokers the transfer of resources between the mother and the mammalian fetus: the placenta. If paternally active genes bent on extracting extra resources from the mother to nourish that male's offspring were going to exist anywhere, the supply line to the fetus would be the place to look, Haig reasoned.[31]

Obstetricians have long referred to the placenta as "a ruthless parasitic organ existing solely for the maintenance and protection of the fetus, perhaps too often to the disregard of the maternal organism."[32] They were aware that maternal options concerning whether or not to terminate a pregnancy decrease markedly once implantation occurs, since the placenta secretes hormones that maintain the pregnancy. Usurping the authority of the maternal pituitary, the placenta synthesizes large quantities of a hormone called chorionic gonadotropin (the hormone whose presence causes the white Cheerio to form in early pregnancy detection kits) and releases pregnancy-maintaining hormones into the mother's bloodstream. In effect, after twelve weeks, the fetus is in a strong position to command the mother's body to carry on with the pregnancy.

As with any well-organized invasion, the next step is to establish supply lines. In some mammals, including prosimians such as the galago, a relatively "noninvasive" placenta absorbs nutrients secreted by glands in the mother's uterus (referred to as "uterine milk"). In others, including mice, bats, armadillos, and "higher" primates (monkeys and apes), the placenta has evolved so as to tap directly into the mother's bloodstream. As the embryo implants within the uterine lining, cells branch out that destroy the wall separating mother and growing embryo and remodel the endometrial spiral arteries, even expanding the diameter of blood vessels. The mother is unable to constrict the vessels supplying the embryo, so that, in Haig's words, the mother "cannot regulate the flow of nutrients to the placenta without starving her own tissues."[33] Thus does the fetus gain considerable control over its own provisioning.

By the eighth week of human gestation, the placenta composes a whopping 85 percent of the total package. End to end, the villi (finger-like projections growing out of the outer membrane of the embryo to increase the absorptive surface of the placenta) stretch about thirty miles. Once this massive infrastructure has secured the supply lines, the fetus itself starts to grow, unless the mother is malnourished. If there are few resources to go around, the placenta still grows large, but the baby remains small. Specialists in fetal development have come to regard a large placenta as the first and most obvious sign of a malnourished fetus.[34]

Maternal-Fetal Negotiations over Time of Birth, etc.

As much as an immature benefits from remaining in the womb, dimensions of the mother's pelvic canal set a limit on how long gestation can last before it imperils both parties. The exact moment of birth can be viewed as a negotiated settlement between embryonic demands and maternal assessments of the environment they will both inhabit. Extrapolating from data on sheep and monkeys, it is assumed that the baby's brain sends signals that increase estrogen production.

Estradiol released from the placenta stimulates the production of the hormone oxytocin and the signaling molecules (called prostaglandins) that promote coordinated muscle contractions, which deliver the baby. It is the mother, however, who ultimately controls levels of oxytocin, and hence the precise time when she gives birth. Among primates, birth begins during the hours when the mother and her group are least active—nighttime for diurnal primates—thus increasing the odds that parturition occurs in a safe spot.[35] (Although they may not know why it occurs, those attending in hospital maternity wards have come to expect this nighttime peak in deliveries.)

Long before sociobiologists raised the specter that the mother's interactions with the fetus developing inside her might be other than harmonious, generations of pathologists had already dipped into the lexicon of warfare to describe what plainly looked to them like an "invasion" of the lining of the mother's uterus by fetally derived cells. In the words of one pathologist writing in the *Journal of Obstetrics and Gynaecology of the British Empire* just around the outbreak of World War I: "The border zone . . . marks the division between the foetal and maternal tissues. . . . It is the fighting line where the conflict between the maternal cells and the invading trophoderm takes place, and it is strewn with . . . the dead on both sides."[36] Haig was not the first to borrow military vocabulary to describe the placenta, but he found the analogy useful.

Haig assumed that the placenta originally evolved in a world where mothers experienced periodic unpredictable famines, such that investment in current offspring might well detract from a mother's future ability to reproduce. Under such circumstances, it's each player for herself. After birth, infant ploys to circumvent maternal economies are played out through tantrums and other behavior; prior to birth, fetal grabs for a bigger share of resources are mediated by chemical messengers sent coursing across the placenta into the mother's bloodstream. Under normal circumstances what enforces the truce, though, is not just the fetus's current and future dependence on its nurturer but, Haig speculates, a "fifth column" established in the fetus by maternally imprinted genes that evolved to counter fetal greed.

Various medical pathologies can be explained by this accumulated history of conflicting agendas. If one side or the other were underrepresented—if, for example, there was a genetic deletion of either the paternal command or the maternal countermand—there could be serious repercussions. Hormones such as human placental lactogen, produced by the placenta, increase nutrients like glucose in the mother's bloodstream. To drown out such hormonal propaganda, the mother is forced to issue even stronger instructions of her own, work orders calling for more insulin. If for some reason the mother's body fails to counter the fetal command on cue, diabetes could result as an unfortunate by-product of the fetus's call for sugar-rich blood. Outcomes detrimental to both mother and fetus, such as the form of pregnancy hypertension known as preeclampsia, could represent a last-ditch effort by a poorly nourished fetus to increase its supply of nutrients. Given that low-birth-weight infants are especially likely to suffer from high blood pressure later in life, this drastic survival-at-any-cost fetal ploy (raising the mother's blood pressure in an effort to avoid starvation of the embryo) might not only take an immediate toll on the mother, but might also take a continuing toll from the embryo after birth, through infancy and into adult life, contributing to a lifetime of high blood pressure.[37] Amazingly, one of the best predictors of hypertension at age fifty is a combined measure of—what else?—placental weight and birth weight.[38]

For anyone puzzled—as I was—as to why a gene involved in placentophagia (eating the placenta after birth) should be expressed when transmitted from father to daughter but not when passed from mother to daughter, the placenta's role in promoting fetal interests is an obvious point to consider. Yet the more I try to understand what is going on with the genetically imprinted mest gene, the more puzzled I become. When would it be more in a father's than a mother's interest for a daughter to continue to be swayed by fetal propaganda after birth? What is still missing in this story? There is a very great deal left to know. Such curious facets of development—undreamed of just ten years ago—are only beginning to be studied. Over the next few decades, the study of imprinted genes is going to open up a whole new window into the dynamics of sexual interaction, and competition between matrilineal and patrilineal interests.

No one is suggesting that outcomes so detrimental to organisms—the unwitting "vehicles" transporting self-serving "replicators" vying among themselves to cancel each other out—are by design. Nor do such findings mean that a mother is inevitably pitted against her mate, or her fetus. For better or for worse, two parents, and to an even greater extent mother and fetus, are self-interested members of the same team. As Trivers cautioned decades ago, "Unlike conflict between unrelated individuals, parent-offspring conflict is expected to be circumscribed by the close genetic relationships between parent and offspring."[39]

As long as circumstances are "for better," all goes well (as when mothers have plenty of food and adequately long spaces between births to recoup). As long as the team is winning, both parties get along fine. But humans evolved in an unpredictable world, where neither party could count on things going well.

Putting Maternal-Fetal Conflict in Context

The pendulum has swung from depicting the most intimate human relationship as a harmonious partnership in which a fetus placidly "shares every breath that its mother takes and every meal that its mother eats" to the post-Trivers view emphasizing a team composed of players who share some goals but not others. At their most sensationalized, these disparate maternal-fetal interests are described as a "war in the womb."[40] Omitted from this catchy sound-bite is the complex dialectical history of maternal, paternal, and fetal measures and countermeasures implicit in Haig's evolutionary explanation for the perils and inefficiencies of gestation. Even a chromosomally defective offspring slated for early spontaneous abortion is not—even metaphorically—in an all-out war with the mother evicting it, although certain of its genetically produced traits might be. For the mother's next, probably more viable, tenant will share many of the same genes by common descent. This is a fairly subtle dispute in which the different parties have different thresholds for aborting or proceeding. At multiple points conflict may be exacerbated, or cooperation enhanced, depending on circumstances.

Just how inevitable is mother-infant conflict? Cautionary voices were not long in sounding, reminding theorists how conditional on social and ecological circumstances such conflicts could be. One of these voices was Jeanne Altmann, the primate sociobiologist studying mothers and infants in the real world. Selfish or parliamentary, she was less interested in genes than in how individuals behave and in situating behavior within "local ecologies." In a gentle rejoinder to Trivers's 1974 paper on "parent-offspring conflict" she sought to shift attention from the genetic relatedness of mother and infant (who share 50 percent of their genes by common descent), back to the "cost" and "benefit" functions that went into the tradeoffs being calculated by a time-strapped mother and a dependent infant.

Although Altmann admired the original way Trivers conceptualized parent-offspring conflict, she wanted more attention paid to the "underlying mysteries"—to development and the particulars of an individual's life. Whereas theoreticians focus on hypothetical effects of genes no one can see, fieldworkers focus on individuals they watch day after day. Unless an animal manages to survive and reproduce, it scarcely matters what any of those unseen selfish genes are up, or stoop down, to.

To Altmann, genes would never be anything more than accretions of molecules, some of them replicated at different rates, depending on which traits they produce. She went out of her way to emphasize instances of mother-infant consensus in response to Trivers weighing in so heavily in the other direction. Some would argue that she went too far, giving short shrift to conflict. But that's one of the ways science (like evolution) proceeds, through inefficient dialectics. Although some of the baboons at Amboseli did indeed throw "really wonderful" tantrums—"just like a [human] two-year-old"—others, she stressed, voluntarily weaned themselves.[41]

Baboons, Altmann knew, had lots of other business besides making babies. A mother ordinarily spends 23 percent of each day walking from safe refuges to feeding places, and another 55 percent of each day picking acacia pods or digging up the underground corms where grasses store their richest nutrients. This leaves only about 20 percent of daylight hours to qualify as "leisure," and that's not counting time needed for resting and grooming, more nearly essential than optional activities. As with most working moms, there is little slack in this "dual career" baboon mother's life. Suppose a baboon baby genetically predisposed to demand more milk did manage to extract it from her—where would the mother find time in the day to scrounge the extra calories to support such a greedy freeloader? Extracting more from her would be like getting the proverbial blood from a stone. Half of all baboons born die within the first two years, and it is not clear what more the mother could have done to keep them alive. Already those mothers who are carrying unweaned infants suffer twice the mortality rate of females unburdened by infants.[42]

For both mother and infant, survival depends on the hypothalamic feedback loop that prevents her from conceiving again while she is still nursing a dependent. Shorten the birth interval, though, and both underwriters of this precarious consensus are in trouble. So, at five months, a mother baboon in this population is in no hurry to wean. Nor does she have the energy to fight about it. With their mothers already pushed to the physiological limit, in a population already characterized by (for baboons) long birth intervals, there exists scant incentive for overt conflict.[43] (Covert conflict, between mother and a fetus in the womb, was outside of Altmann's purview.)

When Infants Volunteer to Be Weaned

Just when to wean is a delicate business. If the mother weans too soon, her infant dies. But if she concedes too much, they both die. This is one reason age of weaning turns out to be surprisingly negotiable. As the mother responds to environmental conditions, information about her condition is communicated to her infant. Similarly, information about the infant's nutritional needs is relayed back to her ovaries through the frequency and intensity of the infant's sucking.

Each party responds to the condition of the other. For example, recent experiments with rats by Montserrat Gomendio and others at Madingley— the same animal facility set up years before to test Bowlby's ideas—reveal that infants prepare themselves for weaning, producing the enzyme sucrase, which is essential for digesting solid food, at an earlier or later age, depending on maternal condition. If the mother is pregnant again, pups start to produce sucrase in preparation for early weaning, although how they "know" remains a mystery.

Infants, it appears, "volunteer" to ready themselves for early weaning. Yet if mother rats are experimentally deprived of food, and in consequence their infants fail to grow properly, mothers compensate by nursing their pups longer. These babies begin producing sucrase later, and are weaned later, but at the same body weight as infants born to well-fed mothers.[44]

Even with infants adjusting to maternal condition, and vice versa, the two parties' knowledge about each other, and about the world around them, can never be anything but imperfect. In general, the mother is more experienced and has more direct access to information about current and future environmental conditions than her infant will. But neither party can know when rains will come, or when acacia trees will blossom into soft and palatable foods suitable for just-weaned young. As Altmann reminded, "A mother is to its infant and an infant is to its mother one of many only partially predictable environmental variables, not a predetermined constant."[45]

No matter how brilliant the insights of the gene theorists, the equivalent of an ecological glass ceiling weighs heavily on the heads of individuals. Or, as Haig acknowledges: "It would be a conceptual error to equate the interests of genes with the interests of individuals. Our genes' purposes are not our own."[46]

———

Altmann situates her vision of negotiated consensus between mother and infant squarely in the local "economy" where baboons make their living, in a world shaped by rainfall, food availability, population density, and travel distances. Her model also applies well to human hunter-gatherers in an unpredictable ecosystem. In a case like the nomadic !Kung, where the optimal birth spacing imposed by constraints upon a gathering mother with no safe place to stash her baby is four years, few women bear more than five children in a lifetime, and only half of those born are likely to survive (a mortality rate similar to that for Amboseli baboons).[47] However, among other hunter-gatherer groups, where allomothers are available back at camp, or where soft and nutritious weaning foods can be had, or where mothers are better nourished, birth intervals are shorter. Even weaned as early as six months, infants can still survive. But abrupt arrival of subsequent siblings meant successive neonatal crises for the mother, and varying degrees of neglect as opposed to indulgence for infants born to her. For once the empowered embryo emerges into a cold, unlubricated world, and once the umbilical lifeline attaching it to its agent the placenta is cut, the infant's control over such essentials as food, shelter, and warmth diminishes. At that point the empowered embryo becomes a supplicant.

By the time the baby is expelled by the uterine muscles, it must be prepared for its exile from gestational Eden. From hormonally empowered, firmly entrenched, fully enfranchised occupant of its mother's body, the baby's status declines to that of a poor, naked, two-legged mendicant, not even yet bipedal, a neonate who must appeal in order to be picked up, kept warm, and be suckled. What the infant encounters in this far less certain world is largely a matter of luck. Yet there are a few tricks a newborn can try. Among them are looking good and engaging its mother from birth.

Why Be Adorable?

This fable of an Ape, which had two children of the which [she] hated the one, and loued the other, which [she] took in her armes, and with hym fled before the dogges. And whanne the other sawe that his moder lefte hym behynde, he ranne and lepte on her back. And by cause that the lytyl ape whiche the she ape held in her armes [prevented her from fleeing] she lete hit falle to the earthe. And the other whiche the moder hated held fast and was saued.

—From Aesop's fables, ca. 620 B.C. (translated from the Greek by William Caxton, 1483)

Anyone who has ever wondered about the expression "bald as a coot," or worried lest a naked pate makes him less lovable, needs to meet the little mud-hens that dive for a living in the shallow ponds near my home, mysteriously submerging when I approach, only to resurface shyly in the reeds. The American coot, *Fulica americana,* ranges from Canada down to Mexico. Profligate in egg production, coots lay anywhere from seven to thirteen eggs on floating nests constructed of dead vegetation.

Sibling Rivalry in Birds

Like many birds, coots routinely lay more eggs than the number of young two parents can, or will, successfully rear. Coots are among those birds ominously termed "brood reducers" by ornithologists. One-third to one-half of the chicks will starve before they fledge. Tough competition at home explains why baby coots are born not necessarily dressed to kill but certainly garbed to take their seats among the elect at a fancy banquet. Babies are born spectacularly different from their drab black parents, who sport nothing flashier than a white bill. The front portion of a baby coot's body is covered with conspicuous orange bristles. The eyes and the tip of the red beak are surrounded by brilliant papillae, clustered like coral beads, and the nearly bald pate shines forth a bright vermilion. Being flamboyant enough to attract parental attention is critical for survival, but it also bears the liability of making newly hatched chicks conspicuous to predators.

As David Lack observed years ago for other brood-reducing species, a

Fig. 19.1 Drawing from a twelfth-century English bestiary. (See epigraph on preceding page.)

mother coot lays her eggs sequentially over a period of days, but begins brooding them right away so that the first-laid eggs get a head start and hatch first. Hatching late has serious drawbacks. In species such as the lesser spotted eagle, in which siblicide is obligate, the first-hatched eaglet pecks its younger sibling to death every time.[1] Presumably, parents laid that extra egg as a form of insurance, in case the first was defective, succumbed to disease, or was lost to predators.

Older chicks benefit the most from staggered hatching. By producing successful progeny, parents benefit as well. Hence parents remain astoundingly nonchalant, never intervening as older chicks peck younger ones to death. The losers are the last-hatched chicks, scarcely in a position to do anything about it beyond waiting for a miracle. (It's remotely possible, even at this stage, that the older might suffer a mishap.)

In some siblicidal species (like egrets) one or more chicks are only *sometimes* killed, if food is short. Death results from older chicks bullying weaker ones, who become so intimidated that they starve. In other siblicidal species, several eggs are laid and all but one of these hatchlings are *invariably* killed. These obligately siblicidal species tend to be large carnivorous birds, such as eagles, pelicans, cranes, boobies, and penguins. To qualify for the scientific definition of an "obligately siblicidal" species, 90 percent or more of all broods produced must be reduced.[2] Normally, the last-hatched chick would survive only if it was reared as a replacement for an alpha chick.[3]

The longer the interval between the laying of the first egg and the last one, the longer an insurance egg will be on hand to pinch-hit if needed, and the longer the first-hatched has to grow larger than the runt. With that head start, termination of the underling by the bully can occur swiftly and, from the point of view of parents and surviving chick, efficiently.[4] But in species where death of laterborns is by no means preordained, if food turns out to be abundant, parents rear the whole brood.[5]

Exaggerate Inequalities? or Minimize Them?

Presumably it was energy constraints and the risk of maternal overcommitment that initially favored mother birds who spaced egg-laying over a span of days. Once established, this staggered laying of eggs that would hatch asynchronously set the stage for the first-hatched offspring to prevail over younger ones. At that point, there was an opportunity for natural selection to favor parental behaviors that either exacerbated or else mitigated the resulting inequalities between siblings. For example, by beginning to brood the first egg right away, the mother ensures that the first-laid will hatch before her other eggs. Among some birds, mothers have been selected to further tip the scales in favor of elder bullies by depositing an extra dose of aggression-enhancing androgens in the yolks of the first-laid eggs. From there, the chicks peck it out to the preordained outcome, while parents remain aloof from the fray.[6]

Other birds, however, like my friends the coots, opt instead *to equalize rather than exacerbate* the odds, permitting smaller chicks to coexist with older rivals. Instead of allowing the stronger to prevail, coot parents compensate for the competitive weakness of later-borns by preferentially feeding them. Among the coot chicks, fancy neonatal plumage metamorphoses gradually, over the space of three weeks, into the less conspicuous and safer regulation-gray adult uniform, providing parents with an accurate index of how old the chick is. By favoring the fanciest, parents make sure of feeding the youngest. As all chicks in the nest gradually dim down with age, turning first Quaker-gray, then black at maturity, parental preferences for neonatal coloration fade away as well.

Coot parents are not the only birds to inadvertently subscribe to such post-Enlightenment ideals, helping out the under-bird. Canary mothers equalize investment in offspring through other means. The mothers add extra testosterone to the yolks of their last eggs. After the eggs hatch, mothers provide more food to the hungriest rather than the oldest chicks.[7]

As soon as sociobiologist Doug Mock read this report by Hubert Schwabl about extra testosterone for later-laid canary eggs, he hypothesized that the cattle egret mothers he studied would manage their broods quite differently. He had been studying siblicide for years, and had never seen egret parents do anything to protect the youngest. Collaborating with Schwabl, Mock and colleagues at the University of Oklahoma discovered that among siblicidal cattle egrets, the *first* egg receives extra androgens. Mock and colleagues speculate that the steroid boost enhances aggressiveness and further facilitates domination of later-born chicks by elder siblings. But could the unequal steroid servings be a coincidence? Mock's observation that mothers provide the same amount of androgen content to each of their sequentially produced eggs in a different egret species not characterized by siblicide makes this possibility less likely.

The implications are unsettling. If parental bolstering of bullies in the siblicidal birds is no coincidence, it suggests that the intensity and outcome of sibling rivalry is indirectly being set up by parents. Even given a physiologically imposed interval between laying her eggs, it's still up to the mother whether she begins brooding the first egg right away or waits till everyone is present so eggs will hatch at the same time. It's also up to the mother to deposit the same or different amounts of aggression-increasing hormones in each chick.

What Appeals to Parents?

Mother Nature's "bad habits" include not just acting ruthlessly, but striking early at the weak. Sometimes parents themselves become the agents of selection, biasing the survival of immatures in line with parental interests. In this way, parents magnify the effects of ordinary natural selection. By preferentially feeding the youngest, however, coot parents (like canaries) are opting for equity—even to the point of compensating for weakness. This creates new selection pressures on the young. With half or more of each clutch succumbing to starvation, any trait that makes it likely that a hatchling will receive more food will be strongly selected for. It may be selected for even if that trait happens to be looking more helpless.

It was a team of three behavioral ecologists—Bruce Lyon, John Eadie, and Linda Hamilton—who discovered that coot parents preferentially deliver food to the youngest-looking (neediest) chicks. But how did parents "know"? Were there particular traits that appealed to parents? They postulated that what must have begun as modest differences between adults and hatchlings (perhaps, the tendency for chicks in all species of coots to hatch sporting a bare spot on their head) combined with a reliable alarm signal emitted by a chick in need (the skin of a chick in any kind of distress flushes red) had become perpetuated and magnified. Parents who responded to such distress signals had a higher proportion of chicks survive. As parents provided more food to chicks with balder, redder heads, redness itself became an attractive trait. Thus chicks that hatched with the baldest, reddest heads were more likely to survive and pass on these traits. Here "survival of the fittest" meant survival of those who looked youngest and neediest, even though they were still vigorous enough to beg. Sure enough, experiments revealed that coot parents preferentially fed the most brightly colored chicks.[8] (See Plate 2.)

Because parents exercise preferences among offspring, offspring are selected to display those traits that catch their attention, leading to such exaggerated signals as the huge, gaping (sometimes elaborately marked) mouths of baby birds. Over generations, in a process similar to the peafowl's preference for a peacock's tail, flamboyant neonates were preferred by parents. The payoff for the flamboyant chick is not greater breeding success for the attractive, as in classic sexual selection, but survival of the attractive chick long enough to show off another day, that other way. More than feathers make the bird. Success depends on parental preferences for a particular look.[9]

Runaway Selection for Infantile Traits

Like the more famous case of sexual selection through female choice of a male with fancy feathers, "parental choice" has the potential to spark "runaway selection."[10] If parents prefer fancy chicks, those parents who preferred them are selected for. Their surviving offspring would be disproportionately fancy, and also grow up to be parents who both prefer and produce appropriately fancy chicks. Selection is simultaneously favoring chicks with the trait and parents who possess preferences for it. I return to this topic of runaway selection when I discuss the possibility that maternal preferences contributed to selection for fatter babies in the course of human evolution.

Natal Attractions in Primates

Early on, William Hamilton proposed that sexual selection has shaped our aesthetic sensibilities. Displaying his own poetic gift (a talent, like singing, that some have suggested evolved in men because of its appeal to members of the opposite sex), Hamilton wrote that sexual selection probably "plays a part in making us all so over-intellectualized as to love our bright and capricious world as we do, so far beyond any utility to us." But if Hamilton is right, adult preferences for specific infantile traits must also have shaped human tastes in colors, shapes, and textures (big-headed, cute, soft, paler skinned, and just generally neotenous, or "baby-like").[11]

It should come as no surprise that birds are not the only ones to enter the world in fancy dress. Some of our primate relatives are also born as if dressed in spun gold or dusted in snow. At the top of this list are the orange-gold babies born to dusky leaf monkeys (*Trachypithecus obscurus*). Glistening golden babies stare out at the world through eyes rimmed by large, chalk-white circles; so remarkable is their costume, that the world can only stare back, agog. (See Plate 3.) Perhaps for this reason, natal coats have more often been a subject of amazement than of study. Little research has been done on primate natal coats.[12]

High in the canopy of Malaysian forests, these resplendent golden *T. obscurus* infants are anything but "obscure," although the dusky gray adults are. The new babies stand out brilliantly against their mothers' bodies, visible half a mile away. More spectacular still are the "crucifer patterns" embellishing the natal coats of their Sumatran relations, the mitered leaf monkey. Neonates are born white with a dark stripe from head to tail-tip, crossed by another stripe between the shoulders, in a cross pattern resembling a heraldic emblem—or, as in Hose's leaf monkey of Borneo, just white with a simple dark stripe. What on earth could such decorations be for?

Most mammals, if they can't hide their babies in burrows, at least camouflage them, giving birth to fawns with spotted coats that disappear in the high grass of their hideaways. Others are camouflaged by blending into their mother's mottled fur, as in the case of baby koala bears, marmosets, and orangutans. Borrowing a bit of the best of both worlds, many cercopithecine monkeys—like baboons and vervet monkeys—are born like lawyers, in discreet black suits, just different enough from adults to give babies a discrete uniform without screaming "Baby on Board"—discreet but discrete. The more terrestrial relatives of the leaf-eaters (Hanuman langurs and proboscis monkeys) also tend to dress down, or at least to limit the broadcast of their more flamboyant signals, so as to bypass the unwelcome attention of predators but still be able to selectively advertise neonativity. Gray Hanuman langur monkeys are born all black except for the flamingo-pink skin of their face, while proboscis monkey babies are colored like their parents except for a Druid-blue face, which they direct toward those whose attentions they desire.

But exactly whose attentions are desired? What possible survival advantage could outweigh the disadvantage of being so visible to hawks, eagles, and other enemies? Could it be that infant primates born impractically ornate evolved this way so as to attract more care from mothers? But since they were born to single-mindedly devoted Old World monkey mothers, why on earth would they need to? In the absence of competition with other similarly endowed siblings, how could the coot-caliber flamboyance of all these snow-white, gold, and blue-faced babies evolve? Given that their mothers bore only one baby at a time, who were the ancestors of all these babies competing with that they needed to dress-to-be-noticed?

Why Be a Flamboyant Baby?

Recall that monkey mothers are paragons of unconditional love. Unlike siblicidal egrets (and some humans), but like coots and canaries (and other humans), monkey and ape mothers compensate for infant disabilities rather than discriminate against the weak. Among the anthropoid primates, only human mothers are the odd-primates-out in this respect, since human mothers do discriminate on the basis of offspring attributes.

Given how unconditional the "love" of mother monkeys and apes typically is, what possible selection pressure to be "flamboyant" could outweigh evident costs to baby colobus and leaf-eating monkeys of evolving conspicuous natal coats? To solve this mystery, we need to widen our field of vision to encompass other females, not just the genetic mother. Unlike that of the coots, flamboyant natal dress in primates did not evolve to elicit care from parents. It evolved as a natal attraction, screaming "New Baby" to all benevolent alloparents in the vicinity.

Recall that most monkey and ape mothers are exceedingly possessive of newborns, but with many exceptions. Prime among these are the "infant-sharing" species, many of them belonging to the subfamily Colobinae. Colobine monkeys are characterized by large many-chambered stomachs—sacculated guts, of the sort cows and other ruminants have—where anaerobic bacteria break down cellulose. This ability to digest large quantities of tough fiber permits these monkeys to live off leaves when flowers or fruits are unavailable. Along with the high degree of relatedness among group members typical of species that live in small, one-male groups, there is a tendency among colobines for female-dominance hierarchies to be less rigid than among macaques or baboons.

Mothers in these easygoing colobine societies feel comfortable turning over their infants within hours of births and allowing their female kin to carry them about—in langur monkeys, for about half of all daylight hours. Primatologists had long suspected that mothers freed to forage would be better fed and better able to produce milk. Their babies would thus be better nourished as well—a "trickle-down" argument familiar (if not always completely convincing) to every ambitious working mother who rationalizes leaving her infant in daycare while she seeks a better life for her family. It was very satisfying, then, when John Mitani (at Yale) and David Watts (at the University of Michigan) used comparative data from a broad array of primates with and without infant-sharing to show that babies born to infant-sharers grow at a faster rate, and that their mothers give birth again after much shorter intervals without compromising their own health or infant survival. Crudely put, mothers with good daycare had the highest fertility rates.

Do Humans Have Natal Coats?

Flamboyant natal coats in primates no longer seem such a mystery. Conspicuously garbed babies do not (like coots) compete with sibs in the same clutch for parental attention, but instead compete with all the other just-born infants in their cohort to attract the attention of allomaternal caretakers. Just as coot parents favor the youngest chick in their clutch, primate allomothers prefer the youngest, most "neonatal-looking" infant in their troop. The younger the baby, the more times per minute allomothers attempt to handle it. The more time they carry it, the more the mother gets to feed. Both mother and infant benefit as a result. It sounds like a system humans could use. *Did* they?

Human babies are indeed adorable, especially to those of us culturally as well as biologically primed to see them so. When I look back at photographs of my own babies right after birth, squench-faced and puckered, I am astounded at how homely they are compared to my recollection that each was the most beautiful creature I had ever seen. The eye of the maternal beholder notwithstanding, nothing about human newborns broadcasts signals nearly so conspicuous at a distance as the bright golden coat of a dusky leaf monkey baby or a snow-white colobus newborn would do.

Human neonatal signals seem more tailored to local consumption. Our dowdy species enters the world covered with barely detectable whorls of fine hairs called lanugo. Otherwise, we are as furless as our parents in whose eyes naked babies are nonetheless luscious. At birth, all Caucasian babies have blue eyes, while the eyes of African babies are almost colorless. !Kung newborns, Mongolians, and the Native American descendants of Mongolian peoples often have clusters of dark pigment at the base of their newborn spines, detectable upon close inspection. Otherwise, the human infant's color scheme is some blend of its parents', albeit often a traumatized purple *right* after birth, or a jaundiced peachy hue. For some months after emerging into the light of day, most babies are paler than adult phenotypes. But for the most part, human primates are born the same color as their parents.

Our newborns are visibly different from grown-ups. But they emit no special long-distance signals (apart from vocal ones, like crying). Human babies are not a bit flamboyant—as if they were born so confident of maternal love that they have no need to attract attention from anyone else. Yet from ethnographies, classical accounts, the archives of foundling homes, down to today's tabloids—humankind's record is far from reassuring. Realistically, babies have few guarantees that mothers will care for them.

Sad truth to tell, millions of babies born to our species could have used some magically irresistible natal garb to attract the kindness of allomothers. There is all too great a risk that the proverbial magnetic attraction of baby primates doesn't always work in *Homo sapiens*. Among monkeys, the younger or more "neonatal" they look, the more attractive they are. Yet *exactly the opposite is the case among humans*. Babies right after birth are at the greatest risk of maternal abandonment or infanticide.

————

Something had to change radically for a primate mother ever to find a very new baby, right after birth, less attractive than an older, weaned, or nearly weaned one. Yet the record is clear: among humans, very new babies are more likely to be abandoned, or even killed, than their older siblings are.

What changed, of course—though we can't say exactly when it began—is that human babies started to be born after shorter intervals. Apparently mothers were routinely receiving a message that was different from the one primate mothers had evolved to read. The message was a decline in her prolactin levels, which the mother read as meaning that her current baby had ceased to suckle (or was suckling less) and was either mature enough to be weaned or else dead, and the signal meant time to replace it. The trouble was, that in the new context in which human mothers found themselves, her baby was neither dead nor mature. Rather, the baby was suckling less because it was partially subsisting on porridge or animal milk, and spending time apart, swaddled perhaps on a board, or rocking in a cradle. Possibly it was held by allomothers. Such an infant might not be taking such frequent swigs at the nipple as previously, but he or she remained very dependent indeed. Thus many mothers were ending up with a new baby on top of a current baby who was still dependent.

When organisms must cope with new situations, all Mother Nature has at hand are the leftovers in the cupboard. Mutations are rarely any use. This is why the most rapid adaptations tend to be behavioral rather than physiological. Individuals use behavior to produce a new phenotype, and subsequently, at a more leisurely pace, selection for traits that complement and enhance the new phenotype can be selected for in the conventional Darwinian way.

In the case of these newly fertile humans, ancient predispositions, left over from a time when mammalian mothers bore litters and gave birth to babies that cost less than primate babies do, re-evolved or were reactivated. Instead of classic primate mothers who care for whatever very new baby they have at hand, these newly hyperfertile mothers, who found themselves caring for a clutch of babies of different ages, became more discriminating. It was more practical to change decision-making by mothers, transforming an uncritical, unconditionally loving mother monkey, than to modernize the way a woman's ovaries register incoming information about the status of her baby's needs. (Basic primate sensibilities still persisted, of course, and that's one reason why many of the cruel choices human mothers must make are so fraught with ambivalence.)

Mothers who tended to be more discriminating about which babies they cared for fared better. During famine and other crises, selective mothers would be the only ones who left surviving offspring. More discriminating, situation-dependent solicitude toward newborns—a common enough trait among litter-bearing mammals—reemerged. (A still unexamined question is whether nonhuman primate mothers in a pinch would respond to hyperfertility in the same way. The evidence for twins suggests that an overburdened mother monkey or ape also makes a choice between infants [see pages 180–81]. But information remains too sketchy to permit any conclusion.)

If we are to believe the record from antiquity (and I do), the behavioral response to inopportune births was to abandon or terminate care. Hence the many human infants who were forced to rely on the "kindness of strangers," as described in chapter 12. Human mothers are far more likely to abandon their infants than are other primates who give birth to one baby at a time.

———

The single-minded, unconditional devotion characteristic of mother monkeys and apes contrasts with the more discriminating solicitude of human mothers in traditional societies and across history. In terms of maternal infanticide and abandonment, humans resemble birds and litter-bearing mammals more nearly than they do other primates. Although utterly unprimatelike, infanticidal mothers are not in the least unnatural. Because we take maternal love for granted, it has rarely occurred even to evolutionists to ask the same sort of questions about humans that we now routinely ask about coots or canaries. What effect has discriminative maternal solicitude had on the survival equipment of human infants, the attributes each infant enters the world with, and the signals it sends? In the next chapter, which I caution readers is even more speculative than this one, I attempt to answer this question.

How to Be
"An Infant Worth Rearing"

*At first I was going to throw [the newborn daughter] away on the tundra,
and then I could not, for it was too dear to me.*

—Reminiscence of a Bering Straits Eskimo father, 1901

*[The lady] had a child in the cradle and she wanted to see it before leaving
[to join a group of heretics]. When she saw it she kissed it; then the child
began to laugh. She had started to go out of the room where the infant lay so
she came back to it again. The child began to laugh again; and so on,
several times, so that she could not bear to tear herself away from the child
[that she would never see again].*

—From a fourteenth-century account of the French Inquisition

*Now the midwife, having received the newborn . . . [and having] examined
beforehand whether the infant is male or female . . . should also consider
whether it is worth rearing or not.*

—Soranus, second-century A.D. Greek physician

N atural disasters just happen, floods sweeping away anyone in their
path, epidemics ravaging populations. Neglect or abandonment of
a child may occur because the mother has little choice. She may be
very young and inexperienced, or she may have too many children already. In
today's world a mother may become a victim of civil war, infected with HIV,
incapacitated by poverty, depression, or "crack." Perhaps her mate has died or
walked out, or strange males have stormed into her group. Such contingencies fall outside of Mother Nature's ruthless sphere, governed by her flighty
sister, the occasionally generous but ever arbitrary Dame Luck.

Nothing about an embryo affects its order of birth in a family. Even sex at
conception is usually a chance outcome of the chromosomal dice. Nor can an
embryo alter environmental conditions that await outside the womb. To this
extent, the hardiness or appeal of a particular baby is irrelevant.

Bad Luck

When a certain Aché hunter was killed by a jaguar, the deceased left behind a young widow big with child. A bachelor named Betapagi, regarded as "ugly" and a lackluster hunter, began to pay court. He convinced the pregnant girl, named Pirajugi, to marry him. When her baby was born in camp a few days later, all the Aché examined it. Young children crowded close to touch the baby as their mothers hastily rebuked them and pushed them away. The only absent band member was Betapagi, who had grabbed his bow and left camp as soon as his new wife went into labor. "The baby was small, and had very little hair on its head," noted ethnographers Kim Hill and Magdalena Hurtado.

> The Ache felt little affection for children born without hair. No woman volunteered to cradle the baby while the mother recovered from the birth. No man stepped forward to cut the umbilical cord. The signs were clear, and it took only Kuchingi's [a respected group member's] verbal suggestion to settle the point. "Bury the child," he said. "It is defective, it has no hair." "Besides, it has no father. Betapagi does not want it. He will leave you if you keep it." Pirajugi said nothing, and the old woman Kanegi began to dig silently with a broken bow stave. The child and placenta were placed in the hole and covered with red sandy soil. A few minutes later the Ache packed up their belongings and Grandpa Bepurangi began to break a trail through the undergrowth with his unstrung bow. Pirajugi was tired, but she had nothing to carry, so she was able to keep up without difficulty. Women cried softly as they walked through the forest. "Ooooooh Kuajy maiecheve. . . ." "our parents and grandparents who took care of us. . . ."[1]

Kuchingi, the hunter who had first brought Pirajugi the news of her husband's death, and who now advised her to bury their son, was himself no stranger to such decisions. As a child, his own life had depended upon such a choice. As Kuchingi explained the circumstances in his own words:

> The one who followed me [in birth order] was killed. It was a short birth spacing. My mother killed him because I was [still] small. "You won't have enough milk for the older one," she was told. "You must feed the older one." Then she killed my brother, the one who was born after me.[2]

Probably any infant known to have been sired by the wrong father, or born at the wrong time, would have suffered a similar fate. Even if born just a bit larger, or with a bit more hair, another infant would have fared no better. But not all mortality is random. How robust an infant is, whether innately resistant to a particular disease—these traits surely matter for survival. But because infants are so peculiarly dependent for survival on the adults around them, selection also acts *through them*. How much, or whether, a mother invests depends in part on her perception of the infant. Mothers who find themselves between a rock and a hard place sometimes do opt for the hard place and keep an infant in spite of their predicament.

The cases I summarize here hint that infant attributes can be exceedingly important, of life-or-death significance. For unlike other primates, humans have a conscious capacity to assess outcomes—to predict what costs a particular infant might impose on the survival of an older sibling, or on family resources or harmony; to predict how a father or stepfather might behave; to know ahead of time that the baby's grandmother is about to die and that there will be no one to help; in short, to mentally run through a store of information concerning what has happened in the past to other infants born with this particular configuration of traits and circumstances and to make a *conscious* prognosis for this one. The cognitive capacities that place human infants at particular risk of maternal neglect or abandonment (compared with other primates) may also be selecting for particular traits in infants. We need to understand how maternal criteria helped shape the environment in which infants evolved.

Close Calls

German anthropologists Wulf and Grete Schiefenhövel document how a mother committed to infanticide changed her mind. The events took place among the Eipo people living on the wet northern slopes of the Central Cordillera of Papua New Guinea. The particular vigor of an unwanted infant girl tipped the decision in her favor.

For decades prior to the arrival of the Schiefenhövels, in 1974, the Eipo population had remained stable. Zero population growth was maintained by small family sizes—an average of 2.6 children per woman.[3] Lack of growth was not due to child mortality, however. For the death rate, fifty per thousand, was far lower than in most developing nations. In spite of the near absence of medical facilities, prolonged breast-feeding (up to three years) ensured that such infants as were kept remained relatively healthy.

The key to the puzzle was the high rate of infanticide, which was occurring right after birth and in private. When infanticide is taken into account, Eipo infant mortality rates rise to 480 per thousand.[4] Aid workers in other developing countries have sometimes speculated that the death of some children improves nutritional status of remaining ones,[5] and this was probably the case among the Eipo.

Whether or not a mother in this remote horticultural community kept her infant was up to her. However, each mother would have been acutely aware of population pressure on the supply of sweet potatoes. A conscious concern for the future would have entered her calculations.

Given that the rate of child mortality in this population was low, and that infanticide accounted for 80 percent of what there was, the oft-cited rationale attributing high rates of infant mortality to maternal indifference did not fit. Mothers emotionally distancing themselves from doomed babies could not explain high rates of infanticide among the Eipo.[6] Rather, the deciding factor was the desire for sons. Mothers who eliminated a daughter were aware that they would be likely to conceive again within the year, possibly giving birth the next time to a son. In the early years of the Schiefenhövels' study, between 1974 and 1978 (before missionaries transformed the lifeways of the Eipo), 41 percent of all live-born infants were eliminated. Of twenty infants not kept, five were male and fifteen were female.[7]

The Schiefenhövels tell the story of one woman who during late pregnancy "repeatedly and openly stated that she would not accept another baby girl (because she already had one daughter and no son)." Prior to birth, then, the mother mentally prepared herself not to form any bond with the child, to remain distant at birth. After the birth of a particularly healthy-looking female, however, "she showed obvious ambivalence."

Still, without cutting the umbilical cord, leaving the placenta still attached, the mother wrapped the newborn in fern leaves laced with rope made from lianas. For a long while, she sat thoughtfully near the bundle. The infant screamed lustily, struggling to live, her pudgy hands and feet bursting out through the leaves. Eventually, the mother departed, leaving the infant where she lay. Yet she did not throw the bundle into the bushes as would be typical in an Eipo infanticide.

Fig. 20.1 This Eipo mother had just given birth to a third child, a girl, that she did not want, considers abandoning, then changes her mind. *(Courtesy of Wulf and Grete Schiefenhövel)*

Two hours later the mother returned, cut the umbilical cord, and took up the baby. Almost apologetically she explained: *this daughter was too strong.*[8]

Arbitrariness of Criteria Defining Human Life

Inspecting the newborn is one of the few behaviors mothers everywhere engage in. Like other primates, a human mother pays special attention to the genitals. She may delicately run her hands over the entire body. Unlike other primates, however, the human mother compares her baby with cultural ideals. Quite arbitrary standards—a blend of traits linked (or once linked) to viability, along with current fashion, family needs, or just whimsy—can have lasting repercussions for the way a mother relates to her infant.

Nowhere is such arbitrariness more evident than in debates over just when life begins. Even groups with similar cultural legacies exhibit big differences. Not long ago a painful international incident occurred when it was revealed in the United States that personnel working in an East German maternity ward defined any neonate smaller than 2.2 pounds as fetus rather than baby, and disposed of them in a bucket of water.[9] In the ensuing controversy, neither side would give way. Leaving aside differences in medical technology and available resources, the problem was that biologically, there *is* no set point where life obviously begins. Answers must always be relevant to circumstances.

The criteria for what mothers consider acceptable or desirable in a baby vary with available resources and probable outcomes. Yet this being acknowledged, there do seem to be traits in newborns that all mothers are especially attuned to. Not surprisingly these are traits that have traditionally been good predictors of survival. Hardiness registers, as do signs of vigor. Assessment begins with the first cries of a just-born baby.

Factors Affecting Maternal Assessment

Some maternal likes and dislikes have an innate component. Certain kinds of cries, for example, given by premature infants seem to be universally disturbing, even aversive, while rounded heads and plump features are universally appealing. The sight of a defective newborn is profoundly disturbing. Even low-birth-weight babies are distressing and generate psychological distress.[10] In many postindustrial Western societies today we suppress these responses. It is considered grossly insensitive to mention them.

Post-Enlightenment sensibilities about children, combined with the old assumption that all normal mothers automatically respond to their infants in a nurturing way (or, as Erich Fromm put it, "Motherly love is by its nature unconditional"), has meant that very little research has been done on which infantile traits engage mothers. We know even less about which ones disturb or repel them.

Yet based on the handful of studies where researchers have looked in this direction, maternal commitment does appear to be sensitive to certain fetal and neonatal attributes. In a 1991 study, Canadian researchers documented a change in pregnant women's attitudes toward impending birth once they were told that the baby they were carrying was normal. By using chorionic villus sampling, which can be done earlier than amniocentesis, an increased sense of attachment to the not-yet-born baby was documented.[11] Further evidence that mothers may be more hesitant to fully commit to babies whose survival is in doubt is suggested by maternal responses to premature babies. The delicate and wizened facial features of infants born preterm are less likely to be rated as "babyish," or attractive, and less likely to elicit nurturing responses than are the rounded heads and plump cheeks of full-term infants. For mothers with plenty of support, such initial reactions are not going to compromise care, but the risk of maternal distancing is still there.[12]

Some of the traits mothers attend to are only indirectly linked to survival prospects. For example, a great deal of attention is paid to whom the infant is thought to resemble. Mothers may be particularly interested in increasing a husband's confidence in paternity, and concerned when physical attributes of the baby threaten to undermine that. Father's perceptions on this point may influence whether or not he stays and continues to invest in a particular wife and her children, how much time he spends near the infant, and how much help he provides. An infant's physical appearance—skin and hair color, and eventually eye color—may raise confidence of paternity or further fuel a "father's" suspicions.

Shakespeare's grisliest play, *Titus Andronicus,* provides a stark example. The empress of Rome has just given birth to a dark-skinned baby who more nearly resembles her lover—Aaron the Moor—than he does his father, the emperor. "Here is this babe," moans the nursemaid, "as loathsome as a toad amongst the fair-faced breeders of our clime." Hence the mother beseeches her lover to destroy the telltale evidence of her infidelity, something (as we will see) Aaron is loath to do.

Turning to a less dramatic real-life example, psychologists Martin Daly and Margo Wilson report the tendency of people (especially mothers) in hospital delivery rooms to spontaneously *tell* fathers that their babies look like them. But even the sociobiologists doing these studies simply assumed that well-wishers were being polite, softening Dad up for future duty. After all, we know that babies at birth resemble other babies (especially their siblings at the same life stage) more nearly than they do either parent. However, a

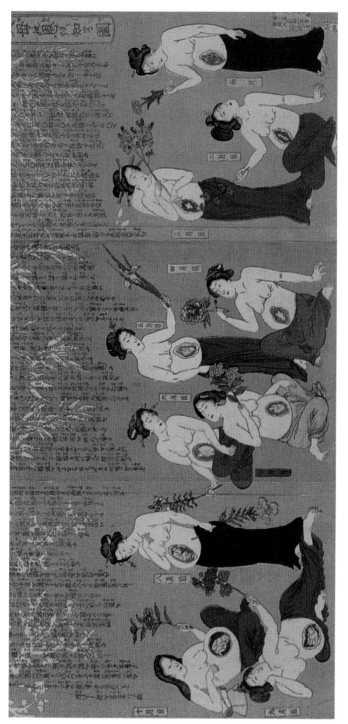

PLATE 1. Long before biologists coined the term *maternal effects* to describe nongenetic traits transmitted between generations, people in Asia were convinced that a woman's conduct during pregnancy affected the embryo. This nineteenth-century Japanese print is titled "To learn about the *on* [obligation to one's parents]." The text reminds viewers that the womb is like a plant that may wither. The mother holds a different flower for each stage of pregnancy, a wild cherry blossom for the third month, a peony in the fourth, an iris in the fifth. Timing of conception and age of the parents were considered critical. In China as well, medical texts advised parents on "fetal education" since what the mother ate or how she behaved could affect the child's qualities. Emphasis on fetal education is, if anything, even more pronounced today owing to the one-child policy. One woman interviewed in 1994 describes what she ate during pregnancy to guarantee a perfect child: "Apples and tomatoes for white skin, and vegetables…for putting more waves in the brain.…" *(From* Conception to Delivery *by Sono Kochi Hasegawa. Courtesy of the National Library of Medicine, Bethesda)*

PLATE 2. When researchers trimmed off the bright orange feathers from one group of chicks but left the decorative feathers on same-aged controls, they demonstrated that coot parents prefer the brightest-colored babies and differentially provision them. *(Courtesy of Bruce Lyon)*

PLATE 3. The flamboyantly golden "natal coat" of this dusky leaf monkey baby attracts the attention of an allomother, who takes it from the mother. *(Photo by Ernst Müller. Courtesy of the Frankfurt Zoo)*

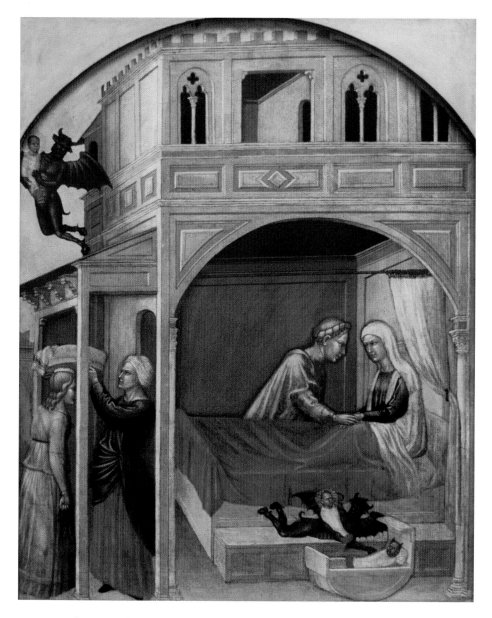

PLATE 4. Christianized versions of changeling myths persisted in Spain and Italy in connection with the lives of particular saints. According to the life of Saint Stephen, depicted here, the saint-to-be is stolen at birth by demons and a devil-baby that fails to thrive is left in his place. *(Painting by Martino de Bartolommeo, fourteenth-century. Courtesy of the Städel Museum, Frankfurt)*

PLATE 5. Tintoretto based his *Origin of the Milky Way* on a Byzantine legend about the life of Hercules (Heracles to the Greeks). According to that version, this star-studded corner of the celestial landscape was created when Zeus sent a messenger to retrieve his bastard son, who had been abandoned at birth by Alcmene. Zeus wanted Hercules to nurse from his own wife, a goddess, and thus become immortal. But when Hera awoke, she shoved the foundling off only to have the force of the baby's herculean sucking send a spray of *gala* (the Greek word for "milk") into the heavens. Thus was our galaxy christened, producing a term for humankind's universe that is, appropriately enough, derived from mother's milk. *(Courtesy of the National Gallery, London)*

"I disagree. I think he looks like his father."

Fig. 20.2 A lot of humor surrounds play on parental concerns over who the baby resembles. *(Robert Weber © 1995 from The New Yorker Collection)*

preliminary report that facial features in year-old babies are more likely to resemble those of their fathers than their mothers raises the intriguing possibility that a baby's facial morphology is especially responsive to genetic input from the father. This finding still needs to be replicated. Should it hold up, genetic imprinting of the kind studied by David Haig would be a possible candidate mechanism to explain it.[13]

Among the handful of researchers to tread—very cautiously—in this highly sensitive, scientifically almost taboo, realm of parental preferences is psychologist Janet Mann. Mann undertook a longitudinal study of fourteen twins born at extremely low birth weights. Eight months after mothers brought the twins home from the hospital, all seven mothers in this small sample were directing more attention to the healthier twin of the pair, even though in several of these cases remedial care for the sicker of two twins was also being provided. Mann, who hastens to qualify her data as preliminary, hypothesized that mothers have an innate template for assessing infant quality, based on physical appearance and cries.

Innate responses to infant appearance cannot by themselves explain the full range of maternal reactions, certainly not the cases where full-term, seemingly perfect babies are abandoned, or the many modern mothers who throw themselves utterly and wholeheartedly into the care of babies unlikely to survive. Relevant factors to take into account include the mother's social and economic circumstances, religious beliefs, as well as learned attitudes about children and about how families "should be."

In contemporary Western society, parents are respected and admired for caring for the same infants that in other societies mothers would be condemned by their neighbors for not disposing of. Some adopting parents in the West go out of their way to select the neediest infants, and commit themselves to years of therapy on behalf of children who will never repay that care in any material sense. Unlike other animals, humans are able to consciously make choices counter to their self-interest. Indeed, much of what we consider "ethical behavior" falls in this category.

Such voluntary behavior counter to all biological self-interest is what qualifies as true heroism, moral heroism of the kind George Eliot had in mind when she distinguished humans from Darwin's "silly animals" and noted that our deeds determine us as much as we determine our deeds. Such acts make it awkward for even the most hard-core materialists to completely discount the existence of free will, and I have no desire to.

Effects of Baby's Signals on Survival

A mother's standards may be higher or lower, depending on availability of resources; her age; what other opportunities to have a child she would be likely to have; on just how much she wants a son; or whether, if she already has three sons, she prefers a daughter. Or she may not want any infant at all at that time.

As a mother scrutinizes her just-born baby, what she sees depends in part on her culture and circumstances, on what she wants and expects. How a mother responds to what she sees, whether her love deepens or she pulls away, in turn affects who a child becomes, whether it prospers or fails to.

Children whose mothers draw back emotionally often fail to thrive. Some literally quit growing. Survival chances fall when mothers distance themselves from preterm or malnourished children. Downward spirals ensue as mothers convince themselves that this small-for-its-age, increasingly apathetic, increasingly isolated child does not desire to eat. On the other hand, displaying *too much* appetite can also be lethal. Among Datoga pastoralists in East Africa, babies who appear to be always hungry risk the diagnosis that their mother produces insufficient milk. Such babies will be weaned posthaste and thereby exposed to potentially lethal new pathogens and stressors.[14]

Growth is not easily disentangled from love. In one recent study of mothers and full-term infants in a low-income, high-risk population in Santiago, Chile, the infants who appeared adequately nourished were more likely to have been classified as securely attached to their mothers (using Ainsworth's strange situation test) than were infants who failed to gain weight normally for their age. Ninety-three percent of children falling behind in growth had previously been classified as anxiously attached.[15]

In the most extreme cases, the way a mother conceptualizes her baby may determine whether it survives at all. This conclusion was not anticipated when the psychiatrist Marten DeVries set out to study temperament among Masai pastoralists in Kenya.

DeVries arrived in Kenya at the height of a ten-year drought. In a pastoralist community such as the Masai's, when pregnant cows and their suckling calves starve from lack of forage, the milk supply will be diminished for at least a year before surviving cows give birth again and lactate. Children and infants are the first in the population to starve, especially in a case like this one, where the majority of children were already malnourished. During this particular famine, infant mortality rose to almost 50 percent. Of thirteen newborn infants registered in the initial study population, DeVries could locate only six of them by the end of his study; the others had all died. Amazingly, only one of the six infants with difficult, "fussy" temperaments had died; five of the seven with "easy" temperaments had done so.[16]

DeVries's results might have been obtained by chance. Or, perhaps the "easy" babies were already weak. But I accept his explanation: under stress from famine, it was easier for the mother to neglect an undemanding baby.

Whenever a Masai infant frets, the mother takes him up and soothes the baby at her breast. Hence the babies who complained the most were put on the nipple more. They stimulated more milk production and fed more. The "squeaky wheels" got the milk.

Worlds away, Western parents find themselves staggered by the staying power (and seeming maladaptiveness) of whining offspring, unquietable beyond reason. Writer Anna Quindlen can even wax witty about it. "Who was it coined that old saw about God making them so cute so we will not kill them? It has particular resonance at 4 A.M."[17] Such parents might do well to contemplate the catastrophic famines and die-offs, the long-ago bottlenecks, narrow as a birth canal, through which some babies just squeezed. In our not so distant evolutionary past, some forms of "obnoxiousness" signaled "vigor."

Still and all, under duress, infants do well to toe a fine line between signaling their distress and appearing *too* needy. Though a certain quantum of distress is likely to trigger a succoring response, signaling distress could prove risky for a baby whose mother is ambivalent. Instead of being classified as "vigorous," or eliciting a rush of compassion, a complaining baby might inspire an aversive reaction in a mother who perceives her baby as overly demanding, or, worse still, who defines her baby as sickly, "doomed," less worthy of further investment.[18]

Directional Selection

Some infants find themselves held less or fed less than others, or left with someone other than their mother to rear. The unluckiest will be abandoned altogether. But how often in the past would such patterns have had to occur to have an impact on the way infants are? There is more than one answer.

When infant mortality in the natural world runs very high and the pattern influencing which babies die is biased in a consistent way—as was true in the example of frog egg clutches that failed to hatch early enough to evade predation—the "blind but efficient . . . daemon" of Darwinian selection will not be far behind. Such "directional" selection can produce evolutionary change even more rapidly than Darwin—and most evolutionary biologists as recently as ten years ago—dreamed possible.[19] But even if, as in the example at hand, predation by snakes had been less common—say, only 5 percent of clutches destroyed—evasive adaptations could have evolved (albeit over many more generations), *so long as selection operated consistently* in the same direction. When catastrophe forces a population through a bottleneck, the whole process of transforming gene frequencies speeds up.

———

Population crashes followed by subsequent expansion and migration, or else migrations followed by explosions, were precisely the conditions attending human evolution. Bottlenecks characterized our expansion in fits and starts around the globe. Traces of prehistoric bottlenecks can still be detected today in the genetic variation (or lack thereof) found in samples analyzed by Henry

Harpending, Alan Rogers, and other human population geneticists striving to reconstruct ancient demographic histories. This is a commonly accepted explanation for why our species is so genetically similar compared to the tremendous genetic diversity found among other widely spread, weedy primates like Hanuman langurs, or the now increasingly rare but once abundant chimpanzee.

In the course of their slow peopling of the planet, human populations would have fluctuated many times. There would have been several steps backward for every three forward. Wherever famine was implicated, the toll would have fallen heavily on those least able to defend themselves and compete for resources—the very young, most especially those just weaned.

In evolutionary terms all that matters is the genetic composition of those who survive to breed *after the last big crash.* Only those who squeeze through the bottleneck contribute to the composition of the future gene pool. Many (most?) of these bottlenecks would have coincided with hard times when mothers did not have the luxury of rearing every baby that they bore.

It is tempting, and certainly much less troubling, to overlook how heavily selection must have fallen on human immatures. When on top of ordinary selection pressures parents are also assessing infants, making conscious prognoses, and investing accordingly, the weight of selection against immatures is magnified in specific ways.

Scrutinizing newborn group members is a primate universal. But consciously *deciding* whether or not to keep a baby is uniquely human. For midwives and educated parents living in the Greco-Roman world of the second century A.D., the Greek physician Soranus was the Berry Brazelton of his day, albeit a Brazelton with an edge.

Gynaecology, Soranus's influential text on the care of the newborn, instructs midwives to announce to the parents the sex of the infant, then to proceed with a physical exam to determine whether or not this newborn was "worth rearing." Soranus wanted to know: Was the mother healthy during the pregnancy? Is the infant well formed? Does it respond well to sensory stimulation? Cry normally? Is the infant full-term and neurologically normal? Two thousand years later, Martin Daly and Margo Wilson independently arrived at the very same criteria for what, *if their only concern was reproductive success,* parents should take into account in their "initial assessment of the newborn's fitness prospects."[20]

Viability Tests

In some cultures, parents go further, subjecting infants to specific tests of viability. Soranus was aware of such practices among European barbarians and vehemently disapproved. Born in Ephesus, and practicing medicine first in Alexandria, then in Rome, Soranus had a broad knowledge of such customs. Among Germans, Scythians, and even some civilized Greeks (which apparently surprised Soranus), newborns were subjected to icy-cold baths to toughen them, and also to test them "in order to let die, as not worth rearing, one that cannot bear the chilling."

In earlier times, Aristotle had recommended chilling as "an excellent practice" to toughen newborns, and "accustom children to the cold from their earliest years." Such customs persisted in parts of Europe through medieval times. Similar traditions arose, apparently independently, in highland South America as well as in more tropical climes. Among some West African tribes, parents bathed babies in cold streams and then swung them to dry in the chill morning air.[21] As Soranus pointed out, only the "most resistant" would be likely to survive such an ordeal. Preterm or underweight babies, those already weak, more than likely would not. Soranus's primary objection was that the test was *too* stringent; even infants "worth rearing" would be put at risk. Nevertheless, neonatal ordeals persisted in medieval Europe for many centuries more, becoming intermingled with Christian rites.

Universally, people use myth to make sense of their world. Myths are especially useful for reconciling discrepancies between a widely held worldview and anomalous things in the real world that don't quite fit into accepted categories, things that would otherwise be disturbing. With the spread of Christianity, longstanding Norse, Greek, Germanic, and Celtic traditions for disposing of unwanted infants had become deeply disturbing indeed. The new religion held it sinful. Since the early Middle Ages, the Christian Church has actively denounced infanticide. From the thirteenth century onward, it became a crime tried by lay judges as well as a sin condemned by the Church. Parents grew increasingly uneasy about pursuing the old customs. Yet they also remained reluctant to rear infants they did not want or that they deemed unlikely to prosper.

Parents were caught between conflicting prescriptions, between newly introduced Christian doctrine and old customs for testing infant viability. Amid such tensions, postpartum scrutiny merged with a peculiar set of folk

Fig. 20.3 Nineteenth-century illustration of peasant mother dipping her newborn in cold water, captioned "*La Mortalité des Enfants en Bas Age.*" (La Source Miraculeuse, *from* L'Illustration, December 12, 1874)

beliefs about "changelings." Sickly babies were reclassified by parents as something other than human—demon or goblin babies exchanged for real, human ones.

Anyone whose bedtime stories ever included Maurice Sendak's *Outside Over There* will be familiar with how goblins arrived one day, "pushed their way in and pulled baby out, leaving another all made of ice," a haunting tale derived in part from an eighteenth-century story collected by the Brothers Grimm:

> Fairies stole a mother's child from its cradle and in its place laid a changeling with a big head and staring eyes who wanted to do nothing but eat and drink. . . . The neighbor told her to take the changeling into the kitchen, set it on the stove and make a fire and boil water in two egg-shells; this would make the changeling laugh, and if he laughed, that would be an end of him.[22]

At the surprising sight of water boiling in an eggshell, the baby laughs, so the fairies are forced to return the human baby and take back its changeling substitute. In Sendak's version, the baby, "who lay cozy in an eggshell, crooning and clapping as a baby should," is identifiable as the real human baby.

In real life, as in folklore, people convinced themselves that sickly babies were impostors left by goblins in place of healthy ones. The infant left behind became an *enfant changé* in France, a *Wechselbag* in Germany, in England a "fairy child" or changeling. In the best known versions from northern Europe, changelings were left overnight in the forest. If the fairies refused to take it back, the changeling would die during the night—but since it was not human, no infanticide could have occurred. If by a miracle the exposed baby

Fig. 20.4 Goblins steal the real baby and leave a changeling "with a big head and staring eyes. . . ."
(Drawing by Michelle Johnson)

survived, it meant that the original healthy human child must have been returned.

In peasant communities, ancient superstitions seeped into "Christian" practices. Among the best known examples is the twelfth-century French cult of the Holy Greyhound that grew up around Saint Guinefort, the healer of children, who was worshiped till quite recent times. Shrines to Saint Guinefort were forested locations where babies suspected of being changelings could be left overnight.

The cult of the Holy Greyhound grew up around the legend of a loyal family dog unjustly slain by its master. In one version, the master misconstrues events when he finds blood on the dog's muzzle. But the dog was no predator. He had been bloodied while protecting the man's baby son from a snake. The father killed the dog, only to find his baby unharmed. Afterward, mothers brought children who failed to thrive to special locations in the forest associated with the burial site of this martyred dog. They left their infant overnight in hopes that the sick baby would either die without further suffering or recover fully. "À Saint Guinefort, pour la vie ou pour la mort," went one French incantation to the saint—if not a healthy life, then death. Such cults persisted into the early twentieth century despite efforts by Church authorities to suppress them.[23]

According to one thirteenth-century definition, a changeling was "a child who exhausts the milk of several wet nurses but to no avail, for it does not grow, and its stomach remains hard and distended. . . ."[24] In his 1405 *Treatise of Superstitions,* Nicolas of Jawor drew on material set down by the Bishop of Paris two centuries earlier to explain that: " 'changelings' . . . are said to be thin, always crying and unhappy, and so thirsty for milk that no quantity, no

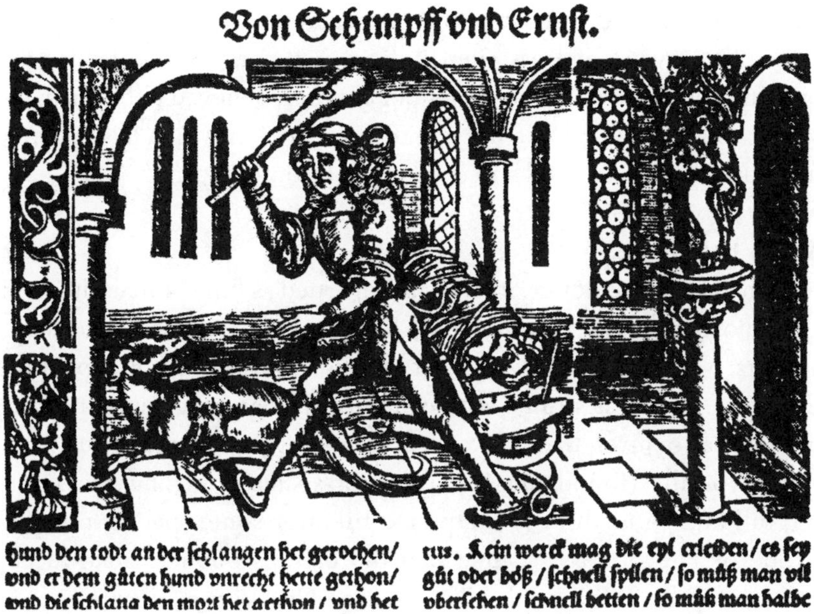

Von Schimpff vnd Ernst.

hund den todt an der schlangen het gerochen/ tus. Kein werck mag die epl erleiden/es sey
vnd er dem guten hund vnrecht hette gethon/ gut oder böß/schnell spilen/so muß man vil
vnd die schlang den mort het aethon/vnd het vbersehen/schnell betten/so muß man halbe

Fig. 20.5 In this sixteenth-century illustration of the legend of Saint Guinefort, the viper still lurks near the baby's cradle as the knight clubs the baby's innocent protector. *(Courtesy of the Bibliothèque Nationale, Paris)*

matter how abundant, could ever satisfy even one of them." It is impossible to know whether such descriptions apply to children who are malnourished and dehydrated from illnesses like diarrhea, or perhaps to unwanted infants who fail to thrive because they are neglected. In her study of "maternal selective neglect" among impoverished mothers in northeastern Brazil today, anthropologist Nancy Scheper-Hughes provides detailed descriptions of such "doomed" and stigmatized babies, changelings in a contemporary Third World context whose desperate mothers cannot afford the luxury of trying to save them. They are said to be "difficult to rear" and to suffer from "child sickness." As one Brazilian mother told Scheper-Hughes, "They come into the world with an aversion to life. They are overly sensitive and are soon fed up" with what food is given them. Another mother commented on how "It hurts the mother to see her baby delay so in dying."

As late as the seventeenth century, French children "born thin," "rather smaller than the others," or "who wear out three nurses without getting plumper and cry when one handles them" were thought to be the products of insemination by the devil. Folk beliefs intersected with reality. Between 1850 and 1895 there were at least eight judicial inquiries from various parts of Europe into the death or mistreatment of children believed to be changelings. In America, cases like the one in New York in 1877 in which Irish immigrant parents burnt a suspect child still crop up from time to time.[25]

Birth by Stages

In modern Western societies, infants are regarded as fully human the moment they are born, tagged, blood-tested, footprinted, granted citizenship and legal rights. Mothers are encouraged to bond with their infant almost immediately and regarded as strange if they do not. It is an efficient and admirable scheme. It is also, given the full range of human experience, unusual.

More typically (found in 86 percent of societies in one cross-cultural study), full human identity is postponed till after some specific postpartum milestone or rite of passage in which the baby is publicly acknowledged, given a name, receives a soul.[26] The ancient Greek *amphidromia* is a classic naming ritual. Alloparents pick up the baby and carry it around the hearth, then hand the newly named baby back to the parents, thereafter a full-fledged member of their community. In an Indian village in Rajasthan, some time after birth neighbors and kin will gather at a "cradling ceremony" to joyously sing songs and sew clothes for the baby. *Hijras* (sexually ambiguous castrated men dressed as women) show up to entertain people and bless the baby—especially if it is a boy. Prior to the late nineteenth century, infanticide would have been sanctioned before the ceremony, but not after.

In hunter-gatherer societies, the transition point is early and very simple. By custom, a !Kung mother goes into the bush alone to give birth. If she comes back with the baby, it is recognized and protected as a group member. However, if she abandons the baby before returning, she is not regarded as someone who has killed a person.

Birth rituals may occur right after birth, or months later. They may be celebrated simply, or with all the fanfare of a cathedral christening. In the majority of cultures, full rights as a group member are postponed till after inspection and acceptance of the baby by parents or alloparents who have committed to rear it. Criteria for acceptance can be very arbitrary, and open to interpretation. Recall that in some South American tribes, too much or too little hair is considered a sign of maternal misconduct, dooming the neonate—but not necessarily. "This is not a human being, this child has no

Figs. 20.6 One consequence of a belief in changelings was that sickly babies would not be reared. Here Saint Stephen—stolen as a baby (see color insert, Plate 4) and supposedly suckled by a white doe—returns to challenge the devil-baby left in his stead. After all those years, the baby had never grown—proof that it was not human. To make the demons take back the impostor, the changeling is threatened with fire, perhaps burned. *(Painting by Martino di Bartolommeo, fourteenth century. Courtesy of Städel Museum, Frankfurt)*

hair," Elena Valero was told by other women after she gave birth among her Yanamamo captors. "Kill him at once," they said, only to be contradicted by the husband who had taken Elena. He merely said, "Let her bring him up, even if he has no hair . . ." and told the women to go away.[27]

In parts of Africa, South America, and in the Pacific Islands, babies born with teeth, multiple births, or breech births can be viewed as "ill omens" that prejudice parental acceptance of the baby. Such traits are often susceptible to interpretation, as illustrated by the case of a West African father who demanded that his baby, who was born with six fingers, be killed. The midwife, however, ridiculed him for misreading the signs of witchcraft; to her experienced eye, polydactyly signaled prosperity.[28]

Passages to personhood may be viewed as more gradual. Among the Ayoreo, the critical milestone is relatively late. No child is considered completely human until he can walk.[29] In other societies, milestones range from taking food, beginning to smile (around five weeks), laughing (four to five months), or cutting teeth. Passage through universal developmental phases hints at the infant's future potential and also attests to how much investment has gone to getting an infant to this stage. The mere fact that an infant has survived long enough to pass these milestones further indicates that someone has taken responsibility for the child.

Grief displayed by bereaved parents is often greater if the baby dies *after* naming or after one of these developmental milestones has been passed, signifying both the emotional bond that has intensified over time, and also the child's perceived worth. DeVries was struck by the contrast between how detached, almost nonchalant, Masai mothers were if an infant succumbed before being named, as compared to the extreme grief—screaming, self-mutilation, and chaotic racing about—exhibited if death occurred afterward.[30] Apparently, passage through the "second birth" marks not only a public decision to keep the infant but also a change in the emotional tone of the relationship between mother and infant.

When United States senator Daniel Patrick Moynihan declares that he is pro-choice but against late-stage abortion because it is too much like infanticide, he is restating the anthropologically obvious. Almost all infanticide in traditional societies occurs right after birth, and is conceptually identical to late-stage abortion.[31] Neonaticide is favored over abortion simply because infanticide is safer for the mother. (Although traditional societies do have ways to induce abortions, such crude methods as the pregnant woman asking someone to jump on her stomach, they are neither effective nor anything approaching safe.) The situation is reversed for societies with Western medicine. Abortion—especially in early stages of pregnancy—is safer for the mother than giving birth is. No one with other options chooses infanticide.

———

In no other primate do mothers appear to distinguish between a new baby and a new baby that will be kept. Nothing is known about when in the course of hominid evolution such distinctions came to be emphasized. My own guess is that mothers grew more discriminating as they were increasingly called upon to simultaneously provide for multiple offspring of different ages. If this supposition is correct, mothers faced with the prospect of provisioning a staggered clutch of slow-maturing, highly dependent offspring would already have been somewhat discriminating prior to the Neolithic. After it, mothers would have become more fastidious still as settled living brought with it birth intervals shorter still. Such mothers would have been among the earliest intellectuals, enlisting natural history, myth, and ritual to explain anomalies, justify their actions, and reconcile necessity with emotions. Then, as now, combining survival, maternity, and work confronted mothers with chronically irreconcilable dilemmas. Emerging belief systems made such dilemmas easier to bear, as intelligent and increasingly compassionate creatures invented stories they could live with.

The way a mother thought about her fetus would have become integrally linked with her view of the world generally. Practices such as assigning names to individuals came linked to a mother's dawning awareness of death and history. Increasingly, mothers would be able to imagine and articulate what the future might hold, including what the impact of a new birth might be for her other children. To a mother giving birth during this dawn-time of humanity, it would have been critically important not to regard a neonate as having equal standing with older children.

One of the earliest reference points for any emerging cosmology would have been the all too obvious connection between childbirth and mortality. Birth must have seemed a dangerous life stage, rendered more so by the pain women experience at that time.

Birth is a perilous time, when the alien, ghostly world of fetuses comes in contact with the human world. If things go badly during passage of the large-headed baby through the birth canal, such a ghost might easily claim the mother's life. It is the neonate that mediates between these worlds and must remain in limbo after birth till rendered safe, till post-birth rituals mark the incorporation of something previously nonhuman into the "civilized," as opposed to supernatural, world.[32]

Belief structures devised to cope with and help explain these experiences took on a life of their own. Historians, psychologists, and primatologists have all called attention to the role played by the physiological lag between the onset of lactation and the mother's emerging sense of unity with her infant. Some delay in the onset of lactation is built into being a primate. Whether or not this window is any longer, or any different, in humans is unknown. What is obviously different, though, is the way cultural concepts have been used to elaborate this period of grace between a woman post-partum and a full-fledged lactating mother.[33]

In eighth-century Holland, for example, among the still-pagan Frisians, infanticide was permissible, but only so long as the child had not yet tasted "earthly food."[34] This was a common pattern. It is surely no accident that so many culturally recognized milestones about becoming human specify intake of nutrients. Once lactation is established, with all its attendant hormonal changes, the mother is physiologically and emotionally transformed in ways that make subsequent abandonment of her baby unthinkable.

———

Anomalies that fall between traditional categories are everywhere supernaturally charged. Centuries after the Age of Reason, hard-to-classify individuals (a man in drag; an infant speaking in a deep voice) can still make us uncomfortable. Aware of this, makers of horror movies often design aliens that look like fetuses.

With startling regularity, the fetus has been regarded as dangerously anomalous, a symbolically charged entity. In the belief systems of many Maya-speaking people—contemporary descendants of the ancient Maya still living in southern Mexico and Central America—there exists a complicated symbolic analogy between making tortillas, their staple food, and making a baby. Among the Tzotzil Maya of Chiapas, a warm and gracious people whose folklore I used to study, menstrual blood is thought to be waste, bodily substances *not* used in making babies.

In this symbolic framework, menstrual blood is equivalent to the waste from the special lime-water mixture that is used throughout Central America to soak cornmeal before patting the mush into tortillas to be cooked. Like tortillas, babies are viewed as essentially raw and uncivilized until cooking (gestation, and application of civilizing ingredients like salt after birth) brings them into the realm of human culture. Given this symbolic framework, such puzzling customs as applying salt and chile to the lips of a just-born baby suddenly make sense. Salt is applied again at baptism, to seal in the baby's soul. Prior to such ministrations, babies are viewed as anomalous, still linked to the supernatural world, and less than human.[35]

In modern societies, science and medicine have removed much of the physical risk from the birth process. So far as genes and tissue are concerned, embryo-fetus-baby represents a biological continuum. No distinction can be other than arbitrary. In this sense, the transition between fetus and person-

Fig. 20.7 In between alive and not-alive, human and not-quite-human, fetuses fall somewhere between "nature" and "culture." Aware of their anomalous status, filmmakers routinely use fetuses as templates for aliens from outer space. *(Alien from* Men in Black, *1997 © Columbia Pictures Industries, Inc. All Rights Reserved, courtesy of Columbia Pictures)*

hood is no less ambiguous today than it was a hundred thousand years ago. The ambiguity persists in spite of quite conscious, even anthropologically sophisticated, propaganda intended to humanize the fetus by, for example, displaying photographs taken with fiber-optic cameras that show a near-term fetus inside the womb sucking its thumb. Ultrasound and wide-ranging neonatal technologies, along with medical advances that constantly increase a fetus's ability to live outside the womb, not to mention the increased care and staggering parental and societal investment required for the exterogestate fetus to survive, are carrying us into novel terrain. There are no precedents—emotional, conceptual, legal, or otherwise. This is totally uncharted territory.

And Baptism?

In the Christian world, dipping a baby in cold water, or sprinkling him with water, or, in the case of the early Christian Church, total submersion in cold water, marks the transformation of a newborn into a community member. The rite of baptism represents a form of rebirth, or social and spiritual birth, after the baby's initial debut into the world.

Nothing is known about the rite of baptism prior to the first century A.D. Most scholars assume that baptism has to do with purifying, primarily because ritual immersion was traditionally important in Judaism, as in the mikvah, a ritual bath taken by Jewish women to purify themselves after menstruation. Others have suggested that baptism was a substitute for circumcision in Judaism. French social historian Jean-Claude Schmitt offers an alternative explanation. Baptism, he argues, derives from the early Celtic-Germanic practice of exposing children to cold or immersing them in cold water. If so, baptism would have derived from a rite originally used by parents to set a threshold of viability below which an infant would not be reared.[36]

———

The question of what determines a woman's reaction to her baby is often deemed too sensitive to discuss. For some the idea that there might be more than one possible response is abhorrent. Yet from the little we have learned, it is clear that people are more drawn to look, or to look longer, at some babies than others. One typical response would be to say that interest was elicited because the baby was "cute." But what does that mean?

What "cute" usually means is: a rounded head, big eyes, and plump cheeks—all prime criteria toward Soranus's determination of "full term." These same neotenous traits persist in humans longer than in any other mammals. To some, this suggests that childish appeal coevolved along with maternal ambivalence, as a sort of sweetener, inducing a discriminating mother to commit.[37] Everywhere, one trait parents pay close attention to is how plump the baby is.

A Matter of Fat

IT'S A BOY!

Alexander Wilson Taylor

August 16, 1991

7 lb 10 oz

Out of the havoc of a collapsed apartment building in the 1985 Mexico City earthquake, rescue workers, who had virtually given up hope, extracted a small miracle. After a week of entombment someone was still alive. Neonatologists may be less surprised than most to learn that the lonely survivor was a newborn baby. For unlike most mammals, and unlike any of the other 175 species of primates, humans are born with such substantial stores of white adipose tissue that even a newborn starved for a week can maintain blood sugar levels still within a normal range.[1] The baby in the earthquake would have been at dire risk of dehydration and hypothermia, but stores of fat made him less dependent on being fed promptly than almost any other mammal would have been. Whereas newborn rats or piglets are nearly devoid of fat at birth, and would starve after just one day of fasting,[2] human milk production (apart from colostrum which is present in the first twenty-four hours) does not even get under way for two or three days after birth, sometimes longer.

Why Are Human Babies So Big and Fat?

Adipose tissue of a full-term human fetus accounts for 16 percent of its weight at birth. This is proportionally four to eight times more fat than is found in a baby monkey. Although length of gestation is about the same in gorillas and humans, the gorilla newborn weighs just over half what a human baby does. Chimp babies (whose gestations are shorter) are similarly small.[3]

Comparisons of the weight of human newborns relative to that of other primates indicate that humans are indeed off-scale; they fail to conform to predicted correlations between neonatal and adult body weights.[4] As the great primate anatomist Adolph Schultz put it, "Most human babies are born well padded with a remarkable amount of subcutaneous fat, whereas monkeys and apes have very little, so that they look decidedly 'skinny' and horribly wrinkled."[5] It would appear that human babies are particularly fat, presumably because they need fat for something. But what? The puzzle is only compounded when we recall that human mother's milk has no more fat in it than does the milk of monkeys and apes. Composition of human milk is the same low-fat formula found in virtually every other primate. Like theirs, our milk is dilute, adapted over tens of millions of years for infants that remain in more-or-less continuous contact with their mothers and suckle at will. By contrast, the handful of mammals that are born with stores of subcutaneous fat comparable to humans, such as seals and walruses, produce incredibly rich milk. The milk of a harp seal mother is up to 23 percent fat at the beginning of lactation, 40 percent fat by the end; the milk of a gray seal is as high as 53 percent fat. Compare that to the paltry 3 to 4 percent fat content in the milk of monkeys, humans, and other apes.[6]

Aquatic Origins Hypothesis

It was the curious convergence of subcutaneous fat in humans and marine mammals (along with the seal-like sleekness of "naked apes") that years ago suggested to Sir Alister Hardy—and to the best-selling author Elaine Morgan, who wrote *The Aquatic Ape*—that our ancestors evolved to spend large amounts of time in water. As in blubbery marine mammals, they argued, subcutaneous fat would render human waders and swimmers warm and buoyant. The "aquatic origins" hypothesis has received little support from anthropologists, yet the puzzle that inspired the speculation still persists.

As with any bodily curiosity—the functional wings on a flightless bird, for example—the most obvious possibility is that the fat deposits are remnants

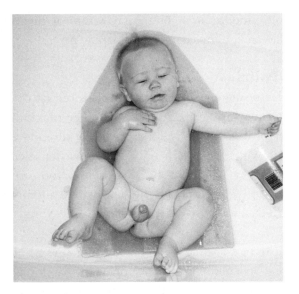

Fig. 21.2 Compared to other primates, human neonates are born inordinately fat. *(Courtesy of Peter Rodman)*

of some ancestral form. Yet human fat deposits have no known homologues in the order Primates, meaning that this trait is most likely a special, derived characteristic that has evolved since the last common ancestor humans shared with chimps and bonobos. No known feature of fetal metabolism can explain the last-minute buildup of fat during the final weeks of pregnancy, a buildup accounting for over half the total energetic costs of pregnancy at that point. Neonatal fat has all the earmarks of a peculiarly hominid adaptation. But for what?

Insulation for Naked Neonates

Apart from the aquatic origins hypothesis, four additional (not mutually exclusive) hypotheses have been suggested to explain the unique development of neonatal fat deposits in human babies. First there is the "insulation hypothesis." In 1998, the Polish anthropologist Boguslaw Pawlowski proposed that fat provided *insulation* to protect babies from nighttime cold in a species that had recently moved to the savanna and slept "naked" and "on the ground."[7] It is not clear, however, why hominids would not be able to solve this problem behaviorally at least as well as monkeys and other apes do: langurs, for example, huddle at night to stay warm at high altitudes in the Himalayas; gorillas and chimps construct sleeping nests.

Insurance Policies

Secondly, it has been suggested that a larder of stored fat could represent an *insurance policy.* If a new mother were indisposed, her infant might have to wait for her to recover, or if she died, wait for a lactating allomother to come along and adopt him. But why should humans be under heavier selection pressure in this respect than other primates? Baby monkeys, too, are orphaned and adopted, and we know of cases from the wild where allomothers who were not lactating carried an infant until its sick mother recovered enough to nurse it herself.[8]

If baby fat is insurance against emergencies, monkeys would also have benefited from putting on a blubbery life vest prior to abandoning their mother's womb. If fat was useful insurance, other primates should have evolved it long ago. Yet fat deposits in primate fetuses have never been documented except in humans. Counterintuitively, hominid babies evolved to be plumper just as selection pressure was making their births an increasingly tight squeeze. Narrowing of the pelvic canal was one of the tradeoffs quadrupedal, knuckle-walking ape mothers made in exchange for the energetic savings to be had from walking upright on two legs, replacing the knuckle-walking shamble of a chimp with an easy two-legged stride.[9]

If fat provided useful insurance for neonates, it is surprising that this trait never evolved before. Compared with humans and monkeys, the pelvic opening of Great Ape mothers has ample room to accommodate a bigger head or fatter shoulders. (For relative proportions refer back to fig. 7.8.) Great Apes have room to spare, yet there apparently has been no selection pressure for ape neonates to stockpile fat prior to birth the way human neonates do.

Food for Thought

This brings us to the third hypothesis, which relies on a uniquely human need: *neonatal fat provides a stockpile, or larder, critical for the development of fast-growing, lipid-guzzling human brains.* Over the past four million years, hominid brains have expanded from around 450 cc in apes and australopithecines to a ponderous 1,400 cc in *Homo sapiens.* The bulk of our braininess evolved in the last three hundred thousand years, long after apes became bipedal. By the time brains were expanding, *Homo erectus* females had already evolved to be bigger mothers.[10] Nevertheless, the combination of a narrow birth canal and big brains cost mothers dearly.

Brain tissue is very expensive to produce and maintain. As anthropologist Leslie Aiello of University College London has shown, the demands of this single greedy organ account for more than 50 percent of the total basal metabolic rate of a baby.[11] Gargantuan as it seems (especially to the laboring mother, whose cervix the compressed cranial case of her baby just manages to squeeze through), the newborn's brain is only a quarter of what it soon will be. Continuing to grow at a rapid, almost fetal, rate during the first months of life, the brain attains nearly 70 percent of its final mass within a year of birth.[12] Hence, according to the "food for thought" hypothesis, big-brained hominids needed extra fat to grow on, like an extra candle on a birthday cake.

Birth weight turns out to be the single best predictor of infant survival. In a study done fifty years ago, before many of the interventions that are available today were in use, infant mortality during the first months was still around one out of twenty births. Those below the optimum weight (about eight pounds) and those above it were less likely to survive. Today, this remains a textbook case illustrating ongoing stabilizing selection.[13] Babies born at low birth weights are also more prone to neurological impairment, though caution is required in interpreting this because such babies are also more likely to be born to mothers in poor condition.[14] Birth weight also turns out to be a moderately good predictor of certain aspects of health in adulthood,[15] and of subsequent mental development.[16]

Advertisement for Myself

In spite of their long, nine-month gestations, human infants are born at an earlier phase of development than other primates, probably to accommodate their disproportionately large heads.[17] Yet even with this early expulsion from the womb, cranial plates still have to compress. To accomplish this, Mother Nature takes advantage of the sutures already present in the brain cases of young vertebrates.[18] Slabs pass over one another like drifting continents as the skull squeezes through the birth canal, leaving newborns with a fontanel, a temporary "soft spot" where the quadrants of the brain case have not quite fused.

Other primates are born precociously able to latch on to their mothers and support their own body weight within hours of birth; the wiring of their central nervous systems is more nearly complete. Not in humans. Compared with those of monkeys, human bones are incompletely ossified and still quite malleable at birth. The cartilaginous hands of human babies are not nearly sturdy enough to cling for long. For all these reasons, human neonates—those "exterogestate fetuses"[19]—are particularly helpless by primate standards and require full-time carriage by a caretaker to survive.

The fact that intelligence is probably useful for human success provides compelling motivation for a mother to "finance" the energetic requirements of an organ as costly as the brain. How well she supplies her baby's fat stores may constitute one of the mother's more enduring maternal effects. Emerging evidence suggests that the enhanced heritability of IQ found among twins reared apart is partly due to their *shared prenatal environment*.[20] Interestingly, the slightly lower average IQ scores of twins as a group compared to singleton births (not more than seven points), is often attributed to the fact that twins must share maternal resources in the womb, and hence weigh less at birth.[21]

But none of these observations explains why babies put on so much fat *before* birth, when they could more safely fatten up *after* birth. Why should the baby stockpile fat *prior* to its treacherous trip through a bipedal mother's narrow pelvic canal? So perilous is the prospect of birth that women living in areas of both West and East Africa deliberately starve themselves during pregnancy. In Nepal, women are aware that smoking stunts the baby's size at birth yet continue, hoping to have a smaller baby and a safer, less painful delivery. The problem is serious enough and widely reported enough that the United Nations commissioned an advisory group to study this counterintuitive practice of "eating down" during pregnancy.[22]

Surely both infant and mother stand to benefit from waiting until after passage through the birth canal before building up infant fat stores. Why not develop a lay-away plan whereby the mother puts on even more extra fat so as to accommodate the needs of the developing brain by delivering lipids in safe installments after birth? Richer milk has evolved independently in various mammals, from tree shrews to bears, under a wide range of selective constraints. So why not in humans?

In defiance of so reasonable an arrangement, near term the fetus builds up a thick layer of fat between the shoulder blades and lays down fat around the arms and on pudgy hands. This explains why my own babies arrived with a crease around their wrists that looked like someone tied a string there to remind them it was time to be born. This sequence seems all wrong, like assembling a model ship and raising the sails *before* putting it in the bottle.[23]

There is, however, a simple solution to the riddle of this masochism, a mother who makes life difficult for herself by delivering large quantities of nutrients and fattening up her baby just prior to delivery. We've been watching the wrong operative: The mother is not responsible for the buildup; the fetus is.

As David Haig points out, once placental supply lines are laid down, the mother no longer controls the flow of nutrients to her fetus. Between gestational weeks 24 and 38, the fetus takes advantage of access to maternal food stores to stockpile fat. Its body composition shifts from 89 percent water and 1 percent lipids, to 76 percent water and 10 percent lipids, so that by the time of birth, fatness is a fetal *fait accompli*.[24] Such fatness might provide insurance, a larder for the developing brain to draw upon.

This brings us to the fourth hypothesis, albeit one not mutually exclusive with the other three, because it depends upon the assumption that fat babies are more viable. From there, the "self-advertising" hypothesis goes a step further to include the mother's perception of how viable her baby is. According to this last hypothesis, *neonatal fat not only helps a neonate survive, it advertises what a good bet this baby is to survive, be healthy, and enjoy full neurological development.* Hypothesis four (or five if we count "aquatic origins") includes the mother's assessment of her baby among the selection pressures upon infant attributes.[25]

When immatures must compete for nurture (in this case, newborns would be vying for preference with past and future prospects, older siblings already born and siblings that might yet be born), selection to appeal to parents takes on a life of its own. Think of the old notion of a "bonny" babe. As the *Oxford English Dictionary* reminds us, a "bonny babe" is not just a physically attractive and appealing one. In its original meaning, the defining feature of a bonny babe is fine size. Remnants of such folk wisdom can be found in the contemporary practice of sending out birth announcements that inform relatives and friends not only of the baby's sex and date of birth but of the infant's weight at birth as well: "It's a boy, seven pounds, ten ounces!"[26]

The inclusion of the infant's weight on a birth announcement no longer seems so strange or arbitrary, merely very quaint and downright ancient. Birth weight has always provided humans with the readiest available index of a neonate's prospects for survival and healthy development, including brain development. Even without the benefit of modern medicine, many societies are aware that low birth weights (especially when associated with prematurity) are indicative of poor survival prospects.[27] As the second-century physician Soranus observed, "The infant which is suited by nature for rearing will be distinguished by the fact that his mother has spent the period of pregnancy in good health . . . [and] by the fact that it has been born at the due time, best at the end of nine months."[28] Longer-term consequences of low birth weight might also be recognized, as when an observant midwife or shaman noticed that those born scrawny were prone to respiratory infections (bronchitis and pneumonia) and even to die young from adult diseases like heart attack.[29] Where local knowledge informs parental preferences, culture joins hands with natural selection and magnifies her effects: parental preferences for bonny babies make it more likely they will prosper.

If infants could be certain of maternal investment, as almost all Old World monkey and ape infants born to fully mature mothers could, it would be far more practical for both parties (in terms of the hazards of delivery) for a mother to stockpile fat and transfer it to her infant in the form of richer milk after birth, and give birth, as every other primate does, to streamlined babies. Instead, mothers expel their big-brained and fat babies at a far less developed stage of development than is the case in any other primate.

Unable to count on mothers to commit to them as reliably as newborn chimps and other primates can, hominid fetuses are under pressure to convince mothers to do so. The more a fetus extracts from its mother prior to birth, the more pregnancy has already cost her. Fat stores diverted to this fetus, although nothing close to the final tally, cannot be recouped. Neonatal fat becomes the equivalent of a coot's fancy natal costume, adipose advertising in place of colorful signals of neonativity. The message is about irretrievable costs that carry with them the promise of a well-endowed product.

If fat were not actually advantageous (either as survival insurance or as evidence of the sufficient wherewithal to provision a greedy brain), plumpness would probably never have become the object of parental preference. But once selection for butterballs was under way, the advantage of appearing bonny took on a life of its own, propelled by parents who preferred fat babies. Parental preferences enhanced favored offsprings' prospects of surviving. All four factors—insulation, emergency stores, wherewithal for greedy brains, and self-advertisement—may be implicated in the puzzle of plump neonates. Indeed, one could even speculate that parental choice favoring plump babies selected for babies with *even more* fat than was essential for brain growth, relieving some of the constraints that otherwise might have prevented the evolution of larger and larger brains in the course of human evolution.

In this still imaginary scenario, uniquely human neonatal fat deposits testify to the intensification of parent-offspring conflict (and with it, maternal-fetal conflict) in *Homo sapiens.* Like the life-historical changes that occurred hundreds of thousands of years before (namely, the longer lifespans that brought with them longer childhoods), this later contribution to humanity's unique intellectual capacity, all that gray matter, could have had origins more coincidental than pride in our own brilliance normally leads us to imagine. Oh yes, having more fat on hand is useful for making better brains, but payoffs from cleverness were not the only selection pressure leading to fatter babies. Some babies were already fatter than they needed to be because some mothers were discriminating about which babies they cared for.

The erotic appeal and sheer deliciousness of a baby's soft, plump flesh, perhaps also such potent neonatal equipment as a new baby's fleeting "fairy" smiles, may be viewed as the product of countervailing selection on infants, to compensate for Mother Nature's discriminating human mothers.

The Infantile Equivalent of Sex Appeal

Almost nothing a baby does so delights or engages its mother as the baby's first laugh. The baby catches her mother's eye and gives a sociable chortle. And yet how many of us have really examined, really thought through the sources and rationales that lie behind that unique and fairly public testimonial to neurological development—that little laugh—and the long evolutionary history that produced in us the powerfully pleasurable sensation on hearing it, the rush of pride and warmth and adoration? How can creatures so helpless take our breath away, appall us with their vulnerability, and seduce us to their bidding? Some of their allure, no doubt, is in the eyes of the beholder. After all, parents themselves have been selected to *find* their babies luscious. But it is not so simple.

If mothers were instinctively nurturing, if they evolved to care for any infant born (as essentialists argue), why should infants be selected to expend so much metabolic energy making certain that a mother does so? Any dull, calorie-conserving lump of rapidly developing tissue would suffice. (Let this grub do its learning at some later stage—no need to seduce at birth.)

No, something more than just selection to survive has been at work. Calories that could otherwise be channeled to somatic growth (for example, growing quickly past the most vulnerable phases of infancy) are channeled instead into such extravagant diversions as early cognitive capacity, imitation, and engaging the mother with that odd arrangement of the cheek muscles we call a "smile," an elaborate reconfiguration of the ancient primate open-mouth invitation to play.[30] With all the various constraints on what is physically or evolutionarily possible, human infants have been selected to be activists and salesmen, agents negotiating their own survival. I am convinced that there must have been stringent selection among our ancestors for this infantile equivalent of sex appeal.

But none of these extravagant powers babies use to massage maternal sensory systems matter unless infants get to first base. That means being picked up and held near the breast. Thus, newborns must appeal on sight. But once a baby's status as "a keeper" is assured, is there anything special about a baby's first hours after birth? Probably only in the sense that a mother close to her baby after birth is that much closer to recognizing that baby as her own so that the two are off to a promising beginning to a long relationship. In the next chapter I evaluate some of the current advice about "bonding" right after birth.

Of Human Bondage

Ultimately, the most liberating piece of information a woman could have is that her infant can attach to anyone.

—Erik Hesse, 1996

When our first child was born in the late 1970s, no doctor or nurse questioned our desire for my husband to be present at the birth or to have the baby "room in" rather than disappear into a centralized nursery. This was a marked departure from earlier policy in the United States and many other Western countries.

By the end of the nineteenth century, administratively convenient centralized depots where nurses could monitor their newborn charges had become a standard feature of American maternity wards. Mothers waited in their rooms for a chance to hold their babies during scheduled visits. Friends and family milled around a pane of glass, straining to catch a glimpse of the baby, asking each other, "Which one is it?" It took more than half a century for medical professionals to question the hygienic as well as the humanitarian wisdom of communal nurseries.

Newborn mammals did not evolve to cope with exposure to the pathogens of so many different individuals. Apart from a handful of colonial breeders, like the Mexican freetailed bats, among whom thousands and thousands of creched babies hang from the ceilings of nursery caves, most newborn mammals are sequestered among close kin, and many—wild apes, for example—remain exclusively in contact with moms. It is a big advantage for newborns to have their skin and the insides of their nostrils colonized by the same bacteria strains as the mother's. This benign colonization reduces the opportunity for disease-causing forms such as *Staphylococcus aureus* to infect these areas later.[1] Similarly, intestinal and other germs contracted from the mother will be the same ones her milk provides defenses against. For a newborn, whose own immune system will not really function for another six months, there are merits to keeping infections in the family.

During the 1950s, a series of nursery-related diarrhea epidemics occurred in U.S. hospitals.[2] These outbreaks, together with pressure from attachment theorists, caused medical personnel to rethink the advisability of centralized nurseries. A further impetus for allowing babies to remain in the same rooms as their mothers arose with the publication in 1972 of a report in the *New England Journal of Medicine* purporting to demonstrate a "critical period" in the minutes and hours just after birth when mothers imprint on their infants.[3]

"Critical Periods"

Par for the course, an excellent reform movement, intended to make the birth process in the United States more humane, swiftly turned evangelical. What had been presented to me at the birth of my first baby as a welcome option, had by the births of numbers two and three been elevated to "advisable." At the extreme end of this movement, some new mothers (whether they felt like it or not) were instructed to engage in set amounts of postpartum flesh-against-flesh intimacy, beginning right after birth, to ensure that "bonding" took place.

In this fiasco, "bonding" quickly became conflated with the much slower attachment process Bowlby and coworkers were talking about. By the 1990s, the pendulum had swung so far that in some hospitals new mothers were being rated: "bonded" or "not bonded." Nor was Madison Avenue far behind. The "Right Start Catalog" advertises a "Baby-n-Momerobics Video" to guide mothers as they "cuddle and bond" with their infants, simultaneously exercising their "bonding techniques" while getting back into shape—"for use with infants up to 25 lbs."

"Bonding" is the modern version of the old ethological process of imprinting. The term *imprinting,* from the German *Praegung,* derives from the idea of a coin, which can be easily shaped for a brief molten interlude, after which the metal becomes unstampable. Konrad Lorenz and other early ethologists used it as a metaphor for the instinctive attachment of just-hatched geese to the first moving creature they see. The possibility that such imprinting occurred in humans had intrigued Bowlby, but only briefly. Ethologist Robert Hinde quickly convinced him that there might be "sensitive periods" for learning, but that nothing so mechanistic as "imprinting" was likely to evolve in a primate.

Attachment versus Bonding

Whereas Bowlby focused on how *infants* become attached to their mothers in the months after birth, "bonding" posited a rapid process whereby *mothers* form an emotional attachment to infants in the hours right after birth. Scientific evidence for this Velcro-style attachment derived from studies of sheep and goats, whose mothers imprint on their smells while licking off the amniotic fluid right after birth (see pages 157–61). None of the research on bonding came from primates, who are far more flexible in this respect. In stark contrast to primates, sheep are intransigently discriminating about which babies they will accept. Unless the baby smells just like the baby whose scent the mother imprinted on moments after she gave birth, she rejects all overtures. Sheep virtually never adopt orphans. Nevertheless, people continued to extrapolate from sheep to women, proposing a ewe-style "critical period" for humans.[4]

For the most part, studies comparing human mothers who spent extra minutes with their babies after birth with "control" mothers who did not were inconclusive *except in the case of mothers who were already at risk of abandoning their babies.* For mothers at risk, contact in the hours after birth could indeed make a difference. In the case of mothers generally, however, not having contact with their infants immediately after birth had no measurable effects on the security of mother-infant attachment a year later.[5] But at the time, even research scientists aware of the fallacy of comparing herd-living, precocious lambs with human babies and their mothers, initially held their tongues. None wanted to interfere with welcome efforts by pediatricians to "naturalize" the birth process. Those who did speak out confined their criticisms to research journals.[6] As usual in such cases, a conspiracy of well-intentioned silence backfired—in spades.

Time spent with her infant after birth probably does stimulate a mother's desire to continue this rewarding activity. The process of recognizing her baby, learning its smell, and, especially, the initiation of lactation, change a mother's physiological state. By several days after birth there is an emerging relationship between mother and infant. It would be difficult for a mother who has allowed herself to grow close to a baby to change course at that point. This is why close physical contact right after birth reduces the likelihood that *mothers at risk for distancing themselves from their baby* will do so, as we saw in the "experiment" historian Rachel Fuchs documented at Paris's Mater-

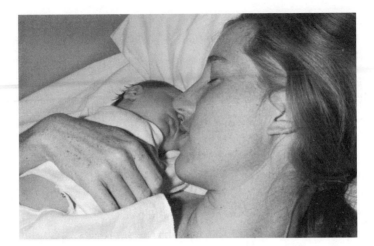

Fig. 22.1 Newborn "rooming in" right under her mother's nose. *(Daniel B. Hrdy / Anthro-Photo)*

nité (see page 315). But in terms of subsequent mother-infant attachment, these first hours and days are more nearly a nice idea than a "biological need."[7] Should a mother otherwise committed to caregiving miss this rendezvous with her infant, and begin to commune the next day, there are no measurable ill effects—provided someone else keeps the infant warm and safe in the interim. Unlike sheep, bonding right after birth is by no means essential for the development of love. It can, however, facilitate the process.

Published critiques of the notion of a "critical period" immediately after birth were available to anyone who took the trouble to look them up.[8] But the idea of a quick-fix that guaranteed a nurturing mother proved irresistible. Thus an interesting (if improbable) hypothesis picked up steam, in spite of the fact that, if critically examined, it would have been quickly disproved. For the "bonding" movement generated its own peculiar mythologies and potent afterbirth rituals. In the popular imagination, in the media, and in maternity wards the promise of a simple rite that could guarantee "good" mothers spun out of control.

Backlash Against Bowlby

When inevitably the backlash came, it, too, arrived in excess, with a vengeance. "Infant Bonding: A Bogus Notion," proclaimed the headline in the *Wall Street Journal*. Well-meaning pediatricians, the "medical establishment,"

scientists generally, baby advocates, and anyone connected to attachment theory (even though its focus was on how infants attach to mothers, rather than
how mothers bond to babies) were indiscriminately lumped together and
made to sound preposterous. Balanced scholarship was sacrificed to the political expedient of exposing an anti-mother "conspiracy" between "the medical
establishment" and the "baby gurus."[9] To a certain faction of feminists, the
"bonding" mandate was just another instance of looking to the animals for
"natural laws" that happened to validate patriarchal interests.

With books like *Mother-Infant Bonding: A Scientific Fiction,* angry advocates
on behalf of mothers, especially mothers who now felt guilty because they
had "never had a chance to bond," declared open season on the whole idea of
any special biologically based relationship between a mother and her infant.

Many feminists had even more compelling reasons than the bonding fiasco
to dismiss attachment theory, an unwelcome message that went beyond what
a mother was supposed to do right after birth. For attachment theory seemed
to prescribe what mothers should do for years to come.

———

One fine spring day in 1977, John Bowlby, seventy years of age and innocent
of much of the trouble brewing, was belatedly awarded an honorary degree
by Cambridge University. At what should have been an uncomplicated celebration, august Cambridge dons found themselves squinting in disbelief as
a small band of women materialized on King's Parade and marched along
the road in front of the gray stone quadrangle of King's College. To the
dons, honoring a "baby man" scarcely seemed controversial. Why were these
women demonstrating?[10]

It was no fluke. Growing numbers of women were coming to regard
attachment theory as anathema. Rarely mentioned in feminist circles,
Bowlby's name, when it did come up, was uttered with derision. Why might
women have a stake in discrediting research ostensibly focused on infant
well-being? Having panicked often enough myself over whether I could live
up to the stiff responsibilities of motherhood, I understand why.

Irreconcilable Dilemmas

About the time Bowlby was being honored, back in the other Cambridge, the
one in the United States, I was being interviewed for an article on the new

field of sociobiology. One angle to the story had to do with women who combine motherhood with demanding careers, a reasonable line of inquiry since my daughter had just been born (and, thanks in part to an unseen Bowlby over in England, immediately turned over to her father to hold). Robert Trivers— the brilliant young sociobiologist whose theories about parental investment are so fundamental to this book—was asked to comment on my work. No doubt Trivers spoke straight from the heart when he told the reporter: "My own view is that Sarah ought to devote more time and study and thought to raising a healthy daughter. That way misery won't keep traveling down the generations." Needless to say, these off-the-cuff remarks, which cut straight to the heart of feminist wariness of evolutionists, were prominently published.[11]

Here was not just a reference to intergenerational transmission of bad mothering, what the Bowlbian Selma Fraiberg had termed "the ghost in the nursery." This was a reference to the ghosts in *my* nursery. Was I offended? After all, my daughter's father was an infectious-disease specialist working long hours, and Professor Trivers, also a father, was just as consumed by his work as I was by mine. Clearly, in addition to the assumption about what infants need, there was an additional assumption being made about precisely *who* should meet those needs. Progenitors of one sex only needed to realign their priorities in order to prevent "misery" passing down generations.

At the time, though, the unfairness of this logic was far less riveting than my own *nagging anxiety lest Trivers was right*. Worse than sciatica, these are nerve spasms that never subside. Long after a child is grown and has left the nest, if she makes a single misstep, painful signals are triggered in a maternal body far away, as if from a phantom limb.

By endowing human infants with a long mother-centered primate heritage, by envisioning infants with special needs and mothers as creatures especially designed to satisfy them, and by situating an infant's sense of security and self in the availability of this specially equipped mother, evolutionists like Bowlby and Trivers appeared to be imposing on women painful choices no man need ever make: her aspirations versus her infant's well-being; vocation *or* reproduction. Twenty years later, I still return to this topic with trepidation.

————

A century before de Beauvoir warned about the "enslavement" of mothers, George Eliot had produced a timeless portrait of the conflict between a talented woman's drive to reach the pinnacle of a demanding profession and her awareness of her baby's needs. In her novel *Daniel Deronda*, Eliot tells the story of an opera singer, Princess Alcharisi, who gives her baby son away.

Here is Eliot's portrait of a woman torn between maternity and vocation. Late in life, the renowned singer is reunited briefly with her son, now grown. Her talent waning, and at a postmenopausal time in life when such a woman might well suffer regrets (the age when the childless de Beauvoir was rumored to have said she tasted "ashes" in her mouth), she tries to explain why, in spite of abundant resources, she chose not to rear her son.

"You are not a woman," Princess Alcharisi tells her son, reliving the old decision.

> You may try—but you can never imagine what it is to have a man's force of genius in you, and yet to suffer the slavery of being a girl. To have a pattern cut out for you . . . "this is what you must be; this is what you are wanted for . . ."[12]

Motherguilt

Not just great artists, but ordinary women who simply aspire to chart their lives their own way, as individuals rather than as nursemaids captive to infant needs, fear that an infant attached means the mother enchained. From this perspective, the flip side of mother-centered attachment models is the arrival of a baby who like a bailiff enters where a mother lives and "attaches" her life. Like early European mothers convincing themselves that a "changeling" child is not really human, professional mothers today can be found persuading themselves that infant needs are as negotiable as an employment contract. A mother, she tells herself, can make up next month for the business trip she needs to take this month, atoning for ten-hour disappearances five days a week and an ever-fluctuating roster of baby-sitters with "quality time" after dinner or during the hours outside her overtime work on weekends.

Alternatively, a mother can dismiss attachment theory altogether as another facet of patriarchal oppression. This explains the popularity of books like *Motherguilt: How Our Culture Blames Mothers for What's Wrong with Society* that urge women to ignore the "din of mother bullying which has been midwifed by the baby gurus T. Berry Brazelton, Penelope Leach, and Benjamin Spock, among others." Attachment critic Diane Eyer blames their "psychobabble about attachment and bonding" for a "backlash against working mothers in the courts and in the workplace." This was only one of many broadsides against the "attachment canon." Its author explains how attachment theory was "first developed in the 1950s . . . [and rests] on the bedrock assumption that only mothers can provide the caring, nurturing relationship their children need to survive."[13]

An unguarded remark to a reporter by Professor Trivers summons all the phantoms that press in upon ambitious mothers, or mothers who must work, unleashing a cascade of self-doubts. Just who comforted her infant when she was otherwise occupied? Were they loving? What will be the long-term psychological consequences of repeated separations? For a mother who has just wrung herself free from the desperate grip of a toddler, frantic and unwilling to stay at daycare while Mom gets to her shift at work, such questions are sickening. They remind her of irreconcilable dilemmas. At that point, she is in no condition to patiently and clearheadedly tease apart all the disparate facts, ideas, distortions, and prejudices tangled there in a mare's nest of conflicting demands. How much more expedient to simply reject any theory that legitimizes infant needs. This is why women were demonstrating against John Bowlby in 1977 and why polemical "get mother off the hook" books are selling well.

The quickest way out, in the short term at least, was to distort and caricature what evolutionists and developmentalists were saying about infant needs and then conflate disparate literatures—the scientific literature on how infant primates come to attach to caretakers with a different literature on the critical need for ungulate mothers to "bond" with their infants in the first hours after birth. Likewise, the horrendous experiences of unwanted children warehoused in Romanian orphanages were lumped together under the heading "maternal deprivation" right along with children spending some hours each week in daycare. A little pseudoscholarly sleight of hand and, presto, Bowlby's observation that primate infants evolved so as to form a strong attachment to their mother was recast into a strict prescription for healthy emotional development in human young that depends on an early and exclusive association with the mother. Once rendered preposterous (since it is obvious that infants reared on other regimens can do fine), what a relief to deep-six the whole attachment enterprise and replace it with new superstitions about innately flexible and resilient ready-formed personalities waiting to emerge, for whom "good enough" care suffices in a world where it is considered crass to ask anyone to define what "good enough" means.

" 'At risk,' 'attachment,' 'bonding': It's a wonder mothers don't break out in a rash every time they hear these psychobabble terms," writes the author of *Motherguilt*. "And they hear them constantly from the professional child-rearing advisers. . . . Thou shalt worry that anyone but yourself that takes care of your children will shame you and damage them. Thou shalt see your husband, babysitter, neighbor, day care provider, aunt and grandmother as a threat to your standing as a Good Mother."[14] The same newspapers and Sunday supplements that had provided a willing forum for the folklore about mother-infant bonding now unselfconsciously championed the "new scholarship." Rarely has that overused phrase "throwing out the baby with the bath water" applied so well.

But wait a moment. Somebody pick up the baby while we take a timeless tip on scholarship from George Eliot: "It seems to me much better to read a man's own writing than to read what others say about him," she wrote to a friend, "especially when the man is first-rate and the 'others' are third rate."[15]

Mothers and Allomothers

What did Bowlby actually say? He said that primate infants, including humans, are born immobile and vulnerable. This is true. He pointed out that they respond very poorly to being left alone, or otherwise being made to feel insecure, which is also true. Human infants have a nearly insatiable desire to be held and to bask in the sense that they are loved. To this extent, the needs of human infants are enormous, and largely non-negotiable. The question is, what are the implications of this for their mothers? How warranted is the deep suspicion that if evolutionary perspectives are allowed a foot in the door, the bailiffs representing "natural law" will march in and impose the notion that the organism that bears this baby, and she alone, must bear responsibility for caring for it? How inevitable is this tyranny, which tells a mother that "*this*" (to quote Princess Alcharisi) "is what you are wanted for"?

Part of the problem of course is that there is little agreement about whose interests are to be maximized in a world where conflicting self-interests—between parents and offspring, between mothers and fathers, within families, between families—are endemic. What goal are we trying to achieve? Secure adults? Good citizens? Independent ones? Self-starters focused on fast tracks? Satisfied mothers? Reproductively successful family lines (in the recent past usually patrilines)? Maximized human potential? And if so, whose? Typically, those with the least power, and of course no vote, are infants. Understandably, perhaps, those most threatened by acknowledging infant needs—mothers with aspirations to do things other than mother—were the ones who felt most compelled to downplay infant needs.

———

"DENIAL IS NOT A RIVER IN EGYPT" reads the message on a musty placard in an archaeology museum where I once worked. Denial of infant needs runs like an invisible and insidious countercurrent through publications purporting to correct the "river of mother-blame" coursing through our society. Yet one of the many ironies in the bonding saga is that even if someone were so rash as to succumb to the naturalistic fallacy and imagine that what evolved is necessarily what should be, there is remarkably little in the primate evolutionary record that turns a female's sex into a precharted destiny of full-time stay-at-home caretaking. Nor is there anything that rules out a mother sharing or delegating caretaking to others. Female primates have always been dual-career mothers, forced to compromise between maternal and infant needs. It is precisely for this reason that primate mothers, including human foragers, have always shared care of offspring with others—*when it was feasible.* Acknowledging infant needs does not necessarily enslave mothers.

Sad to say, however, that acknowledgment does not solve the basic problem of surrounding developing infants with people who feel committed to them. Demonstrating that in the evolutionary past the answer to the question "Who cared?" was more often a matter of circumstances does not instantly make adequate alternatives available. The politician who proclaims that "every child from the moment of conception [should be] protected by law and by love" is merely mouthing words.[16] Parental emotions cannot be legislated.

An updated evolutionary perspective cannot solve such problems, but it can at least focus attention on the real issues. It is not a matter of "motherhood versus vocation." Rather, the question is: Under what circumstances can a mother safely afford to delegate care to allomothers? And, additionally: How can allomothers be motivated to care? Obstacles mothers face when they seek to go off to forage or to work are real enough, particularly in environments (workplaces) where infants are not welcome and mothers lose credibility for considering their needs. These problems have less to do with an infant's need for *exclusive* care by its mother than with values and attitudes in modern workplaces, and the scarcity of reliable and willing alloparents.

Even Bowlby, from his very first published paper in 1944, had specified "mother" *or* "mother figure." Critics like Margaret Mead had seized on Bowlby's concern with maternal deprivation to claim that he meant that mothering "cannot be safely distributed among several figures." And Bowlby continued to be associated with this essentialist posture (only the mother would do) even after he specifically objected that "No such views have been expressed by me."[17]

So many people were so passionately divided over the question "Who cares?" that the point of the enterprise—defining infant needs—was lost. Bowlby's personal opinions on the subject of new mothers working did not help matters. On this score, Bowlby was all too outspoken:

> This whole business of mothers going to work, it's so bitterly contro-versial, but I do not think it's a good idea. I mean women go out to work and make some fiddly little bit of gadgetry which has no particu-lar social value, and children are looked after in indifferent day nurs-eries. *It's very difficult to get people to look after other people's children.* [Italics added.] Looking after your own children is hard work. But you get some rewards in that. Looking after other people's children is very hard work, and you don't get many rewards for it. I think that the role of parents has been grossly undervalued. . . . All the emphasis has been put on so-called economic prosperity.[18]

What is important to keep in mind, however, is that even if Bowlby's personal shortsightedness caused him to be selective about the evidence he cited, it did not invalidate the central premise of his model: infants seek secure attach-ments and need a secure base for healthy emotional development. Bowlby focused on a !Kung-like ideal of an all-indulgent mother who provides exclu-sive care for her infant for the first four years of life. Alternatively, he might have chosen a less indulgent Hadza mother, who weans after two years, or even looked at infant-sharing primates instead of totally possessive chimps. Today we have much more extensive data on people like the Aka and the Efé, where infants from birth are passed among multiple caregivers with whom they become very familiar and are quite at ease. Far from growing up less secure, such infants are if anything more so.

The central reason Bowlby gave for opposing the notion of new mothers working outside the home was his purely practical observation that it is "very difficult to get people to look after other people's children."This, alas, is true. It is the crux of the matter.

————

Bowlby's personal views about working mothers do not undermine the validity of his theory about *how* and *why* infants become attached to their caretakers any more than Charles Darwin's blindness to the sexual assertiveness of females in some species invalidates the theory of sexual selection. Aspects of the model may need to be revised, but the underlying logic holds true so long as the assumptions and observations essential to the model hold.

No doubt patriarchal bias played a role in Victorian (and earlier) ideas about motherhood. And, yes, such biases permeate the writings of evolutionists, from the nineteenth century onward. It was an all-male club and, unwittingly, Darwinians accepted the biased assumptions handed them on a platter by their predecessors who were more nearly moralists than scientists.

Bowlby was not immune. The same biases were still very much in evidence in 1975 when Wilson published his pioneering work on *Sociobiology,* which included a notoriously inaccurate description of foraging societies that claimed that "During the day the women and children remain in the residential area while the men forage for game or its symbolic equivalent in the form of barter and money."[19] A Victorian (and a 1950s suburban) ideal of mother tending the hearth was substituted for the actual life of a highly mobile Pleistocene gatherer.

But Wilson, let's recall, was an entomologist and had to give himself a crash course in ethnography in order to write the chapter on humans for *Sociobiology.* Perhaps more telling, professional anthropologists themselves failed to register this whopper—*even* anthropologists *who had actually helped collect the data* indicating that a woman in a hunter-gatherer society might travel a full 1,500 miles in a year while carrying a year-old baby. The error was simply overlooked because it corresponded with expectations about how the world *should* appear.[20]

Mothers at home, caring for kids—here was an ideal so powerful that it led researchers to overlook the significance of their own important findings. Does this mean that scientists are inevitably too mired in their personal biases to ever be objective? No, absolutely not. Locating and correcting biases is a normal feature of the often slow and inefficient progress of scientific understanding. Science, to paraphrase Winston Churchill, is the worst way to get at the truth, except for the alternatives.

There is no doubt about it: biases informed early definitions of female nature. Where research touched on who cares for babies, the biases took even longer to correct than is usually the case. No doubt, too, the lag had more to do with men's evolutionary histories than with women's. It derived from all the things that it is rarely in male interests to perceive.

For millions of years male reproductive success has depended upon viewing females as individuals to be coerced, defended, and constrained. Changing such ancient attitudes does not come easily to men, nor is it a subject that I as a woman can feel dispassionate about. I still remember the first time I heard a male anthropologist giving a seminar about "exchanging women between groups." His casual depiction of women as chattel enraged me. I scribbled fiercely in my notes: "This is what it must feel like to be black and go to a lecture on the Klu Klux Klan." As a woman, I had an evolutionary past and a history of my own.

Long socialized for subordinate roles, women may be more inclined to look at the world from more than one perspective, male as well as female, dominant as well as subordinate. For those accustomed to the perquisites of patriarchy, however, it would less often be useful to see the world from the point of view of those female subordinates whose reproductive potential they sought to coopt for their own ends. And few men—without guidance and extra effort—seem eager to do so. Once alerted to the problems, however, and with the opening of professions like "naturalist" and "evolutionist" to encompass a wider range of genders, old biases began to be corrected (although I need to be careful here not to overstate progress on this score).

The existence of past biases in science, then, is no grounds for abandoning the enterprise altogether; but it certainly is grounds for introducing better safeguards against the all-too-human propensity for self-serving self-deceptions. So, back to mothers. Does attachment theory "attach" the lives of mothers? Not necessarily, but for mothers who care about their infants' well-being it often comes to the same thing.

Childcare Problems

The chief problem facing mothers is the one Bowlby identified years ago: a scarcity of motivated alloparents. How do mothers who go off foraging keep their immobile and vulnerable infants safe and secure? In many species, fathers are the co-caretakers of choice, as well as caretakers by default—at least of older infants. A survey of the primatological and ethnographic evidence also provides a spectrum of tolerable solutions, not all of them equally palatable to infants. Among anthropoid primates, species-typical sex roles are not quite as fixed as many assume. Solutions vary with species, as we saw with titi monkeys. They also vary with circumstances, as we saw with the "Silas Marner response" among gorillas and langurs, and as we see every day in the lives of the men around us.

Unarguably, the majority of hunter-gatherer societies are characterized by close mother-infant proximity for at least the first several years. Care of infants by fathers is unusual, and care by male alloparents rarer still. But the recipe is not carved in stone. When ecological circumstances permit (or require), mothers readily avail themselves of help from a father, grandmother, niece, nephew, or sibling. Female primates have always entrusted infants to willing allomothers *whenever a mother could be confident of safely retrieving them.* Some of these relationships were reciprocal, but for the most part, allomothers have been fathers or kin, or, with the emergence of stratified societies, desperate subordinates with little choice.

Infant-sharing in other primates and in various tribal societies has never been accorded center stage in the anthropological literature. Many people don't even realize it goes on. Yet where studied, the consequences of cooperative care—in terms of survival and biological fitness of mother and infant—turn out to be all to the good, or at least on balance preferable to alternatives.[21] In infant-sharing primate species, infants grow faster, and (as in the case of parents using wet nurses) mothers produce more surviving infants in a lifetime. Shared vigilance helps reduce infant mortality. Mothers not burdened by infants forage more efficiently, and their infants cultivate wide social networks, and possibly suffer less from sibling rivalry for maternal attention. For an alternative notion of what it means to be born human, consider the Agta foragers of the Philippines, a very unusual hunter-gatherer society because women participate along with men and dogs in hunting wild pig and deer. Fifty-seven percent of their meat comes from hunts in which women participate. When an Agta is born,

> The infant is eagerly passed from person to person until all in attendance have had an opportunity to snuggle, nuzzle, sniff, and admire the

newborn. . . . A child's first experience, then, involves a community of relatives and friends.[22]

Late in the first year of life, an Agta mother may leave her infant in camp with an older sister, grandmother, or the father. If necessary, she carries her infant along with her, even though there is no doubt that her burden constrains her freedom of movement and diminishes her efficiency. Mothers may carry infants as young as six months out hunting over treacherous and thorny terrain. The prospect of injury is high for all concerned, and 30 to 50 percent of Agta children die before puberty.[23]

Even when mothers gather rather than hunt, allomothers can provide an important service, and even a few unusual perks. Not only do infants among the Efé pygmies of Central Africa grow up in famously gentle company, but these newborns get to taste breast milk sooner than babies anyplace else in the world. Why? Although breast milk does not normally come in until a few days after birth, allomothers are quite casual about suckling another mother's baby, including a newborn being passed from woman to woman.

In the only systematic survey of multiple caregiving in traditional societies, anthropologist Barry Hewlett identified three circumstances that facilitate it: flexibility in schedules and considerable leisure time, which may be more characteristic of foraging peoples than herders, farmers, or wage-earners; living in close quarters, so that many people will be familiar with the infant (and vice versa); and, perhaps most importantly, high adult-to-child ratios, meaning that there will be more eager caregivers than there are infants.

Still there is a gap between, on the one hand, these isolated cases where males are active participants in infant care, and, on the other hand, the rest of the world. Since the women's movement revolutionized the way that we talk about male caretakers and paternal roles, there has been a marked change. It would not be honest, however, to pretend that this represents a revolution. For example, fathers in the United States spend much more time today directly caring for children than they did at the beginning of the twentieth century, but the change is still measurable in *minutes* per week, not hours. Even in the most progressive countries in this respect, such as Sweden, where either parent has the option to take paternity leave, few fathers take advantage of the policy.[24]

Fig. 22.2 An Efé girl passes a baby back to its mother. By four months of age, an infant in the Ituri forest will spend an average of 60 percent of each day away from its mother, passed from caretaker to caretaker like a baby langur, at a rate of eight times an hour, to as many as fourteen different allomothers in a day.[25] *(Steve Winn/Anthro-Photo)*

In the Environment of Evolutionary Adaptedness Bowlby had in mind, the birth mother has the greatest chance of being on the spot, lactating and motivated to hold and carry *this* baby. As discussed in chapter 9, the mother is likely to be more sensitive than the father to signals of infant need. This may be due to innate temperament, or to "the company she has kept," for a mother would have spent more time near other women and small infants and be more practiced in caretaking.[26] Once a relationship between mother and baby is forged, the mother becomes the person emotionally most primed to extend her tour of duty for the requisite number of years to ensure survival.

Rather than some magical "essence of mother," what makes a mother maternal is that she is (invariably) at the scene, hormonally primed, sensitive to infant signals, and related to the baby. These factors lower her threshold for giving of herself to satisfy the infant's needs. Once her milk comes in, the mother's urge to nurture grows stronger still. Furthermore, compared to the father (who also shares at least half of his genes with this infant by common descent), there is a good chance that this infant represents a higher proportion of her reproductive prospects than of his (though not necessarily, if she has several, and this is the only child he ever sires). These factors make the mother the likeliest candidate to become the primary caretaker. *But they do not constitute an unyielding prescription.*

Mammalian babies eagerly suck milk wherever they learn to obtain it. Of itself, this observation speaks volumes about the inherent slack in the infant's motivational system for seeking nurture; human babies can recognize but are not specifically imprinted on their mothers' breasts, and seek nurture wherever they are comforted and rewarded. Note, for example, how easily infants figure out that rubber nipples deliver milk faster than breasts do, and although it's harder on their digestion, still learn to prefer bottles over breast.

Once formed, of course, there is no more powerful emotional glue than an infant's attachment to a favorite caretaker, and yet even this bond, if broken, can be reforged with someone else (though there are limits, which vary from individual to individual, to the elasticity of learned trust). But what predisposes an infant to attach to one person rather than another? And why is it that, hands down, the mother so often emerges as the frontrunner for infant affections? It's not *just* a matter of her milk.

It is not preordained that the infant's primary attachment will be to the mother. There are other options. But when you consider that the mother usually has a lower threshold for responding to infant needs, and is the first to respond to a cry of discomfort, with her face, her voice, her breasts, the satisfying sweet milk they spurt in the baby's mouth—these attributes make the mother the likeliest prospect for her infant to form a primary attachment to. But second-best has proved adequate often enough. In fact, when the allomother is more committed than the mother, second-best may be superior.

Maternal propensities interact with infant needs in ways that make certain preferences highly probable; the infant takes it from there. The mother's sex may not be her destiny. But from the perspective of a newborn, staring blurry-eyed at the world, there are attributes to a mother that make her an easily acquired taste. For reasons that have less to do with innate properties of mothers than they do with how effective infants are in achieving their first choice, babies in the majority of primates are found in the exclusive possession of their mothers, and vice versa. Once initiated, infants develop a passionate preference for this arrangement.

Why Don't Mothers Use Allomothers More?

Statistically an observer is far more likely to see a mother carrying her infant and caring for him than witness the fleeting moment when an infant is abandoned. Everywhere in primatedom, mothers are typically seen carrying babies about, with infant and mother mutually "attached" to each other. As a result, mothers were assumed to be innately and uniquely qualified for "mothering." It was an easy conclusion to reach. In the context of this sort of tautological thinking, no one asked: *Why* are mothers carrying their babies?

The question was finally addressed when anthropologists became acquainted with peoples like the Efé and individuals other than the mother *were observed* mothering. Only then did researchers give more thought to multiple caretaking. So long as it had been mothers carrying babies, as expected, alternative styles of caretaking weren't considered.

Given the manifest survival and reproductive advantages of relying on allomothers, why weren't mothers giving their babies to someone else to hold while they foraged (or turning infants over to their subadult daughters for as long as those daughters remained in the vicinity)? If in nonhuman primates, infants born to mothers who hand their babies over to allomothers grow faster, and the mothers reproduce at shorter intervals without having to make tradeoffs in survival chances; if, as seems clear, maternal fitness can be enhanced by using allomothers, why haven't mothers been selected to use allomothers more?

Consider the case of !Kung mothers. Why don't these mothers leave their babies with prereproductives, the children who are hanging out in a !Kung encampment when their mothers go off to forage? According to Patricia Draper and the other ethnographers of the !Kung, older children are doing things like cracking open mongongo nuts already brought back to camp. Why does this supply of baby-sitters go untapped?[27]

Think back to the discussions in chapters 4 ("Unimaginable Variation") and 11 ("Who Cared?") for reminders about why nonhuman and human primates might not use allomothers more. A chimpanzee mother who turns her baby over to its older sister risks having her infant eaten by a predator or killed by an infanticidal conspecific. When a mother chimp rebuffs a daughter eager to hold her little sister, it is not so much this would-be caretaker she is worried about as the daughter's ability to keep her infant safe from these others. Since chimpanzee mothers almost never give up infants to eager older siblings, observers never have a chance to find out what would happen if they did. Unsuspecting mothers in another species have, however, unintentionally provided natural experiments. These are the rare and unfortunate instances in which African women left babies unattended only to find that a passing wild chimpanzee seized and ate them. Depending on the circumstances, a chimp may adopt an unattended baby—or kill it. A competent caretaker is essential.[28]

In the case of human foragers, mothers have trouble finding someone both willing and qualified to fill this role. Where there are no infertile women or women past childbearing age, there will typically be a dearth of adult volunteers; and unless camp is a very safe place, using immature "childminders" is risky. A range of considerations may make allomothers unsafe or unpractical, depending on conditions. Among the desert-dwelling !Kung, for example, keeping the baby hydrated is a big constraint, since infants have no other safe drinking source than mother. Unable to safely leave infants at camp, mothers carry them long distances.

My guess is that alternative childcare systems—like the Hadza's, or even the Efé's—have been more prevalent than commonly supposed. Even before the Neolithic, allomothers were employed *when safe allomaternal options were available*. It is beginning to look like the prevalence of exclusive and long-lasting mother-infant relationships might be an artifact of the harsh environments where hunter-gatherer lifestyles have persisted long enough for anthropologists to study them, an artifact of mothers needing to travel long distances to find food and water, of predators lurking nearby, and of settlement patterns where the mother's kin live far away.

What *is* certain is that as soon as an infant begins to be weaned and a wider pool of baby-sitters becomes eligible, they are enlisted. Among the Hadza, where food sources are available near camp, children as young as three are left in multi-age groups to roam about and combine play with food gathering. It would not be unusual for a ten-year-old Hadza child to combine baby-sitting with a quick pick-me-up in the form of 200 calories' worth of makalita berries casually picked. Such snacks are shared with younger children, who gather round begging from their baby-sitters.[29]

"Psychological Thalidomide"?

Identifying *infant needs* is separate from deciding *who should provide for them*. Monkey and ape mothers tend to be possessive of their infants. The exceptions include those species where benign caretakers (usually kin) are available so that mothers can be confident of retrieving hale and hearty babies. When these conditions are met, primate mothers readily delegate infants to allo-mothers. Sufficiently desperate, even model monkey mothers drop their standards and consign their infants to caretakers who are less than ideal. (Recall the ousted males with whom langur mothers deposit babies when infanticidal males threaten them "at home.") Critics of daycare who condemn it as "psychological thalidomide"[30] need to ask themselves: Compared with what? For those who find it unnatural, to ask: Compared with when?

The real constraint on working mothers has little to do with imagined ideals of Pleistocene motherhood, far more to do with locating and enlisting reliable, motivated, long-term allomothers. Human infants need mothers and allomothers to keep them warm, safe, mobile, stimulated, clean, fed, hygienically hydrated, and, most important, to communicate tenderly and responsively their commitment to go on caring. Hour by hour, supplying this kind of care is tedious. In addition to deep-seated predispositions so well summarized by Hamilton's rule, caregiving is also more or less onerous, depending on personal tastes, training, expectations, and working condi-tions. A vast divide separates the ideal of childcare from its realities.

Shortage of Alloparents

Many mothers find the smell of their breast-fed infants' excrement entirely congenial. (In other mammals, like dogs, special molecules in the excrement motivate mothers to eat it, thus keeping their burrow clean.) Few other indi-viduals of either sex, however, can honestly say that they share this taste, or that they enjoy changing dirty diapers. For truth in advertising, consider this job description for a modern allomother:

> WANTED: Someone to turn his/her life over to the whims and needs of a smaller, weaker, often unreasonable individual for a period of months or years. Low pay. Little prestige. No security or long-term mutual obligations; faint prospect that any relationship formed will be maintained over time. Caution: if the infant comes to love you best, the mother may grow jealous and terminate the relation-ship early.

Not surprisingly, there are far more mothers seeking nannies and daycare spots, more activists calling for universal subsidized daycare, and politicians promising it, than there are reliable allomothers signing on to provide it.

Fig. 22.3 Older children entertaining a toddler on a swing near a !Kung encampment. Environmental constraints prevent mothers in foraging societies from relying on young baby-sitters even more than they do. *(Marjorie Shostak/Anthro-Photo)*

Those who seek one-on-one infant care, but are not lucky enough to have kin to provide it, would do well to make it emotionally and materially worthwhile to their potential caregiver, that is, to beef up the "benefits" component in Hamilton's equation.

Wherever affirmative action and/or equal rights in the workplace have created new job opportunities for women, a shortage of female caregivers follows in its wake. Nevertheless, the same optimism that convinced John Boswell and other historians that infants abandoned in antiquity were gathered up and reared by "kind strangers" leads contemporary policy-makers to imagine that if we can just throw some money at the problem, we can create adequate childcare. Somehow caretakers both willing and qualified will emerge.

In a recent radio broadcast, a politician voiced the opinion that we needed to hire a few more case workers to "shake the bushes" to find homes for foster children. Once we did, suitable caregivers and suitable homes for all unwanted children would just come tumbling out. Among the infants he particularly had in mind were "boarder babies" born to heroin-addicted mothers

—babies who are among the most challenging in the world to care for. With just a bit more effort, he asserted, all of New Jersey's "boarder" babies could easily be placed in foster homes. He reminded me of Catherine the Great and the other great "progressives" who were convinced they could solve the problem of unwanted babies by building handsome edifices.

———

Scarcity of local alloparents often forces working parents to look beyond their own neighborhoods. But grouping infants together—like bats in a communal nursery—for a certain number of hours every day under the supervision of *paid* alloparents who are not kin, but who are expected to act as if they are, is an evolutionary novelty, completely experimental. Already there are greater numbers of children in daycare than ever before in human history. The burgeoning demand for *good* daycare far outstrips availability, just as its cost outstrips most budgets.

Given the rapid spread and highly experimental nature of paid communal daycare for very young infants (under two years), it is unfortunate that the many complex factors involved (different temperaments of the infants, their variable experiences at home, the different personalities of the providers, and so on) make it impossible to give definitive answers to parents' most pressing questions. Evidently most children in daycare survive and develop normally. But will they grow up as secure as those reared in continuous contact with mothers? Are there children whose temperaments are ill suited to daycare? Or parents whose caretaking styles are incompatible with using daycare? Will children placed in daycare from an early age grow up to be less able to form strong relationships, or be less caring toward others? And will they be more, or less, qualified for life? (Keep in mind that it could turn out that those *less* securely attached and those *less* capable of forming close relationships are the "more qualified" for life in the modern world.)

So far as the best small-scale daycare centers are concerned, recent studies are mostly reassuring. Even for infants younger than age two, there are no detectable bad consequences, provided babies were in high-quality care, had a good relationship with their parents at home, and the daycare hours were limited—rather daunting specifications.[31] Results for very young infants who spend more than thirty hours a week in the more institutionalized settings, where a few caretakers struggle to meet the needs of many infants, or for children who bounce from one facility to another, are less encouraging. They make open-minded mothers with options queasy and those without options miserable. Others will simply continue to deny and contest the results as some antifeminist conspiracy.

Not only can effects be seen in the way infants respond to their mothers, but also in the way mothers respond to their babies, who are already harder to soothe. Mothers who used daycare more than thirty hours a week tended to be less sensitive with their six-month-olds, more negative with fifteen-month-olds, than mothers who used daycare ten hours a week—an effect that one suspects will only become magnified through time as the child finds its mother less comforting to be around and therefore complains more.

Few topics are more passionately debated by developmental psychologists. Yet beneath the furor lie areas of near consensus, which—at my peril—I summarize here.

Experts differ over just how flexible, how adaptable, human infants might be, yet no one is saying that human adaptability provides a carte blanche for indiscriminate care. A strong advocate of daycare, psychologist Sandra Scarr, professor emeritus at the University of Virginia, recommends monitoring each child's response and calibrating hours in daycare to temperament; how a child copes with daycare should be a matter of continuing concern and vigilance. Even the iconoclastic grandmother Judith Rich Harris, the author of *The Nurture Assumption,* touted by her publishers as a book showing how "parents matter less than you think," notes that

> In order to complete its development the [newborn baby's] brain requires certain inputs from the environment. . . . The developing brain "expects" the baby to be taken care of by one person, or a small number of people, who provide food and comfort and who are around a lot. If this development is not met, the department of the brain that specializes in constructing working models of relationships might not develop properly. . . .[32]

Fundamentally, neither of these two views differs very much from those of their supposed opponents, such as Jay Belsky, the developmental psychologist who became identified as a critic of daycare, a researcher self-described as one who "violated the XIth commandment of developmental psychology: thou shalt not speak ill of daycare." Belsky, who feels his reputation was dam-

Fig. 22.4 In France, all children between ages three and six attend state-sponsored *écoles mater-nelles,* even children whose mothers do not work. In the United States, good daycare programs have long waiting lists and are expensive. The highly regarded daycare program in this picture, open to children ages two to five, costs $365 per month for a maximum of five hours per day, with an adult-to-child ratio of 1:7. Places in infant daycare (ages six weeks to eighteen months) are even scarcer and can cost twice as much. *(Photo by Karen Froyland)*

aged because he raised questions about a sacred cow, is nevertheless at pains to balance his main message that "Mothers matter. Early experience is influential . . ." with a caveat: "But it's not the only thing that matters."[33]

Virtually every researcher in this field concedes that when parents are particularly inattentive, even "indifferent daycare" is preferable to home neglect. Tucked away within the polemic, even a rabid debunker of attachment theory and an outspoken feminist advocate of daycare like Diane Eyer will allow that "all [babies] certainly need the consistent loving care of a few people"[34]—which is what I took Mary Ainsworth, Mary Main, and the other attachment theorists Eyer criticizes to have been saying all along. *Caretakers need not be the mother, or even one person, but they have to be the same caretakers.*

The Importance of Behaving Like Committed Kin

The numbers are staggering. The majority of mothers in postindustrial countries use some form of daycare. Many are working mothers with babies under
two. Given Bowlby's personal opposition to new mothers working, it is
ironic that attachment theory today provides the most useful theoretical
model we have to make daycare better, to adjust allomaternal care to the
needs of baby primates.

When I was looking for infant daycare in Cambridge, Massachusetts, back
in the seventies, the best place I found was the Harvard Yard Daycare center,
whose program was designed by Berry Brazelton and other pediatricians
heavily influenced by Bowlby. These pediatricians, identified by the critics of
attachment theory as special enemies of working mothers, were among those
who made it possible for me to continue to work part-time even while my
children were infants. I could confidently leave a six-month-old baby in the
care of two allomothers for part of each day. The center was set up in the
basement of what had once been Harvard's ROTC* building—a case of
swords beaten into plowshares—near the campus and a block from my
home. The nursery was three shabby rooms occupied by two extraordinarily
patient, allomaternally gifted women who brought supreme calm wherever
they went. A condition of leaving a baby there was volunteering to work in
the nursery, so that in addition to the very permanent staff (each of them
stayed for years) there was a fluctuating contingent of family members in a
tiny village-like setting, with a high ratio of adults to infants.

If multiple caretakers *are* involved early in life, it frequently happens that
the infant selects one with whom to form a primary relationship. Yet children
are rather flexible in this respect. All early caregivers become the emotional
equivalents of kin. Any caretaker is capable of communicating the message
infants desperately seek—"You are wanted and will not be set aside"—the
message that elicits an infant's sense of security, a double-edged sword when
an allomother suddenly disappears.

Looked at this way, Bowlby's insights invite fathers to engage more with
their newborn babies and stick with it. They lend urgency to reform movements to place children in adoptive homes as early as possible. They do not
discourage mothers from using daycare, but they do provide tremendous
impetus to make daycare resemble families with a stable cast of characters

---

*Reserve Officer Training Corps.

Fig. 22.5 John Bowlby transformed an empathetic concern for infants into a new way of understanding infant needs. This photo was taken on the Isle of Skye three years before his death in 1990. *(Courtesy of Erik Hesse)*

and an atmosphere that provides infants with a sense of belonging. From a mother's point of view, there is a big difference between what evolution designed as the likeliest outcome (infant attaches to mother) and what under other circumstances serves perfectly well.

From a newborn's point of view, however, little has changed. Mothers are rarely as close or consistently available as the infant desires. Babies have no way of knowing that the mother who went out of town on business is not dead, that sabertooth tigers are extinct, jaguars scarce, abandonment illegal, or how few modern mothers would, in fact, contemplate it. For babies are designed to proceed as if no baby bottle had been invented, no laws had ever been passed. I turn now to the question of just *why* this sense of belonging matters so much.

Alternate Paths of Development

After Alcmene had brought forth the babe, fearful of Hera's jealousy she exposed
[the baby Heracles] . . .

—Diodorus Siculus, first century B.C.

How shall I be in this life, like a flower or a twig? Bellowing and dominant like a fully developed male orangutan, or a diffident and self-effacing smaller fellow? Trusting and empathetic, or a self-centered loner? Different morphs of caterpillars and orangutans provide a metaphor for what some evolutionary-minded specialists on infant development suspect might be going on in our own species. Depending on the input infants receive from the world around them, genetically very similar individuals embark on quite different developmental pathways, growing at a different rate, developing a different sense of self to confront the world, assigning different priorities to the well-being of others, even reproducing in a different way. Individuals as closely related as full siblings may be slated for different microenvironments, not only in infancy, but potentially through life. The question is, why? Recently people have begun to suspect that mothers and early experience have something to do with it. This has even led many people to "blame" mothers if a child turns out badly. This way of thinking is very new.

Blaming Mothers for Sociopaths

In previous times and places, mothers were blamed if they failed to produce offspring, or produced offspring of the wrong sex, but they weren't blamed if their children grew up to be criminals. Among the ancient Greeks, victims of heinous crimes railed at a particular god, or the Fates. In India, people blamed bad luck, karma, or turned to the malefactor's clan for restitution. Orientals lamented lapses in the education of an unfilial child. In Africa, witchcraft was suspected. Another favorite candidate is "breeding," or genes that produce a certain kind of person. Think of Oliver Twist. If mothering mattered more than breeding, how could Dickens's orphaned hero, adopted by a clan of thieves, have turned out so well?

The idea of blaming the behavior of a grown-up sociopath on insecure attachment to his mother has only entered our culture since Freud, and especially since Bowlby. In a short space of time, it has become widely accepted that a serial killer's early relationship with his mother can account for subsequent sociopathic tendencies. ("One false move and our precious bundle of joy will turn into an ax murderer" is how psychotherapist Shari Thurer parodies this stereotype.)[1]

Hence, when *Newsweek* ran its obligatory story on the "Blood Brothers"—Theodore Kaczynski, the suspected "Unabomber," and his brother, David, who turned him in—the magazine also included the obligatory portrait of a third person, their mother, Wanda Kaczynski. Yet the most damning attribute the magazine could unearth was that Mrs. Kaczynski "encouraged her sons in school."

Newsweek was sheepish enough to include the caveat that "Blaming one's mother is the oldest [though of course it really isn't] and least original excuse in history. . . . It may not be fair to hold Wanda Kaczynski, who is described by her neighbors as a sweet old lady, accountable for turning her son into a possible serial killer. . . ."[2] Still, *Newsweek*'s editors couldn't resist including an aunt's offhand observation that perhaps Ted had initially embarked on the path that grew lonelier and lonelier by way of a defensive response when the birth of his younger brother displaced him from the center of attention.

Bowlby as a young researcher had been among the first to suggest that the sort of young delinquents "who seemed to have no feelings for anyone and were very difficult to treat were likely to have had grossly disturbed relationships with their mothers during their early years. Persistent stealing, violence, egotism, and sexual misdemeanors were among their less pleasant characteristics." He noticed that the rare human outcomes who become "sociopaths" were disproportionately drawn from the portion of the population that had been "avoidantly attached" to their mothers. Bowlby assumed that their behavior was due to a disturbance in their internalized working model of the world that made it hard for them to learn to trust others.[3]

———

Like many mothers, I was immediately indignant at *Newsweek*'s presumption in putting the spotlight on Mrs. Kaczynski. Six weeks later, however, I felt chastened when the *Washington Post* broke an amazing and deeply moving story about the hospitalization of Ted Kaczynski at the age of nine months:

> The image still haunts Wanda Kaczynski. She can still see the photograph of her baby son, pinned down on his hospital bed. It offered what she now sees as a clue into how her oldest son grew into the troubled man he would become. . . . [The infant was strapped down,] terrified, spread-eagled so doctors could examine what they believed was a severe allergic reaction. His naked body was blotched with hives. His eyes, usually normal, were crossed in fear.[4]

During that week in the hospital, the infant who would grow up to be a strange, vengeful man who shunned human contact, and lashed out at the world in murderous ways had no reason to assume otherwise than that an infant's worse-case scenario had befallen him: he was abandoned. Such accounts reinforce one's conviction that Bowlby was on to something. There were legitimate reasons for journalists to ask whether the perception of abandonment might not have disrupted neural connections in the peculiarly fragile psyche of Ted Kaczynski, diverting him to some alternate, antisocial course of development. No one could blame Mrs. Kaczynski if her son was born with an inborn propensity for schizophrenia. Yet schizophrenia, a disease with a known genetic component, depends on context for its expression. Whether or not and how the disease is expressed is rarely separable from context. But even if diversion to alternate tracks—ranging from mild to extremely delusional—occur, why might this happen? Must we presume pathology? Surely humans, the most flexible of all primates, should be able to overcome early deficits, as many children obviously do. Even if a disproportionate number of people who grow up to be sociopaths were avoidantly attached, one has to wonder why their "internalized working model" didn't do a better job of correcting itself when things improved.

The Nurture Assumption

Bowlby was the first modern psychologist to examine infant development in an evolutionary context. Nevertheless, his training was in psychoanalysis, not biology. He did not keep up with new developments as ethology merged with ecology, population genetics, and life history theory to become the more inclusive study of the biological bases of behavior known as sociobiology. Bowlby was apparently not aware that there is no one species-typical, one-size-fits-all pattern of development.

Individuals differ in innate temperament, whether they are, for example, shy or extroverted.[5] They differ in appearance, endurance, in what foods they can digest, in susceptibility to diseases, both physical and mental, and myriad other attributes known to be influenced by genes. Individuals also experience different parental effects, including many areas where a mother's or father's social status, locale, or kin networks affect an offspring's prospects.

In all primates, juveniles enter their social world with a great deal of innate and acquired baggage that shapes their prospects, including their social status relative to others in their same world. A baboon daughter, for example, inherits her mother's rank in the female hierarchy, and if she is the youngest daughter, is assisted by her mother to rise in rank above her older sister. Compared to the daughter of a lower-ranking female, she will reach menarche and breed at an earlier age—a reproductively important milestone. Similarly, a young male orangutan who enters a habitat already patrolled by a full adult, dominant male, assesses that situation and indefinitely suspends further maturation, remaining for years as a "subadult," opting to be a "Peter Pan" morph who does not grow up until such time as it is safe and advantageous for him to do so. These individuals, entering the social world with certain advantages and disadvantages, somehow assess their prospects, match their phenotypes to their available options, and develop accordingly. When that social world changes, so does the developing organism. Why should humans be any less opportunistic or less flexible? Indeed, there is every indication that they are more so.

In *The Nurture Assumption* Judith Rich Harris correctly points out that genetically inherited traits have been given short shrift by social psychologists in their explanations of how children develop. At the same time, however, Harris underestimates—although she never entirely discounts—the extent of parentally mediated effects. ("Parents matter less than you think and peers matter more," the book promises.) What she leaves out is the extent to which input from parents, especially from mothers, influences the developing individual as he or she begins to negotiate his or her place in the world. Demographic profiles bottom-heavy with young, and social institutions of developing nations in the late twentieth century with their schools and gangs, mean that this world is largely inhabited by peers, but this has not always been so.

It is the individual child or teenager who is the active agent in this transaction. Yet this young person's phenotype is already very much shaped by what parents and kin have done. In earlier human environments, the assistance of kin would continue to be very relevant. A young person's entire reproductive future would depend upon what group members (mostly kin) could and would provide.

The mother-infant relationship already has factored into it a great deal of relevant input from the surrounding world, including how much support the mother can expect from her mate and kin around her. Depending on whether we are talking about a matrilocal or a patrilocal society, either the mother's or the father's status and kin ties affected the status of the child. For example, status of kin might determine such a major life circumstance as whether or not he or she remained in that group or migrated out of it.[6]

To support her argument that peers (not parents) socialize children, Harris cites case studies from two ends of the social spectrum. She asks us to consider the son of Polish immigrants who learns to speak idiomatic, unaccented English just like his peers, and the son of a British baronet who spends his first eight years being tended by governesses with a variety of accents, then goes to preparatory school, and from there on to Eton. In spite of having virtually no contact with his father, she notes, the boy learns to speak and act just like him, providing the young man "his membership card for belonging to the upper class."[7]

But there is an important difference between these examples: the immigrant child has a big incentive to adopt the language of peers. This child could not have failed to notice how poorly treated and disenfranchised his parents were. By contrast, the son of a baronet saw everyone around deferring to his father. Each child then took the course open to him for attaining "cultural success" in the world around him. In nonhuman primates, the object of status can be fairly narrowly defined (gaining access to desired resources), and is almost always correlated with reproductive success. This is no longer necessarily true for humans. Nevertheless, the aristocrat's son will be predisposed to emulate his socially prestigious father, to mimic the clipped linguistic habits of his peers, rather than learn to speak in the Irish lilt of his nannies. Even if he subsequently goes on to school in America, the baronet's son—

new peers' accents notwithstanding—will continue to talk like his dad so long as people respect him more for doing so.

In an example closer to home, when my sister went to live in west Texas, her son picked up a strong twang—essential for a manly identity in that part of the world. However, my nieces—who went to the same schools, with peers from the same background—continued to talk just like their mother, with only the faintest trace of a Texas accent. I assume this was because speaking without a drawl seemed more refined, which would be a desirable trait for a young lady. Between their natal homes and their peers, the children were active agents, consciously and unconsciously deciding whom to emulate.

In all these examples, parental effects (which include children's perceptions of how others treat their parents) are critically important. This would have been even more true in small foraging communities. In worlds where it would be uncommon in the extreme for children to find themselves in large classes of same-age peers, and where technology and culture changed at an infinitesimally slow pace, whether one was an orphan or the son of a skillful hunter, related to no one or related to many, would shape that child's life choices. From the perspective of an infant in humanity's environment of evolutionary relevance, there was only one really meaningful question to ask about nurture—would it be forthcoming? "Will I be protected and provided for, or not?"

Surviving Detachment

Mothers vary in what they are able to provide to infants, and infants vary in viability. But mothers also vary in how much they are willing to commit to give to each infant. This variation in maternal investment produced selection pressure on infants to elicit as much care and commitment from their mothers as they could.

To discourage maternal retrenchment, all human infants seek to maintain a close and, if possible, continuous association with their mothers. Any serious deviations from this "normal," species-typical pattern (namely, frequent or prolonged absences) lead to the responses Bowlby called "insecure attachment"—a state he regarded as an abnormal outcome.

If separation from the people infants are attached to occurs frequently enough, and persists long enough, Bowlby feared that resulting insecurities would lead to pathological development. In the 1950s, when he was proposing this model, the idea of an Environment of Evolutionary Adaptedness was itself new. No attention was paid to the possibility that there might have been fluctuating circumstances, or more than one adaptive path, depending on the conditions during infant development. Insecure attachment to the mother was assumed to be maladaptive, period.

It never occurred to the first generation of attachment theorists that high proportions of insecurely attached infants (ranging as high as 30 to 80 percent in some populations studied) could be anything other than aberrations produced by unnatural rearing conditions in the modern world. But the fact is, no one knows what proportion of insecurely attached infants existed during the most relevant periods of human prehistory, when the genetic makeup of our ancestors was being forged. Were a fraction of orphans survivors? Were some abandoned infants rescued? How many survived? Were there "espositos" with character traits especially suited to surviving separation from their mother and growing up among alloparents?

Survivors of Abandonment in Prehistoric Times

By definition, we have no history of what happened during prehistory. But we have tiny windows into this murky past that provide hints for interpreting evidence from historical times. We know, for example, from DNA analysis of the human genome, that the human species passed through at least one major and prolonged population bottleneck. In some areas, populations experienced multiple population busts and booms. Furthermore, we also know that such crashes almost always strike disproportionately the old and the very young. So which infants survived after such a demographic calamity?

What more can we know? The effort requires taking some risks. Myths are slippery sources of information, yet ancient mythology is often the best window we have into the experiences and customs of people who lived four thousand and more years ago. Mothers (we know from Boswell) were already abandoning infants in significant numbers. It is not far-fetched to think that some of these children would have survived, as did the *espositos* whose many descendants can be found today by flipping the pages of a phone book.

Heracles—better known by his Roman name, Hercules—is perhaps the most famous *exposito* of all time. According to the first century B.C. account in Diodorus Siculus, he was abandoned at birth, found, briefly suckled, and returned to his mother by Hera because he seemed such a robust baby. Heracles grew up to be a risk-taking loner who wandered about Greece killing his enemies with a club and his bare hands. In another age, such a man would be viewed as more nearly a sociopath than a hero. And even in the ancient world people found appalling the story in which clouds descend upon his mind and Heracles murders his nephew, six of his own sons, and two other boys who happened to get in the way. Yet it never occurred to any of the ancients to blame the behavior of this sociopathic loner on his mother. Heracles, whose name literally means "fame of Hera," is linked for all posterity not with the mother who abandoned him but with the goddess Hera, who plotted against him.

I take the Heracles myth to be more than a fanciful story. Although the vast majority of abandoned children would have died, a few would have survived the dismal rearing conditions of a foundling. Based on what we currently know, there is no basis for assuming that all children in the past who survived and reproduced were necessarily securely attached.

Since Bowlby formulated his attachment theory, anthropologists and historians have learned a great deal more about just how often mothers deviated from Bowlby's Pleistocene ideal. Many infants *were* left with allomothers for extended periods of time. Others were neglected to varying degrees, even abandoned. Indeed, as we have seen, during some periods of human history, children were neglected and abandoned on a massive scale. And a broad spectrum of maternal distancing resulting in outcomes less drastic than abandonment must have been even more common.

It requires far-fetched, special pleading to argue that what went on ten thousand years ago was completely different from what goes on among nomadic foragers and semisedentary forager-horticulturalists today, or among those chronicled in a mythology over three thousand years ago. In prehistoric as in historical times, some fraction of infants would have been neglected, even abandoned, and some of these would have been adopted by others as substitute children, or else exploitively reared for their labor, or (in the case of females) their reproductive potential. What if children with some traits were more likely to survive less-than-ideal child-rearing than others?

No doubt infants without committed mothers usually succumbed to predation or starvation, as Bowlby assumed. But there is no reason to think that evolutionary provisions for extreme neglect exist. Even when neglected infants in post-Neolithic worlds do manage to subsist "protected" (or warehoused) inside walled buildings called orphanages, these infants, left untended, scarcely touched, rocking themselves day after day, do not develop basic human sensory, motor, linguistic, or emotional potentials. Nevertheless, over a vast stretch of unrecorded history (ten? twenty? fifty thousand years?) some percentage of children survived rearing conditions that fell somewhere between lethally inadequate and the Pleistocene ideal.

Precise data on infanticide have never been easy to come by. It is even more difficult to estimate, and impossible to measure, the effects on infant survival of mothers who cut back to varying degrees in maternal investment. An infant who died in an epidemic might not have done so had its mother nursed him a month longer. Or perhaps the toddler would not have stumbled into the fire if she had carried him with her that day. Unfortunately, such contingent possibilities don't make for hard data. The behavior of mothers who provide for their infants *less well than they might* (surely an immeasurable quantity!), who delegate care to others with some unknowable prospect that their children will be adequately cared for, are literally "off the record." Nor is there any way to measure the impact on subsequent survival of children of the various tactics parents might use to reduce the costs of rearing them.

What we can be certain of is that infanticide (ranging from rare to common) has characterized human societies throughout history and prehistory. Less-than-fully-committed mothers and mothers who delegated care to others were even more common.

Whether a mother is single-mindedly attentive, a bit distracted, overwhelmed, or absent altogether conveys a great deal of information to an infant, not only about what he or she can expect during infancy, but—barring some miracle—after weaning as well. Almost by definition, a mother well provided for by her mate, or surrounded by supportive alloparents, will be investing more in her infant, and her infant can expect to rely on this network through childhood, adolescence, and perhaps into adulthood.

Hamilton's rule figures just as strongly in humans as it does in other animals, and affects us both in body (our physiological capacity for recognizing the smell of kin) and in mind (the emphasis on kinship in all human cultures). There is no good, intuitive reason to expect humans to differ from other animals in this respect. It would behoove a man no less than a baboon to pursue different developmental and reproductive strategies according to social circumstances.

Scientifically, the case has so far been impossible to prove, because human lives are shaped by so very many variables. The task is rendered even more complicated by cultural customs. For example, gender- and age-specific

clothing and age-graded segregation of children in schools may make children seem more similar than they are. It is important, then, to make a clear
distinction between what has been demonstrated in other creatures and
what, beginning in the 1980s, a handful of evolutionary-minded social scientists began to *suspect* might be going on in humans. A novel question was being
asked: Did human children evolve different ways of coping with the varying
social environments they encountered?

The Post-Bowlby Era of Attachment Theory

Patricia Draper was a graduate student in anthropology at Harvard when she
first went, in 1968, to the field to study the !Kung San. At that time the
!Kung still lived as nomadic hunter-gatherers, and Draper planned to do a
Ph.D. thesis on how their social structure affected relations between adults
and children. Scholarly attention was still focused on the question of whether
prior to the seventeenth century people even *had* a concept of childhood.
Like many anthropologists who had studied child-rearing in traditional societies, Draper found such ideas hard to accept. Still, she wanted to know
more about just what it meant to be a child in a hunter-gatherer environment.

At first she focused on how mother-infant relations were influenced by
ecology and social structure. Gradually her interest expanded to include
emotional development.

Over the next twenty years, Pat Draper, following the lead of Richard
Lee, studied the transition of the !Kung from a nomadic Pleistocene lifestyle
to settled living amidst other African villagers, where older children play a
major role in rearing younger ones. By that time, Draper had become a professor and had three children of her own. For Draper, Bowlby's concept of an
Environment of Evolutionary Adaptedness remained important, but she no
longer expected that environment to be everywhere the same. Her environments of evolutionary relevance were typically arduous locales, characterized by fluctuations of drought and plenty. For children, resource availability
not only varied with the climate and the availability of game, but with their
social resources, as people came and went, as kin networks were decimated
or expanded.

"A problem for scientists who view modern humans from the vantage
point of the adaptive legacy," Draper told me, "is that they do not know very
much about the way people lived 30,000 or 40,000 years ago, when anatom-

ically modern humans were spreading in such a rapid way over much of the globe." For a child growing up in one of these fluctuating environments, differences could be as great as night and day, depending on whether the father stayed around or left, whether the mother struggled on her own or turned to family, or what kind of baby-sitters were available.

Bowlby had assumed that drastic departures from the circumstances of the Environment of Evolutionary Adaptedness (such as separation from mother in infancy) would lead to aberrant adjustments. Draper was not so sure. After long discussions, she and her then husband, biological anthropologist and geneticist Henry Harpending, hypothesized different developmental trajectories that would be more or less adaptive for a child, depending on circumstances.

Both were struck by similarities and differences between children's life experiences in middle-class American families and those in various parts of Africa. "There were all these assumptions that derive from our own Victorian past concerning what children need and what women need," Draper told me.

> Our own culture has been characterized by intense emotional bonds between mother and offspring. Children look up to their parents as providers. But if children are weaned earlier, or cared for by peers and other caretakers nutritively and psychologically, and farmed out into larger, more diffuse and less diffuse groups of child-rearers, it will be a different experience. Children will come to different conclusions about what the goodies are and where to get them.[8]

Draper and Harpending hypothesized that children would mature at different rates and develop different styles of coping with the social world, depending on whether they had learned to rely on their parents or on peers. They speculated that a child's perception of the availability of resources, and the most effective way of obtaining them, elicited different ways of responding in social situations, investing in reciprocal relationships, or trying to cajole or manipulate others into helping them. This was learning, but learning of a special kind. As in the acquisition of language, children are born with "learning biases" that at some life phases cause them to learn some tactics and skills more easily than others. "We now realize that the individual organism does not come equipped with a generalized capacity to learn anything in the way of response to cues that are coupled with different rewards and punishments. Instead," they wrote, "there is a growing understanding that natural selection has shaped the central nervous system to promote the 'ease of learning,' certain responses which themselves are conducive to survival and ultimately the reproduction of the individual."[9]

Increasingly concerned about the extent to which psychologists were idealizing Bowlby's "Environment of Evolutionary Adaptedness," an emerging group of sociobiologically influenced researchers in human development— including Draper and Jay Belsky, Michael Lamb, Jim Chisholm, and Mary Main—accepted Bowlby's evolutionary framework for how infants attach to their caretakers but began to diverge from Bowlby concerning how uniform ancient worlds, and how variable human adaptations to them, were. "[I]n light of current theory in evolutionary biology," declared a 1985 consortium of scholars, including child psychologist Lamb and the life-history theoretician Eric Charnov, "it is not easy to designate some behaviors as adaptive, others maladaptive. There is no species-appropriate pattern of behavior against which all other patterns can be evaluated." To them, infants come equipped with a flexible repertoire, depending on the specific environment in which they live. Viewed from this perspective, it became critical to specify *how* alternative patterns might be adaptive under *what* caregiving circumstances. This challenge marked the beginning of the post-Bowlby era of attachment theory. It brought with it a new awareness of naturally occurring variation within the same species.[10]

This marriage of sociobiology and child development is redefining what we mean by "normal," replacing the older notion of species-typical development with an awareness of "developmental trajectories" that are adaptive for coping with the circumstances at hand. This new perspective is already leading to novel interpretations.

As early as 1981, Mary Main was already beginning to wonder about infants who avoid looking at their caretakers. Earlier researchers took for granted that avoidance must be pathological. Now, Main postulated that avoidance might be an adaptive tactic for infants to cope with parents who routinely reject them. She used the "strange situation" to identify particular infants who on reencountering the person to whom they were most strongly attached would avert their gaze and perhaps move away, avoiding precisely the person who common sense tells us should be the foremost object of desire and concern. Such cases had always seemed puzzling and counterintuitive.

Fig. 23.1 In his *Biographical Sketch of an Infant,* the loving but typically distracted Charles Darwin described the development of his firstborn son, William (two years and eight months old in this photograph with his father, taken in 1842). Darwin's account of William's "unconscious shyness" is probably the first published description of the childhood behavior termed "avoidance": "I saw the first symptom of shyness in my child when nearly two years and three months old: this was shown towards myself, after an absence of ten days from home, chiefly by his eyes being kept slightly averted from mine; but he soon came and sat on my knee and kissed me, and all trace of shyness disappeared." *(Courtesy of Darwin Museum, Down House, and English Heritage Photographic Library)*

Main hypothesized that children avoid looking directly at a wayward parent in order to conceal angry reactions that might jeopardize their ability to maintain this all-important relationship. She called the response "avoidance in the service of attachment."

Birth Order and Personalities

The most ambitious effort to date to actually test how evolutionary ideas apply to the development of different human personality types is the book

Born to Rebel by Frank Sulloway. Although Sulloway never mentions "poly-phenisms," he is in fact focused on the same dichotomy as Erick Greene in his research on early- and late-hatched caterpillars: the development of quite different morphs in response to the different resources available to early-born versus later-born offspring. The difference is that Sulloway's morphs involve different personal styles.[11]

Sulloway argues that firstborn children are advantaged early on by their monopoly on parental attention, and later on by their greater size and maturity, which allows them to dominate later-born siblings. When magnified by parental preferences for older children or by such institutional elaborations of their advantage as preferential inheritance by firstborn offspring, firstborn children have every reason to identify with parents and with the authority of the social system, while the later-born have every reason to feel aggrieved, to chafe and rebel at the injustice of this resource allocation—to question authority rather than identify with it.

In a very large sample of people from different domains throughout history (primarily scientists but with similar patterns documented for political figures), Sulloway was able to show distinctive patterns in which firstborns were more likely to identify with and defend the status quo, while later-borns (those "born to rebel") were more likely to challenge the status quo and endorse revolutionary ideas. Siblings become different in the course of development by internalizing quite different working models about how they expect (and would like) the world to work. This internalization of a model of the world based on early experience is Bowlbian to the core.

A wanted child would learn to expect very different treatment, especially different levels of support, compared to a less wanted child. Anthropologist Nancy Levine, who worked among Tibetan-speaking farmers, herders, and traders in northwestern Nepal, tells the story of a woman who had nine children, two sons and seven daughters. Both of the sons died, while three daughters survived. The last-born of all her children was a girl. The experience of bearing another daughter after losing her two sons made the mother "so furious" that at first she refused to feed her. But the baby did not die, and relatives and friends finally prevailed upon the mother to feed her. The mother's bitterness, however, resurfaced in her refusal to interact with the child or even name her. The girl was given a series of abusive epithets. One of her nicknames was "Ready-to-die," which in fact she did not do. Instead, she matured into a healthy, if rather quiet, child.[12] The point of this sad story is that a wanted son and an unwanted daughter might experience a mother's care *very* differently, as differently as if they had been born to different species, or born in quite different habitats. Yet somehow, some unwanted children (like "Ready-to-die") do manage to survive, albeit stunted to varying degrees, emotionally and physically. Whereas the first-born son would grow up with a working model of a world in which those around him seek to help him out, the last-born girl would feel just the opposite.

This sensitivity to intraspecific variation is reflected in a recent summary statement on child-rearing by Jay Belsky and colleagues at Penn State: "Meeting children's emotional and social demands in a supportive responsive manner fosters a social orientation that values mutually beneficial interactions and relationships, whereas patterns of rearing that are negative, inconsiderate and coercive lead children to behave in ways that are self-centered."[13] Children with "negative, inconsiderate and coercive" caretakers will be those born to parents who did not want them, or else reared by exploitive alloparents. Such children could not realistically grow up to expect much help from others, or to find an advantageous niche for themselves within local kin networks.

In a foraging context, it might be highly adaptive for an avoidantly attached individual to learn to downplay love, to dismiss the importance of close human relationships, which would not in any event be forthcoming. Rather than rely on those around him, the most advantageous course open might be for the child to become self-reliant and to avoid developing empathetic feeling for others around him, unlikely to behave like committed kin.

"Wretched" Developmental Courses

The term *wretched* derives from the Old English *wrecche,* for exile, literally a person without kin nearby, dehumanized by circumstances. No one could be more kinless or in this sense more wretched than an abandoned infant. Yet heroic tales from antiquity abound with stories about such wretches. Famous foundlings include Heracles, Oedipus, Paris, and Heracles' own abandoned son, Telephos (see Plate 5).

Just after birth, Telephos's mother, a priestess to Athena, hid him in a thicket, where he was suckled by a doe. For such a child, growing up like a foundling, brought up by strangers, it might well be adaptive for him to develop into an aggressive loner. When, as a young adult, Telephos is

Fig. 23.2 In his *Retour de Nourrice* Jean-Baptiste Greuze (1725–1805) depicts the weaned child's return from the wet nurse. As the mother reaches for the child, he avoids her gaze. My guess is that Greuze's model for the child in this portrait was a child who had been separated from his mother for only a relatively short period, for he does not treat the mother like a complete stranger. Or perhaps there has been enough contact between mother and infant while the child was at the wet nurse for him to have the sense that she is a special figure. The painting remains a poignant evocation of the effect on children of ruptured attachments. *(Courtesy of Bibliothèque Nationale, Paris)*

wounded, he is told that only King Agamemnon can heal him, whereupon Telephos takes the king's son hostage to extort his help. For a kinless human, little is likely to be proffered, and goods must be snatched, secretly stolen, won by guile or daring heroic exploits. Such human relationships as might be available are likely to be on terms exceedingly disadvantageous to the kinless wretch. There would be no kin to arrange a marriage or provide wherewithal for a bride-price. Even mating would be out of reach, except perhaps through rape. In terms of Hamilton's rule (Cost to donor must be less than benefit to a recipient related by r, $C < Br$), there might as well be no r, no degree of relatedness, in his internalized equation, only perceived benefits and calculated costs. For a human, especially a child, it is the most dismal imaginable situation. Thus a certain type of sociopathic personality, forged out of desperate circumstances, though highly undesirable from society's point of view, might from the developing child's perspective be a way of making the best of an appalling situation.

To date, the hypothesis that the personalities of infants who are avoidant and inner-directed, or who develop to be manipulative, exploitative, and self-centered, might once have been adaptive for coping with wretched circumstances has failed to generate predictions that are testable—at least not testable in any way sufficiently ethical to get past human subjects review boards. We know little about how avoidant attachments are played out in men's reproductive adaptations, even less about women's, and are only beginning to learn how these histories play out in the formation of adult relationships. Documenting the emergence of a personality deficient in empathy—what we often mean by "sociopathic"—is problematic. Long-lived and multifaceted, humans are difficult to study, and there are daunting ethical constraints (which is as it should be).

There are numerous practical reasons why we know so much more about alternate developmental states in caterpillars than we do in children—reasons that go far, far beyond the greater complexity of the latter. Given how undesirable "sociopaths" are from society's point of view, there is a real risk that studying certain types of people might lead to stigmatization of certain children.

What Makes Us Human?

By the third trimester a fetus can hear noises beyond the womb, can process affective quality of the speech, and differentiate whether mother or someone else is speaking. This provides the fetus his first clues about the world. It marks the beginning of feeling "embedded" in a social network and the sensation of belonging that gradually develops, after birth, into the capacity to experience empathetic feelings for others. The capacity to combine such feelings with our uniquely human ability to guess what someone else must be

Fig. 23.3 Abandoned at birth, Telephos, like his father before him, survives and grows up to be a terrorist. Suffering from a festering wound, he has no kin to help him. Depicted as scrawny man (on left), he attempts to force King Agamemnon to help him heal his wound by holding the king's son hostage. According to Diodorus Siculus, the desperate Telephos exclaims: "I will kill your son unless you cure me!"—a credible threat, coming as it does from a wretch unrelated to his defenseless victim.[14] *(Lucanian krater attributed to Policoro, ca. 400 B.C. Courtesy of the Cleveland Museum)*

thinking and feeling is the main difference between humans and other animals. As Harvard psychologist Marc Hauser likes to sum it up, what distinguishes humans from other apes is our "ability to put ourselves emotionally and cognitively in someone else's shoes."[15]

It is this ability to imaginatively construct the way others think and feel that takes us beyond mere empathetic awareness (say, of an offspring's grief at the death of its mother) and beyond mere strategic manipulation of another's behavior. A chimp, for example, is quite capable of giving a false food call that leads another animal away from a hidden food source, allowing the deceiver to return later for the plunder. Chimps, as well as many animals that hunt for a living, show an uncanny ability to read intentions and guess what other animals will do. Chimps are quite capable of consciously calculating certain kinds of costs and anticipating benefits. They can even anticipate the cost-benefit decisions other animals are likely to make.[16]

But humans go a step further. They combine these analytical capacities with new ones—like being able to imagine the future. Even more important, they are able to translate hunches about how another animal will react into full-scale speculation about what others are thinking, and articulate their concerns both to themselves and to others. In this way, humans transform ingenious capacities of observation into the sophisticated capacity to care what happens to others, even those they have never met.

From whence such gifts? Unfortunately, in the case of minimal levels of human empathy, we know the answer. As with all primates, infants who grow up without social contact, with no one to touch, hold, cuddle, with neither mother nor allomother to reassure them of their commitment to the infant's well-being—these infants fail to proceed along the developmental pathway that is the essential first step for development of this uniquely human sensibility. Such people may grow up brimming with analytical abilities, even uncannily able to anticipate what other humans will do. But they lack the capacity to hook up the cognitive and emotional components of human potential.

In many cultures the word for one's own group is the word for "humans." Humans are, by definition, those people like oneself, and also likely to be those people one is most closely related to. People not like oneself, especially enemies with widely divergent self-interests, are defined as "other-than-human."

There is a variety of preconceptions about what makes us human, and people tend to hold very tightly to their own views on this subject. My own favorite effort to encapsulate in words the features that make us human appear at the conclusion of *Ishi in Two Worlds*, Theodora Kroeber's poignant account of the life of the "last surviving wild Indian in North America."

On August 29, 1911, a starving Native American materialized outside a slaughterhouse in Oroville, California. Although still a young man, he had outlived every other person in his tribe, most of whom were murdered by those who wanted to rid the country of people of his race. Yet somehow, in spite of inestimably cruel and inhumane treatment of his people, Ishi navigated this "sudden, lonely, and unmitigated change-over from the Stone Age to the Steel Age" with his humanity intact to demonstrate that *Homo sapiens,*

> whether contemporary American Indian or Athenian Greek . . . is quite simply and wholly human in his biology, in his capacity to learn new skills and new ways as a changed environment exposes him to them, in his power of abstract thought, and in his moral and ethical discriminations. It is upon this broad base of man's panhumanity that scientists and humanists alike predicate further progress away from the instinctual and primitive subhuman strata of our natures.[17]

Confronted with the most appalling possible life circumstances, isolated with neither kin nor peers, Ishi exhibited a well-honed understanding of the social contract, a scrupulous sense of fairness and concern for others. Against all odds this dignified and nobly mannered man with the "flashing, friendly smile" adapted to an alien world by forging friendships that were characterized by genuine concern for the feelings of people who belonged to a very different race and era. The source of his abiding nobility was his own internalized moral values, personal standards of correct versus incorrect behavior.

Such sensibilities could scarcely be attributed to the world around him. Most who met Ishi would see in him little more than a freak from another era. Where rehabilitation through selective kindness has been tried on people Ishi's age, the results are rarely positive. So what made Ishi so special? Simply put, he was orphaned late enough. His adult life was spent hiding from genocidal cowboys who sought to kill him, who had already killed his immediate family and remaining people. But prior to that terror, his early experiences included being part of a community. What made Ishi human derived from his early sense of belonging to a group of kin committed to his well-being.

Practical Implications

No doubt some people are born innately deficient in the potential to develop compassion. But what interests me here is the possibility that some portion of noncompassionate people are *not* born that way. Rather they have responded adaptively to signals in their early upbringing. These cues alert the infant that one of Mother Nature's worst-case scenarios is about to befall him: he will have no mother or adequate substitute for one, and no supportive kin network. Under these circumstances, lack of compassion might well be an adaptive (or once adaptive) response. This hypothesis is consistent with evidence that secure attachments early in life have long-lasting effects, not so much on temperament or cognitive capacities as on degree of empathy for others.

At present we are only making educated guesses, yet the stakes seem terribly high, as countries like the United States pour money into prisons while investing proportionally far less in early development programs designed to convince infants and children that they belong to a community that will treat them like kin.[18] These are the circumstances that signal a developing human organism that it will be worthwhile to grow up caring about what happens to others, and to behave as if those others might be kin.

Devising Better Lullabies

Baby, baby, naughty baby,
Hush, you squalling thing, I say.
Peace this moment, peace, or maybe
Bonaparte will pass this way.

Baby, baby, he's a giant
Tall and black as Rouen steeple,
And he breakfasts, dines, rely on't,
Every day on naughty people.

Baby, baby, if he hears you,
As he gallops past the house,
Limb from limb at once he'll tear you
Just as pussy tears a mouse.

And he'll beat you, beat you, beat you,
And he'll beat you all to pap,
And he'll eat you, eat you, eat you,
Every morsel snap, snap, snap.

    —Nineteenth-century English lullaby

The image of a cannibalistic, strange male can scarcely be an image little apes find soothing. Modern readers would of course never sing anything so grisly to their children. But pause. Recall the words from Mother Goose that English-speaking children hear many times:

Hushabye Baby, on a tree top,
When the wind blows, the cradle will rock.
When the bough breaks, the cradle will fall.
Down tumbles Baby, cradle and all.

That message may not be very comforting either. (Indeed, "Hushabye," for today's more common "Rockabye," may be a distant corruption of an old French warning that a wolf is near: "Hé bas, là le loup").[1]

Childhood Fears

Most parents reading this book would regard lightning striking as more probable than that they might neglect or place in danger a baby in their care. ("The parental instinct drives us to nurture our children to the utmost of our ability," reads a just-published medical text.)[2]

The assumption that ambivalent parents (especially mothers) must be abnormal is one reason selection pressures on infants to elicit their commitment have received so little attention. It may also be the reason maternal ambivalence has for so long remained a preserve of psychoanalysts (who study the odd) rather than of evolutionists (who study the natural).

"If anyone had suggested, when I started my psychoanalytic practice some twenty-five years ago that my patients would fear that I—or their parents— might kill them, I think I might have responded with shock. I might even have rejected it as somebody's nightmare," wrote psychoanalyst Dorothy Bloch. Bloch drew on her entire professional experience to document her preposterous-sounding claim that all the children she treated feared being killed by their parents and employed a wide range of fantasies and self-deceptions to make themselves feel safe.[3]

I doubt that all, or even most, children live in a state of chronic fear. However, I am convinced that Dr. Bloch provides insights into the self-protective fantasies devised by *insecurely attached* children (perhaps the young patients most likely to end up in her office). Such infants have caught whiffs of ambivalence, detected hints that their mother might (figuratively or really) withdraw or go away. The promise of current and future care, something more than just maintaining proximity, is a central preoccupation of immatures; it is also critical for social and physical growth.

From an infant's perspective, there is a critical distinction between being cared for by multiple familiar, as-if-kin caretakers, and being abandoned to the less-committed care of strangers—or, worst of all, abandoned altogether. Herein lies an important practical tip for mothers, what I think of as Bowlby's first law of maternal freedom: convince your toddler that *he* is the one who wants to leave *you* to play with someone else, not vice versa; or see to it ahead

Fig. 24.1 *Time* magazine used a woodcut of Hansel and Gretel being led into the forest by their father and stepmother to illustrate their review of Dorothy Bloch's *So the Witch Won't Eat Me.* Bloch read into children's fantasies her "patients' struggle to win their parents' love." To her, this was their "primary defense against their fear of infanticide."[4]

of time that baby is comfortable with the baby-sitter, that he sees the sitter as surrogate kin, part of his extended family. The extra time it takes will be more than repaid by ease of the transition.

A securely attached infant is an infant secure about his world in general, present *and future.* A secure infant is far more comfortable, even in his mother's absence, than an infant in doubt about his mother's commitment. It is not turning an infant over to allomothers that has harmful repercussions. It is failing to convince him or her that abandonment is out of the question.

First Steps to Security

Yes, there *is* bondage here. But the nature of the conspiracy is different from the one women were demonstrating against down King's Parade when Bowlby received his honorary degree. From an infant's point of view, the period between birth and the onset of lactation, some days later, remains fraught with significance. Not because there is a "critical period" during which a mother either bonds with her infant or fails to, but because her proximity is an essential first step to lactation and the ensuing processes that forge powerful ties between mother and baby. A mother closer on day one is more likely to want to remain close on day two, be driven to remain close on day three, and so on.

The past few decades have witnessed a long, pointless, and ill-informed debate over whether or not women have "maternal instincts." Given the historical context, the battle lines were understandable. The early literature on the biology of motherhood was built on patriarchal assumptions introduced by earlier generations of moralists. What was essentially wishful thinking on their part was substituted for objective observation. It has taken a long time to correct these errors and revise old biases, to "raise Darwin's consciousness" and widen the evolutionary paradigm to include both sexes. But by the time this happened, feminists, social historians, and philosophers were already convinced that they knew what evolutionists had to offer, that it was necessarily flawed, determinist, and uninsightful. Natural selection, and with it the most powerful and comprehensive theory available for understanding the basic natures of mothers and infants, was rejected, as social scientists and feminists took another route. That path, which led away from science, led them to reject biology altogether and construct alternative origin stories, their own versions of wishful thinking about socially constructed men and women, and infants born with more nearly a desire for mothers than a need. Mother love could safely be interpreted then as a "gift" consciously bestowed, or as a by-product of changing fashions in sentiment. In the meantime, biologists were developing a more multifaceted view of mothers, featuring flexible actors whose responses were contingent on circumstances. But how were feminists—long ago embarked upon an alternate itinerary—to know?

Lost in the shuffle over what it was natural for mothers to do and dust-ups over "bonding" and mother love, was the infant's often noisy two cents' worth: "No matter who gives it, I need it. And I need it now. Not when grown-ups sort out where it comes from. But now. Don't care if it's quality time or some other time. NOW."

More than anyone before him, it was John Bowlby who, at a theoretical level, addressed this need. Bowlby provided scientific legitimacy to the anxiety, distress, terror, and, finally, desolation that infants experience when they fail to detect "the meeting eyes of love." What is it infants so dread? he asked.

Many of the dangers turn out to be different from those Bowlby initially envisioned. Maternal alternatives to caring for their own infants were more varied than he realized. But Bowlby's central explanation of how and why infants become attached to their caretakers was on target.

In a modest addendum to Bowlby, then, I would argue that infants strive for attachment to their mothers not only to stay close lest something pounce, but also to stave off the possibility of maternal retrenchment, and forestall it in any of the manifold forms it takes, ranging from mild negligence to abandonment. Babies are geared to making sure that maternal care is forthcoming and ongoing. Among humans, protection from predators comes as a bonus from maintaining this even more crucial attachment.

Toward this end, infant connoisseurs of mothering are designed as they are. Every trait, every nuance to a trait that made this more likely, was selected for. Over tens of thousands of years, any infant whose lusciousness was detectably less, also proved ever so slightly less likely to survive. Robustness, plumpness, cuteness were not just useful physiological attributes. They were signals to the mother that the *benefit* function in Hamilton's equation was well worth whatever *cost* it might entail for a mother to care for her infant. That infants, if they are to get the care they need, must be so plump and adorable is a reminder of the dreadful bottlenecks and close calls that the survivors who became our ancestors passed through.

In addition to these bonny inducements to get themselves picked up, infants also need innate diagnostic skills. Once proximity is achieved, it is their job to make the most of it by rooting for the mother's nipple, latching on, and sucking. Being attached to one's mother, not incidentally, also initiates and then maintains lactation, with the attendant cascade of physiological consequences in the mother, suffusing her body with a sense of well-being, overriding inhibitions she might feel about her unfamiliar passenger.

The Sensual Responses Babies Produce

As pursed lips clamp tightly onto her nipples and tug, the little head gives a jerk, like a fish on a line whose movement secures the hook. But in this instant, just who is it that is being caught? Within moments, maternal cortisol levels subside; oxytocin courses through her veins. As if she were getting a massage, the mother's blood pressure decreases, oxytocin suffuses her (if all goes well) in a beatific calm that lulls her normal inhibitions about being so close to a stranger. If she allows it to happen, close contact with this baby does indeed transform her mindset and produces in her a need to be close to her baby, a need to smell her baby that, for some women (true in my case), borders on addiction. Whether or not this constitutes a gift given—and reciprocated—or bondage, depends on what else a mother wants to be doing, and who is there to help.

Once nursing begins, bondage is a perfectly good description for the ensuing chain of events. The mother is endocrinologically, sensually, as well as neurologically transformed in ways likely to serve the infant's needs and contribute to her own posterity. As her mammary glands go into production, it will be a long time before she is again emotionally and physiologically so at liberty to cut bait.

From that point on, a mother (especially one without a breast pump) lives on a mammary leash. Levels of prolactin are raised, emotional bonding has begun, aided and abetted by that inhibition-lowering, affiliation-inducing, calming surge of oxytocin. Absence may or may not make her heart grow fonder (though it usually will). What absence will always do, though, is cause her nipples to itch, tingle, and eventually ache.

Does a mother consciously desire to suckle her baby? Many probably do. But once the milk supply builds up (and Mother Nature has designed babies to ensure that it does) a mother, whatever her emotional state, will be adversely conditioned, tortured really, if she fails to. She becomes addicted to the act of nursing as well as to the baby herself. At the first cry of hunger, a lactating mother immediately recognizes her own infant's cry and warm secretions drip from her mammae, as if she were some long-ago platypus wanting nothing more than to hole up in a den someplace and let this baby suckle. As the baby sucks, the mother experiences an exquisite relief from the pressure of milk that has built up in her glands. The baby sucking on the receiving end of this let-down reflex brings with it pleasurable sensations, bordering on and blending into the erotic.

Whether or not the earth moves, these are powerfully conditioning sensations. To classify maternal sensations as "sexual," and therefore in puritanical minds to condemn them, is to privilege sexuality in a very nonpuritanical way, implying that sexual sensations are more important than equally powerful sensations that reward women for caring for babies. We might just as logically describe various orgasmic contractions during lovemaking as "maternal." These responses by a lactating mother mammal to her baby's sucking long antedated sexual responsiveness to breast stimulation in heterosexual (or any type of intimate) contacts. Propaganda from androcentric sources like *Playboy* notwithstanding, the feelings we identify as sexual were originally maternal. This is why, even in contemporary women, it should not surprise us that erotic arousal in the mother during breast-feeding can be correlated with increased milk ejection.[5] For maternity and sexuality are inseparably linked in ways that just are not true for paternity and male erotic experience. Sexual desire in men and other male primates evolved because copulating with females increased the chances their sperm would fertilize an egg. Having eggs fertilized, however, is just one of the ways that copulating serves a woman's reproductive ends. Others include the various roles men play, both positive and negative, in keeping her baby safe and fed.

Under the circumstances, then, who is say which came first? The eros of suckling or the erotic sensations of heterosexual adults coupling? I would guess the former. Not counting gestation, when the infant is still in utero, mammalian reproduction entails three types of relationships: copulation, parturition, and nursing. Each one involves participation by two individuals and involves intense psychophysiological responses to vaginal stimulation or uterine contractions, and in the case of lovemaking and lactation, significant breast stimulation, all deeply sensual experiences.

Maternity is inextricably intertwined with sexual sensations, and it is an infant's business, through grunts and coos, touches and smells, to make the most of Mother Nature's reward system, which conditions a woman to make this infant a top priority. Evolutionary logic is firmly on the side of mothers who enjoy the sensual side of mothering for its own sake.

Each moment in proximity enhances the prospect that there will be a next one, increasing the probability that the mother will remain accessible to receive the neonate's adorable signals, gradually guaranteeing the mother's attachment to the increasingly familiar creature nestled against her. Satiated, drugged by warmth and just a nip of oxytocin lacing the breast milk, the relaxed infant nestles against mother, their comfortable relationship feeding upon itself and growing into love.

We can discard an erroneous notion about how mothers bond to infants in some critical period right after birth. The window for bonding is fairly open-ended. Children can be adopted days after birth and loved just as intensely, as ferociously, as those babies whose bodies passed through their mother's birth canal and then pressed against her right after birth. It's the window for rela-tively painless (or only tolerably painful) termination of investment that is brief—usually in the first seventy-two hours.

Once established, the influence babies exercise over mothers does not easily fade. There is a conspiracy, all right, but the enslavement at issue predates by millions of years the medical and scientific establishments that feminist writers on the subject of the "bonding fiasco" complain about.

What Babies Seek

The intense, often highly sensual feelings babies produce, *and* the guilt mothers experience when they notice them, as well as the guilt mothers feel when they are absent, or when they feel blank or even negatively toward their babies, have caught the attention of psychoanalysts. Acutely aware of the ambivalence mothers feel about their slavish devotion to infants, psychoanalysts have devoted much attention to the expression of such feelings in sadistic lullabies and fairy tales.

Yet the same psychoanalysts tend to start out with a baseline presumption of mother-infant harmony. "For all of us," writes the psychoanalyst Alice Balint, "it remains self-evident that the interests of mother and child are identical and it is a generally acknowledged measure of the goodness or badness of the mother how far she really feels this identity of interests."[6] But the *cost* function of Hamilton's rule calls this presumption of harmony into question. As Robert Trivers, David Haig, and others make clear, the interests of mother and infant are only identical when they don't differ.

Instead of viewing it as an abnormality, a pathology to be treated, sociobiologists accept some degree of maternal ambivalence as inevitable. This insight, together with an awareness of the underlying sources of maternal ambivalence, helps make sense of some of the very odd things that mothers say and sing to children.

When they were young, I called all my children "sweet potato," "muffin," "cutie-pie." I'd say, "You're so adorable I could eat you up." I actually spoke to them this way. So have many other parents. What on earth could be the point of all this talk of eating? In retrospect, after a bit of Darwinian self-analysis, I suppose I really did have flesh on my mind. Soft as rose petals, such delectable new tissue, so healthy, not to mention parasite-free. But, however delectable, I am positive that I never had any inclination whatsoever to eat my children (although I acknowledge desiring more control over them, body and soul, than their own strong wills were ever inclined to grant). Furthermore, like the fry of mouth-brooding fish who dart confidently right in and out of mother's maw—and do note that, unlike humans, most fish, size permitting, are highly cannibalistic—my own small fry never seemed the least concerned about my culinary endearments. They giggled.

For, as designed by Mother Nature, the delectability of infants seduces to quite different ends. My children's deliciousness rendered *me* more willing to be consumed by *them,* to give up bodily resources, and in my own contemporary example, most importantly, time—time, time, time, right down to the last syllable of allocatable time (when I could get no allomother to substitute)—and so to subordinate my own aspirations to their desires so we could all (more or less) contentedly take our places at posterity's table.

But around the world, lullabies intended to calm babies often seem to fall short of the mark. Lurking predators, not to mention strange males (like "Bonaparte"), or the prospect of maternal ambivalence, are precisely the threats little primates ought to fear most deeply. What infants yearn for is the reassurance that they will never lose their caretaker's love, that no matter what, she (or he) will keep them safe from any lurking hazard. Although female caretakers are more fluent in high-pitched babyese, and have the advantage of being able to nurse, and a head start toward becoming the infant's primary object of desire, fathers should not sell short their own ability to reassure—or harm. Far better than most, Aaron the Moor, the Shakespearean dad desperate to keep his lover from killing their child, understood the message infants want to hear from anyone who matters when he announced that "this [babe], before all the world will I keep safe."

From an infant's point of view the desired message is best summed up by an Egyptian incantation from the sixteenth century B.C., chanted to forestall evil spirits covetous of the child, spirits who might otherwise approach under the cover of dark:

Hast thou come to kiss this child?
I will not let thee kiss him!
Hast thou come to silence him?
I will not let thee set silence over him!
. . . .

I will not let thee injure him!
Hast thou come to take him away?
I will not let thee take him away from me![7]

"Ah, there, *that's* what I call a lullaby. That's the message I want to hear, see, smell, and touch," the infant doesn't precisely think but, in the old mammalian recesses of his brain, emotionally processes to himself as he nestles against his caretaker's breast, master of his galactic empire, and falls securely asleep.

Notes

Preface

1. See Genevie and Margolies 1987:5 for a broad-scale survey asking U.S. women how they feel about being mothers. More than half of the mothers expressed some ambivalence, and 20 percent were outright negative and wished that they had not had children. As will become clear in the course of this book, ethnographic and historical evidence makes it unlikely that the proportion of ambivalent mothers documented for the postindustrial U.S. is an aberration.

Chapter 1. Motherhood as a Minefield

1. This particular query, by Karl Zinsmeister, appeared in the *Sacramento Bee,* October 8, 1988.
2. Thomas 1998; Eyer 1992b.
3. According to Dr. Allan Rosenfield of the Columbia School of Public Health, fewer than *1.5 percent* of U.S. abortions are performed after 20 weeks, even fewer after the 23rd week of pregnancy, when there would be around a one-in-four chance that the fetus would be viable outside the womb. An even smaller fraction would ever be performed using the contested procedure, perhaps one tenth of 1 percent (Seelye 1997a).
4. Stolberg 1997a; there are no agreed upon figures. I accept the estimates cited by Katharine Seelye of the *New York Times* (1997c). Her numbers are consistent with those that appeared in the most recent *Journal of the American Medical Association* editorial on this subject where Grimes (1998) reported that 0.8 to 1.7 percent of all induced abortions performed in the U.S. occurred after 21 weeks and hence might be called "late term." Of these, only a fraction would be dilation and extraction. This type of abortion is so rare that the reason for focusing on it can only be to galvanize public sentiment against abortion in general.
5. The National Right to Life Committee, for example, opposes not only abortion, but also the pill, Depo-Provera, the IUD, and Norplant, because they flush the fertilized egg out of the uterus prior to implantation (Conniff 1998).
6. See Seelye's reporting on the "partial-victory abortion vote" (1997b, 1997c).
7. Santorum wrote a May 4, 1997, commentary for the *Philadelphia Inquirer* describing his own experience in order to illustrate that "I've gone through what everyone on the Senate floor said I had no understanding of. . . ." My source was Jesdanun 1997.

8. Stolberg 1997b.

9. Rasekh et al. 1998.

10. Shostak 1981; especially pp. 206–8, 309, 326; Draper and Buchannon 1992.

11. Judge and Hrdy 1992; Hrdy and Judge 1993 (Table 1).

12. For the most extensive documentation of this widely noted generalization see Betzig 1986, 1993; for a brief summary of studies showing a correlation between male rank and reproductive success in tribal societies, see Irons 1998.

13. This phenomenon is best documented in research by the behavioral ecologist Monique Borgerhoff Mulder among Kipsigis herders in Kenya (e.g., Borgerhoff Mulder 1998; Luttberg, Borgerhoff Mulder, and Mangel 1999); for Japan see Associated Press 1998; for India see Srinivasan 1998.

14. Gilibert 1770:257–58. I am indebted to Elisabeth Badinter (1981:156ff.) for this translation, and for her account of eighteenth-century French attitudes on motherhood.

15. Schiebinger 1995; see especially her fascinating chapter "Why Mammals Are Called Mammals." Gilibert's translation of Linnaeus's 1752 anti-wet-nursing pamphlet was titled "La Nourrice marâtre, ou dissertation sur les suites funeste du nourissage mercenaire" (The unnatural step-mother, or a dissertation on the deadly consequences of mercenary wet-nursing). For a related discussion see also Yalom 1997:108–11.

16. Spencer 1873:32.

17. From Spencer's autobiography (vol. 1, p. 395), cited and discussed in Paxton 1991:17–18; here and throughout I am indebted to Nancy Paxton for her superb analysis of the intellectual ties between Eliot and Spencer.

18. Spencer's views on female inferiority became stronger over time. In the *Principles of Biology,* two volumes published between 1864 and 1867, he waxed even more authoritatively concerning the physical and intellectual inferiority of women, now (thanks to Darwin!) "proven." See Paxton 1991:118; Russett 1989:12ff.

19. Supposedly the pejorative term *bluestocking* was originally given to an eighteenth-century lady, Mrs. Stillingfleet, who always wore blue stockings to the meetings of her women's literary group. Subsequently, the term was extended to any woman of letters who seemed either pedantic or ridiculous. Spencer's quote from his *Principles of Biology* (vol. 2, p. 486).

20. Spencer 1873:32. For critiques of nineteenth-century characterizations of women see Hubbard 1979; Russett 1989; Shields 1984; Sayers 1982. See esp. Gould 1981; Tavris 1992.

21. Eliot 1859:285–86.

22. Eliot ended her only surviving love letter to Spencer with the provocative observation that "I suppose no woman ever before wrote such a letter as this—but I am not ashamed of it." With characteristic courage, she laid bare her feelings to a degree remarkable in any age, much less for a woman in a Victorian one: "I want to know if you can assure me that you will not forsake me, that you will always be with me. . . . If you become attached to someone else, then I must die, but until then I could gather courage to work and make life valuable if only I had you near me. I do not ask you to sacrifice anything—I would be very glad and cheerful and never annoy you. But I find it impossible to contemplate life under any other conditions. If I had your assurance, I could trust that and live upon it" (from a transcription of the original quoted in Karl 1995:146).

23. Spencer 1859:395.

24. As Spencer put it years later in his autobiography, "the lack of physical attraction was fatal." He described Eliot as of "ordinary feminine height [but] strongly built" with a "physique" exhibiting "a trace of that masculinity characterizing her intellect" (cited in Paxton 1991:17).

Note that genes had not yet been discovered and Spencer was under the misapprehension that women got looks from their mothers, brains from their fathers. By comparison with either Spencer or Darwin, Eliot's intuitions about how heredity works seem brilliantly prescient. She correctly guessed that physical and mental traits can derive from both parents and sort themselves out in unpredictable ways. Barred from science, Eliot used her novels as a laboratory in which to set up fictional matings, experiments that would yield unanticipated personal and genetic outcomes, ranging from quite comic to tragic—novels that one way or another always expanded essentialist conceptions of male and female natures. In *The Mill on the Floss,* for example, Eliot has Maggie Tulliver's father follow Spencerian advice to a tee. He married a woman specifically for her looks and lack of brains, only to find himself the disappointed progenitor of a clever daughter and a handsome but plodding son. "Did you ever hear the like on't?" asks Mr. Tulliver as he laments what a pity it is that his daughter was so clever since, as he put it, "I picked the mother because she wasn't o'er 'cute [overly acute]—bein' a good-looking woman too . . . but I picked her from her sisters o' purpose 'cause she was a bit weak, like. . . . But you see when a man's got brains himself, there's no knowing where they'll run to; an' a pleasant sort o' soft woman may go on breeding you stupid lads and 'cute wenches till it's like as if the world was turned topsy-turvy. It's an uncommon puzzling thing" (Eliot 1860:18).

Darwin himself explicitly parted company with Spencer on this one. As far as he was concerned, "men who succeed in obtaining the more beautiful women will not have a better chance of leaving a long line of descendants than other men with plainer wives . . ." (Darwin 1874:580). Still, the criteria Spencer used when he turned George Eliot down are worth pondering, not because of what they tell us about Herbert Spencer's sexual preferences or personal love map (who cares?) but because of the discrepancy between the theoretical reason Spencer offered (eugenic ones) and what were more probably the real ones. Was it really Eliot's genetic qualifications for motherhood that bothered Spencer, as he said? His stated reason bears closer examination. Thirty-two when they met, Eliot would have still been fertile, qualified by robust health, intelligence, good character, and excellent earning power (better than Spencer's and absolutely remarkable for her time). Eliot's more likely deficit had to do with a heavy-set body and a constellation of facial features—fairly evidently inherited from her father, to whom she bore a striking resemblance—that made her look both masculine and mature beyond her years.

25. Darwin 1874:558. For the subsequent debunking of this misguided view of male versus female intelligence, see Gould 1981; Tavris 1989.

26. Eliot 1871–1872:183, 301. This tendency of men to define female worth in the terms that suit them best drove Eliot to counter with a roster of beautiful but destructive and nonmaternal heroines (Rosamond Vincy in *Middlemarch,* Gwendolyn Harleth in *Daniel Deronda,* and Hetty Sorrel in *Adam Bede*). Eliot contrasts these sterile, destructive beauties with the plain but nurturing Mary Garth of *Middlemarch,* who made a wonderful wife and mother.

27. Blackwell 1875:13–14.

28. Fraisse 1985. Quotes here come from Royer's suppressed manuscript, in the appendix to Joy Harvey's biography (1997): *Clémence Royer on Women, Society and the Birthrate*, pp. 193–203.

29. According to Harvey (1987:161), Royer supported all forms of birth control, including abortion, and was intrigued by what she heard was going on in America. Note that even though the population was much smaller, the rate of abortions in the U.S., per woman, was as high or higher than it is today because there were no other reliable means of birth control (Reagan 1997). Darwin, who was opposed to birth control (see Desmond and Moore 1991:627–28), would have had a different view.

30. Cited in Harvey 1997:194.

31. Ernest Renan, cited in Harvey 1987:165.

32. The ideological conflicts of the 1990s—what many journalists referred to as "the Science Wars"—grew out of the bitter residue left from these early failed dialogues. Any semblance of constructive discussion was replaced by absurdly polarized debates over Nature vs. Culture. See detailed accounts in Morell 1993b; Begley 1997; Macilwain 1997b; Gross and Levitt 1994; Segerstrale 1997. See Harding 1986 for the immediate precursors.

33. Consider, for example, this passage from a 1976 book: "The double moral standard which punishes an adulteress severely while often condoning the man can be defended on biological grounds. It increases a man's reproductive potential and it might be added that those who indulge in extramarital activities are those who are the 'fittest' and most deserving to be biological fathers as they must possess a high degree of cunning and initiative, and often physical agility" (Burton 1976:155). For further discussion, see Hrdy and Williams 1983; Horgan 1995.

34. Cowley 1996.

35. Pinker 1997:480.

36. From *Daniel Deronda* (Eliot 1876:132, 645). Darwin himself was marvellously outspoken in his opinions about how stupid it was to arbitrarily leave estates to the eldest surviving representative of the father's line regardless of merit. "Oh, what a scheme is primogeniture for destroying natural selection!" he wrote to Alfred Russell Wallace in 1864. He was admirably even-handed to his sons, but his daughters fared less well (Hrdy and Judge 1993).

37. Buss 1994b:114.

38. De Beauvoir 1974:51.

39. The most cogent summaries of this argument are to be found in Badinter 1981.

40. Kempe et al. 1962; for a broad overview see Korbin 1981.

41. Rich 1986:217.

Chapter 2. A New View of Mothers

1. Eliot 1990d.

2. Jay 1963.

3. Piercy 1986:78.

4. The text actually reads "the normal mother always is a mother," but I assume that the word female was intended. From Jay 1963:44. Such structural-functionalist views are typically associated with Durkheim and, within anthropology, with the influential anthropologist A. R.

Radcliffe-Brown. For a brief overview see Goldschmidt 1996; for a discussion of the effect of Radcliffe-Brown's social theory on primatology, see Hrdy 1977:7–11, 246, 276.

5. In one late 1950s study of wild langur monkeys, data were collected for as many minutes as the monkeys behaved "normally." If the behavior of the animals seemed "abnormal" (and by "abnormal" was usually meant aggressive behavior toward another animal), the group was deemed to be in a state of social "disequilibrium." At that point, the collection of data was halted, to ensure that only "normal" behavior was recorded. Not surprisingly, the resulting reports depicted remarkably pacific societies, with every individual a team-player, and mothers confirmed in their roles as nurturers (Jay 1962: chapter 8).

6. See the classic 1963 volume *Maternal Behavior in Mammals* edited by Harriet Rheingold. See classification of "normal" versus "abnormal" mothers in Calhoun 1962. Mothers and infants were similarly compartmentalized in studies of infants, e.g., Rheingold and Eckerman 1970.

7. Lack 1941.

8. Mock and Forbes 1995. For a comprehensive overview and update, see Mock and Parker 1997.

9. Many of us were then impressed by a series of experiments with rats at the National Institutes of Health published in a paper on "Population density and social pathology" (Calhoun 1962). Calhoun, who apparently had the emerging global population crisis and "a chilling possible end for Man" on his mind, began by citing Malthus on vice, misery, and the limits on population growth that nature would inevitably apply to those who outbred their capacity to adapt. Calhoun described the behavior of "normal" rat mothers, who constructed nests and cared for their young, and the pathological behaviors of "abnormal" mothers who failed to build nests and neglected their young. These same studies would be interpreted very differently today. However, they remain of historical importance.

10. At Mount Abu, where I studied langur monkeys intermittently between 1971 and 1979, repeated male invasions accompanied by infanticide occurred in one unusually small troop. Because so few infants were allowed to grow up, this troop grew smaller still—so small that males who usurped it were soon tempted to try to take over the larger troop next door as well, attempting to control both troops at once, and in the process made the small troop especially vulnerable to yet another takeover (Hrdy 1977). Chronically vulnerable to a new male coming in and attacking the infants, the group still existed when our observations ended in 1979, but it seemed to be on its way to oblivion. The neighboring group was expanding at its expense. Infanticidal behavior was on average advantageous to individual males, but quite detrimental to this group.

11. Hausfater and Hrdy 1984; Parmigiani and vom Saal 1994; for the most up-to-date review see van Schaik, van Noordwijk, and Nunn 1999. For red howler monkeys in Venezuela, see Crockett and Sekulic 1984. For gorillas at the Karisoke research center in Rwanda, see Watts 1989. By contrast with the langur or howler monkey cases, gorilla males do not oust the resident male and take over his troop. Rather, by attacking a young mother and killing her infant, the killer essentially demonstrates that the mother's current mate is unable to protect her offspring. She subsequently decamps and follows the killer.

12. Summarized in Sommer 1994.

13. Note that most savanna baboon troops have a number of simultaneously resident and breeding males, and these males may help protect infants. Palombit, Seyfarth, and Cheney (1997) argue that the unusually high rates of infanticide among savanna baboons in Botswana can in part be attributed to the fact that there is often only one resident male doing most of the breeding.

14. Hrdy 1977. For the best data on this point see Sommer 1994. At Jodhpur, only 5 percent of 55 cases in which invading males attacked infants involved infants that *might* have been sired by them. Recent DNA data collected by Borries and colleagues from langurs in Nepal indicate that, in their study, none of the infants killed by invading males were sired by them (Borries et al. 1998a, 1998b).

15. Frankel 1994; Batten 1992; for an elegant treatment of the history of ideas about female choice, see Cronin 1991.

16. Trivers 1972:173. Subsequent researchers would find it difficult to quantify the time, energy, and risk that mothers commit to reproduction and to compare costs from one domain to the other, and practically impossible to measure how investment in one offspring detracts from the ability to invest in another. For this reason, anyone planning research in this area should consult Clutton-Brock 1991, which deals at length with these problems. Nevertheless, one of the continuing merits of Trivers's original definition of parental investment theory is how well it seems to match parental psychologies in humans, the species of special interest in this book.

17. Petrie, Doums, and Møller (1998) show that across populations of socially monogamous birds females in the populations with the most genetic variation among males are those most likely to avoid being monopolized by their partner and seek extra-pair copulations; see also Gowaty 1996.

18. Petrie and Williams 1993; Petrie 1994.

19. Møller 1992a; Thornhill and Gangestad 1994.

20. For the theory see Hamilton 1982. For specific studies of parasite load in organisms where working measures are available for determining how far organisms deviate from perfect bilateral symmetry (known as degree of fluctuating asymmetry), see Møller 1992b. For correlations between performance and low levels of fluctuating asymmetries see Downhower et al. 1990 for fish; Thornhill 1992a for insects; Møller and Hoglund 1991 for birds. For discussion of an important alternative hypothesis, see Johnstone (1994) and Enquist and Arak (1994), who raise the possibility that female preferences for symmetrical males may be an artifact of neural infrastructure and the way females process information in the course of recognizing mates.

21. For an overview of this research by two of the pioneers, see Gangestad and Thornhill 1997. For an important cautionary note, however, see Jones 1996. Jones looked at how well measures of fluctuating asymmetries correlated with perceived attractiveness among subjects from five populations: Brazilians, Americans (U.S.), Russians, and two South American tribal groups (Aché and Hiwi). His correlations were in the same direction as Gangestad and Thornhill's (i.e., people with lower levels of fluctuating asymmetries were deemed more attractive); however, the results were not significant, leading him to conclude that "these results are consistent with fluctuating asymmetries being a component of attractiveness, but not a very important one. It is possible that fluctuating asymmetry is more important as a component of attractiveness in populations under heavy stress from nutrition and pathogens" (111).

22. This study by Furlow et al. 1997 is cited in Blinkhorn 1997.

23. Trivers 1972; Burley 1977; Thornhill 1979; Eberhard 1996; Gowatay 1996.

24. The consortium of biologists working with Gowaty on the "free female choice project" is funded by a $360,000 grant from the National Science Foundation, an unusually large grant in the field of animal behavior and testimony to how important research on female choice—including research inspired by Gowaty's explicitly feminist ideas—has become. Their goal is to evaluate the criteria used by females to choose mates in a range of insects, rodents, fish, and birds. Through examining female preferences and their biological consequences they hope to answer questions such as: Does female mate choice matter for "outbred vigor"? For "good genes"? Are there detectable differences in the viability of offspring depending on whether or not females have more freedom to select which males they mate with? The suspicion of nineteenth-century feminists like Eliot and Charlotte Perkins Gilman that breeding systems constrained by males are not only dystopic but dysgenic is now for the first time being subjected to scientific inquiry. As Gowaty points out, it would have saved a lot of time and misguided effort if we had asked such questions earlier. But the important thing is, we are asking them now.

25. Rice 1996. Quotes from William Rice are from the News Release published by the University of California, Santa Cruz, on May 10, 1996.

26. Evans, Wallis, and Elgar 1995.

27. Altmann 1980:1.

28. Jeanne Altmann's 1974 paper "Observational study of behavior . . ." would eventually (1986) be recognized as the most cited paper in the field of animal behavior.

29. Altmann 1980:6.

30. Altmann, cited in Walton 1986.

31. Day and Galef 1977; Gandelman and Simon 1978; Tait 1980; Packer and Pusey 1984.

32. Andersson 1994:186–88. Among certain fish, for example, males have been shown to prefer large and fecund females (Downhower and Brown 1980).

33. See Ralls 1976; Harvey, Martin, and Clutton-Brock 1987; Lessells 1991.

34. McHenry 1996; Kramer 1998.

35. Frank, Weldele, and Glickman 1995; Frank 1997.

36. Of fourteen females born at Gombe, as of 1995, six remained, five transferred to new communities, and three disappeared. Of eleven adult females present in 1995, five were natives, six were immigrants (Pusey, Williams, and Goodall 1997, note 22).

37. Wallis and Almasi 1995; Pusey, Williams, and Goodall 1997; Anne Pusey, personal communication, January 14, 1998, concerning fates and final tallies of Fifi's progeny. Similarly, the earliest birth ever recorded for a wild gorilla female was also documented for a female who managed to remain in her natal area (Harcourt, Stewart, and Fossey 1981:267). The moral is clear: if you are female and can pull it off, philopatry pays. According to Wallis (1997) Fifi had successive surviving births just 3.26 years apart, compared to the more usual five- to eight-year intervals characteristic of wild Great Apes. No unprovisioned wild chimps had been known to breed that fast.

38. Quote from Hrdy 1981:109. See Sherman 1981 and Digby 1994 for case studies; see Digby, Merrill, and Davis (in prep.) for general review. At the time of the first reports of infanticide and cannibalism by female chimps, Goodall (1977) was convinced that the killer was deranged. However, what struck a sociobiologist examining patterns of infanticide across a wide array of animal species was how similar the general outlines of the chimp cases were to patterns of infanticide in other animals (Hrdy 1981a:108–9). By 1997, however, even Goodall and her colleagues were convinced that infanticide by chimp females was "a significant if sporadic threat, rather than the pathological behavior of one female" (Pusey, Williams, and Goodall 1997:830), which I take to be the latest word on the subject.

39. Blackwell 1875:22.

40. Women fieldworkers, especially primatologists, would play key roles in articulating the special problems confronted by females. See Fedigan 1982; Haraway 1989; Strum and Fedigan 1996; Norbeck et al. 1997. Within science, coverage of these issues has been broad and balanced, if belated (e.g., see Morell 1993a).

Biases in primatology may have been more blatant, but were scarcely confined to this one taxon. Why, then, were women primatologists more sensitized to the problem of bias? Many of them came from backgrounds in the social sciences, and hence were more exposed, and exposed earlier, to seepage from feminist writings. But there is another reason. The more nearly animals resemble ourselves, the more difficult it becomes to replace preconceptions about females with empirically derived observations, but also, I suspect, the greater will be the likelihood that researchers identify with the interests of members of their same sex (Smuts 1985; Hrdy 1986a). Recently Holmes and Hitchcock (1997) surveyed the topics that animal behaviorists choose to study. Their findings lend support to this generalization. In general, women behaviorists are no more likely than men to study topics related to females or young animals, although women, it turns out, *are* more likely to study mammals than they are to study insects or fish. Among primatologists, however, women are significantly more likely to study females, men to study males.

Symposia proceedings and other publications marking inclusion of female perspectives in evolutionary biology generally include Lancaster 1973; Wasser 1983; Small 1984; Rosenqvist and Buglund 1992; Gowaty 1997. For overview see Batten 1992; Liesen 1995, 1997; Rosser 1997; Schiebinger 1999. See also Gowaty, ed., 1997.

41. Eberhard 1990:263, 1996. Some biologists have queried whether this new "female perspective" was "a female or a scientific triumph" (Cunningham and Birkhead 1997). The question seems almost mischievous. Obviously, benefits from correcting long-standing biases accrue to science. If the question is rephrased, however, to ask if women researchers played disproportionate roles in attempting to correct the biases, a hundred and fifty years of largely unwritten history indicates that the answer is yes.

Chapter 3. Underlying Mysteries of Development

1. Wilson 1971a:171–76.

2. Seger 1977.

3. M. J. West-Eberhard, May 4, 1998. "The flexible phenotype," lecture at University of California, Davis. For a comprehensive treatment of these themes see West-Eberhard, in prep.

4. See Gowaty 1995.

5. Cited in Paradis and Williams 1989:84–85.

6. Description of the queen's message from interview with M. J. West-Eberhard, May 5, 1998.

7. West-Eberhard 1986.

8. Bourke and Franke 1995.

9. Anyone who imagines that this is a new gloss on sociobiology, added after the fact to reburnish a field ideologically tarred and feathered in the seventies, should go back and read Wilson on "behavioral scales" (1971b).

10. Hamilton 1963, 1964. These articles are available in volume 1 of his collected papers, *The Narrow Roads of Gene Land* (1995).

11. Number of eggs per day is for domestic bees (Mairson 1993).

12. Instead of a 50-percent chance of getting certain genes from their father, which is the case in diploid reproduction (the way humans do it), in haplodiploid organisms all genes that sisters derive from their father are the same. Normally (under diploidy) the average degree of relatedness between full siblings is one-half (they have a one-quarter chance of receiving the same genes from their mother plus another one-quarter chance of receiving the same genes from their father). Provided the queen mates only with one male, the average degree of relatedness between full sisters in a haplodiploid system is three-quarters, since the half of their genes received from their fathers is always identical. However, queens mate with one male less often than initially supposed.

13. Behavioral development in the drones and timing of life events (such as first flight) are mediated by the same endocrine mechanisms as they are in the worker bees, but the outcomes are very different (Giray and Robinson 1997).

14. Hamilton 1963, 1964. Darwin himself had arrived at the same logical explanation for the devotion of sterile workers but without knowing about genes was unable to work out the problem as Hamilton did. Instead, Darwin relied on a common analogy. He compared the distinctive attributes of individuals belonging to a sterile caste to "a well flavoured vegetable" that, even though destroyed when it is eaten before going to seed, has nevertheless induced "the horticulturalist [to sow] seeds of the same stock" so as "to get nearly the same variety" (cited in Hölldobler and Wilson 1994:97). Technically, Hamilton's rule is expressed as $K > 1/r$ but the form I use here is the one preferred by those teaching Hamilton's rule to students, as it is easier to understand.

15. West-Eberhard 1967.

16. For the *pièce de résistance* confirming Hamilton's rule see Mock and Parker 1997.

17. For evidence that people treat close kin preferentially, think how people allocate their resources, especially after they die. Even when donors target a particular sex—the sex most likely to successfully translate resources into long-term inclusive fitness for a family line—or discriminate by birth order, they channel wealth to kin (Alexander 1979; Betzig 1992; Hrdy and Judge 1993). Even when people leave wealth to spouses rather than blood kin, which was not always the case but has become the rule in the United States over the past 150 years, wives avoid leaving wealth to husbands presumably because a widower may marry again and divert wealth to other offspring, while widowers, who need worry less about reproduction (by mostly postmenopausal wives) do not bias against spouses (Judge and Hrdy 1992). For evidence of nepotism in personal favors see Essock-Vitale and McGuire (1980, 1985a, and 1985b).

18. For a now-classic treatment of Hamilton's rule among humans, see Alexander 1979.

19. Hamilton 1963; 1995 reprint, p.7.

20. For almost everything I say about ants I rely on Hölldobler and Wilson 1990.

21. Apologies to wasp geneticist David Queller for taking some poetic license with his fine idea; see Queller (1994 and 1996) on the advantages of extended care for maintaining eusociality.

22. Cited in Hrdy and Bennett 1979:28.

23. Hamilton 1995:355.

24. Werren 1988:69.

25. Short 1977b:27.

26. Alberts et al. 1994:1083.

27. Lloyd 1975; Eisner et al. 1997.

28. For more on the recent explosion of interest in maternal effects, see Pennisi 1996; Fox, Thakar, and Mousseau 1997; and, for a detailed overview, Rossiter 1996; Mousseau and Fox 1998.

29. West-Eberhard 1989.

30. Greene 1989; 1996.

31. Mackinnon 1979 (esp. 269–70); Galdikas 1985a and 1985b; Kingsley 1982; and especially Maggioncalda et al. 1999.

32. West-Eberhard, in prep.; McNamara and Houston 1996.

33. For unusually clear discussion of how this works see Boyd and Richerson 1985.

34. The term *meme* was coined by Richard Dawkins (1982) to describe units of cultural information transmitted by imitation or teaching, rather than inherited, like genes.

35. Williams 1966b:15.

36. Hostetler 1974:203, 290–96.

Chapter 4. Unimaginable Variation

1. Daly and Wilson 1978:59. Through two editions, Martin Daly and Margo Wilson's *Sex, Evolution and Behavior* would stand as the best available and most widely used textbook in this area, the one I chose for my own classes. I am thus not citing a "weak link" but the standard view.

2. For one of the best recent papers on this subject see Altmann 1997. For a classic example in humans of when "counting cops" starts to break down as even a reasonable estimator of male reproductive success, see Perusse (1993) for modern industrial populations where people use artificial birth control.

3. Cronin 1980:302. See also Abernethy 1978:129, 132.

4. Primatologists in Japan (e.g., Kawai 1958) recognized the significance of female ranking systems far earlier than their counterparts in the West perhaps because they bypassed nineteenth-century Darwinian thinking and started by observing the wild macaques in their own backyards.

5. Silk 1983 and 1988.

6. Altmann, Hausfater, and Altmann 1988.

7. Known as "the Bateman paradigm" after Angus John Bateman, who reported in 1948 that 21 percent of drosophila males in his lab colony failed to breed, whereas only 4 percent of the females did. A successful male fruit fly could thus produce nearly three times as many offspring as the most successful female. From these results, Bateman extrapolated to humans. He stressed that the difference between the most successful and the least successful male, what today is known as the variance in reproductive success, was always greater than the variance in female reproductive success. Bateman's observation became central to sexual selection theory after it was picked up by Robert Trivers (1972), who elaborated and refined it by incorporating the concept of parental investment. The basic observation was correct, but old paradigms left over from the nineteenth century shaped the way these results were interpreted, and overinterpreted, to mean that selection operates more powerfully on males than females. In fact, what Bateman's experiments showed was that in fruit flies traits that helped in mate competition were more important for males than females.

8.   Even today, data on reproductive success over the female's entire lifespan are available for relatively few species. For some of the first attempts to address the problem see Wasser, ed., 1983, and Small, ed., 1984, and especially the compilation of relevant field studies brought together in 1988 by Clutton-Brock; for an exemplary experimental case study see Honig 1994. For examples from the human literature, see Essock-Vitale and McGuire 1985a and 1985b; Voland 1990; and Boone 1986.

9.   Summarized in Wrangham 1993, from data collected by Caroline Tutin, Michael McGinnis, and Jane Goodall. For early theorizing about why we see this "polyandrous" component to primate breeding systems, see Hrdy 1981a and 1986a; for an updated overview see Small 1990 and 1994; and Wrangham and Peterson 1996.

10.   Gagneux, Woodruff, and Boesch 1997; comparable genetic studies from Gombe and from Sugiyama's study site at Bossou corroborate the Tai results (Gagneux, Woodruff, and Boesch 1999).

11.   Hiraiwa-Hasegawa and Hasegawa 1994.

12.   For DNA evidence, see Borries et al. 1998a, 1998b; for similar data from a larger sample, but without DNA evidence to back it up, see Sommer 1994; for the proportion of infant mortality due to infanticide at Ranagar, South Nepal, see Borries 1997; for naturally occurring experiments in which langur males ignore unrelated infants kidnaped from other troops see Hrdy 1977:225 and 280. These males do not attack strange infants so long as they are carried by a familiar female.

13.   Cantoni and Brown 1997; see also Gubernick, Wright, and Brown 1993.

14.   Huck 1984; Storey 1990.

15.   Bruce 1960. Even after biologists realized that infanticide was probably implicated in the Bruce effect, the first explanation invoked selection on males to somehow cause females to reabsorb the fetus (Wilson 1975:154). In fact, it must be the case that selection operated *on females* to reabsorb their own fetuses. Although losing her fetus would never be advantageous, it would be less disadvantageous than continuing to invest in a doomed litter. Since at that point her best option is to conceive again, the mother's and the infanticidal male's interests suddenly coincide—a subtle distinction, of interest primarily to those tracing changes in the way evolutionary theory was applied to the two sexes (Labov et al. 1985). Some of the best work on this topic has been done with voles, which have a somewhat different breeding system from mice, and a different form of male-induced pregnancy disruption that apparently continues to operate later into the pregnancy than is the case in mice (Storey 1990).

16.   This brief treatment does not do justice to the enormous variation both within and between species of rodents. For an overview see Parmigiani and vom Saal, eds., 1994.

17.   This applies to the Mongolian gerbils Elwood studies (1994) as well as to various kinds of mice; see Labov et al. 1985; Soroker and Terkel 1988; Elwood and Kennedy 1990.

18. For savanna baboons see Pereira 1983; for gelada baboons see Mori and Dunbar 1985; for Hanuman langurs see Agoramoorthy, Mohnot, and Sommer 1988 and Sommer 1994 (esp. 174–75 on "Abortions during takeovers," and observations itemized in Appendix III: "Abortions in connection with male change"). Spontaneous abortions under similar circumstances have also been reported among lions (Packer and Pusey 1984) and wild horses in the western United States (Berger 1983).

19. Barbara Konig of the Theodor-Boveri-Institut, in Würzburg, has shown for house mice that survival of young to weaning constitutes 46–64 percent of the total variance in lifetime reproductive success among females (Konig 1994). The study was made possible by the relatively short lifespan of mice (six months) and because it is feasible to simulate naturalistic social conditions in the laboratory.

20. At present there is no exhaustive overview of all the animals known to have cooperative breeding systems. For excellent case studies, see Solomon and French 1997 (*Cooperative Breeding in Mammals*) and Stacey and Koenig 1990 (*Cooperative Breeding in Birds*). For the first theoretical overview, attempting to lay out decision rules employed by cooperatively breeding animals, see Emlen, Wrege, and Demong 1995; for mole rats see Lacey and Sherman 1997.

21. For review of animal literature see Packer, Lewis, and Pusey 1992; for humans see Hrdy 1992; for an experimental study that actually compares the reproductive success of similar subjects under communal and solitary rearing conditions see Konig 1994; for bats see Wilkinson 1992; for elephants see Lee 1987; for cebus monkeys see O'Brien 1988. For overviews of communal suckling, see French 1997, Tardif 1997.

22. The oldest female was always behaviorally dominant over other females; even though subordinate females are more numerous, alpha females accounted for 219 of 302 pregnancies. In all but one of the packs studied, the alpha female could dominate the alpha male, who was also the oldest of his sex in the group (Creel et al. 1992; see also Creel et al. 1991).

23. Digby 1994; Rasa 1994; review of earlier literature in Hrdy 1979.

24. Hoogland 1994, 1995:150–53.

25. Hubert 1998; see Lock 1990 for another such case.

26. Bianchi et al. 1996. I am indebted to David Haig for calling my attention to this phenomenon. See esp. Artlett, Smith, and Jimenez 1998.

Chapter 5. The Variable Environments of Evolutionary Relevance

1. Bowlby 1972:301.

2. Bowlby 1972:319; for updates see Tooby and Cosmides 1990; Pinker 1997.

3. Coss and Goldthwaite 1995:89.

4. See Prechtl 1965; Jolly 1972a.

5. First described by Heinz Prechtl in the Netherlands (1965), this functional explanation, based on the survival value of the Moro reflex for primate ancestors, is now found in standard references about child development, e.g., Brazelton 1969:27–29 or my own favorite, Konner 1991:54.

6. Bowlby 1972:86.

7. Borrowed from Hartmann 1939; see Bowlby 1972 (esp. chapter 4; for quote see 91–92).

8. Tooby and Cosmides 1992.

9. Pinker 1997:21, 207.

10. Did humans in the Pleistocene live like the desert-dwelling !Kung people that Bowlby would have known a bit about, whose offspring are born as far apart as five years? Or were Pleistocene foragers breeding at more nearly two- to three-year intervals, as some forest foragers with more meat in their diet routinely do? Did plant foods gathered by women provide the staple food, or was meat more important? When people relied upon shellfish and other aquatic food, or hunted with nets, were both sexes participating? Answers to such questions have implications for how dependent on men mothers were for food to feed their children, and would have influenced whether women always lived among their husband's kin or spent more time among their own matrilineal kin. How much did fathers do to provide and help care for offspring? Who else helped? Answers depend not only on which hunter-gatherer society in what part of the world, but also on demographic features of the group such as which individuals were on hand to help, discussed in chapter 11.

11. For concise overview of this literature see Foley 1996. Paleontologists like Robert Foley along with behavioral ecologists like Kaplan and Hill and cultural anthropologists like William Irons have been at the forefront of those shifting the focus away from the Environment of Evolutionary Adaptedness (EEA) to "decision rules" humans and other primates use to cope in different environments. Irons first proposed replacing the EEA with the phrase "Environment of Evolutionary Relevance" (Irons 1998).

12. A mother who remains near her kin in "matrilocal" arrangements retains more autonomy than if she travels a long distance from her natal place to breed and lives among strangers, a generalization that applies across all well-studied species of primate, as well as human societies where one or the other sex customarily leaves their natal place at marriage. See Schelegel 1972; Quinn 1977; Hrdy 1981a; Smuts 1995.

13. Kolata 1982; Coss and Goldthwaite 1995:89; Vogel 1998.

14. Retinas from guinea pig eyes also secrete melatonin—and even isolated from their owners, kept alive in tissue culture, they continue to secrete this compound on a twenty-four-hour, now-it's-night, now-it's-day cycle (Morell 1995). Research on day rhythms was done with hamsters (Weaver and Reppert 1986).

15. Wood 1994:368–70 (and fig. 8.39).

16. Sieratzki and Wolf 1996.

17. Guppies provide one of the best examples of rapid evolution. These brightly colored freshwater fish can be found in rivers on the island of Trinidad, where they are preyed upon by voracious cichlids. In response to this heavy predation pressure, Trinidadian guppies evolved so as to live life in the fast lane, maturing as fast as possible so as to reproduce before being eaten. Further upstream, however, a series of waterfalls prevented both guppies and their predators from penetrating upland streams. When experimenters released guppies in these low-predation habitats, there was a rapid transformation. In less than eleven years, after fewer than eighteen generations of breeding under reduced predation pressure, the guppies evolved a completely different "life history." They grew bigger, bred later, and lived longer. Mothers' slower reproductive pace was matched by higher infant survival.

When captured and bred in the lab, the slower-breeding guppies pass on their slow-maturing life-history schedules to progeny who, like their parents, also grow bigger, breed later, and produce smaller litters containing larger individuals. Faster-breeding guppies from the pre-dated stock downstream, however, continue to breed at their same rapid pace. Thus, differences between the two stocks involved genetic changes in the wild populations from which these spec-imens had been taken. See Resnick et al. 1997; Rice 1996; Holland and Rice 1999 for additional examples; for overview of recent studies, see Svensson 1997. For similar examples in life-historical changes related to the genus *Homo* see pp. 284–87 of chapter 11.

18. E.g., Glass et al. 1985.

19. Wiley 1994.

20. Example calculated by University of Utah geneticist Jon Seger.

21. Gray et al. 1998.

22. Fowke et al. 1996.

23. Simoons 1978; Bodmer and Cavalli-Sforza 1976; and especially Durham 1991:226–85 for an in-depth case study. In many parts of the world, milk is soured or fermented into yogurt or aged cheese (a process discovered thousands of years ago [Kosikowski 1985:88]) or curdled into yogurt before being fed to people, reducing the concentration of milk sugars to be digested, an example of how people use culture to adapt to new conditions.

24. For more on this Darwinian approach to diseases, see Nesse and Williams 1994.

25. Thurer 1994:287; Hays 1996.

26. For the first demonstration of correlation between female rank and reproductive success among rhesus macaques, see Drickamer 1974. For case studies see Silk et al. 1981 and Altmann 1997. For overview see Silk 1987c. In Hrdy 1981a:96–130, I emphasized how little was known about female-female competition in humans. There is now an emerging, albeit still slim, litera-ture: Essock-Vitale and McGuire 1985a and 1985b; Wasser and Isenberg 1986; see Campbell 1993; Campbell and Muncer 1994; see esp. Cashdan 1996.

27. Shostak 1981:172–75.

28. Many newspapers followed the story of Wanda Holloway of Channelview, Texas; for an account of her sentencing, see the *L.A. Times* (September 10, 1996).

29. In fact there are many modern workplaces. A sea change *is* taking place in the ecology of some of these, making it more practical for some women to combine work with infant care.

30. This is not an area where we have a great deal of information, but see David et al., eds., 1988 (*Born Unwanted: Developmental Effects of Denied Abortion*), which provides longitudinal studies of children born to women in Sweden, Finland, and Czechoslovakia who had sought abortions dur-ing pregnancy but been denied. The studies document long-term emotional deprivation with the effects magnified through time, and—where there was social support—in some cases amelio-rated.

The most dismal prognosis came from the Prague cohort, based on a 1983–84 follow-up of children born in 1961–63 to Czech women twice denied abortion requested for the same preg-nancy. A pair-matched controlled study revealed developmental risks reflected in behavior disor-ders, learning disabilities, and difficulties forming long-term relationships that widened over time, "influencing quality of life in adolescence and young adulthood, and perhaps even casting a shadow on the next generation" (Dytrych, Matejcek, and Schuller 1988:102). Boys were more vulnerable than girls (Matejcek, Dytrych, and Schuller 1988:72).

Chapter 6. The Milky Way

1. David Macdonald, ed., 1984:4, 94–95; Pond 1977.

2. Based on field reports by M. Yamada, cited in Jolly 1972a:62.

3. For leptin connection see Angier 1997; for fat see Frisch 1988 and Lancaster 1986.

4. Pond 1978:559.

5. Rose Frisch first proposed this "lay-away" plan for in-depth discussion of steatopygia in the context of comparative fat deposition among mammals; see Pond 1978:559; caloric estimates from Prentice et al. 1996; Lawrence et al. 1987; Dewey 1997. It is possible that this Hottentot woman was photographed without being asked for her "informed consent." If so, it provides an invaluable record acquired by methods no longer defensible.

6. See summary of practices in Miller 1981:98ff.

7. For review, see Mascia-Lees, Relethford, and Sorger 1986; for "deceptive hypothesis," see Low 1979; for "honest hypothesis," see Caro and Sellen 1990; for "symmetry hypothesis," see Manning et al. 1997; Møller, Soler, and Thornhill 1995; Scutt et al. 1997.

8. Ben Shaul 1962; Blurton-Jones 1972; Patino and Borda 1997.

9. See Low 1978 for the classic treatment comparing marsupial and placental mammals. Milk for the older joey is about 4 percent higher in fat content than milk produced for the neonate (Ealey 1963; Oftedal 1980).

10. Ealey 1967; Frith and Sharman 1964 (both cited in Pond 1977).

11. Hayssen 1995.

12. It is not yet known why male Dyak fruit bats lactate (Francis et al. 1994). Under unusual circumstances, male goats and men have also been known to lactate. For further discussion on this topic see Daly 1979 and Diamond 1997a.

13. Dixson and George 1982. Soon new analytical techniques allowed scientists to assay hormone levels using urine and fecal samples. Hence data collection became far less invasive. Ziegler and Snowdon 1997; for more see Gubernick and Nelson 1989.

14. According to sea horse specialist Amanda Vincent, this might explain the curious practice of eighteenth-century British women who bought patent medicines made from ground-up sea horses to increase their milk supply.

15. Ziegler and Snowdon (1997). This study was done with *Saguinus oedipus,* the cotton-top tamarin, in an effort to replicate Dixson and George's results with *Callithrix jacchus.* Both species are cooperatively breeding New World monkeys.

16. For a recent case study showing high prolactin levels in helpers among cooperatively breeding Florida scrub jays, see Schoesch 1998.

17. John Wingfield's quote from personal communication, June 10, 1998. Cowbird research by Rissman and Wingfield is summarized in Nelson 1995:298.

18. Example from Schoesch 1998:74.

19. Riddle and Braucher 1931; Nicoll 1974. Crop milk is described for emperor penguins by Prevost and Vilter 1963, for flamingos by Studer-Thiersch 1975; for pigeons by Griminger 1983. See also Desmeth 1980; Desmeth and Vandeputte-Poma 1980.

20. Rousseau 1977:393 (from *Émile,* originally published in 1762); Westmoreland, Best, and Blockstein 1986.

21. Darwin (1859:171) was following his correspondent St. George Mivart on this; Blackburn,

Hayssen, and Murphy 1989; Hayssen 1995. Note that Darwin's nutritional hypothesis and this antibacterial one are not mutually exclusive; both could be right.

22. Human colostrum contains 1.5 calories per ounce compared with 30 calories in the milk a woman expresses later.

23. Fildes 1986:85–88 (esp. fig. 2.3).

24. Gillin, Reiner, and Wang 1983; see also Newburg et al. 1998.

25. Short 1984:41; for specific example see Merino, Potti, and Moreno 1996; for a classic study of immune properties of mother's milk see Pittard 1979; for an up-to-date overview see Cunningham 1995.

26. Insel 1992; Insel and Hulihan 1995 and references therein.

27. Tense women are less likely to experience uninhibited let-down reflexes or, in general, the extraordinary sense of calmness being extolled here. They are not abnormal, simply anxious, a state that may make breast-feeding harder to establish and maintain due to failure of the let-down reflex. Since milk was not spurting out, the baby sucked harder, producing sore or cracked nipples. In such cases, exogenous oxytocin (administered in a spray) may help, although not so much as just relaxing. This book is not intended as a how-to guide, but today there are awfully good ones available, like Sheila Kitzinger's *The Experience of Breastfeeding* (1980). For all aspects of breast-feeding see Stuart-Macadam and Dettwyler (1995).

28. Newton 1955; Uvnas-Moberg 1997. For early speculations along these lines, see the pioneering writings of Niles Newton (1977:82). Masters and Johnson (1966:161–63) reported that women who are breast-feeding become sexually responsive sooner after giving birth than do non-breast-feeding women. A few nursing mothers report sexual stimulation to plateau levels, and even orgasm, induced by suckling their infants. For an authoritative recent overview of oxytocin and sexual responsiveness, see Carter 1992; also Carter, Izja, and Kirkpatrick, eds., 1997.

29. Carter and Getz 1993; Carter and Roberts 1997.

30. Keverne, Martel, and Nevison 1996; Dunbar 1992; see also Byrne and Whitten, eds., 1988.

31. Keverne, Martel, and Nevison 1996; Gibbons 1998a.

32. For review of the literature and empirical documentation from ground squirrels see Holmes and Mateo 1998.

33. This is an old idea, lent new credence by Eric B. Keverne and his colleagues at Cambridge. See Keverne, Martel, and Nevison 1996; Keverne, Nevison, and Martel 1997; Dunbar 1992; Byrne and Whitten, eds., 1988.

Chapter 7. From Here to Maternity

1. One headline promised "clues to the nature of nurturing" (*New York Times,* 1996a). The "essence of mothering" quote is from *The Harvard Gazette* (October 31, 1996). In *Science* (Cohen 1996) the headline asked: "Does nature drive nurture?"

2. Brown et al. 1996; personal communication from Jennifer Brown, March 11, 1998.

3. Brown's quote is cited in Cohen 1996:577. Although Brown's description mentions pups scattered about the cage, this photo shows pups aligned equidistant from the nest. This raises the possibility that the mother herself pushed them there, as if she failed to process the still-living pups as creatures that belonged in her nest.

4. I have relied here on both the *Cell* article (Brown et al. 1996) and an interview with members of Greenberg's lab reported in the *New York Times* (1996a).

5. Cohen 1996:578.

6. For an exhaustive overview of the physiology of maternal behavior see Numan 1988.

7. Terkel and Rosenblatt 1968.

8. Onset of postpartum aggression is faster in some wild strains than in others. But even in the extraordinarily aggressive wild Canadian *Mus domesticus* used by vom Saal and colleagues, rates of attacks against intruders did not peak until day three after birth (vom Saal et al. 1995).

9. Parmigiani et al. 1994; McCarthy and vom Saal 1985.

10. Konig, Riester, and Markl 1988. In studies of *Mus musculus,* 75 percent of pups killed by infanticidal males were killed in the first three days (Manning et al. 1995). Stephano Parmigiani and colleagues (1994:349–50) have speculated that postpartum maternal aggression might additionally function to test qualities of different potential mates in animals with postpartum mating.

11. Svare and Gandelman 1976. Note that far stricter animal welfare guidelines exist today than when these studies were done.

12. Parmigiani et al. 1994:342–43.

13. For example, see Bekoff 1993.

14. Portions excerpted from Hrdy and Carter 1995. For placentophagia, see Kristal 1991.

15. Based on work with voles (Insel and Shapiro 1992); for overview, see Carter 1998.

16. Keverne 1995.

17. See especially Bridges et al. 1985.

18. The most famous example of socially deprived mothers who learned to improve is provided by the rehabilitation of Harry Harlow's infamous "motherless mothers" (Suomi and Harlow 1972). Apes born in zoos also sometimes fail to care for firstborns. When zookeepers then have to remove and rear these babies by hand, the problem may be perpetuated. Some zoos now actually have remedial "mothering" programs: future gorilla mothers are given dolls to play with! Scientifically, however, such programs are hard to evaluate, since some mothers may improve with age as a matter of course.

19. Gibber 1986.

20. Higher mortality rates for firstborn infants are widely documented for both captive and wild primates (Harley 1990; Drickamer 1974; Silk et al. 1981). The failure rate for wild howler monkeys is probably the highest on record; in some small samples, none of the firstborn infants survive (Glander 1980).

The same J-shaped pattern (i.e., mortality highest among infants born to very young and older mothers) is typical of human populations (Srivastava and Saksena 1981; World Health Organization 1976). Some of this increased mortality is probably due to the physical limitations of a very young mother, as well as inexperience.

21. *Time,* September 2, 1996. Primatologist Frans de Waal made a stab at restoring order to delirium with a *New York Times* op-ed piece titled "Survival of the kindest: A simian Samaritan shows nature's true heart" (August 22, 1996).

22. Kendrick, Levy, and Keverne 1992.

23. Honig 1994.

24. Porter 1991; Fleming, Corter, and Steiner 1995; Formby 1967.

25. "Switched at birth," *USA Today* (April 17, 1998), 2A.

26. Silk 1990; Hrdy 1976.

27. "Kidnaping" resulting in starvation of the infant has been reported for various species of cercopithecine monkeys (rhesus and Japanese macaques, baboons, and guenons). Reports are very rare, but it is also rare for mothers in these species to permit another female to take her new infant. When I first identified the phenomenon, Ed Wilson had not yet coined the term *allomother,* so I referred to it as "aunting to death"; see section on "Incompetence, kidnapping and aunting to death" in Hrdy 1976:125–28. For modern reviews of "allomaternal abuse" and kidnaping see Silk 1980 and the up-to-date overview of allomaternal care and abuse of infants by Nicolson (1987).

28. The "freedom to forage hypothesis" was first proposed in Hrdy 1976. Subsequently Whitten 1983 showed that wild vervet monkey mothers do obtain feeding benefits from turning over their baby to a group mate. See also Fairbanks 1990 for captive vervets; Stanford 1992 for wild capped langurs. In 1997 Mitani and Watts published a comparative analysis across primates showing that allomaternal care is correlated with faster infant growth and more rapid reproduction. See Ross and Maclarnon 1995 for similar finding.

29. Hrdy and Hrdy 1976; McKenna 1979; Borries, Sommer, and Srivastava 1991.

30. Although what looks like hormonal "priming" for motherhood occurs among pregnant monkeys, both in captivity and in the wild, I have been unable to locate any endocrinological research for primates providing definitive evidence that hormones are responsible in the same way that Terkel and Rosenblatt were able to show for rats. For data on wild langurs see Hrdy 1977:198–241.

31. Hrdy 1977:214–27. Long ago I wondered if the mother's milk in infant-sharing species might not be richer than the dilute milk typical of most primates. But when, thanks to Dr. Jill Mellen at the Washington Park Zoo in Portland, Oregon, I was finally able to get samples of milk from hanuman langurs and black and white colobus monkeys, Bo Lonnerdahl of the Department of Nutrition at the University of California, Davis, found that their milk was not significantly higher in protein and lipids than rhesus monkeys' milk—a species with no infant-sharing.

32. Trevathan 1987:59–60; see also Jordan 1985 and 1993. Initial indifference was also reported among half of all first-time mothers in a British sample from the late 1970s by Robson and Kumar (1980). For Machiguenga, see Johnson 1981.

33. Their sample contained 153 primiparous women, and only 40 multiparae (Robson and Kumar 1980).

34. Jordan 1993:107; Trevathan 1987:59. See also Newton and Newton 1962.

35. Initially it was thought that Great Apes did not lick their babies; better observations confirmed that they do. Reviewed in Lindburg and Hazell 1972.

36. Stewart 1984 and personal communication.

37. "They do not wash and bathe a newly born child, but the mother licks it as soon as it is born" (Rockhill 1895:231).

38. Lindburg and Hazell 1972; Soranus (1956 translation).

39. Details of !Kung's birth come from Shostak 1981:194–95.

40. Astonishingly little cross-cultural research has been done on birth in humans, a point made by Wenda Trevathan and Brigitte Jordan, whose books are the best sources of information.

41. Cox 1995.

42. Asch 1968 discusses this theory.

43. Pinker 1997:444; Hazen 1996.

44. Fleming et al. 1988.

45. For example, see case study presented in Ahokas, Turtiaien, and Alto 1998.

46. See Howell 1979 and discussion in chapter 8 ("Family Planning Primate-Style").

47. The average age for mothers suffering psychotic reactions associated with childbirth is twenty-eight (Herzog and Detre 1976).

48. The lactational aggression hypothesis was proposed by a team of psychiatrists in Italy (Mastrodiacomo et al. 1982–83, cited in Numan 1988:1607).

49. Mastrodiacomo et al. 1982–83.

50. Fleming et al. 1990.

Chapter 8. Family Planning Primate-Style

1. Participants included 367 women and 180 men of high social status; and 1,454 women and 877 men of low status, the mean age was around eighty. No comparable relationship between children and lost teeth was found for twins who were men (Christensen et al. 1998).

2. Lancaster and Lancaster 1983.

3. Charnov and Berrigan 1993; Galdikas and Wood 1990.

4. As always, there are a handful of exceptions, species of primates where other group members continue, like humans, to provision immatures after weaning. Tamarin dads, for example, provide weanlings the occasional grasshopper, while bonobo allomothers may also provide food to the offspring of a female they are friendly with. A common chimpanzee mother may allow offspring as old as eight years of age to cadge an occasional termite or nut meat. Much of the "sharing" occurs while youngsters fumble alongside their mothers, beginning to master the art of poking stripped twigs into termite holes to fish for them, or cracking open hard-shelled nuts. None of this, however, is provisioning on the same scale human children require.

5. Kaplan 1994:760.

6. Kalish 1994.

7. Berkson 1973; Fedigan and Fedigan 1977.

8. Daly and Wilson 1988.

9. *Never* is a risky word to use about behavior. But in years of doing research in this area, I have not heard or read of any exceptions apart from those found among captive primates. It is well known that monkey mothers deprived of opportunities to develop socially, or to learn to mother, can be abusive to their infants. For some of the most extreme cases see Harlow et al. 1966. A well-studied but poorly understood case of abuse by a wild-born but now captive Japanese macaque mother at the Rome Zoo comes closest to being an exception. The lethally abusive treatment of her infants by this macaque mother has been described (Troisi et al. 1982). A sample of "abusive" mothers has been identified at the Yerkes Primate Center in Atlanta (Maestripieri 1998). However, these rhesus macaque mothers—mostly low-ranking or of unusually anxious temperaments—seem more nearly overly restrictive and controlling than murderous. Dearth of reports of abusive mothers in the wild could mean either that abuse does not occur, or that the abused infant dies before observers have an opportunity to document it. Given the hundreds of thousands of hours that primates have been studied in the wild, however, the latter is unlikely.

10. For maternal preference for her own over a borrowed infant see Hrdy 1977 (chap. 7). For quantitative evidence on "cost" of being a twin and maternal preference for own over adoptive infant twin, see Ellsworth and Andersen 1997. Anecdotal cases involving twins are relevant but difficult to interpret, for example the abandonment of the sicker of two twin daughters born to a wild gorilla mother (Watts and Hess 1988).

11. Endangered in the wild, *Saguinus oedipus* breeds well in captivity. The following data are for 659 infants born over two decades at the New England Primate Center. For some pairs, one or both parents already had caretaking experience. For them, what mattered most was the availability of help. Of 65 liveborn young who had older siblings present to help rear them, only 12.3 percent were rejected. But if no older siblings were present, 57.4 percent of 148 liveborns were rejected (Johnson, Petto, and Sehgal 1991). Abandoned marmoset and tamarin infants have also been observed in the wild, but no rates are available (personal communication from Leslie Digby).

12. I am not counting the benign, and nearly universal, maternal practice of threatening—or scolding—an older sibling when a younger one squeals, or when an older, weaned juvenile attempts to displace its infant sibling on the nipples.

13. In the langur case, the attacker was known to be the mother's new consort; in the gorilla case, the researcher David Watts suspected that he was. Watts has observed eight cases where young gorillas (aged 2.8 to 4 years) were left behind when their mothers transferred to groups containing new males. In almost all cases, the young gorilla attached himself to a brother or a former consort of its mother (Watts 1989; also "Karisoke orphans," n.d., courtesy of D. Watts). See Moore 1985 for additional accounts of mothers leaving nearly weaned offspring with all-male bands. For prepubescent female soliciting male see Hrdy 1977:269–71, 278.

14. Henry Harpending and colleagues at the University of Utah use patterns of gene differences among contemporary humans to reconstruct "ancient demographic events" (Harpending et al. 1998). There is some evidence that sub-Saharan African populations may have begun to expand within Africa prior to the diaspora "out of Africa"; see Relethford 1998.

15. Gene Hammel (1996); see Cohen (1995) for current growth rates.

16. Darwin 1874:586; Birdsell 1968. See discussion and wider review in Lee 1979:317–20.

17. Hrdy 1992:table 1; Howell, 1979, chaps. 3–5; Scrimshaw 1984.

18. Frisch 1978; Huss-Ashmore 1980.

19. Angier 1997.

20. Goodall 1986:81, 443. Some have argued that adolescent subfertility also prevents a girl from giving birth before her pelvis is developed enough to permit safe passage of the baby. However, this is unlikely to explain the occurrence of adolescent subfertility among chimps, who do not suffer from the tight squeeze giving birth that human females do.

21. Wrangham (1993:55) estimates that on average, a female chimpanzee will copulate around 6,600 times in her life, and that around 60 percent of this sexual activity will occur during the recurring cycles before she conceives for the first time—an estimated 3,600 copulations. For age of first pregnancy, see Wallis 1997.

22. Tutin 1975.

23. For Nepal and highland Papua New Guinea see Wood 1994:37–38. In one of the most careful of such studies, Strassmann and Warner (1998) used detailed interviews and hormonal profiles to document peak fecundability between ages 26 to 29 among married Dogon women in West Africa. The key variables were age, parity, and breast-feeding status.

24. It is difficult to detect early spontaneous abortions among wild primates. But in captivity, young females appear more prone to abort than parous mothers (e.g., Graham 1970 for chimps). For general discussions see Lancaster 1986; Anderson and Bielert 1994.

25. Dahl 1998.

26. Rates of teenage pregnancy in the United States are dropping from the highpoint in 1957—partly because of abstinence campaigns, but especially as a result of new, easily used, and reliable contraceptives such as Norplant and DepoProvera (Lewin 1998).

27. Overpeck et al. 1998:1215 (and references therein).

28. Danker-Hopfe 1986.

29. Surbey 1990, 1998.

30. Draper and Harpending 1982. For developmental aspects of this hypothesis see Belsky, Steinberg, and Draper 1991. For the classic paper on the relationship between income predictability and promiscuity in humans, see Weinrich 1977.

31. Belsky, Steinberg, and Draper 1991; Walsh 1998; for the quote see Rossi 1997.

32. For discussion see Moffitt et al. 1992.

33. Bugos and McCarthy 1984; Daly and Wilson 1988: 62-63.

34. For overview see Lancaster and Hamburg, eds., 1986. A recent series of papers by sociologist Arline Geronimus (e.g., 1996) analyzes some of the benefits of giving birth early to women with poor employment or marriage prospects. Among other things, these teenage mothers are more likely to have mothers and grandmothers still alive to help them.

35. Since vervets are infant-sharing monkeys, even first-time mothers would have had opportunities to practice mothering prior to birth, whether they gave birth early or at an age more usual for wild vervets (Fairbanks 1995; Fairbanks and McGuire 1995).

36. For how one gets from simple, predictable interactions between a couple of monkeys to the full-fledged cercopithecine system, see the experiments by Canadian primatologist Bernard Chapais (1988; Chapais et al. 1991).

37. Hrdy 1981a:111–12. For a well-documented case study see Fairbanks and McGuire 1986; Fairbanks 1988. Fairbanks's study of vervet monkeys showed that a primiparous female's risk of losing her first baby was always high, but far higher if her mother was no longer in the group to offer support. This grandmaternal effect was most pronounced if the daughter was very young at the time.

38. This pattern is documented in North and South America and in Africa. See Lee 1979; Hawkes n.d; Kroeber 1989:203, 242ff; Voegelin 1942; Hill and Hurtado 1989.

39. According to Murdock (1967), 71 percent of cultures worldwide are patrilocal—so the wife leaves her natal place to live with husband's kin. Among contemporary foraging societies living more nearly as our Pleistocene ancestors would have, however, roughly half (56 percent) lived in patrilocal groups; most others are matrilocal or bilocal—moving opportunistically between his kin and hers (Ember 1978). This statistic is obtained by deleting hunting and gathering people who use horses, or live primarily on fishing from boats—recent innovations of the past ten thousand or so years. For Aché case, see Hill and Hurtado 1996:234ff.

40. Claus Wedekind and colleagues at the University of Bern, Switzerland, are pursuing this research on histocompatibility complexes; see Furlow 1996.

41. Wollstonecraft 1978:315. Sadly and ironically Wollstonecraft herself died of puerperal fever (or, "milk fever") in 1797 shortly after giving birth to her only child.

42. Lewis 1986:212ff, n. 63.

43. Ellison 1995.

44. Ad from Pocket Microscope Company, *Harper's Bazaar,* October 1997, p. 132.

45. For a brief introduction to the physiology of this feedback loop see Vitzthum 1997. For a history of how we came to understand the role of breast-feeding in fertility control see Ellison 1995. For more exhaustive accounts, see Wood 1994, and Ellison in press. For detailed examination of the role of cultures, see Panter-Brick 1989; Stallings, Panter-Brick, and Worthman 1994; Vitzthum 1989.

46. Stern et al. 1986.

47. Interview with Peter Ellison, in Ackerman 1987:626; see also Ellison 1995.

48. This model, first proposed by Lee (1979:442) was tested by Blurton-Jones (1986).

49. Blurton-Jones 1993; Hill and Hurtado 1989.

50. Lancaster 1978; Klein 1992; for division of labor in chimps, see McGrew 1979.

51. It has long been assumed that the demographic repercussions of the Neolithic included faster rates of population growth, but only in the past few decades have demographers had life tables for hunter-gatherers to test assumptions underlying this received wisdom. Eugene A. Hammel (1996) examined demographic repercussions of female and infant mortality and birth spacing in foraging and settled populations. See also Pennington 1996 for the impact of sedentary living on infant and child mortality and population growth; for quotes, see Lee 1979:330-32.

52. In December 1996, Japanese archaeologist Syichi Toyama presented radiocarbon dates for rice grains from sites along the Yangtze River, documenting their cultivation in central China more than 11,500 years ago, even earlier than millet cultivation known from northern China—as early as 7,800 years ago (Normile 1997:309); see also Smith 1997; Bogucki 1996.

53. The earliest fired pottery from the Near East, France, and the Sudan almost always includes the little vessels used to boil grain into a gruel for use as weaning foods (Molleson 1994; personal communication with Theya Molleson, Department of Palaeontology, The Natural History Museum, London, November 25, 1997; Fildes 1986:328-50).

54. According to archaeological wisdom, cultural traits rather than genes diffused as people adopted useful innovations. But in their "demic expansion hypothesis," Ammerman and Cavalli-Sforza (1984) proposed that as agricultural populations increased they expanded into new regions, causing diffusive gene flow between Neolithic farmers and the Mesolithic groups they encountered at the same time that they changed lifeways and language. This hypothesis is now generally accepted for Europe (Jones 1991).

55. The longest naturally occurring interval between births of wild Great Apes is eight years (Galdikas and Wood 1990). Human birth intervals are shorter, but as Pennington (1996) points out, the correlation between birth interval and lifestyle is variable. I follow demographer Eugene Hammel (1996): in general, there *is* a tendency for birth intervals to become shorter as nomadic people became sedentary. See also Campbell and Wood (1988); Kaplan, Hill, Hurtado, and Lancaster in prep.

56. Lummas et al. 1998.

Chapter 9. Three Men and a Baby

1. Eliot 1854, reprinted 1990a:8.

2. Eliot 1861:168.

3. 1996. "Gorilla cradles injured child." *Laboratory Primate Newsletter* 35(4):9, Primate Behavior Lab, Psychology Department, Brown University.

4. A lovely example would be the enormous silverback gorilla male who sleeps nestled next to an adopted infant, the only other gorilla ever allowed to share his sleeping nest (Stewart 1981). David Watts records eight cases when gorillas aged 2.8 to 4 years were left behind when their mothers transferred to other groups. Almost all "orphans" were able to attach to an older brother or to a former consort of the mother (Watts 1989 and n.d.).

5. Absence of a fos-B gene, for example, has the same obliterating effect on male retrieval of young that it has on the mother's. This suggests that the wiring for parental care (in this case, retrieval) is similar in both sexes (Brown et al. 1996).

6. From data collected among the !Kung by Mel Konner, cited in Hewlett 1992 (table 2).

7. Wilson 1978:129−32.

8. See Mendoza and Mason 1986 for the full story of division of labor in these little South American monkeys belonging to the genus *Callicebus*. The figure 93 percent is for the first two weeks of life. For an introduction to a vast literature on paternal care, see Whitten (1987) for primates, and Hewlett (1992) for humans.

9. Mason 1966.

10. There is a rich literature on these "special friendships": see Ransom and Rowell 1972; Strum 1987; and Smuts 1985. Smuts added an extra twist to the baboon story. She proposed that, sometimes, male baboons forged "special friendships" with females and looked out for infants, not because her infant might be his, but because his intimate relationship with the mother improves a male's chances of mating when the mother weans her infant and becomes fertile again.

11. The savanna baboons at Moremi are "Chacma baboons" (*Papio ursinus*). But the differences between these baboon populations and those studied in Kenya are probably due as much to the different habitats as to these "strain" differences (Palombit, Seyfarth, and Cheney 1997). As with langur monkeys, or house mice, the amount of intraspecific variation is enormous. The terminology and basic concepts in animal behavior are really not well suited to describe it. It is our failure to have recognized the extent of intraspecific variation in "weedy" and highly adaptable species like these that led anthropologists for so long to imagine that intraspecific variation was uniquely human. It is not.

12. For pronghorns see Byers, Moodie, and Hall 1994; for galagos see Lipschitz 1992; for reviews see Hrdy and Whitten 1987; Wallen 1995. This topic has been recently summarized in a highly readable account by Jared Diamond (1997a: chap. 4).

13. For a case history see Hrdy 1977:137ff.

14. Alexander Harcourt et al. (1981) plotted average testes size for all primate species for which they could obtain the information on one axis, and number of breeding males per female on the other axis; the resulting correlation allowed them to reliably distinguish primates with one-male breeding systems (like monogamous titi monkeys or harem-dwelling gorillas) from those most often found in multimale ones, like chimps or savanna baboons, where a number of adult males reside permanently in the troop and a female in estrus mates with many or all of them within a period of a few days. Even the seeming outliers turned out to be more nearly exceptions that confirm the rule. Tamarins, for example, had testes way too big for their traditional designation as monogamous primates. Then new field studies revealed that tamarin females in fact breed with multiple males—who all thereafter help care for offspring. For mammals generally see Kenagy and Tromulak 1986.

15. Parish and de Waal 1992; Dahl 1985; Parish 1994, 1996; de Waal and Lanting 1997. For food-sharing I relied on Parish and Voland 1998.

16. Stern and McClintock 1998.

17. Matteo and Rissman 1984; Worthman 1978, 1988.

18. Measurements were made by asking volunteers to wear pedometers (Morris and Udry 1970). Hampson and Kimura (1988); Furlow (1996) reviews emerging evidence that women may be using smell to assess certain immune properties in males related to the major histocompatibility complex (MHC) and to use MHC cues to avoid mating with close kin.

19. Worthman 1978. For more on primate origins of human female sexuality see Wallen 1990; Small 1994; Hrdy 1997. Human generalizations here are informed by Adams et al. 1978; Grammer 1996; Matteo and Rissman 1984; Slob, Ernste, and van der Werff ten Bosch 1991; Stanislaw and Rice 1988.

20. For the old view see Morris 1967; Pugh 1977:248. For reviews of some of the research that convinced primatologists that female orgasms occur in other primates, see Slob, Groeneveld, and van der Werff ten Bosch 1986; Slob and van der Werff ten Bosch 1991.

21. Masters and Johnson 1966; Baker and Bellis 1995.

22. For example of disfigurement ("nose-biting as retribution for adultery"), see Okimura and Norton (1998). For role of sexual jealousy in wife-beating and spousal homicide see Daly and Wilson 1988: chap. 9 ("Till Death Do Us Part").

23. Clark and Hatfield 1989.

24. Kenrick et al. 1997.

25. Researchers used both subjective (interview) and objective (temperature of the woman's labia) methods to measure arousal (Slob et al. 1996); see also Slob, Ernste, and van der Werff ten Bosch 1991.

26. Kleiman and Malcolm 1981; Taub, ed., 1984; Katz and Konner 1981:181; Hewlett 1992.

27. Among the Aka, one-fifth of all infants born die before the end of the first year, mostly due to infectious and parasitic diseases (Hewlett 1992:161).

28. Soffer et al. 1998. However, all such dates are hard to interpret and, even if correct, can mean only that net hunting was at least that old.

29. Barnett and Rivers 1997; Greenberg and Morris 1974.

30. For case study see Hames 1988. For a general overview on the topic of fitness tradeoffs see Hill and Kaplan 1988.

31. Cited in Hodder 1996.

32. Kaplan et al. 1984 for Hadza; Marshall 1976 for !Kung; Hawkes 1991; Hawkes, O'Connell, and Blurton-Jones n.d.

33. For more on risk in stochastic environments see Winterhalder 1986; Cashdan 1985. For more on the "selfish origins" of sharing, see Blurton-Jones 1984; Moore 1984.

34. For an up-to-date summary of a vast literature on the meat-for-sex exchange in both nonhuman primates that hunt (primarily savanna baboons and chimpanzees) and especially across human hunting-and-gathering societies see Chris Knight's 1991 book on how mothers through human history and prehistory have attempted to force men to hunt for them.

35. Biesele 1993:1. "Women like meat" is the title of Megan Biesele's lively account of these themes as they play out in the folklore of the Kalahari Ju/'hoan people.

36. The study was undertaken by the London ad agency Lowe Howard-Spink and the results reported by E. S. Browning in the June 13, 1993, *Wall Street Journal*. Even though (as readers by now will realize) I don't believe there ever is a gene "for" anything, I thought the title was amusing: "Somewhere in Every Man's DNA There May Be a Gene for Sirloin."

37. I rely here on material summarized by Judith Bruce, The Population Council, New York; see especially Bruce 1989:985. Elsewhere, Dwyer and Bruce (1988) discuss conflicting interests within the household as it pertains to family planning and child well-being. According to current statistics from the U.S. Department of Health and Human Services, parents shirk court orders to pay child support in four out of five cases, and custodial parents (and this usually means mothers) collect less than 20 percent of what they are owed by the father.

38. For more on universal features of romantic love, see Fisher 1992.

39. Shostak 1981:175 and 197.

40. Although a topic outside the scope of this book, there are actually data for modern Americans suggesting that economics, income level, and life prospects are better predictors of marriage duration than professed ideas about family values or religion are (Whelan 1998).

Chapter 10. The Optimal Number of Fathers

1. Aberle 1961:680. For best available overview see Schneider and Gough 1961.

2. Shostak 1981:211.

3. Hill and Hurtado 1996:434.

4. Daly and Wilson 1988, 1995.

5. For a balanced report of both claims and critiques about "the evolution of child abuse" see Wray 1982. More recently, I have been dismayed to learn of proposals to funding agencies to use monkeys as models for the "evolution of child abuse." The money would be far better spent on daycare and other social services for families in need.

6. Daly and Wilson (1980; 1995) attribute the man's indifference to "discriminative parental solicitude," which is another way of summarizing at the psychological level the outcome of Hamilton's rule (discussed above, page 62–65). They treat this topic in depth in their classic monograph on human homicide in evolutionary perspective (Daly and Wilson 1988).

7. Daly and Wilson 1980.

8. For a review of this literature see Wrangham and Peterson 1996.

9. Most scholars who adopt this position are loath to mention it, but Herbert Spencer was among the first to arrive at this conclusion, and a surprising consensus has emerged among Darwinians, Marxists, feminist historians, cultural ecologists, and classicists that raiding for women was a primary objective of primitive warfare. The fact that Spencer was prescient on this point does not of course mean that we have to agree with him on every point.

10. Darwin 1874:556–57; Engels 1884:126. For a sociobiological interpretation in the same vein see Ridley 1993:205.

11. Chagnon 1972; quotes from Elena Valero in Biocca 1971:34–35.

12. For chimps see Hiraiwa-Hasegawa 1987; Hiraiwa-Hasegawa and Hasegawa 1994. For statistically significant bias in number of male victims in langurs see Sommer 1994.

13. Wrangham and Peterson 1996; "Do humans have evolved homicide modules—evolved psychological mechanisms specifically dedicated to killing other humans under certain contexts?" asked evolutionary psychologist David Buss on John Brockman's "Edge" Web site. Cited in a December 30, 1997, *New York Times* article entitled "In an on-line salon, scientists sit back and ponder"; Buss and Duntley 1998; Duntley and Buss 1998.

14. Amy Parish 1996. See also de Waal and Lanting's beautiful book on bonobos (1997).

15. For data on genetic relatedness see Goodman et al. 1997. For discussion of the relation between short male tenure time and the evolution of infanticidal behavior in males, see Hrdy 1974. For male-biased infanticide in langurs see Sommer 1994.

16. The anthropologist William Irons has argued that when men pursue culturally defined goals, what has been selected for is not specific behaviors but a general desire to strive for recognition, prestige, control, and well-being within the social context that the individual grew up in, among people who share the same values (Irons 1979:258).

17. This famous study, titled "International preferences in selecting mates: A study of 37 cultures" (Buss et al. 1990), is cited by evolutionary psychologists as evidence that women universally value men's earning capacity more than men value women's, and that women have an innate, species-typical capacity to seek men with resources. Since most of the data came from interviews with university students in countries or cultures where a wife's well-being and status depends on her husband's earning capacity, the result is scarcely surprising. Nevertheless, financial prospects consistently ranked below measures of reliability. For any reader who has trouble locating the published study, Boyd and Silk (1997:645–47) provide a summary and reanalysis.

18. Hill and Hurtado 1996:442; Beckerman et al. 1998.

19. For tradeoffs between fidelity and promiscuity from the point of view of an Aché woman see Hill and Kaplan 1988: esp. 298–99; Crocker and Crocker 1994:32, 83–84; for survival data see Hill and Hurtado 1996:444.

20. Crocker and Crocker 1994. Even though this particular ceremony no longer involves group sex, all these tribes are at appallingly high risk of AIDS.

21. Wife-sharing is reported for one-third of all human cultures (Broude 1994:334). It is most often mentioned by Africanists in the context of "sexual outlets" for unmarried younger men. But group cohesion and shared protection and provisioning of offspring may also be relevant. See Schapera 1933; Middleton 1973; LeVine and LeVine 1979; Llewelyn-Davies 1978; Leakey 1977, vol. 2:810; Temple 1965:103; Wiessner 1977: 359–60.

22. For the Canela case see Crocker and Crocker 1994:186; for instability of matrilineal social systems after contact with other groups, see the essays in Schneider and Gough, eds., 1961.

23. Engels 1884:54–55; Boster et al. 1999; Baker and Bellis 1995:200.

24. Scheper-Hughes 1992; Schuster 1979; Guyer 1994. Some experts on African family life have proposed an alternative explanation based on cultural diffusion. They view mother-headed households and reliance on kin networks to help rear children as "signs of an African heritage" (e.g., Miller 1998). Obviously if unpredictability of resources provided by fathers also pertains in the West African cases, these two explanations are not mutually exclusive. Also, cultural diffusion offers little help explaining why people with the same cultural background set up nuclear families when husbands get better jobs. Unquestionably, the unpredictability of resources provided by fathers is a factor. See Ilsa M. Glazer Schuster's *New Women of Lusaka* (1979). This is part of an important emerging literature on the lives of women, especially poor women, in modern third world countries. For entry into it, with special reference to the social strategies of mothers, see the extraordinary account of women's lives in the shantytowns of Brazil by Nancy Scheper-Hughes (1992). The quote is from Schuster 1979:9. For more on "Female-headed families by choice and default" see Batten 1992.

25. Graglia 1998.

26. Diamond (1997a:281ff.) provides readable overview.

27. Witkowski and Divale 1996. Today, matrilineal societies are found mostly in Africa, the Pacific, and in tribal South and North America. Half of them are still also matrilocal, and 56 percent are horticultural. Matrilineal societies are rarely found among pastoralists, or where extensive irrigation systems or plows are used. As Aberle (1961:680) puts it: "the plow is the enemy of matriliny, and the friend of patriliny."

28. Robert Trivers interviewing Irv DeVore for *Omni* magazine in 1993.

29. See Emlen and Oring 1977 for overview of original nonhuman models; for application to humans see Hartung 1982 and Borgerhoff Mulder 1990.

30. Gamble 1894:72.

31. Strassmann 1993, 1997 and personal communication (1998).

32. Costs to wives of being polygynously married to the same man are complicated by wives having different rights. First, or "senior," wives often have customary or legal rights that give them first access to family resources, or the right to extract labor from women their husbands subsequently marry. See Isaac 1980 and especially Borgerhoff Mulder 1992b.

33. Strassman and Hunley 1996. Similar accusations of witchcraft against cowives are also documented for East Africa (LeVine 1962).

34. The best documented example comes from Australian Aborigines studied by anthropologists James Chisholm and Victoria Burbank (1991). Monogamously married women became pregnant more often, gave birth to more infants, and their offspring enjoyed higher survival rates than did polygynously married women. Detrimental effects of polygyny on motherhood were ameliorated, however, when the cowives were sisters.

35. Tim Clutton-Brock, Geoffrey Parker, Patricia Gowaty, Barbara Smuts, and Robert Smuts have begun to explore the implications of male coercion on female freedom of movement, foraging, and mate choice in a range of birds and mammals (Smuts and Smuts 1993; Clutton-Brock and Parker 1995a, 1995b; Gowaty 1996, 1997; Hrdy 1997). But there are also human populations in which females sequestered first in their father's home, and then in seraglios, have never had anything approaching a choice among their mates. When this goes on for many generations, are there genetic consequences for men? For women? We have not the slightest idea.

36. In Crocker and Crocker's (1994:37ff) account of the Canela, complete nudity still requires fixed standards of modesty. Women must keep knees together so as not to expose the labia minora; a man would be disgraced if any other man ever saw his glans penis; and so forth.

37. Chapman 1982 and 1997 for Selk'nam; cf. Blaffer 1972 for *h'ik'al*.

38. Cited in Reeder 1995a:126.

39. Reeder 1995b:299; for descriptions of lion matings see Bertram 1975. Lions have a breeding system similar to langurs except that males take over groups as packs of brothers, and all the brothers remain subsequently. Male lions are exceedingly infanticidal (Pusey and Packer 1994). Hence, like primates, female lions may have evolved polyandrous mating with multiple males to confuse paternity and discourage infanticidal impulses.

40. Stewart 1995; Reeder 1995b:300. The quote is from the nineteenth-century medical authority William Acton, cited in Stone 1977:676, n. 6.

41. Mann 1997:26; see also pp. 27–29, 56.

42. These words are spoken by Princess Alcharisi to her son, Daniel Deronda, as part of her explanation for why she gave him away as an infant (Eliot 1876:694).

Chapter 11. Who Cared?

1. Darwin 1874:778.

2. For example, Lovejoy 1981 on "The origin of man."

3. Comparisons of fossilized teeth from early hominids with those of chimpanzees indicate that the "delayed maturation" often associated with tool use and division of labor does not emerge until fairly late in hominid evolution; there is, for example, no evidence for delayed maturation from teeth of australopithecines (Conroy and Kuykendall 1995). For *Homo erectus* see Smith 1993; Walker and Shipman 1996: chapter 9.

4. Hill and Kaplan 1988:282–83; Hawkes, O'Connell, and Blurton-Jones n.d.

5. According to the best available estimates, the larger *Australopithecus afarensis* morph, assumed to be male, weighed around 45 kg. The smaller morph, assumed to be female, weighed 29 kg (McHenry 1992: table 1).

6. See Tronick, Morelli, and Winn 1987; Gillogly 1983.

7. At this point some readers might enjoy additional verses from Trivers's anonymous student-author of "Allomothah . . .":

I suppose that Cousin Mabel
Did the best that she was able.
She's not really grown up either;
I can't blame her that she threw me at a cheetah.

Take me back, my darling mother
Take me back; I want none other.
You're such fun, you always make me laugh;
And besides, our relatedness is half.

8. Greaves 1996; Gragson 1989: table 6.9.

9. For analysis of !Kung, Hadza, Aché, and Hiwi data see Kaplan 1997. See also Blurton-Jones et al. 1997.

10. Hilton and Greaves 1995; personal communication from Rusty Greaves, University of New Mexico.

11. For elephants see Macdonald, ed., 1984:457. Age-specific pregnancy rates recorded for pilot whales showed that none of 76 slaughtered females older than age 36 was pregnant (Marsh and Kasuya 1986). Kim Hill and Magdalena Hurtado (1997) review and evaluate such evidence, including evidence from hunter-gatherers.

12. For endocrinological data on nonhuman primates see Graham 1986; Walker 1995; Caro et al. 1995. Although we don't have the same hard physiological measures for wild monkeys, that female baboons, mangabeys, and macaques quit cycling around twenty-four years (Hrdy 1981b; Packer, Tatar, and Collins 1998). Some argue that reproductive changes in other primates are more idiosyncratic than among humans, but I am far more struck by the similarities between humans and chimps. Not all wild chimps reach menopause before they die, but they would if they lived long enough. Menopause in women, although guaranteed to happen, is also not invariant in time of onset. For example, in a large sample of American nurses not taking hormone replacements, 8 percent still occasionally menstruate at age fifty-five (Alice Rossi, personal communication, August 4, 1997).

13. Packer, Tatar, and Collins 1998; Sherman 1998.

14. Williams 1957.

15. Reviewed in Sherman 1998; Nesse and Williams 1994. Another version of the "stopping early" or "grandmother hypothesis" proposed that menopause itself evolved so that old females would be free to help their kin. So far it has received little support. Based on work by Kim Hill, Alan Rogers, and Magdalena Hurtado, most researchers in this area have been dissuaded from continuing to argue that inclusive benefits to grandmothers from helping kin might be sufficient to cause females to cease to produce offspring themselves in order to devote themselves to offspring of others. That is, women *may* be doing this, but this is probably not why menopause evolved. For the first attempt to test this idea empirically, see Hill and Hurtado 1996:427–34.

16. Amundsen and Diers 1970. For admirable overviews, see Pavelka and Fedigan 1991; Lancaster and King 1992. In a provocative early review Weiss (1981) argued that menopause in humans is an artifact of how much longer women live today. Although it is true that *more women* *today* live longer, demographic data from foraging peoples, as well as fossil evidence, indicate that some of our ancestors lived long lives as well.

17. For East Africa see LeVine et al. 1996:25–26, 110; for village India see Miller 1981: table 6.

18. Hrdy and Hrdy 1976. Borries, Sommer, and Srivastava (1991) have confirmed this pattern for langurs at Jodhpur.

19. Packer, Tatar, and Collins 1998.

20. Fairbanks 1988; Fairbanks and McGuire 1986.

21. Note that this as yet unsupported hypothesis is distinct from the still viable, if unproven, hypothesis whereby a prudent mother, confronted with inevitable senescence of her ovaries, is selected to stop cycling early in order to increase her chances of pulling her infant through. Unfortunately, though, these two quite different propositions both get referred to in the popular literature as "the grandmother hypothesis." The muddle is aggravated by the fact that still a third proposition is also referred to under the catchall phrase. This third version is not being used to explain menopause per se, but to explain why females in species such as humans go on living so long after menopause.

Hence, in a recent news account, zoologist Craig Packer was widely quoted to the effect that "menopause isn't adaptive. It has no function" (see, for example, Gibbons 1998b). Packer's data on baboons and lions suggested that females did not stop reproducing early to help others' infants. However, the same data are consistent with the "prudent mother stopping early" version of the "grandmother hypothesis." Readers were left to figure out for themselves which of the three possible versions of the "grandmother hypothesis" had been debunked: (1) stopping early to help own offspring as well as grandchildren; (2) stopping early to help offspring of others; or (3) given that menopause is inevitable, females live on afterward to help kin. (It was only "grandmother hypothesis" number two that Packer's data excluded.) Readers seeking clarification should see Sherman 1998.

22. Judge and Carey (1998) analyzed available data for female body weights, brain weights, and lifespan across thirty-five anthropoid genera in order to construct a regression that would allow them to project what the lifespan of *Homo sapiens* should be. The number they came up with, 72 years, was substantially longer than the 52–56 years estimated for *Homo habilis* or the 60–63 years for *Homo erectus* (Hammer and Foley 1996). The *cebus* monkey is the only other primate with a lifespan quite so off-scale. Interestingly, lactating female cebus monkeys routinely allow other females' infants to suckle, suggesting that there may be a valuable service older kin can provide (Perry 1996; see discussion below, in chapter 13).

23. For genealogy of oldest Frenchwoman see Robine and Allard 1998; for age of menopause among twins see Treolar et al. 1998 (and references therein).

24. Biesele and Howell 1981.

25. Hawkes et al. 1998: table 1.

26. Of eight women in a sample of postmenopausal Hadza helpers, two were mothers of new mothers, two were maternal aunts, one was a maternal great-grandmother, two were the fathers' mothers, and one was a more distant relative. No nursing mother in the group at the time of the study lacked a postmenopausal helper (Hawkes, O'Connell, and Blurton-Jones 1997). This picture differs from that reported for the !Kung (Shostak 1981:323).

27. Quote from Biesele and Howell 1981: 77–79; data from Hawkes, O'Connell, and Blurton-

Jones 1989 and 1997. Older kinswomen's help in provisioning mothers could explain the tendency toward matrilocality in societies where gathering is especially important. It might also explain a tendency for women to remain near their own kin *where hunting is important.* So surprising was this correlation to anthropologists, who had long assumed that hunting must be correlated with male philopatry and bands of males sticking together as "men in groups," the first response was to discount correlations between subsistence and residence patterns as not meaningful. Hence in a recent review of residence patterns among hunter-gatherers, Witkowski and Divale write: "When hunting is especially important . . . this primarily male activity does not lead to patrilocality as expected; instead there is actually a weak tendency toward matrilocality. This suggests that the modest association between subsistence contribution and unilocal residence in food collecting societies is not a meaningful one" (1996:674).

But there is another possibility, namely that correlations between subsistence and residence patterns are indeed meaningful, but that they tell anthropologists something they had not expected: that whenever they can, women remain near their female relatives because in foraging societies a mother's mother and sisters are more reliable allomothers than the father and his kin are. The fact that in non–hunter-gatherer, post-Neolithic societies patrilocal residence has become the rule (not counting large-scale modern societies in which people do not necessarily reside near kin) may reflect the fact that women started to have less say over where they lived. See Witkowski and Divale 1996 for general overview.

28. Amoss and Harrell 1981; Hill and Hurtado 1996:54, 156–57, 236–37; Sharp 1981.

29. Hill and Hurtado 1996:236–37.

30. Statistics from the U.S. Census Bureau, cited in Gilbert 1998a:B8; for front-page headline, see De Parle 1999; for case studies, see Stack 1974.

31. Charnov and Berrigan 1993; for full technical treatment of the "invariant" rules of life histories see Charnov 1993. For application to primates, I would recommend Harvey, Martin, and Clutton-Brock 1987. For the best available demonstration of how to test hypotheses about human life-historical traits, such as reproductive senescence, see Hill and Hurtado 1996 (esp. chapter 13: "Kin effects on life history").

32. Kaplan 1997; Hawkes et al. 1998.

33. For nutritional status of *Homo erectus* see Wrangham et al. in press. Hawkes's quote was a personal communication, September 7, 1998.

Chapter 12. Unnatural Mothers

1. Wille and Beier 1994.Psychiatrist Philip Resnick proposes that the term "neonaticide" be used to distinguish "the killing of an unwanted neonate within the first few hours of life" from "filicide" (the killing of child after its role in the family is more firmly established) because the motives and disposition of the mother differ. The latter is more likely to involve psychopathology than the former (Resnick 1970; d'Orban 1979). In the United States being born to a mother under seventeen or being the second infant born to a mother under nineteen are the highest risk factors (Overpeck et al. 1998:1213). For comprehensive treatment see Daly and Wilson (1988), chapters 3 and 4.

2. Eliot 1859:497.

3. The experimental stimulus was the cry of a three-month-old girl receiving a vaccination; the control was an artificial noise at the same sound level (Bleichfeld and Moely 1984).

4. Overpeck et al. 1998.

5. For years 1902–1927, see Hopwood 1927.

6. For example, Pryce (1995) defines mothers with "good maternal attributes" as those who "score highly on extraversion, agreeableness and interpersonal affect and low on neuroticism."

7. Hetty Sorrel is convicted, but her death sentence is commuted to transportation. Her modern counterpart, Melissa Drexler, received a fifteen-year sentence in October 1998.

8. Matthews Grieco 1991:44; Oliver St. John, Esq. F.R.S. in *Philosophical Transactions,* no. 412, November 30, 1731. Cited in Fildes 1986:196. Fildes mentions 4,000 infant deaths attributed to overlaying between 1701 and 1776.

9. Langer 1972:96.

10. Associated Press 1998a.

11. The study appeared in the November 1997 issue of the journal *Pediatrics,* described in Hilts 1997; for details of the Waneta Hoyt case, see Judson 1995; Firstman and Talan 1997.

12. Quote from Linda Norton, cited in Firstman and Talan 1997:64.

13. Asch 1968.

14. From the nursing records for Molly Hoyt, and interview with senior nurse working in pediatric ward of Upstate Medical Center at the time; cited in Firstman and Talan 1997: 472 and 260.

15. Dolhinow 1977.

16. Vogel 1979; Schubert 1982; Aronson 1995; Mestel 1995; Sussman et al. 1995; Hrdy, Janson, and van Shaik 1995; Dagg 1999.

17. See, for example, Bugos and McCarthy 1984.

18. Proceedings of this workshop at Ettore Majoranà Centre were published in Parmigiani and vom Saal, eds., 1994.

19. Rosenthal 1997.

20. Elizabeth Marshall Thomas titled her famous book about the bushmen of the Kalahari *The Harmless People,* providing a cultural stereotype that stuck—even though, as ethnographer Richard Lee points out, the murder rate among the !Kung is about the same as in Detroit. Meanwhile, Napoleon Chagnon's monograph about the Yanamamo of Venezuela, titled *The Fierce People,* gave them a different image, for which Chagnon was much criticized. In fact, though, as Chagnon has been at great pains to point out in some of his writing, and as others like Irenäus Eibl-Eibesfeldt and Mattei-Müller (1990) have stressed, these people can be very caring as well as very fierce. But in terms of how these tribal groups were treated by their neighbors, neither epithet made much difference: their land was appropriated and they became second-class citizens.

21. Wissow 1998. For Erice Statement see Parmigiani and vom Saal 1994: xvi–xvii.

22. Some speculate that Mrs. Hoyt suffered from "Munchausen by proxy" syndrome, where people use the illness of someone close to elicit sympathy for themselves, a then nameless syndrome described in the 1930s in Thomas Mann's fictional "A Man and His Dog."

23. Boswell 1988:4.

24. Boswell 1988:134–35, n. 161, citing *Controversiae* 10.4.10, 429.

25. McLaughlin 1989; Boswell 1988:160.

26. Trexler 1973a.

27. I am indebted to David Kertzer for reviewing and clarifying this information on Italian foundling homes from Kertzer 1993:139.

28. Ransel 1988, and comments by Ransel in Tilly et al. 1992: 21.

29. Part of this high mortality was due to a smallpox epidemic. See Ransel 1988:45–46 (and esp. his appendix) for admissions and deaths for the entire period between 1764 and 1913.

30. Ransel 1988:194–95, and chapter 10 about networks of exchange between town and village.

31. This conference on anthropological and historical studies of child abandonment, organized by Catherine Panter-Brick and Malcolm Smith, was code-named "Nobody's Children," Durham University, September 25–30, 1995; see Panter-Brick and Smith, in press.

32. There is growing literature, beginning with Richard Trexler's account of the foundlings of Florence (1973b). Of particular value are Delasselle 1975; De Mause, ed., 1974; dos Guimaraes 1992; Ransel 1988; Sherwood 1988; Sussman 1982. Several studies trace a gradual improvement in these institutions; see Fuchs 1984; Corsini and Viazzo, eds., 1993; Kertzer 1993.

33. Kertzer 1993:72, 80–81; 141–142; table 1.1 and personal communication from David Kertzer.

34. Scheper-Hughes 1992.

35. Kertzer 1993 (chap. 5) traces the rise and fall of the wheel in Italy; see p. 156 for quote from Lamartine.

36. See Boswell 1988, and especially Kertzer 1993:120–21 for the derivation of the name Colombo; Boswell counted 328 Espositos for New Haven (432, n. 2).

37. See Trexler 1973a for Renaissance Italy; statistics for France indicate that as many as 92 percent of children in some foundling homes died prior to their eighth birthday (Dupoux 1958); Herlihy and Klapisch-Zuber 1985:147.

38. For this exchange from *Le Nouvel Observateur,* see Francine du Plessix Gray's foreword to Badinter 1981.

39. Shorter 1975:168. For discussion see Pollock 1983; Kertzer 1993.

40. Aries 1962; Shorter 1975; Stone 1977. For a more recent overview see Suransky 1982.

41. De Mause 1974: 51–54.

42. Cited by Nancy Scheper-Hughes (1992:401), who says of the reporter: "[he] may have overstated my case. But in effect I suppose it is close enough to what I am saying."

43. Scheper-Hughes 1992:354, 400–401.

44. French social historian Philippe Aries (1962) argued that the concept of childhood has a culturally determined life of its own. It emerged gradually in the West between the Middle Ages and the eighteenth century. Aries does not concern himself with the vast ethnographic literature for other peoples.

45. Reviewed in Badinter 1981.

46. See review of this literature in Scheper-Hughes 1992.

47. For much more on this see Kertzer 1993.

48. See Langer 1972; for a definitive recent treatment see Cohen 1995:42–45.

49. Scheper-Hughes 1992.

50. Shostak 1981:309.

51. As social historian Linda Pollock (1983) cautions, these diaries are written by a select, literate sample, and mostly fathers. Nevertheless, Pollock demonstrates how poorly such accounts support the proposition that parental emotions in the past were qualitatively different from today.

52. Cited in Ross 1974:183.

53. Cited in Ross 1974:198–99.

54. For quotes from Rousseau see Cranston 1997:182–183; see also Badinter 1981:138.

55. Badinter 1981:99–100; Madame d'Épinay, quoted in Lorence 1974.

56. Bugos and McCarthy 1984:512.

57. For New Guinea, see Eibl-Eibesfeldt and Mattei-Müller 1990. The best-documented ethnographic evidence for increasing maternal solicitude with age remains Bugos and McCarthy (1984). Daly and Wilson (1988) use police statistics to document the same pattern in contemporary North America.

58. Feldman and Nash 1986. As discussed in chapter 11, an older mother has a different threshold for investing. David Nathan, a physician at Harvard Medical School, provides a moving account of a woman who has devoted decades to keeping alive a son born with a lethal genetic defect. The mother confides that in her youth, had she foreseen the difficulties, she would have aborted the pregnancy. But—in a different life phase and now past menopause—she says she would opt to go ahead with the pregnancy even if she had known (Nathan 1995). For differences in the way women of different ages mourn lost pregnancies, see Fein 1997.

59. Scheper-Hughes 1992:459.

60. Fuchs 1987.

Chapter 13. Daughters or Sons? It All Depends

1. This headline appeared over Nicholas Kristof's story in the *New York Times* (1991a).

2. Visaria (1967) reported sex ratios in Korea as high as 116 boys per 100 girls, leading some to conclude that Koreans may be genetically predisposed to produce sons. This interpretation received tentative support when Morton, Chung, and Mi (1967) reported higher than average sex ratios among Hawaiian children whose fathers were of Korean descent. Subsequently, Park (1983) proposed that these sex ratios were the outcome of parental decisions to continue giving birth after daughters are born, stopping with sons. Mathematically, however, the probability of producing a son or a daughter at each birth should remain the same. Hence, others argued that instead of "stopping rules," parents were somehow actively intervening through selective termination of female fetuses. There the matter stands—unresolved.

3. Although officially forbidden for use in sex-selective abortion, prenatal sex determination is widely available in China. Ultrasound imaging is the most common technique, but amniocentesis permits parents to learn the sex by the sixteenth week of pregnancy. Results from analyzing placental tissue are available by week eleven. This technique, called chorionic villus sampling (CVS), was pioneered in China, and became available there long before it was in the West.

4. This account derives from published interviews with the demographer Susan Greenhalgh and with William Lavely, a specialist in Chinese demographic trends (Herbert 1994), and an unpublished report presented by Lavely and colleagues at the Universit0y of California, Davis, 1992. See also Johansson and Nygren 1991. More detailed analyses for the lowerYangtze region confirm both missing girls and intensification of the sex preference over time (Skinner and Jianhua 1998).

5. Kristof 1991a; for complete analysis of sex ratios in live births for Chinese census data from 1981, 1986, and 1987, see Johansson and Nygren 1991: table 2.

6. Smaller families mean parents invest more in each child, as described by Kathy Chen (1994).

7. Skinner and Jianhua 1998.

8. See summary of sex ratios among populations of children under sixteen analyzed by demographer Ping-ti Ho (1959:8–13, 56–69, table 14). The high sex ratio of 154 cited here comes from a sample of 158,310 people in Shensi Province in 1783. Although incomplete reporting or fostering out of daughters may have been factors, I follow William Skinner and others in assuming that parental manipulation of offspring sets at birth is the main cause of skewed sex ratios.

9. See Hull 1990 and references in Dickemann 1979a. Rarely did families that already had two daughters allow additional daughters to live (Geddes 1963:12–17; Ho 1959:217ff; Smith 1899:308–9; Gordon-Cumming 1900:134–37:272–76; Martin 1847:48–49; for persistence into the twentieth century see Fei 1939:33–34, 51–53; Lang 1946:150–51).

10. Smith 1899:308–9.

11. Cited in Zhao 1997: n. 13; see also p. 749.

12. Lavely, Mason, and Ono 1992; Skinner and Jianhua 1998.

13. When demographer Ansley Coale calculated the number of women who should have been alive but did not appear on censuses for India, Pakistan, Bangladesh, Nepal, West Asia, and Egypt, his results led to another *New York Times* headline: "Stark Data on Women: 100 Million Are Missing." Some demographers argued that the numbers were too high; other scholars, like Nobel Prize–winning economist Amartya Sen, that they were, if anything, too low (Kristof 1991b); see below, chapter 20.

14. Minturn and Stashak 1982.

15. Quoted in Lewis 1985.

16. Ramanamma and Bambawale 1980:107. See also Miller 1981 for a summary of the anthropological literature on son preference. The phrase "a mania for sons" was introduced by sociologists A. Ramanamma and Usha Bambawale at Poona University in India.

17. There are roughly fifteen bachelors for every ten unmarried women in China (Shenon 1994). For market in women, see Faison 1995.

18. The ratio of male to female suicides in China was 0.8 men for every woman, compared with the United States, where suicide is four times likelier for men than women. (These data, from Michael Phillips at Harvard Medical School, were reported in Neal 1998.)

19. Madan 1965, cited in Ramanamma and Bambawale 1980.

20. This study by R. P. Ravindra, a Bombay social worker, was cited in Rao 1986; see also Ramanamma and Bambawale 1980; Jefferey and Jefferey 1984. A letter to prospective clients from one sex determination clinic specifically addresses itself to "couples in quest of a male child" and specifies that sex determination is done for parents who already have female children (cited in Jefferey and Jefferey 1984:1212).

21. Medical doctors and geneticists in the West are divided over the morality of prenatal sex diagnosis, though some resistance lessened with the introduction of amniocentesis, which permits sex determination in the first 20 weeks of pregnancy (Kolata 1988). Even so, many doctors in countries such as Britain and the U.S. are unwilling to perform prenatal sex determinations if they suspect parents are using the test to select one or the other sex. But it is not illegal (Perera 1987).

22. Burns 1994; Jayaraman 1994; WuDunn 1997; Miller 1981; Das Gupta 1987.

23. Mull 1992 and personal communication from D. Mull.

24. Skinner and Jianhua 1998. I know of no literature analyzing the role of mothers in discriminating against daughters in this extreme way. A few studies touch on parallel topics, such as how mothers feel about performing clitoridectomies on their daughters (see Cloudsley 1983).

25. Wyon and Gordon (1971:235) and especially Dyson and Moore (1983:51).

26. WuDunn 1997. Survival of patriline is put before any individual (Mull 1991; Pettigrew 1986).

27. Scrimshaw 1984:462; Neel 1970; Dickemann 1975. Statements like Scrimshaw's have sometimes been used out of context as antiabortion propaganda (e.g., the recent editorial "Will Infanticide Follow Abortion as 'Acceptable Behavior'?", Morris 1996). For this reason, I want to make it very clear that the scholars I cite seek to study, describe, and explain the world as they find it. Neither I nor any of those I cite advocate infanticide.

28. Kristof 1991b.

29. Levine 1987: n. 11.

30. Caine 1977.

31. For general model see Dickemann 1979a, 1979b.

32. Parks 1975, I:59; cited in Miller 1981:50.

33. Panigrahi 1976, cited in Miller 1981:50–51. Reeves 1971.

34. For strategies of heirship among parents in a French peasant community, see Bourdieu 1976; for African pastoralists, see Mace 1996a and Borgerhoff Mulder 1988 and 1990; for colonial Americans see Hrdy and Judge 1993 and references therein. In a slightly different context, Cosmides and Tooby (1989) analyze the human penchant for evaluating social interactions and contracts—very much part of the strategic game families play with one another over time, where the ultimate "gift" exchanged between families is often a wife.

35. For one of the best available discussions of culture and the evolutionary process, see Boyd and Richerson 1985.

36. Stenseth 1978; usually, an XO individual would be infertile but this is not the case in the XY wood lemming; see Fredga et al. 1977; Gileva et al. 1982.

37. Gosling 1986:784; Morris Gosling, personal communication, October 13, 1987.

38. James 1983; literature reviewed in Hrdy 1987.

39. Clutton-Brock and Iason 1986; Austad and Sunquist 1986; Symington 1987.

40. Alternatively, offspring of one sex may impose fewer costs on parents to rear—the case documented by Ann Clark (1978) for galagos. Researchers studying wild dogs proposed a similar model, which they expanded to include behaviors by sons or daughters that augment the value of local resources (Frame, Malcolm, and Frame 1979). In 1985, Gowaty and Lennartz proposed the term "local resource enhancement" and provided the model used by behavioral ecologists analyzing human sex preferences (Sieff 1990). It is very close to, and often indistinguishable from, the more traditional "rational actor models," long in use by social scientists (Hrdy 1990).

41. Komdeur et al. (1997). For overview and perceptive commentary see Gowaty 1997.

42. Altmann, Hausfater, and Altmann 1988. At roughly the same time that the Amboseli results on sex ratios began to emerge, Silk (1983, 1988) was finding almost identical patterns among captive bonnet macaques. These studies are important because females in the best condition overproduce daughters. Since males tend to be the more vulnerable sex, the fact that deer, possum, and coypu mothers in good condition tend to overproduce sons could be attributed to the fact that sons die off at higher rates when mothers are stressed. Clearly, though, this would not explain cases where mothers in the worst condition produce the highest proportion of sons.

43. Tactics used by high-ranking females to harass low-ranking ones range from knocking babies off nipples to forcibly taking them from their mothers and refusing to give them back to nurse—in effect, kidnaping them. Montserrat Gomendio (1990) attributes longer birth intervals after the birth of a daughter among captive rhesus macaques to such interference. Because higher-ranking females targeted mothers with daughters, their babies were knocked off the nipple. To get fed, they had to reattach, stimulating nipples in the process and causing prolactin levels to rise. Gomendio hypothesized that this is why low-ranking mothers who bear daughters take longer to resume ovulating and conceive again than do low-ranking mothers of sons. For extreme interference leading to the infant's death from starvation, see Silk et al. 1981.

44. Paul and Kuester 1987, 1988; Meikle, Tilford, and Vessey 1984; Rhine, Wasser, and Norton 1988; studies summarized in van Schaik and Hrdy 1991.

45. Fisher 1930. For a clear introduction to the subject see Trivers 1985: chapter 11.

46. Trivers and Willard 1973; Betzig 1995; Chagnon, Flinn, and Melancon 1979; Hill and Kaplan 1988.

47. The case most often mentioned is that of Orthodox Jews. Guttentag and Secord 1983:98; Hrdy 1987:119–23; Harlap 1979.

48. Williams 1979:578.

49. For toys, see Pereira 1994. Growth intervention data provided by Genentech, Inc., in advertising for "Protropin."

50. For explicit statement of this view, see Cucchiari 1981.

51. Heightened discrimination against daughters among the rich can be documented in other Asian contexts. During times of famine, discrepancies in the amount of food given to sons versus daughters become *more* pronounced in rich families, as was recently documented for Bangladesh (Bairagi 1986).

52. When Mildred Dickemann (1979a) first used the Trivers-Willard hypothesis to explain female infanticide in the case of hypergamously marrying Rajputs, she knew how dowries worked at the top of the hierarchy but only guessed how they worked at the bottom, predicting there would be bride-prices. The year her paper was published, Parry (1979) confirmed the pattern.

53. Wherever there are saturated habitats and recurring ecological crises, the landless and those with the smallest holdings are most vulnerable. See Smith 1977:117–25; Low 1991; Hrdy and Judge 1993.

54. Bamshad et al. 1998.

55. For Gypsies see Bereczkei and Dunbar 1997; for Slovakia see Sykora 1998; for the ancient brothel see Faerman et al. 1997.

56. Manzoni 1961:135.

57. Boone 1986; for more on this use of convents, see Hager 1992. Data for 3,700 individuals from the top twenty-five patrilines in Portugal were obtained from the *Peditura lusitana,* a seventeenth-century compilation of genealogies (Boone 1988a, 1988b).

58. See Cronk 1993 for overview on societies that prefer daughters. Cronk does not rule out the possibility that the daughter-biased sex ratios are biased at birth, since a survey he undertook yielded 13 sons and 32 daughters born in the previous year; for update on Mukogodo case, see Cronk 1999.

59. Clark et al. 1995.

60. Dullea 1987.

61. Skinner and Jianhua 1998; differential survival not just by sex but by birth order was first documented for the Punjab by Das Gupta (1987). Muhuri and Preston (1991) for Bangladesh. Numbers are from a 1993 paper by Pradip Muhuri and Jane Menken, cited in Skinner 1997:72.

62. Turke 1988.

63. Skinner 1997:76.

64. Skinner 1993.

65. Turke 1991; see also Turke and Betzig 1985; Alexander 1979:68. Drawbacks from precommitting to sons versus daughters are discussed in Hrdy and Coleman 1982.

66. Gilbert 1998b.

Chapter 14. Old Tradeoffs, New Contexts

1. Paul Robinson (1981) has questioned LeNoir's statistics, but so far as I can tell, LeNoir knew what he was talking about. I follow Bandinter 1981 and Sussman 1982, source of the quote.

2. Sussman 1982, esp. pp. 22–23 for the breakdown of LeNoir's estimates.

3. Cited, with discussion, in Sussman 1982:80 ff.

4. Cited in Sussman 1982:124.

5. For a detailed case study see Kertzer 1993.

6. From an 1865 pamphlet by Dr. Alexander Mayer, cited in Sussman 1982:122.

7. From court records cited in Sussman 1982:123.

8. Eighteenth-century English critics of wet-nursing like William Roscoe also come close to equating wet-nursing with infanticide (Foote 1919).

9. From Mayer's pamphlet, cited in Sussman 1982:122.

10. Piers 1978:52.

11. Smith 1984:64c.

12. Most exceptions involve adoption of relatives (reviewed in Hrdy 1976; Thierry and Anderson 1986).

13. This is the case among cebus monkeys. Very rarely, one sees an older langur infant successfully "pirate" milk in this way.

14. Mothers in cooperatively breeding mammals like mice or lions may combine and communally suckle their litters. But among primates, such shared nursing is only found among litter-bearing, nesting prosimians—never in simian, or "higher," primates. Such allomaternal suckling as can be found in monkeys has been most often studied in cebus monkeys. Since mothers among these South American monkeys don't wean infants until long into their next pregnancy, they are almost perpetually lactating. Typically, older, already mobile infants who find themselves temporarily separated from their mothers demand milk from their mother's relatives—especially new mothers who are subordinate to them (O'Brien and Robinson 1991; Perry 1996). As primatologist Susan Perry points out, it's energy-consuming to refuse. Costs to the lactating female are small compared to benefits for a youngster temporarily separated from its mother.

15. For reciprocal altruism see Trivers's 1971 essay.

16. Fildes 1986:178.

17. "Can Adoptive Mothers Breast-Feed?" (1985 editorial, *Lancet* 2:426–27). For ethnographic examples of induced lactation among grandmothers see Wieschhoff 1940; for additional early European accounts see Fildes 1986:53.

18. Auerbach 1981; Auerbach and Avery 1981.

19. Kleinman et al. 1980.

20. Wieschhoff 1940; personal communication from Napoleon Chagnon for Yanamamo; Roth 1896 (cited in Fildes 1988:266).

21. Auerbach and Avery 1981:341.

22. For Arunta, see Murdock 1934; for Solomon Islanders, Gillogly 1983.

23. Van Lawick 1973.

24. This account comes from Wallis Budge 1925, cited in Fildes 1986:6. Here and elsewhere I rely on the histories of wet-nursing by Valerie Fildes (1986 and 1987).

25. Jasper Mortimer, Associated Press, "Tomb of Tutankhamen's Wet Nurse Found," December 8, 1997.

26. Fildes 1988:3–4.

27. Origo 1986:200–201; Trexler 1973b provides similar accounts.

28. Trexler 1973b:270.

29. Fildes 1986:6–7.

30. Sussman 1982; Klapisch-Zuber 1986.

31. Fildes 1988, esp. pp. 257–60.

32. This was the Duchess of Leinster (Lewis 1986:123–24).

33. See Marilyn Yalom's *A History of the Breast* (1997), esp. pp. 71–73 on Gabrielle d'Estrées.

34. For example, in nineteenth-century Sicily aristocratic families using wet nurses produced seven children, on average, born at two-year intervals, compared with four children born at 4.3-year intervals in poor families (Schneider and Schneider 1984).

35. Assuming that the nurse in the home cost 18–20 fiori annually, compared with 8–15 for a nurse in the country, parents are paying more for sons. The longer stint with wet nurses for those sons sent out often came in the form of a "bonus" period that allowed the infant a more leisurely transition to weaning. Of 283 privileged Florentine children sent out of the house to wet nurses, 82 percent of the boys and 84 percent of the girls survived. It is assumed that the sons kept at home to be nursed did even better. Did the extra payments made for those sons who were sent out help them survive? Not according to these statistics, unless one assumes that boys are more vulnerable in infancy and the bonus milk helped compensate for it (Klapisch-Zuber 1986:136).

36. Badinter 1981:65.

37. For Talleyrand's own sense that he suffered parental neglect see Cooper 1986:12–14, and for his biographer Cooper's sense that he did, see p. 73.

38. For an intimate account of pregnancy avoidance and family pressure among eighteenth-century aristocrats, see Lewis 1986, esp. pp. 212 ff.; for the victimization of women, see Shorter 1982.

39. Postpartum sex taboos can be found among the !Kung and Herrero in southern Africa, the Mende of western Africa, the Yanamamo and Nambiwara of South America, and the Sioux of North America, to name only a few. For more on a widespread custom, see Schoenmakers et al. 1981.

40. The most plausible explanation for decreased family sizes among elites was "status anxiety" as proposed by demographer Sheila Johannson (1987). But there must have been other factors, not yet understood. Some kind of "demographic transition" becomes apparent in Europe between 1750 and 1850 and starts to reduce completed family sizes by the beginning of the twentieth century. Human behavioral ecologists show that people voluntarily reduce family sizes as they opt for economic well-being over producing large numbers of offspring. See especially Barkow and Burley 1980; Luttbeg, Borgerhoff Mulder, and Mangel 1999; Kaplan et al. 1995; Borgerhoff Mulder 1998.

41. For preserved Greek contracts see Pomeroy 1984:133.

42. Christiane Klapisch-Zuber (1986), using diaries and household records to reconstruct family life in Renaissance Italy, shows that wet-nursing was initially confined to prominent families, but after about 1450, began to spread to families of more modest rank, eventually becoming the norm for all but the poorest.

43. Hufton 1974:14.

44. The relevant case study is for Lyons, but this is probably also true generally (Garden 1970:137).

45. Garden 1970:12–16 and tableau VII; 95–97.

46. Delasselle 1975; Sussman 1982:66–67; Hrdy 1992a: figure 1.

47. Data from the National Center for Education Statistics (www.acf.dhhs.gov/programs/ccb/faq/demogra.htm), May 9, 1998.

48. Swarns 1998.

49. According to Kertzer (1993:100–101) women in Sardinia had greater equality of inheritance, broader property rights, and generally more autonomy and higher status.

50. Caldwell and Caldwell 1994. I am indebted to a number of Africanists—especially Bob Bailey, Nadine Peacock, Monique Borgerhoff Mulder, and Dan Sellen—who took the time to discuss these issues. For overviews see Laesthaeghe et al. 1994, and Robert LeVine et al. (1996). They detail the importance of children for status, labor contributions, and ritual significance. At the heart of patrilineal pronatalist sentiments is the preoccupation of fathers with "the possibility that those with *fewer* children will be unable to protect their own interests in competition for scarce resources" (LeVine et al. 1996:107). For more on matrilineal societies, including comparisons with those in Southeast Asia, see Skinner 1997.

51. Draper and Buchannon 1992.

52. I have relied for this characterization on Bledsoe 1994, Caldwell and Caldwell 1990, and Robert LeVine et al. 1996:103−10; Mende quotes from Bledsoe 1993:181 and 1994:115 and 124.

53. Gregor 1988.

54. Howell 1979:119−20.

55. Based on interview data reported in Isiugo-Abanihe 1985, one out of three Ghanian women and 40 percent of Liberian women between the ages of fifteen and thirty had a child living in another household. Forty-six percent of Sierra Leonean women aged thirty to thirty-four had fostered children out. See Page 1989 (table 9.1) for Cameroon, Lesotho, and Ivory Coast; Bledsoe and Isiugo-Abanihe 1989 for general discussion. Slightly lower levels of fostering can be documented for Kenya, Nigeria, and Sudan (Goody 1984; Silk 1987b). Significantly higher survival rates for children with maternal grandmothers were calculated by Mace and Sear (1999) using data collected in rural Gambia between 1949 and 1975.

56. One significant outcome is that the availability of fostering reduces incentives for parents to seek reliable methods for birth control (Bledsoe 1990). Among Herrero pastoralists in South Africa, 50 percent of all children born to unmarried mothers were fostered out, compared with 32 percent of those born to married parents. Often too, a mother will divest herself of children by fostering them out before going to a new husband. This may be why unmarried women under twenty were far more likely to foster out children than were older women (Pennington 1991: table 6). Most commonly, mothers send children to live with their own mothers (René Pennington, personal communication, April 29, 1998). For similar patterns in West Africa, see Isiugo-Abanihe 1985.

57. Researchers from the Department of Social Medicine at the Federal University of Pelotas, Brazil, looked at the effects of bottle feedings on infant survival in a large controlled study of two urban areas in southern Brazil. Breast-fed infants who received even supplemental feedings with formula were at four times greater risk of dying. Risk of mortality jumps to fourteen times greater for infants who receive no breast milk at all (Victora et al. 1989, summarized in Scheper-Hughes 1992:316−17).

58. Bledsoe 1993; Pennington 1996.

59. LeVine and LeVine 1981; Bledsoe and Brandon 1992; Bledsoe 1993 and especially Bledsoe's 1994 essay on the Mende metaphor of "little bamboo shoots."

60. In Uganda on the order of 1.7 million children under the age of fifteen have lost one parent to AIDS. Roughly 40 percent will have lost both parents, and some proportion of the children will themselves be infected either at birth or from their mother's milk. UNAIDS "Report on the Global H.I.V. (AIDS) epidemic" (cited in Daley 1998).

61. Kaplan 1997.

62. *Torn Apart* is the British title of Parker's 1995 book published in the United States as *Mother Love / Mother Hate: The Power of Maternal Ambivalence.*

63. Editorial: "Teen Moms," *Wall Street Journal,* May 8, 1998, A14.

64. Figures on female-headed households from Judith Bruce at the Population Council, New York, and Bruce 1989:988.

Chapter 15. Born to Attach

1.   Personal communication from Dr. Mary Main, March 1997.

2.   For example, Desmond and Moore 1991, or Brown 1995.

3.   Karen (1994) speculates that Bowlby was drawn to Darwin's case by the sense that something on the maternal front might have been missing in his own upbringing. I find Karen's explanation plausible, yet I wonder if Karen, a psychoanalyst himself, underestimated just how large Darwin loomed for Bowlby for purely intellectual reasons, as argued by Darwin scholar Frank Sulloway (1991:30).

4.   Bowlby 1990a:71.

5.   See Sulloway 1991.

6.   Cited by Colp (1977:114), who details Darwin's symptoms. For an account of Darwin's perseverance in the face of these symptoms see Brown (1995:445), who describes the curtain across the alcove in Darwin's study in Down House, supposedly there so that Darwin could continue working even if he felt like retching.

7.   That Darwin took a dim view of "society" is apparent from his autobiography, especially the famous notes to himself on the pros and cons of marriage. One of the merits of celibacy was "Choice of society and *little of it*." I thank Frank Sulloway for referring me to Fabiene Smith's 1990 article about Charles Darwin's allergies.

8.   See Sulloway (1991:30) for Darwin quote.

9.   The term "hyperventilation syndrome" was still considered a valid diagnosis at the time Bowlby invoked it in 1990, but is now passing out of medical usage (see Hornsveld et al. 1996). Nevertheless, the hypothesis that Darwin suffered from some sort of anxiety disorder (what is known today as "panic disorder") which among other things affected his breathing, remains plausible (Barloon and Noyes 1997).

10. Darwin's account was preserved in an early, fragmentary version of his autobiography, cited in Bowlby 1990a:58–60.

11. Darwin 1887:22.

12. Cited in Bowlby 1990a:78.

13. Darwin 1887:22.

14. Raverat 1952:244–45. In his own footnote, Bowlby (1990a:469, n. 17) apparently forgot where he had first read this anecdote and Raverat's interpretation. The note simply reads "Raverat."

15. Blass et al. 1989; Oberlander et al. 1992.

16. Dates mentioned here are deliberately vague, because although I know human birth intervals would have grown shorter, I don't know when, or whether the changes were gradual or even always in the same direction. Just as a wild guess, we can posit that by a million or two years ago *Homo erectus* mothers were too hairless for babies to cling to. The improved efficiency with which *Homo erectus* parents and their helpers were extracting calories from their environment (discussed pp. 285–86) would already mean shorter birth intervals. But the real prelude to the Neolithic probably did not begin until around 50,000 years ago, when paleoarcheologists who study stone technologies and cave art hypothesize that something cognitive and important changed in *Homo sapiens* (see Klein 1992).

17. Quoted in Jessica Benjamin's endorsement for Rozsika Parker's *Mother Love / Mother Hate:The Power of Maternal Ambivalence* in an ad in *The NewYork Review of Books,* April 4, 1996, p. 3.

18. King 1998; fifty gene estimate cited in Wade 1998.

19. Lewes 1877:71–72; cited in Paxton 1991:184.

Chapter 16. Meeting the Eyes of Love

1. See, for example, the August 28, 1995, cover of *Time,* which announces: "the new science of evolutionary psychology finds the roots of modern maladies in our genes."

2. Bowlby 1973:403, cited in Sulloway 1991:29.

3. Karen 1994:21–24. See also Robertson and Bowlby 1952 and James Robertson's 1953 film *A Two-Year-Old Goes to Hospital.*

4. The critique by S. R. Pinneau (1955) and the surrounding storm are reviewed in Karen 1994:120–21.

5. Bowlby 1991:301.

6. Interview with Bowlby, January 14, 1989, cited in Karen 1994:46–47, and videotape of Bowlby lecturing at the University of Virginia, May 1986, courtesy of Mary Main and Erik Hesse; remarks by Anna Freud published in the journal *The Psychoanalytic Study of the Child* following Bowlby's 1960 paper "Grief and Mourning in Infancy" are cited in Karen 1994:116.

7. Bowlby 1991:303.

8. Hinde informed Bowlby about current debates in ethology. That mentoring would have included the ongoing debate concerning whether selection operated at the level of the group— as argued by V. C. Wynne-Edwards (1959, 1962)—or at the level of the individual, as argued by Hinde's close colleague David Lack (1966). A key reason for Bowlby's ideas having stood the test of time so well was that all of his arguments about attachment are phrased in terms of survival value to the individual infant, rather than (as most social scientists would have done at that time) being phrased in terms of benefits to the group or family into which the infant was born.

9. From Freud 1926, cited in Bowlby 1969:259.

10. For discussion see the chapter on "Metaphors into Hardware: Harry Harlow and the Technology of Love" in Haraway 1989. Because Harlow's "motherless" infants grew up to be sexually inept adults, for reproduction to occur, the experimenters "resorted to an apparatus affectionately termed the rape rack which we leave to the reader's imagination" (cited in Haraway 1989:238). Reviewing this painful history, Haraway concluded: "Misogyny is deeply implicated in the dream structure of laboratory culture; misogyny is built into the objects of everyday life in laboratory practice, including the bodies of the animals, the jokes in the publications, and the shape of the equipment" (1989:238).

11. Bowlby saw in the use of nonhuman subjects a way to do controlled experiments unthinkable with human infants. Likewise, for many of us, experiments on maternal deprivation are now unthinkable for monkeys as well since the case for simian despair has been so conclusively demonstrated; Robert Hinde 1996, and interviews with Hinde, June 14–24, 1996.

12. Spencer-Booth and Hinde 1971a.

13. The researchers called them "mother goes to hospital" experiments, in explicit reference to

Robertson's famous 1953 film. See esp. Hinde and McGinnis 1977. Even in "infant-sharing" species such as langur monkeys, cold-turkey separation from the mother (with no other female holding the infant) is extremely stressful. All langur babies complain when separated from their mothers. But when experimenters working with a colony of langur monkeys at Berkeley actually removed mothers from the group, some, but not all, infants managed to attach themselves to foster mothers; those that did not, died (Dolhinow 1980).

14. Spencer-Booth and Hinde 1971a:116–17; see also Capitanio and Reite 1984.

15. All monkey and ape infants are distressed when deprived of physical contact with a caretaker. Only among the prosimian primates are there some cases (like *Varecia*) where mothers leave their infants in nests. Almost invariably, monkey, ape, and human infants complain when they are first separated from their mothers. Nevertheless, there is considerable intraspecific as well as interspecific variation in the willingness of a mother to allow other individuals (so-called allomothers) to hold and carry her infant, as well as interspecific variation in the tolerance of infants for strange caretakers. Scientists are just beginning to study the sources of these inter-individual differences in temperament. See Suomi 1999 for monkeys; and for human babies, Spangler and Grossmann 1993.

16. For a marvelously readable summary of the slow but steady accumulation of evidence substantiating Darwin's theory of natural selection, see Jonathan Weiner's *The Beak of the Finch: A Story of Evolution in Our Time* (1994); for attachment theory see Cassidy and Shaver, eds., 1999.

17. Ainsworth 1967.

18. Spencer-Booth and Hinde (1971b) noted that: "The intensely 'clinging' behaviour shown by the rhesus infants in the immediately post-separation period is also characteristic of many human children, though some of the latter after severe separations show instead 'detachment' and a difficulty re-establishing affectional relations which we did not see in these monkeys" (1971b:191). They were struck by the absence of such avoidant responses among macaques. To this day, I know of no published report of an avoidant response for nonhuman primates.

19. Ainsworth's remark from her introduction to videotaped lecture (see note 6, above).

20. Ainsworth and Wittig 1969; see also Main and Weston 1982.

21. Main and Solomon 1986.

22. Main and Hesse 1990. Most people were not aware of child abuse as a medical problem until after 1978, when pediatricians Ruth and Henry Kempe published their book *Child Abuse*. The term *battered child* entered the medical literature in the United States with Kempe et al. 1962. Bowlby credited his own "enlightenment" to the Kempes. The relationship, if any, between child abuse and disorganized attachment states is not well understood, but in some instances it is difficult not to suspect one.

23. Van IJzendoorn 1995; 82 percent of German infants classified as secure were classified the same at age five (Wartner et al. 1994). Lower levels of concordance are reported by Belsky et al. (1996) for a sample of 200 American children.

24. See especially Sroufe and Fleeson 1986; Main 1994:8.

25. Bowlby 1973:323, cited in Bretherton 1992:767.

26. Fraiberg, Adelson, and Shapiro 1975:387–88. For overview, see Fonagy et al. 1993.

27. Main 1991; see also Fonagy et al. 1993; for overview see Cassidy and Shaver, eds., 1999.

28. Belsky and Cassidy 1994; Cassidy and Shaver, eds., 1999.

29. Van IJzendoorn and Kroonenberg 1988; see also Kagan, Kearsley, and Zelago 1978 (esp. fig. 6.2: "Avoidant, secure and resistant infants across cultures").

30. Sagi et al. 1994.

Chapter 17. "Secure from What?" or "Secure from Whom?"

1. Bowlby 1972:229–31. See Bowlby 1972:258 for use of "Jim" in an epigraph, pp. 229–31 for discussion.

2. Hill and Hurtado 1996:424. Outside of foraging societies, the same pattern can be documented during famine: see, for example, Rosenthal 1998:A12 on the current famine in North Korea. Children without mothers are at special risk of starvation.

3. As discussed in chapter 11, wild apes breed even more slowly than humans do. At the extreme end of ape birth intervals, common chimpanzees in the Kibale Forest of Uganda space births up to eight years apart (personal communication from Richard Wrangham, 1997). Dettwyler (1995) hypothesizes that some humans in prehistory suckled as long as seven years. What this means for our ancestors is that *if conditions were harsh enough* they, too, would have nursed that long. When conditions were good, two to four years (Dettwyler's norm) would have been more like it.

4. Parker 1995:xi. Parker's book looks at mother-infant relationships from a psychoanalytical rather than an evolutionary perspective, yet her work is full of such convergences, which is hardly surprising. If psychoanalysis has merit, and evolutionary theory has merit, different agendas for mother and infant should be the first prediction of both.

5. Bowlby 1972:319. For research done since, see Tooby and Cosmides 1990; Pinker 1997.

6. Stern 1990:47–49. For updated overview of research see Chung and Thomson 1995.

7. Johnson et al. 1991; Langlois et al. 1987; Lewis 1969.

8. Schaal et al. 1980; Fernald 1985; personal communication, from Marc Hauser, Psychology Department, Harvard University.

9. De Casper and Fifer 1980. Although very young babies can learn to discriminate between father's voice and that of a stranger, infants were not similarly motivated to hear it (De Casper and Prescott 1984).

10. Weinberg and Tronick 1996 demonstrated that babies confronted with a blank face exhibited increased heart rate and also a vagal response as in the flight-fight reflex.

11. Bowlby 1972:276.

12. Meltzoff and Moore 1977. Not long after, these results were replicated by Field et al. (1982). For updates, see Meltzoff 1993; Meltzoff and Moore 1997.

13. Bowlby 1972:276.

14. Dickemann 1984:430. For the first formal publication of such a model see van Schaik and Dunbar 1990; note that this "bodyguard" hypothesis is distinct from a "paternal provisioning" hypothesis.

15. Freedman 1974:43–46, and personal observation by the author.

16. Mallardi, Mallardi, and Freedman 1961, cited in Freedman 1974.

17. Hill and Hurtado 1996 (table 5.1: "Causes of death during the forest period").

18. When my daughter Katrinka was just beginning to crawl, I asked an old friend, herpetologist José Rosado, to stop off for a glass of wine on his way home from work and bring a friend—a four-foot python. Placed near the snake on the carpet, Katrinka's main response was to try to crawl under the couch after the snake. It was not until we were all talking together later and the python tried to slither underneath my red velveteen skirt that Katrinka *learned* to be afraid of snakes when I inadvertently shrieked.

19. Prechtl 1950; Mineka 1987; Hinde 1991.

20. Personal communication from R. Palombit; Plooij 1984:88–89.

Chapter 18. Empowering the Embryo

1. From a letter by Charles Darwin to Asa Gray, cited in Weiner 1994:49.

2. Charles Darwin, 1859:67.

3. Lipson and Ellison (1996) found that average concentrations of estrogenic hormones during the mid-follicular phase were significantly higher in cycles where a woman conceived than in the cycles where she had unprotected sex but failed to conceive. This was true both for comparisons across women in the sample and when the same individual was monitored across her own cycles. These high-estradiol cycles were in turn correlated with increases in her body weight. For large-follicle advantage, see Miller, Goldberg, and Falcone 1996.

4. Gosden 1996; Dong et al. 1996. For general reference see Wood 1994, quote from p. 123.

5. Geoffrey Parker and Robert Trivers were primarily responsible for calling attention to the relevance of anisogamy for understanding relative parental investment and mating systems in animals. Recently, however, these concepts have been borrowed by social scientists with little attention to the underlying biology. Aware that anisogamy did not constitute some hard and fast rule about how males and females would behave, Trivers (1972) was fascinated by all the exceptions he could find, cases where females competed among themselves for access to males, as in phalaropes and jicanas. Even though the jicana's egg is bigger than the male's sperm, jicana females invest less per offspring than males do, hence the "sex reversed behavior."

6. The fruit fly experiment was published in Bateman 1948, reviewed in Hrdy 1986a. For generalizations from it, see Symonds 1979:23.

7. Rates of fetal loss after implantation was confirmed, were 16.4 percent for rhesus macaques and 21.7 percent for bonnet macaques at the California Primate Center. Rates of fetal loss calculated from retrospective data collected in these colonies were similar to the often cited figure of 18.7 percent reported by Van Wagenen. Rates went up during times of stress, as during an epidemic (Hendrie et al. 1996). For humans see Wood 1994. For extreme cases, see Roberts and Lowe 1975. Some recent estimates for humans report losses as high as one-third of conceptions (e.g., Cross, Werb, and Fisher 1994).

8. See Wood 1994 (table 6.5) for summaries of rate of fetal loss by age of mother for nine populations from North America, Europe, Asia, and the Caribbean.

9. The species name is *Agalychnis callidryas.* For a brief popular account of the species' life history and of Warkentin's work see *Science* 1995; my discussion is based on Warkentin 1995; interview with Karen Warkentin, November 1, 1996.

10. Sih and Moore 1993.

11. Quoted from the opening of Trivers's 1974 paper "Parent-Offspring Conflict."

12. Bingham (1980) describes Trivers's life during this period.

13. Trivers 1985:147.

14. Trivers 1974:249.

15. Shostak 1981:54, see also 58.

16. Trivers 1985:148, quote from 155.

17. Jane Goodall relates a case in point drawn from the life of one of the Gombe chimps, Fifi's four-year-old son, Frodo, who was in the process of being weaned, and well on his way toward perfecting the art of the tantrum: "After [Frodo] had twice tried to climb on his mother's back and twice been rejected, he followed slowly with soft hoo-whimpers. Suddenly he stopped, stared at the side of the trail, and uttered loud and urgent-sounding screams, as though suddenly terrified. Fifi, galvanized into instant action, rushed back with a wide grin of fear, gathered up her child and set off—carrying him. . . . Three days later, . . . the entire sequence was repeated. And, a year later, I saw the same behavior in a different infant . . . who was also being weaned. . . . I am of the opinion that [these infants] were intentionally manipulating their mothers" (Goodall 1986:576–77, 582).

18. Trivers 1985:157.

19. Bowlby 1972:315; Trivers 1985:160.

20. Bingham 1980:66.

21. Trivers 1974:259.

22. Interview with David Haig, November 7, 1996.

23. *The Selfish Gene* (1976) became one of the most famous of all evolutionary metaphors. Trivers wrote the preface for Dawkins's book. By the time Haig arrived at Oxford, W. D. Hamilton had joined Dawkins—making it a mecca for gene-centered interpretations.

24. By "parliament of genes," Leigh (1971) meant genes sometimes pulling together, sometimes pushing apart, other times neutralizing each other's effects. By the terms of the metaphor, other genes could band together against a gene that threatens survival of the legislative body. See Haig 1997: 284.

25. Haig 1996a:232.

26. Haig and Westoby (1989) first proposed that embryonic genes derived from the father were selected to extract more resources from maternal tissue than were genes of maternal origin, while maternally imprinted genes would have been selected to counter or nullify such paternally produced growth factors. For evidence consistent with this, see Kalscheur et al. 1993.

27. Leighton et al. 1996.

28. Angier 1996.

29. Lefebvre et al. 1998; for discussion see Bridges 1998.

30. Haig 1993:496.

31. Haig 1993; see also Kanbour-Shakir et al. 1990. Haig himself explored simple mechanisms for genes to recognize themselves at the maternal-fetal interface of viviparous organisms in his 1996b article, "Gestational Drive and the Green-Bearded Placenta."

32. The quote (from Page 1939:292) is cited in Haig 1993:516.

33. Haig 1995.

34. Macdonald 1984:6; Haig 1993:500. It was long assumed that variations in fetal size were largely determined in the second half of pregnancy. New evidence points to effects as early as the first trimester (Smith et al. 1998). Robinson 1992:15.

35. Nathanielsz 1996; Jolly 1972b.

36. This 1914 quote is from pathologist R. W. Johnstone, cited by Haig (1993:500). By "trophoderm" he meant the outermost layer of cells of the blastocyst phase of the embryo-to-be.

37. For example, small size at birth (i.e., a birth weight under 5.5 pounds) can result in a 35 percent higher rate of coronary death in later life, as well as a sixfold increase in the risk of diabetes or impaired glucose metabolism (Barker 1994). See especially Nesse and Williams in *Why We Get Sick: The New Science of Darwinian Medicine* (1994:197–200) for medical implications of work by Haig and others.

38. Barker et al. 1992 (esp. 183).

39. Trivers 1974.

40. Haig 1993:497. "What's Behind Odd Idea of War in the Womb?" *New York Times,* August 7, 1993.

41. Altmann 1980; quote cited in Walton 1986.

42. Altmann 1980:33–36; 42–63.

43. Altmann, Altmann, and Hausfater 1978.

44. Gomendio et al. 1995.

45. Altmann 1980:178–186, quote from page 186.

46. Haig 1996a:232.

47. Blurton-Jones 1986; refer back to chapter 8.

Chapter 19. Why Be Adorable?

1. Meyburg 1974; see especially review in Mock 1984.

2. Anderson 1990.

3. Mock and Forbes 1995.

4. Hahn 1981; Evans 1996.

5. When Doug Mock and coworkers artificially synchronized egret broods, making all chicks equal in size, parents spontaneously began to deliver 30 percent more food. Whereas earlier theorists following David Lack had assumed that parents are unable to obtain enough food for all, Mock's results raise the possibility that parents hold back, saving themselves. That is, they could, if they would, pull later-hatched chicks through more often than they do. Personal communication from D. Mock, November 9, 1996.

6. Schwabl, Mock, and Gieg 1996.

7. How do parent birds know which chick is hungriest? Begging chicks display brightly colored mouths, and the emptier their stomachs, the redder the display, apparently because more blood bypasses the stomach and flows through this area. Experiments by Rebecca Kilner (1997) using canaries demonstrated that parents provided more food to chicks whose mouth color was artificially reddened; for the extra testosterone in the yolk, see Schwabl 1994.

8. I am indebted to ornithologist John Eadie for helping to construct this scenario for the evolution of bright natal coats in American coots. For key experiment see Lyon, Eadie, and Hamilton 1994; for discussion see Pagel 1994.

9. Mary Jane West-Eberhard was about ten years ahead of her time in identifying the effects of "social choice," a generalized version of female choice (1983, esp. p. 160). John Eadie credits her 1983 paper with inspiring their interpretation of coot natal plumage.

10. In 1915, R. A. Fisher pointed out how sexual selection could take on a life of its own, so that preferences for a particular trait could "run away." For an excellent recent overview see Cronin 1991:201–4; also Thornhill and Alcock 1983:390–91.

11. W. D. Hamilton 1975. Regarding neoteny, the Austrian ethologist Konrad Lorenz used to illustrate his lectures with a slide of rounded snub-nosed heads of puppies, baby seals, and human babies to make this point. The makers of Disney cartoons made profitable use of such ethological insights. So far as skin color goes, even people with dark skin produce babies that are a lighter shade than most adults are.

12. The taxonomy of monkeys in the subfamily Colobinae is currently in flux. Until very recently, dusky or "spectacled" leaf monkeys (*Trachypithecus obscurus*) were known as *Presbytis obscurus,* and many still call them that. For overview on primate natal coats, see Hrdy 1976.

Chapter 20. How to Be "An Infant Worth Rearing"

1. Hill and Hurtado 1996:3. Note that the aversion to babies born without hair is widespread in South American tribal societies (e.g., Gregor 1985:88–90).

2. Translated from an interview with the man they call Kuchingi (Hill and Hurtado 1996:375).

3. Schiefenhövel 1989. Assessments of Eipo demography prior to Schiefenhövel's arrival were calculated from aerial surveys of Eipo country over the previous thirty years.

4. Schiefenhövel 1989 (fig. 10.7).

5. De Meer and Heymans 1993.

6. For the argument that "a high expectancy of child death is a powerful shaper of maternal thinking," leading in some cases to delayed attachment and emotional distance between mother and infant, see Scheper-Hughes 1992:340.

7. These figures, based on forty-nine live births, twenty of which ended in infanticide, are recalculated from Schiefenhövel 1989 (fig. 10.8). After 1978, under missionary influence, infanticide rates dropped to 10 percent.

8. For information on the Eipos, see Schiefenhövel 1989:185–86; Schiefenhövel and Schiefenhövel 1978. The account, based on a filmed record made by the Schiefenhövels, is quoted from Eibl-Eibesfeldt 1989:193–94.

9. Associated Press 1992.

10. Investigation so far has involved interviews where subjects assess photos or drawings of different babies. For example, Alley 1981; Hildebrandt and Fitzgerald 1979; Maier et al. 1984; Frodi et al. 1978. Similar data from traditional societies do not exist. However, the ethnographic record is fairly consistent on this point; see, for example, Hill and Ball 1996. As Daly and Wilson (1988:72–73) have pointed out, long-term parental responses to defective infants would not have been an issue in earlier times since such children were not kept, see also Singer et al. 1999.

11. Women were evaluated using a modified version of the "Cranley Maternal-Fetal Attachment Scale" administered before and again after the results of prenatal diagnosis were made known to them, between 16 and 21 weeks for the amniocentesis group, and between 10.6 and 15.7 weeks for the chorionic villus sampling group (Caccia et al. 1991:1122).

12. Maier et al. 1984; Frodi et al. 1978.

13. Daly and Wilson 1982; Christenfeld and Hill (1995) asked subjects to match photographs of year-old babies with photos of their parents. They were more readily able to match baby to father than to mother. Long suspected, it is now firmly documented that American husbands are more likely to invest when they are confident of the paternity of offspring ascribed to them (Anderson, Kaplan, and Lancaster 1999).

14. Sellen 1995. For additional examples see McCarthy 1981; Scheper-Hughes 1992.

15. Valenzuela 1990. Whereas 66 percent of children in Ainsworth's "normative" sample in Baltimore were found to be securely attached, only 50 percent were found securely attached in Valenzuela's quite high-risk sample of poor mothers (p. 1992). Such failure of infants to thrive, even in the presence of adequate nutrition, led Pollitt and Leibel (1980:196) to postulate that disturbances in the relationship between mother and infant lead to "a reversible growth hormone deficiency associated with emotional deprivation." See also Drotar 1991.

16. DeVries 1984. Unlike the Eipo case, malnutrition is the major cause of infant mortality among the Masai; when the Masai were first studied in 1910, infant mortality was 300 per 1,000 live births. Using Fisher's exact test, DeVries (1987) calculated that the probability of obtaining these results linking temperament to survival by chance was seven in one hundred.

17. From Anna Quindlen's text accompanying Nick Kelsh's *Naked Babies* (1996:29).

18. For an example of an aversive reaction, see Frodi et al. 1978; for classifying a baby as lacking "will" to live, see Scheper-Hughes 1992:314 ff.

19. In a remarkable series of natural experiments, Jonathan Losos, Kenneth Warheit, and Thomas Schoener (1997) were able to demonstrate evolutionary change in just fourteen years, some hundreds at most rather than thousands of generations, among lizard populations introduced to novel environments on various Caribbean islands. For more on rapid evolution, refer back to chapter 5, esp. note 17.

20. Daly and Wilson 1995: esp. 1281–82.

21. Aristotle, *Politica* VII 17 1336a 12–18. For West Africa see Dorjahn 1976:80; McKee 1984:97–99. McKee describes a similar "test of strength" (subjecting newborns to cold air) practiced in the past in traditional Central Andean communities in South America.

22. From the Grimms' tale "The Changeling," cited in Haffter 1968:56.

23. Schmitt 1983:114.

24. From Jacques de Vitry's early-thirteenth-century definition of a *chamion,* cited in Schmitt 1983:75.

25. Scheper-Hughes 1992:364–73, esp. 368. Boguet 1610:376ff, cited in Marvick 1974:280. The European cases were from what were then West Prussia, Posen, and Schleswig-Holstein, as well as Scotland and Ireland (Haffter 1968:60). See also Schmitt 1983:72 (and n. 11). According to Haffter (1968:57), European traditions for coercing elves to return a child included treating the changeling so badly that the fairy parents felt sorry for it. Hence, people should threaten to dip the changeling in boiling water, or place it on a coal shovel as if it were about to be thrust into the fire. In none of these cases would parents believe they were abusing a "human" baby.

26. Generalizations were derived from a sample of fifty-seven different cultures. Conservatively estimated, infanticide was socially sanctioned in at least 53 percent of these societies, and mothers were typically the agents or primary decision-makers (Minturn and Stashak 1982).

27. In China, families do not traditionally celebrate a new birth until the baby is a month old, a custom attributed to historically high infant mortality (Thurston 1996). Among the Bemba of Central Africa, "the taking of the child" by the father and mother, or *ukupoko mwana* ceremony, is performed at four months; this ritual also marks the postpartum point when husband and wife can resume intercourse (Richards 1939). For Elena Valero's account, see Biocca 1971:162–63.

28. See Hill and Ball 1996 for review; see Sargent 1988 for West African case study.

29. Bugos and McCarthy 1984:508.

30. DeVries 1987:173.

31. Minturn and Stashak 1982.

32. Such beliefs fall into the realm of human imagination and experience explored by structuralists seeking to understand "the savage mind." It is a world of curious trickster figures and unclassifiable animals, a weird bestiary of anomalous and supernatural creatures (like fetuses) that mediate between the dead and the living, the raw and the cooked, the civilized and the uncivilized, Nature and Culture. See especially Lévi-Strauss's renowned series of books on the science of mythology, beginning with *The Raw and the Cooked* (1969).

33. Langer 1972; Hrdy 1984; Daly and Wilson 1988.

34. Coleman 1974. The version cited here comes from a German account cited in McLaughlin 1974 (esp. p. 155, n. 99).

35. Blaffer 1972:114–15; 92. Soranus as well recommended that newborns be cleaned with salt. Salt was also part of the Hebrew birth ritual and eventually became a token of baptism. This is why throughout Catholic Europe salt was left with abandoned children to show that they had been baptized (Boswell 1988:322).

36. According to an eighteenth-century Scots account, "The moment a child is born it is plunged into cold water, though it should be necessary first to break the ice." By 1894, the Reverend J. Vaux, who recorded this, does not question his source but suggests that "This is carrying out the principle of the survival of the fittest to rather an extreme length" (Vaux 1894:74).

37. These ideas, originating with Lorenz, have been further elaborated in Alley (1981) and Maier et al. (1984), and are reviewed in Bogin 1996 (esp. 14–15).

Chapter 21. A Matter of Fat

1. Elphick and Wilkinson 1981. These stores of white adipose tissue are distinct from the brown fat that can be quickly mobilized by infants in thermoregulation. For generally low fat content in infant mammals, see Widdowson 1950.

2. Girard and Ferre 1982: esp. 522, 538.

3. Girard and Ferre 1982, cited in Haig 1993:500ff. Study of six human neonates revealed a range from 11 to 28 percent body fat, with the average falling at 16 percent (Widdowson 1950).

Like most physiologists, Girard and Ferre (1982: table 3) lump a great deal of diversity in primates together in one generic category, "the monkey." According to their measurements, one would expect to find 20 grams of fat per kilo of body weight in a newborn monkey weighing 500 grams, compared with 160 grams of fat in a 3,500-gram human. Data for three cebus infants yielded an average fat content of 4.5 percent of carcass weight (Ausman et al. 1982). Although Girard and Ferre put human neonatal fat mass at 16 percent of total birth weight, a subsequent study suggests 14 percent (Catalano et al. 1992)—still high by the standards of other mammals.

4. For cross-species comparison, see Harvey, Martin, and Clutton-Brock 1987. Chimp and orang neonates weigh around 1.7 kg or less (following Schultz 1969:152).

5. Schultz 1969:152–53.

6. Blurton-Jones 1972 for continuously suckling primates; Macdonald 1984:284; Ben Shaul 1962, for other mammals.

7. Pawlowski 1998:65.

8. Reviewed in Hrdy 1976.

9. Rodman and McHenry 1980.

10. As discussed in chapter 11, bigger mothers may be linked to mothers living longer as well as greater food availability. Such life-historical changes, including longer childhoods, may have set the stage for bigger brains to become valuable, but as I pointed out, this does not mean that payoffs from being the cleverest ape were sufficient by themselves to compensate organisms for taking two decades to mature, or to compensate mothers for giving birth to dangerously fat babies.

11. Aiello and Wheeler 1995; Aiello 1992.

12. Martin 1995; Devlins, Daniels, and Roeder 1997.

13. For a concise overview see Jones 1992.

14. Detected by Hardy and Mellits (1977) using the Bender Gestalt test of visual perception.

15. Barker 1994; see also Brody 1996.

16. A linear correlation exists between weight at birth and subsequent scores on Weschler Intelligence Scale Tests. This linear relationship was similar in two large samples (12,315 white and 13,352 black children), even though birth weights were lower in the black subpopulation (Hardy and Mellits 1977). Low birth weight does not necessarily mean poor mental performance. Nutritional supplementation after birth can play an important compensatory role. See also Devlin, Daniels, and Roeder 1997.

17. Rosenberg and Trevathan 1996.

18. As Darwin pointed out, sutures occur in the skulls of a variety of young animals, including birds who hatch from eggs and have no problem traversing a narrow birth canal. Darwin concludes that this aspect of cranial engineering resulted from "the laws of growth," though it subsequently was "taken advantage of in the parturition of higher animals" (Darwin 1859:197).

19. See well-referenced overview of "exterogestation" (a term coined years ago by Ashley Montagu) in Trevathan 1987:143–45.

20. McGue 1997.

21. Devlin et al. 1997.

22. Anthropologists Beverly Strassmann and Catherine Panter-Brick have both called my attention to concern by pregnant women in West Africa and Nepal lest their fetus grow too large. "Pregnant women attempt to become as emaciated as possible in order that the birth may proceed more easily," writes Marten DeVries of the Masai. "During the last 3 or 4 months of pregnancy the woman abandons her normal diet and exists on a near starvation diet, consisting primarily of a broth of lungs, liver, and kidneys cooked with a bitter bark. The last month, she drinks only milk" (1987:170). See also Brems and Berg 1998 on " 'Eating Down' During Pregnancy: Nutrition, Obstetric, and Cultural Considerations in the Third World."

23. Ramsay and Dunbrack 1986.

24. Ziegler et al. 1976.

25. Hrdy 1996a.

26. Mann 1995.

27. Hobcraft, O'Donald, and Rutstein 1985.

28. Soranus 1956:80.

29. Barker and Robinson 1990:154–55, 165–86.

30. Van Hoof 1962.

Chapter 22. Of Human Bondage

1. Klaus and Kennell 1976:75–76; Daschner 1984; Ewald 1996. As Paul Ewald points out, "the disagreement about the value of rooming-in as infection control can be attributed to a failure to distinguish the prevalence of disease organisms from the prevalence of disease" (1996:254).

2. McBryde 1951.

3. See Klaus and Kennell 1976 for an overview by two of the prime movers in this reform movement who were also among the authors of the original *New England Journal* report.

4. Klaus and Kennell 1976; Klopfer 1971 ("Mother-Love: What Turns It On?"). But see Klopfer's subsequent essay (1996), essentially a retraction.

5. Rode et al. 1981; see also Grossman et al. 1981.

6. Lamb 1982.

7. Ironically, the strongest and most visible claims about "Velcro-style" bonding come from those debunking it. See, for example, Diane Eyer (1992a:2).

8. For a review of this literature see Myers 1984.

9. Eyer 1992b, 1992a.

10. See Karen 1994:429. It was Robert Hinde who arranged for Bowlby's degree, and who shared with me his recollections of that day (interview, June 23, 1996).

11. Litwack 1979:56.

12. Eliot 1876:694.

13. Eyer (1996:xiii, 11). Eyer goes on to describe how "the big guns of psychology were trotted out again to curb women's enthusiasm for the workplace. . . . The bullying of mothers may be the necessary tactic to draw attention away from the fact that their free domestic labor and cheap commercial labor currently subsidize American business. . . ."

14. Eyer 1996:72.

15. Eliot in a letter to Sarah Hennell, Nov. 28, 1865, reprinted in Cross 1885.

16. *New York Times,* September 14, 1997 ("Forbes Puts on Anti-Abortion Mantle," National Report, p. 10).

17. Bowlby quotes from Mead's 1962 essay "A Cultural Anthropologist's Approach to Maternal

Deprivation" and replies to her in the second edition of his *Attachment* (1972:303, note 1). See also Ainsworth 1990:193–94.

18. From an interview with Bowlby in 1989, cited in Karen 1994:325–26. Note that Bowlby's assessment was based on women's job opportunities in Britain at the time.

19. Wilson 1975:553.

20. Wilson's colleague DeVore, a leader of Harvard's pathbreaking study of Kalahari hunter-gatherers, was among those who checked drafts of chapter 27 in *Sociobiology,* where the error about women waiting at camp first appeared.

21. A handful of social historians have wondered out loud about the psychopathologies of wretched infants, wrenched away from wet nurses, but no pertinent case studies have surfaced.

22. Jean Peterson 1978, cited in Hewlett 1991:13; see also Etioko-Griffin 1986.

23. Griffin and Griffin 1992; Hewlett 1991, 1992.

24. Lamb et al. 1987; and personal communication from Michael Lamb, May 10, 1998.

25. Tronick, Morelli, and Winn 1987.

26. According to the "company we keep" hypothesis, first proposed by Beatrice Whiting and Carolyn Pope Edwards, the age and sex of those with whom boys and girls associate provide different learning experiences and elicit different social behaviors. Across cultures, little girls tend to spend more time with women, other girls, and infants than little boys do. For recent review see Edwards 1993.

27. Draper 1976; Draper and Cashdan 1988; Blurton-Jones 1993. In fact, work in progress by Draper and Hames (1999) suggests that older siblings do play a more important allomaternal role among the !Kung than suspected.

28. Goodall 1986:282–85, 351.

29. Hawkes, O'Connell, and Blurton-Jones 1995.

30. Quoted in a *Wall Street Journal* editorial by Andrew Peyton Thomas (1998:A-6).

31. These results are just emerging from a massive study begun in 1991 by the National Institute of Child Health and Human Development, involving twenty-five researchers and 1,300 families.

32. Harris 1998:153.

33. Interview with Jay Belsky, Evanston, Illinois, June 29, 1996. Belsky suggested that "some nonmaternal care arrangement in the first year for more than 20 hours per week may be a risk factor in the emergence of developmental difficulties" (1988:235). For rejoinders, see Clarke-Stewart 1988; see also the exchange between Belsky and his critics in Belsky 1986, and a reply in Phillips et al. 1987.

34. Eyer 1993.

Chapter 23. Alternate Paths of Development

1. Thurer 1994:xiii.

2. Thomas 1996.

3. Bowlby 1990b:36. Bowlby's views emerge in several papers written early in his career and training, e.g., "Forty-four Juvenile Thieves: Their Characters and Home Life" (1944) and "Some Pathological Processes Set in Train by Early Mother-Child Separation" (1953). Today, sociopaths are thought to make up perhaps 3 percent of the male population, 1 percent of the female. See review by Linda Mealey (1995; statistics cited on p. 523). Mealey, who summarizes the controversial literature on this subject, is convinced that sociopaths fall into two categories: those who are congenitally incapable of experiencing social emotions that give rise to conscience and empathy, whom she terms "primary" sociopaths; and "secondary sociopaths," who develop in response to adverse rearing and environmental conditions. Bowlby was concerned with such "secondary sociopaths."

4. Kovaleski and Adams 1996.

5. Kagan 1994.

6. Harris (1998:108) assumes that in the Environment of Evolutionary Adaptedness, all humans lived in patrilocal, male-centered societies. Based on ongoing research by human behavioral ecologists there is reason to expect that early humans were more flexible in this respect (see chapter 5, n. 10; chapter 8, n. 39; chapter 11, n. 26).

7. Harris 1998:203–4.

8. From an interview with Patricia Draper, August 1, 1996. See also Draper and Harpending 1982, 1987.

9. Draper and Harpending 1988:343.

10. Lamb et al. 1985; for overview of this new way of thinking about situation-dependent development strategies among humans, integrating infant development with life history theory, see Chisholm 1996. The other seminal paper was by Patricia Draper and Henry Harpending (1982). They examined how early experiences with caretakers among hunter-gatherers and African villagers would be likely to correlate with future access to resources. These researchers were in turn influenced by key developments in "life-history theory," e.g., Stearns 1977, 1982, 1992.

11. Sulloway (1996:83–118) explains the development of personality differences among first- and later-born siblings by invoking Darwin's "Principle of Divergence." "As children grow up," Sulloway writes, "they undergo adaptive radiation in their efforts to establish their own individual niches within the family" (86). This is essentially *polyphenism*. First- and later-born offspring are different morphs of essentially the same genotype, rather than genetically different organisms belonging to different species. Most of the research on polyphenism focuses on identifying relevant environmental cues (e.g., tannins in the diet of a caterpillar). In Sulloway, the relevant cues are *not* birth order per se, but amount of parental attention and access to parental resources. However, for a historical sample such as Sulloway's, the proxy variable "birth order" is obtainable for large numbers of his subjects (scientists), while measures of parental preferences would not be.

12. Levine 1987:292.

13. Belsky, Steinberg, and Draper 1991:656.

14. Diodorus of Sicily 1935:451; Schraudolph 1996:53–112.

15. Interview with Marc Hauser in Davis, California, October 29, 1997.

16. For more on human versus nonhuman primate intellectual capacities, and the "Machiavellian" intelligence that evolved in response to the need to cope with complex social situations, see Byrne and Whiten, eds., 1988; Cheney and Seyfarth 1990; Hauser 1996.

17. Kroeber 1989:230; see also 132ff.; 139–60.

18. For fuller treatment see Bryan Vila (1997).

Chapter 24. Devising Better Lullabies

1. From Baring-Gould and Baring-Gould 1962, cited in Piers 1978:29–30.

2. Nathanielsz 1996:214.

3. Bloch 1978:3.

4. "Terrible Tales: Coping with Fear of Infanticide," *Time,* November 4, 1978, p. 140; Bloch 1978:229.

5. Campbell and Petersen 1953, cited by Alice Rossi (1973), one of the first to take a biosocial perspective on human maternal emotions. For updates, see Carter 1992; Carter et al. 1992.

6. Balint 1985:116. For further discussion of this passage in the context of modern psychoanalytic theory see Parker 1995:103.

7. Pritchard, ed., 1955:7.

Acknowledgments

There is an old saying, "Sons branch out, but one woman leads to another." Perhaps its author was aware of sex-specific parental effects. In any case, this book owes its existence to my mother, Camilla Davis Blaffer Trammell, and to her mother, Kate Wilson Davis, for passing on to me their dogged temperaments (probably genetic) combined with a love of learning (more likely a maternal effect). Both women were closet bluestockings, who after their respective widowhoods metamorphosed into full-fledged scholars. Like other women of their class and time they were determined to "marry well." How else to achieve an acceptable social status? Alternative options in those days were not obvious. Yet these women imparted to me their love of books and ideas, and stood up to support an iconoclastic kinswoman in her defection from tribal custom.

For my own introduction into a world of expanding professional opportunities for women scholars, I thank the Mayanist Evon Vogt. It was while I was his undergraduate advisee at Radcliffe that I first learned how rich a field anthropology could be and also, to my amazement, that people actually studied animals to learn more about humans. Later as a graduate student at Harvard, I was fortunate to have the opportunity to work under the guidance of three pioneers in the field of sociobiology: Irven DeVore, Robert Trivers, and Edward O. Wilson.

It was also during those years that I first developed a feminist consciousness. Often, I felt like a lonely castaway throwing message bottles from a desert island as I struggled awkwardly to become a better scholar by taking into account female as well as male perspectives. My aim was to correct biases long implicit in evolutionary thinking without simply substituting new biases of my own in the process. It was no easy task, yet as the years passed I was delighted to feel the old metaphor of a castaway nudged aside, to be replaced by a more encouraging vision. I was a woman cast upon the waves, gently uplifted by a pod of supportive dolphins—fellow Darwinians and fellow feminists, themselves drawn forward by the currents of those swimming ahead. My debt to these companions is very great, especially to Patricia Adair Gowaty, Jane Lancaster, Barbara Smuts, and George Williams, who have charted new oceans and drawn others in their wake, including those who sail along blissfully unaware of their debt. As a result, "Darwinian feminism," as Patty Gowaty likes to say, "is an oxymoron no longer."

Many old friends provided information and commented on all or parts of the manuscript. I owe large debts to Jeanne Altmann, Sue Carter, Sandy Harcourt, Robert Hinde, Dan Hrdy, Debra Judge, David Kertzer, Jane Lancaster, Monique Borgerhoff Mulder, Peter Rodman, Jon Seger, Frank Sulloway, Kelly Stewart, Fred vom Saal, and Mary Jane West-Eberhard. My favorite uncle, Dietrich von Bothmer, along with other friends helped with illustrations. I thank especially Sally Landry, Virginia Savage, Dafila Scott, and Michelle Johnson for their wonderful draw ings, and Nancy DeVore for making available her treasure trove at Anthro-Photo.

Two years into the writing of this book, it became apparent that I was gestating two books, not one. Surgery was called for to separate these Siamese twins. The prospect left me paralyzed, but Mary Batten came to my rescue. With a combination of talent, good sense, and good grace, she reorganized and helped to rewrite material in Part One. Mary made many suggestions for the book as a whole that added greatly to the coherence of the finished product. I cannot thank her enough.

As I struggled, I never did come up with a pat answer to the question posed in this book about the optimal number of fathers for children born to mothers in humankind's EEA. However, I did learn that for a vast and unwieldy manuscript (at one point, twice as long as the finished project) the optimal number of editors is four. Along with Mary Batten, I am deeply grateful to Dan Frank at Pantheon for wise counsel and for seeing me through a difficult delivery. Savvy Jennie McDonald at Curtis Brown was unfailingly supportive and constantly reminded me to make this book relevant to the world into which it would be born. For the final push, I was fortunate to have the advice of Jenny Uglow of Chatto and Windus, who strove to make the book live up to its dual citizenship.

In addition to these book doctors, delivery of the manuscript required several midwives. June-el Piper was on hand from the outset. She kept track of references, located sources, and obtained permissions for all the illustrations. It fell to a trusted friend, Gene Miner, to keep everybody coordinated and bring order to chaos. Special thanks are due to Beth Post, Shirley Stewart, Katherine Robinson, Jennifer Weh, and to Elizabeth Knoll at Harvard University Press for valuable guidance early on.

Colleagues and staff in the Department of Anthropology and in the Graduate Group in Animal Behavior at the University of California, Davis provided unfailing support. The book was begun while I was a Guggenheim fellow. I also received support from Wenner Gren, and from the Rockefeller Foundation Gender Roles program.

Scholars belong to an amazing subculture. Practitioners are socialized into a unique worldview that accords highest priority to accuracy and shared knowledge. In addition to those whose published works I cite, the following colleagues, some of them former teachers, fellow graduate students from years ago, or my own former students, provided advice, drew diagrams, scolded, and corrected. I list them by what is often just one of their specializations to acknowledge their assistance, and again to thank each one.

Animal Behavior: Steve Emlen, Lawrence Frank, Sandy Harcourt, William J. Hamilton, Peter Marler, Peter Moyle, Judy Stamps, Mary Towner

Animal Husbandry: Ed Price, Lynette Hart

Anthropology: James Chisholm, Lee Cronk, Peter Ellison, Rusty Greaves, Kristen Hawkes, Kim Hill, Chuck Hilton, Debra Judge, David Kertzer, Jane Lancaster, Anne Nacey Maggioncalda, Lauris McKee, James Moore, Monique Borgerhoff Mulder, William Skinner, Volker Sommer

Cognitive Psychology and Neuroscience: Marc Hauser, E. B. Keverne, Sue Parker

Developmental Psychology: Jay Belsky, Susan Crockenberg, Erik Hesse, Robert Hinde, Michael Lamb, Mary Main, Janet Mann, Carol Rodning, Arlene Skolnick, Emmie Werner

Endocrinology and Behavior: Sue Carter, Steve Glickman, Sally Mendoza, Fred vom Saal, John Wingfield

Entomology: Mary Jane West-Eberhard, Hugh Dingle, Erick Greene, William D. Hamilton, Robert Page, Randy Thornhill, Robbin Thorpe

Etymology: Max Byrd

Genetics: Rick Grosberg, David Haig, Brett Holland, Charles Langley, Jon Seger

Herpetology: Karen Warkentin

History: Dietrich von Bothmer, Susan Mann, Lawrence Stone, George Sussman

History of Science: Joan Cadden, Paula Findlin, Joy Harvey, Frank Sulloway, Margaret Lock

Medicine: Dan Hrdy, Mary Rodman

Nutrition: Kay Dewey

Ornithology: Ann Brice, Nancy Burley, Patricia Adair Gowaty, Caldwell Hahn, Doug Mock, John Eadie, Jamie Giliardi, Bruce Lyon

Paleoanthropology: Henry McHenry, Theya Molleson

Primatology: Jeanne Altmann, Fred Bercovitch, Carola Borries, Lynne Isbell, John Mitani, Anne Pusey, Ryne Palombit, Amy Parish, Peter Rodman, Joan Silk, Meredith Small, Kelly Stewart, David Watts, Lalith Jayawickrma

Translations: Junko Kitanaka, Joy Harvey, Katrinka Hrdy, Volker Sommer

And then there are my greatest debts. To Guadalupe de la Concha I owe heartfelt thanks for providing my children an allomother as committed as their genetic mother. It is to my steadfast and resourceful husband, Dan, however, that I owe the most. By expanding the scope of his parental responsibilities, Dan took up the slack I often left during the years I worked on this manuscript. He and our children have been my mainstays. Families, of course, are what humans are all about. I am immensely grateful for mine.

Bibliography

Aberle, David

    1961 Matrilineal descent in cross-cultural perspective. In *Matrilineal Kinship,* David Schneider and Kathleen Gough, eds., 655–727. Berkeley: University of California.

Abernethy, Virginia

    1978 Female hierarchy and evolutionary perspective. In *Female Hierarchies,* Lionel Tiger and Heather Fowler, eds., 129–32. Chicago: Beresford Book Service.

Ackerman, Sandra

    1987 American Scientist Interviews: Peter Ellison. *American Scientist* 75:622–27.

Acton, William

    1865 *The Functions and Disorders of the Reproductive System,* fourth ed. London.

Adams, D. B., A. R. Gold, and A. D. Burt

    1978 Rise in female-initiated sexual activity at ovulation and its suppression by oral contraceptives. *New England Journal of Medicine* 299:1145–50.

Agoramoorthy, G., and R. Rudran

    1995 Infanticide by adult and subadult males in free-ranging howler monkeys, *Alouatta seniculus,* in Venezuela. *Ethology* 99:75–88.

Agoramoorthy, G., S. M. Mohnot, and Volker Sommer

    1988 Abortions in free-ranging Hanuman langurs (*Presbytis entellus*)—a male-induced strategy? *Human Evolution* 3:297–308.

Ahnesjö, Ingrid, Amanda Vincent, Rauno Alatalo, Tim Halliday, and William J. Sutherland

    1997 The role of females in influencing mating patterns. *Behavioral Ecology* 4:187–89.

Ahokas, Antti J., Saija Turtiaien, and Marjatta Alto

    1998 Sublingual oestrogen treatment of postnatal depression. *Lancet* 351:109.

Aiello, Leslie C.

    1992 Human body size and energy. In *The Cambridge Encyclopedia of Human Evolution,* Steve James, Robert Martin, and David Pilbeam, eds., 45. Cambridge: Cambridge University Press.

Aiello, Leslie C., and Peter Wheeler

    1995 The expensive-tissue hypothesis. *Current Anthropology* 36:199–221.

Ainsworth, Mary D. S.

    1967 *Infancy in Uganda: Infant Care and the Growth of Love.* Baltimore: Johns Hopkins University Press.

1990 Further research into the adverse effects of maternal deprivation. In *Child Care and the Growth of Love,* reprint ed., John Bowlby, part III, 191–235. London: Penguin Books.

Ainsworth, Mary D., and B. A. Wittig
1969 Attachment and exploratory behavior of one-year-olds in a strange situation. In *Determinants of Infant Behaviour,* vol. 4, B. M. Foss, ed. London: Methuen.

Alberts, Bruce, Dennis Bray, Julian Lewis, Martin Raff, Keith Roberts, and James D. Watson
1994 *Molecular Biology of the Cell,* 3rd ed. New York: Garland.

Alexander, Richard D.
1979 *Darwinism and Human Affairs.* Seattle: University of Washington.
1997 On God, and such: The view from the president's window. *Human Behavior and Evolution Society Newsletter* (Spring)VI(1):1.

Alexander, Richard D., J. L. Hoogland, R. D. Howard, K. Noonan, and Paul Sherman
1979 Sexual dimorphism and breeding systems in pinnipeds, ungulates, primates and humans. In *Evolutionary Biology and Human Social Behavior,* N. A. Chagnon and W. Irons, eds., 402–35. North Scituate, Massachusetts: Duxbury Press.

Alley, T. R.
1981 Head shape and the perception of cuteness. *Developmental Psychology* 17:650–54.

Altmann, Jeanne
1974 Observational study of behavior: Sampling methods. *Behaviour* 49:227–67.
1980 *Baboon Mothers and Infants.* Cambridge: Harvard University Press.
1997 Mate choice and intrasexual reproductive competition: Contributions to reproduction that go beyond acquiring more mates. In *Feminism and Evolutionary Biology: Boundaries, Intersections and Frontiers,* Patricia Adair Gowaty, ed., 320–33. New York: Chapman and Hall.

Altmann, J. A., S. A. Altmann, and G. Hausfater
1978 Primate infant's effects on mother's future reproduction. *Science* 201:1028–30.

Altmann, Jeanne, Glenn Hausfater, and Stuart A. Altmann
1988 Determinants of reproductive success in savannah baboons, *Papio cynocephalus.* In *Reproductive Success: Studies of Individual Variation in Contrasting Breeding Systems,* T. H. Clutton-Brock, ed., 403–18. Chicago: Chicago University Press.

Ammerman, A. J., and L. Cavalli-Sforza
1984 *The Neolithic Transition and the Genetics of Populations in Europe.* Princeton: Princeton University Press.

Amoss, Pamela T., and Stevan Harrell
1981 Introduction: An anthropological perspective on aging. In *Other Ways of Growing Old,* Pamela T. Amoss and Stevan Harrell, eds., 1–24. Stanford: Stanford University Press.

Amundsen, D. W., and C. J. Diers
1970 The age of menopause in classical Greece and Rome. *Human Biology* 42:79–86.

Anderson, Dave
1990 Evolution of obligate siblicide in boobies, 1: A test of the insurance-egg hypothesis. *American Naturalist* 135:334–50.

Anderson, Connie M., and Craig F. Bielert
1994 Adolescent exaggeration in female catarrhine primates. *Primates* 35:283–300.

Anderson, Kermyt G., Hillard S. Kaplan, and Jane B. Lancaster

1997 Paying for children's college costs: The parental investment strategies of Albuquerque men. Paper presented at the Meeting of the Human Behavior and Evolution Society, June 4–8, University of Arizona, Tucson.

1999 Differential parental investment: Theory and data. Paper presented at the Annual Meeting of the Human Behavior and Evolution Society, June 2–6, University of Utah, Salt Lake City.

Andersson, Malte

1994 *Sexual Selection*. Princeton: Princeton University Press.

Angier, Natalie

1996 Fighting and studying the battle of the sexes. *New York Times* (June 11):C-1, C-11.

1997 Chemical tied to fat control could help trigger puberty. *New York Times* (January 7): C-1–C-3.

Aries, Philippe

1962 *Centuries of Childhood*. New York: Vintage Books. (Originally published in 1960 as *L'enfant et la vie familiale sous l'ancien regime*.)

Aristotle

1970 *Historia Animalium*. A. L. Peck, trans. Cambridge: Harvard University Press.

Aronson, Debby

1995 Infant killing among primates more myth than reality. Press release distributed by Washington University, St. Louis, Missouri, 3 pages plus attachments.

Artlett, C. M., J. B. Smith, and S. A. Jimenez

1998 Identification of fetal DNA and cells in skin lesions from women with systemic sclerosis. *New England Journal of Medicine* 338:1186–91.

Asch, Stuart

1968 Crib deaths: Their possible relationship to post-partum depression and infanticide. *Journal of the Mount Sinai Hospital* 35(3):214–19.

Ashton, Rosemary

1991 *G. H. Lewes: A Life*. Oxford: Oxford University Press.

1992 *George Eliot: Selected Critical Writings*. New York: Oxford University Press.

1996 *George Eliot: A Life*. London: Hamish Hamilton.

Associated Press

1992 East Germans drowned very premature babies. *Davis* (California) *Enterprise* (February 16).

1998a Heart linked to quick death of infants. *New York Times* (June 11):A-20.

1998b Japanese see declining birth rate as threat to country's future. *Daily Democrat* (Woodland, California, August 3):A4.

Auerbach, K. G.

1981 Extraordinary breast feeding: Relactation/induced lactation. *Journal of Tropical Pediatrics* 27:52–55.

Auerbach, K. G., and J. L. Avery

1981 Induced lactation: A study of adoptive nursing by 240 women. *American Journal of Diseases of Children* 135:340–43.

Ausman, Lynne M., Elizabeth M. Powell, Donna L. Mercado, Kenneth W. Samonds, Mohamed el Lozy, and Daniel Gallina

    1982 Growth and developmental body composition of the cebus monkey (*Cebus albifrons*). *American Journal of Primatology* 3:211–27.

Austad, S., and M. Sunquist

    1986 Sex ratio manipulation in the common opossum. *Nature* 324:58–60.

Austin, C. R., and R. V. Short

    1982 *Germ Cells and Fertilization. Reproduction in Mammals,* vol. 1. Cambridge: Cambridge University Press.

Badinter, Elisabeth

    1981 *Mother Love: Myth and Reality.* Francine du Plessix Gray, trans. New York: Macmillan.

Bairagi, R.

    1986 Food crisis, nutrition, and female children in rural Bangladesh. *Population and Development Review* 12:307–15.

Baker, R. Robin, and Mark Bellis

    1995 *Human Sperm Competition: Copulation, Masturbation and Infidelity.* London: Chapman and Hall.

Balint, Alice

    1985 Love for the mother and mother love. In *Primary Love and Psycho-analytic Technique* by Michael Balint. London: Maresfield (originally published 1952).

Bamshad, Michael J., W. Scott Watkins, Mary E. Dixon, Lynn B. Jorde, B. Bhaskara Rao, J. M. Naidu, B. V. Ravi Prasad, Arani Rasanayagam, and Mike F. Hammer

    1998 Female gene flow stratifies Hindu castes (letter). Nature 395(6703):651–52.

Baring-Gould, William S., and Ceil Baring-Gould

    1962 *The Annotated Mother Goose.* New York: Clarkson N. Potter.

Barker, D. J. P.

    1994 *Mothers, Babies and Disease in Later Life.* London: British Medical Journal Publishing.

Barker, D. J. P., A. R. Bull, C. Osmond, and S. J. Simmonds

    1992 Fetal and placental size and risk of hypertension in adult life. In *Fetal and Infant Origins of Adult Disease,* D. J. P. Barker and R. J. Robinson, eds., 175–94. London: British Medical Journal Publishing.

Barkow, J. H., and N. Burley

    1980 Human fertility, evolutionary biology and the demographic transition. *Ethology and Sociobiology* 1:163–80.

Barkow, J., Leda Cosmides, and John Tooby, eds.

    1992 *The Adapted Mind: Evolutionary Psychology and the Generation of Culture.* Oxford: Oxford University Press.

Barloon, Thomas J., and Russell Noyes

    1997 Charles Darwin and panic disorder. *Journal of the American Medical Association* 277(2):138–41.

Barnett, Rosalind, and Caryl Rivers

    1997 The new dad works the "second shift" too. *Radcliffe Quarterly* (Winter):9–10.

Barzun, Jacques

  1981 *Darwin, Marx, Wagner: Critique of Heritage.* Chicago: University of Chicago Press (originally published 1941).

Bateman, Angus John

  1948 Intra-sexual selection in drosophila. *Heredity* 2:349–68.

Batten, Mary

  1992 *Female Strategies.* New York: G. P. Putnam.

Becher, Hans

  1960 Die Surara und Pakidai: zwei Yanonami-Stamme in Nordwestbrasilien. Publication of
  the Museum für Völkerkunde, Hamburg.

  1974 Pore/Perimbo: Einwirkungen der lunaren Mythologie auf den Lebensstil von drei
  Yanonami-Stammen, Surara, Pakidai und Ironasiteri: Ergebnisse der 1970 durchgefuhrten
  Expedition nach Nordwestbrasilien, Volkerkunde-Abteilung. Hannover: Kommissionsverlag
  Munstermann-Druck.

Beckerman, Stephen

  1999 The concept of partible paternity among Native South Americans. Paper presented at
  the annual meeting of the American Association for the Advancement of Science, January
  21–26, Anaheim.

Beckerman, Stephen, and Paul Valentine (co-organizers)

  1999 Partible paternity: When matings with multiple men lead to many fathers for a single
  child. Symposium at the Annual Meeting of the American Association for the Advancement
  of Science, January 21–26, Anaheim, California.

Beckerman, Stephen, Roberto Lizarralde, Carol Ballew, Sissel Schroeder, Cristina Fingelton,
Angela Garrison, and Helen Smith

  1998 The Bari partible paternity project: Preliminary results. *Current Anthropology* 39:164–67.

Begley, Sharon

  1997 The science wars. *Newsweek* (April 21):54–56.

  1998 The parent trap. *Newsweek* (September 7):53–58.

Begley, Sharon, and Adam Rogers

  1996 "Morphogenic field" day. *Newsweek* (June 3):37.

Bekoff, Mark

  1993 Experimentally induced infanticide: The removal of birds and its ramifications. *The Auk*
  110:404–6.

Bell, Graham

  1997 *Selection: The Mechanism of Evolution.* New York: Chapman and Hall.

Belsky, Jay

  1986 Infant day care: A cause for concern? *Zero to Three* (September):1–7.

  1988 The "effects" of infant day care reconsidered. *Early Childhood Research Quarterly*
  3:235–72.

Belsky, Jay, and Jude Cassidy

  1994 Attachment: Theory and evidence. In *Development Through Life: A Handbook for Clinicians,* Michael Rutter and Dale F. Hay, eds., 373–402. London: Blackwell.

Belsky, Jay, Laurence Steinberg, and Patricia Draper
1991 Childhood experience, interpersonal development, and reproductive strategy: An evolutionary theory of socialization. *Child Development* 62:647–70.

Belsky, Jay, Susan Campbell, Jeffrey F. Cohn, and Ginger Moore
1996 Instability of infant-parent attachment security. *Developmental Psychology* 32:921–24.

Ben Shaul, Devorah Miller
1962 The composition of the milk of wild animals. *International Zoo Yearbook* 4:333–42.

Bennett, Neal
1983 Sex selection of children: An overview. In *Sex Selection of Children,* N. G. Bennett, ed., 1–12. New York: Academic Press.

Benton, Michael
1993 Dinosaur summer. In *The Book of Life,* S. J. Gould, ed., 162. New York: Norton.

Bereczkei, Tamas, and R.I.M. Dunbar
1997 Female-biased reproductive strategies in a Hungarian gypsy population. *Proceedings of the Royal Society of London,* Series B 264:17–22.

Berger, Joel
1983 Induced abortion and social factors in a wild horse. *Nature* 303:59–61.

Berkson, Gershon
1973 Social responses to abnormal infant monkeys. *American Journal of Physical Anthropology* 38:583–86.
1977 The social ecology of defects in primates. In *Primate Bio-social Development: Biological, Social and Ecological Determinants,* Suzanne Chevalier-Skolnikoff and Frank E. Poirier, eds., 189–204. New York: Garland.

Bertram, Brian
1975 Social factors influencing reproduction in wild lions. *Journal of Zoology* 77:463–82.

Betzig, Laura
1986 *Despotism and Differential Reproduction: A Darwinian View of History.* New York: Aldine.
1992 Roman monogamy. *Ethology and Sociobiology* 13:351–83.
1993 Sex, succession and stratification in the first six civilizations: How powerful men reproduced, passed power on to their sons, and used their power to defend their wealth, women and children. In *Social Stratification and Socioeconomic Inequality,* L. Ellis, ed., 37–74. New York: Praeger.
1995 Presidents preferred sons. *Politics and Life Science* 14:61–64.

Betzig, Laura, ed.
1997 *Human Nature: A Critical Reader.* Oxford: Oxford University Press.

Bianchi, D. W., G. K. Zickwolf, G. J. Weil, S. Sylvester, and M. A. DeMaria
1996 Male fetal progenitor cells persist in maternal blood for as long as 27 years postpartum. *Proceedings of the National Academy of Sciences, USA* 93:705–8.

Biesele, Megan
1993 *Women Like Meat: The Folklore and Foraging Ideology of the Kalahari Ju/'hoan.* Bloomington: Indiana University Press.

Biesele, Megan, and Nancy Howell
1981 "The old people give you life": Aging among !Kung hunter-gatherers. In *Other Ways of*

Growing Old, Pamela T. Amoss and Stevan Harrell, eds., 77–98. Stanford: Stanford University Press.

Bingham, Roger

1980 Trivers in Jamaica. *Science 80* (March/April):56–67.

Biocca, E.

1971 *Yanoama.* New York: Dutton.

Birdsell, Joseph

1968 Some predictions for the Pleistocene based on equilibrium systems among recent hunter-gatherers. In *Man the Hunter,* Richard B. Lee and Irv DeVore, eds., 229–49. Chicago: Aldine.

Birkhead, Tim, and Anders Møller

1993 Female control of paternity. *Trends in Ecology and Evolution* 9(3):100–3. Elsevier Science.

Blackburn, Daniel G., Virginia Hayssen, and Christopher J. Murphy

1989 The origins of lactation and the evolution of milk: A review with new hypotheses. *Mammal Review* 19(1):1–26.

Blackwell, Antoinette Brown

1875 *The Sexes Throughout Nature.* New York: G. P. Putnam.

Blaffer, Sarah C.

1972 *The Black-men of Zinacantan.* Austin: University of Texas Press.

Blass, Elliott M., T. J. Fillion, P. Rochat, L. B. Hoffmeyer, and M. A. Metzger

1989 Sensorimotor and motivational determinants of hand-mouth coordination in 1–3 day old human infants. *Developmental Psychology* 25(6):963–75.

Bledsoe, Caroline

1990 The politics of children: Fosterage and the social management of fertility among the Mende of Sierra Leone. In *Births and Power: Social Change and the Politics of Reproduction,* W. Penn Handwerker, ed., 81–100. Boulder: Westview Press.

1991 The "trickle-down" model within households: Foster children and the phenomenon of scrounging. In *The Health Transition: Methods and Measures,* J. Cleland and A. G. Hill, eds., 115–31. *Health Transitions Series,* no. 3. Canberra.

1993 The politics of polygyny in Mende education and child fosterage transactions. In *Sex and Gender Hierarchies,* Barbara D. Miller, ed., 170–92. Cambridge: Cambridge University Press.

1994 "Children are like young bamboo trees": Potentiality and reproduction in sub-Saharan Africa. In *Population, Economic Development, and the Environment,* Kerstin Lindahl-Kiessling and Hans Landberg, eds., 105–38. Oxford: Oxford University Press.

Bledsoe, Caroline, and C. Isiugo-Abanihe Uche

1989 Strategies of child fosterage among Mende grannies in Sierra Leone. In *Reproduction and Social Organization in Sub-Saharan Africa,* R. Lesthaeghe, ed., 442–74. Berkeley: University of California Press.

Bledsoe, Caroline, and A. Brandon

1992 Child fosterage and child mortality in sub-Saharan Africa: Some preliminary questions and answers. In *Mortality and Society in Sub-Saharan Africa,* E. van de Walle, G. Pison, and M. Sala-Daikanda, eds., 279–302. Oxford: Oxford University Press.

Bleichfeld, Bruce, and Barbara E. Moely

    1984 Psychophysiological responses to an infant cry: Comparison of groups of women in different phases of the maternal cycle. *Developmental Psychology* 20(5):1082–91.

Blinkhorn, Steve

    1997 Symmetry as destiny—taking a balanced view of IQ. *Nature* 387:849–50.

Bloch, Dorothy

    1978 *"So the Witch Won't Eat Me": Fantasy and the Child's Fear of Infanticide.* Boston: Houghton Mifflin.

Blunt, Wilfrid

    1951 *Black Sunrise: The Life and Times of Mulai Ismail, Emperor of Morocco, 1646–1727.* London: Methuen.

Blurton-Jones, Nicholas

    1972 Comparative aspects of mother-child contact. In *Ethological Studies of Child Development,* N. Blurton-Jones, ed., 305–28. Cambridge: Cambridge University Press.

    1984 A selfish origin for human food-sharing: Tolerated theft. *Ethology and Sociobiology* 5:1–3.

    1986 Bushman birth spacing: A test for optimal interbirth intervals. *Ethology and Sociobiology* 7:91–105.

    1993 The lives of hunter-gatherer children: Effects of parental behavior and parental reproductive strategy. In *Juvenile Primates: Life History, Development and Behavior,* M. E. Pereira and L. A. Fairbanks, eds., 309–26. New York: Oxford University Press.

Blurton-Jones, N., K. Hawkes, and J. F. O'Connell

    1997 Why do Hadza children forage? In *Uniting Psychology and Biology: Integrative Perspectives on Human Development,* N. Segal, G. E. Weisfeld, and C. C. Weisfeld, eds., 279–313. Washington, D.C.: American Psychological Association.

Bodmer, W. F., and L. L. Cavalli-Sforza

    1976 *Genetics, Evolution and Man.* San Francisco: Freeman.

Boesch, C.

    1993 Aspects of transmission of tool-use in wild chimpanzees. In *Tools, Language and Cognition in Human Evolution,* K. R. Gibson and T. Ingold, eds., 171–83. Cambridge: Cambridge University Press.

Bogin, Barry

    1996 Human growth and development from an evolutionary perspective. In *Long-Term Consequences of Early Environment: Growth, Development and the Lifespan Developmental Perspective,* D.J.K. Henry and S. J. Ulijaszek, eds., 7–24. Cambridge: Cambridge University Press.

Bogucki, Peter

    1996 The spread of early farming in Europe. *American Scientist* 84:242–53.

Boguet, H.

    1610 *Discours des sorciers.* Lyon.

Bolker, Jessica A., Marguerite Butler, Jessica Kissinger, and Margaret A. Riley

    1997 Addressing the gender gap in evolutionary biology. *Trends in Ecology and Evolution* 12(2):46–47.

Boone, James L., III

1986 Parental investment and elite family structure in preindustrial states: A case study of late medieval–early modern Portuguese genealogies. *American Anthropologist* 88: 859–78.

1988a Parental investment, social subordination, and population processes among the 15th and 16th century Portuguese nobility. In *Human Reproductive Behaviour: A Darwinian Perspective,* L. Betzig, M. Borgherhoff Mulder, and P. Turke, eds., 201–19. Cambridge: Cambridge University Press.

1988b Second- and third-generation reproductive success among the Portuguese nobility. Paper presented at the Eighty-seventh Annual Meeting of the American Anthropological Association, November, Phoenix.

Borgerhoff Mulder, Monique

1988 Kipsigis bridewealth payments. In *Human Reproductive Behaviour: A Darwinian Perspective,* L. Betzig, M. Borgerhoff Mulder, and P. Turke, eds., 65–82. Cambridge: Cambridge University Press.

1990 Kipsigis women's preference for wealthy men: Evidence for female choice in mammals? *Behavioral Ecology and Sociobiology* 27:255–64.

1992a Reproductive decisions. In *Evolutionary Ecology and Human Behavior,* Eric Alden Smith and Bruce Winterhalder, eds., 339–74. Hawthorne, New York: Aldine de Gruyter.

1992b Women's strategies in polygynous marriage: Kipsigis, Datoga, and other East African cases. *Human Nature* 3:45–70.

1998 The demographic transition: Are we any closer to an evolutionary explanation? *Trends in Ecology and Evolution* 13:266–70.

Borries, Carola

1997 Infanticide in seasonally breeding multimale groups of Hanuman langurs (*Presbytis entellus*) in Ramnagar, South Nepal. *Behavioral Ecology and Sociobiology* 42:139–50.

Borries, Carola, and Andreas Koenig

In press. Hanuman langurs: infanticide in multi-male groups. In *Infanticide by Males and Its Implications,* Carel P. van Schaik and Charles Janson, eds. Cambridge: Cambridge University Press.

Borries, Carola, Volker Sommer, and Arun Srivastava

1991 Dominance, age and reproductive success in free-ranging female Hanuman langurs (*Presbytis entellus*). *International Journal of Primatology* 12:231–57.

Borries, Carola, Kristin Launhardt, Cornelia Epplen, Jörg Epplen, and Paul Winkler

1998a DNA analyses support the hypothesis that infanticide is adaptive in langur monkeys. Submitted to *Proceedings of the Royal Society* (London). Ms. in the authors' possession, Deutsches Primatenzentrum, Göttingen, Germany.

1998b Males as infant protectors in Hanuman langurs (*Presbytis entellus*) living in multi-male groups: defense pattern, paternity, and sexual behavior. Submitted to *Behavioral Ecology and Sociobiology.*

Boster, J. S., R. R. Hudson, and S. J. C. Gaulin

1999 High paternity certainties of Jewish priests. *American Anthropologist* 100(4):967–71.

Boswell, John

  1988 *The Kindness of Strangers: The Abandonment of Children in Western Europe from Late Antiquity to the Renaissance.* New York: Pantheon Books.

Bourdieu, Pierre

  1976 Marriage strategies as strategies of social reproduction. In *Family and Society,* R. Foster and O. Ranum, eds., 117–44. Baltimore: Johns Hopkins University Press.

Bourke, F. G., and Nigel R. Franke

  1995 *Social Evolution in Ants.* Princeton: Princeton University Press.

Bowlby, John

  1944 Forty-four juvenile thieves: Their characters and home life. *International Journal of Psychoanalysis* 25:19–52, 107–27.

  1953 Some pathological processes set in train by early mother-child separation. *Journal of Mental Science* 99:265–72.

  1960 Grief and mourning in infancy and early childhood. *The Psychoanalytic Study of the Child* 15:9–52.

  1972 *Attachment.* Attachment and Loss, vol. 1. Middlesex: Penguin Books (originally published 1969).

  1973 *Separation.* Attachment and Loss, vol. 2. London: Hogarth Press.

  1982 *Attachment,* 2nd edition. Attachment and Loss, vol. 1. New York: Basic Books/HarperCollins.

  1988 *A Secure Base: Parent-Child Attachment and Healthy Human Development.* New York: Basic Books/HarperCollins.

  1990a *Charles Darwin: A New Life.* New York: W. W. Norton.

  1990b *Child Care and the Growth of Love.* London: Penguin Books (originally published 1953; subsequent edition published 1965).

  1991 Ethological light on psychoanalytical problems. In *The Development and Integration of Behaviour: Essays in Honour of Robert Hinde,* Patrick Bateson, ed., 301–13. Cambridge: Cambridge University Press.

Boyd, Robert, and Peter J. Richerson

  1985 *Culture and the Evolutionary Process.* Chicago: University of Chicago Press.

Boyd, Robert, and Joan Silk

  1997 *How Humans Evolved.* New York: Norton.

Brazelton, T. Berry

  1969 *Infants and Mothers: Differences in Development.* New York: A Delta Special.

Brems, Susan, and Alan Berg

  1988 "Eating down" during pregnancy: Nutrition, obstetric and cultural considerations in the Third World. Discussion paper prepared for a United Nations advisory group.

Bretherton, I.

  1992 The origins or attachment theory: John Bowlby and Mary Ainsworth. *Developmental Psychology* 28:759–75.

Bridges, Robert S.

  1998 The genetics of motherhood. *Nature Genetics* 20:108–9.

Bridges, Robert S., Rosemarie Di Biase, Donna D. Loundes, and Paul C. Doherty
  1985 Prolactin stimulation of maternal behavior in female rats. *Science* 227:782–84.
Brody, Jane
  1996 Life in the womb may affect adult heart disease risk. *New York Times* (October 1).
Bronson, Frank
  1984 The adaptability of the house mouse. *Scientific American* 250:90–97.
Broude, G. J.
  1994 *Marriage, Family, and Relationships: A Cross-Cultural Encyclopedia.* Santa Barbara, California: ABC-CLIO.
Brouskou, Aigle, and Aftihia Voutira
  1995 Borrowed children: The case of child indoctrination as an issue in the Greek Civil War, 1947–1949. Paper presented at the conference "Nobody's Children" held at the University of Durham (September 28–30), Durham, England.
Brown, Janet
  1995 *Charles Darwin: Voyaging.* Princeton: Princeton University Press.
Brown, Jennifer R., Hong Ye, Roderick T. Bronson, Pieter Dikkas, and Michael E. Greenberg
  1996 A defect in nurturing in mice lacking the immediate early gene fosB. *Cell* 86:297–309.
Bruce, Hilda
  1960 An exteroceptive block to pregnancy in the mouse. *Nature* 184:105.
Bruce, Judith
  1989 Homes divided. *World Development* 17:979–91.
Bugos, Paul E., and Lorraine M. McCarthy
  1984 Ayoreo infanticide: A case study. In *Comparative and Evolutionary Perspectives on Infanticide,* G. Hausfater and S. Blaffer Hrdy, eds., 503–20. Hawthorne, New York: Aldine.
Burke, T., N. B. Davies, M. W. Bruford, and G. J. Hatchwell
  1989 Paternal care and mating behaviour of polyandrous dunnocks, *Prunella modularis,* related to paternity by DNA fingerprinting. *Nature* 338:249–51.
Burley, Nancy
  1977 Parental investment, mate choice and mate quality. *Proceedings of the National Academy of Sciences* 74:3476–79.
Burns, John F.
  1994 Ban on fetus-sex tests splits India. *Sacramento Bee* (August 28).
  1996 In India, attacks by wolves spark old fears and hatred. *New York Times* (September 1).
Burton, R.
  1976 *The Mating Game.* New York: Crown Publishers.
Buss, David M.
  1994a The strategies of human mating. *American Scientist* 82:238–49.
  1994b *The Evolution of Desire: Strategies of Human Mating.* New York: Basic Books.
Buss, David M., M. Abbott, A. Angleitner et al.
  1990 International preferences in selecting mates: A study of 37 cultures. *Journal of Cross-Cultural Psychology* 50:559–70.

Buss, D., and D. M. Duntley

1998 Evolved homicide modules. Paper presented at Tenth Annual Meeting of the Human Behavior and Evolution Society, July 8–12, University of California, Davis.

Byers, John, J. D. Moodie, and N. Hall

1994 Pronghorn females choose vigorous mates. *Animal Behavior* 47:33–43.

Byrne, R., and A. Whitten, eds.

1988 *Machiavellian Intelligence*. Oxford: Oxford University Press.

Caccia, N., J. M. Johnson, G. E. Robinson, and T. Barna

1991 Impact of prenatal testing on maternal-fetal bonding: Chorionic villus sampling versus amniocentesis. *American Journal of Obstetrical Gynecology* 165:1122–25.

Caine, M. T.

1977 The economic activities of children in a village in Bangladesh. *Population and Development Review* 13(3):201–27.

Calamandrel, G., and E. B. Keverne

1994 Differential expression of Fos protein in the brain of female mice dependent on pup sensory cues and maternal experience. *Behavioral Neuroscience* 108:113–20.

Caldwell, John

1982 *Theory of Fertility Decline*. New York: Academic Press.

Caldwell, John, and Pat Caldwell

1990 High fertility in sub-Saharan Africa. *Scientific American* (May):118–25.

1994 Marital status and abortion in sub-Saharan Africa. In *Nuptiality in Sub-Saharan Africa: Contemporary Anthropological and Demographic Perspectives,* C. Bledsoe and G. Pison eds., 274–95. Oxford: Oxford University Press.

Caldwell, John C., I. O. Orubuloye, and Pat Caldwell

1991 The destabilization of the traditional Yoruba sexual system. *Population and Development Review* 17:229–62.

Calhoun, John

1962 Population density and social pathology. *Scientific American* 206(2):139–48.

Campbell, Anne

1993 *Men, Women, and Aggression*. New York: Basic Books.

1995 A few good men: Evolutionary psychology and female adolescent aggression. *Ethology and Sociobiology* 16:99–123.

Campbell, Anne, and S. Muncer

1994 Sex differences in aggression: Social roles and social representations. *British Journal of Social Psychology* 33:233–40.

Campbell, B., and W. B. Petersen

1953 Milk let-down and orgasm in human females. *Human Biology* 25:165–68.

Campbell, K. L., and J. W. Wood

1988 Fertility in traditional societies. In *Natural Human Fertility: Social and Biological Determinants,* Peter Diggory, Malcolm Potts, and Sue Teper, eds., 39–69. London: Macmillan.

Cantoni, Debora, and Richard E. Brown

1997 Paternal investment and reproductive success in the California mouse, *Peromyscus californicus. Animal Behavior* 54:377–86.

Capitanio, J. P., and M. Reite
1984 The roles of early separation experience and prior familiarity in social relations of pig-tailed macaques: A descriptive multivariate study. *Primates* 25:475–84.

Caro, T. M., and D. W. Sellen
1990 The reproductive advantages of fat in women. *Ethology and Sociobiology* 11:51–66.

Caro, T. M., D. W. Sellen, A. Parish, R. Frank, and E. Voland
1995 Termination of reproduction in nonhuman and human female primates. *International Journal of Primatology* 16(2):205–20.

Carter, C. Sue
1992 Oxytocin and sexual behavior. *Neuroscience and Biobehavioral Reviews* 16:131–44.
1998 Neuroendocrine perspectives on social attachment and love. *Psychoneuroendocrinology* 23: 779–818.

Carter, C. Sue, and Lowell L. Getz
1993 Monogamy and the prairie vole. *Scientific American* 268 (June):100–6.

Carter, C. Sue, and R. Lucille Roberts
1997 The psychobiological basis of cooperative breeding in rodents. In *Cooperative Breeding in Mammals,* Nancy G. Solomon and Jeffrey French, eds., 231–66. New York: Cambridge University Press.

Carter, C. Sue, I. Izja, and Brian Kirkpatrick, eds.
1997 *The Integrative Neurobiology of Affiliation.* New York: New York Academy of Sciences.

Carter, C. S., J. R. Williams, D. M. Witt, and T. R. Insel
1992 Oxytocin and social bonding. *Annals of New York Academy of Medicine* 652:204–11.

Cashdan, Elizabeth
1985 Coping with risk: reciprocity among the Basarwa of northern Botswana. *Man* 20:454–74.
1993 Attracting mates: Effects of paternal investment on mate attraction strategies. *Ethology and Sociobiology* 14:1–23.

Cassidy, J., and P. R. Shaver, eds.
1999 *Handbook of Attachment: Theory, Research, & Clinical Applications.* New York: Guilford Press.

Catalano, Patrick M., Elaine D. Tyzbir, Scott R. Allen, Judith H. McBean, and Timothy L. McAuliffe
1992 Evaluation of fetal growth by estimation of neonatal body composition. *Obstetrics and Gynecology* 79(1):46–50.

Chagnon, Napoleon
1968 *Yanomamö: The Fierce People.* New York: Holt, Rinehart and Winston.
1972 Tribal social organizations and genetic microdifferentiation. In *The Structure of Human Populations,* G. A. Harrison and A. J. Boyce, eds. Oxford: Clarendon Press.
1979 Mate competition, favoring close kin, and village fissioning among the Yanamamö Indians. In *Evolutionary Biology and Human Social Behavior: An Anthropological Perspective,* N. A. Chagnon and W. Irons, eds., 86–131. North Scituate, Massachusetts: Duxbury.
1988 Life histories, blood revenge, and warfare in a tribal population. *Science* 238:985–92.
1992 *Yanamamö: The Last Days of Eden.* San Diego: Harcourt Brace Jovanovich.

Chagnon, N., M. V. Flinn, and T. F. Melancon

   1979 Sex ratio variation among the Yanamamö Indians. *Evolutionary Biology and Human Social Behavior,* N. A. Chagnon and W. Irons, eds., 290–320. North Scituate, Massachusetts: Duxbury.

Chapais, Bernard

   1988 Experimental matrilineal inheritance of rank in female Japanese macaques. *Animal Behavior* 36:1025–37.

Chapais, B., M. Girard, and G. Primi

   1991 Non-kin alliances and the stability of matrilineal dominance relations in Japanese macaques. *Animal Behavior* 41:481–91.

Chapman, Anne

   1982 *Drama and Power in a Hunting Society: The Selk'nam of Tierra del Fuego.* Cambridge: Cambridge University Press.

   1997 The great ceremonies of the Selk'nam and the Yámana. In *Patagonia: Natural History, Prehistory and Ethnography at the Uttermost End of the Earth,* Colin McEwan, Luis Borrero, and Alfredo Prieto, eds., 82–109. London: British Museum Press.

Charnov, Eric

   1993 *Life History Invariants.* Oxford: Oxford University Press.

Charnov, Eric, and David Berrigan

   1993 Why do female primates have such long lifespans and so few babies? *or* Life in the slow lane. *Evolutionary Anthropology* 1(6):191–94.

Chen, Kathy

   1994 Study this, baby. Chinese fetuses bear heavy course loads: Limited to one child, couples nurture progeny pregnant with worldly potential. *Wall Street Journal* (February 8):A1, A9.

Cheney, Dorothy, and Robert Seyfarth

   1990 *How Monkeys See the World.* Chicago: University of Chicago Press.

Chiara, Susan

   1996 Study says babies in child care keep secure bonds to mother. *New York Times* (April 21):1, 11.

Child Care Bureau, Department of Health and Human Services

   1997 FAQ (Frequently Asked Questions): www.ack.dhhs.gov/programs/ccb/faq Website document dated December 19, 1997.

Chisholm, James

   1996 The evolutionary ecology of attachment organization. *Human Nature* 7:1–38.

Chisholm, James S., and Victoria K. Burbank

   1991 Monogamy and polygyny in southeast Arnhem Land: Male coercion and female choice. *Ethology and Sociobiology* 12:291–313.

Christenfeld, Nicholas J. S., and Emily A. Hill

   1995 Whose baby are you? Scientific correspondence. *Nature* 378:669.

Christensen, Kaare, David Gaist, Bernard Jeune, and J. W. Vaupel

   1998 A tooth per child? *Lancet* 352:204.

Chung, M.-S., and D. M. Thomson

   1995 Development of facial recognition. *British Journal of Developmental Psychology* 86:55–87.

Clark, Ann
1978 Sex ratio and local resource competition in a prosimian primate. *Science* 201:163–65.

Clark, R. D., and E. Hatfield
1989 Gender differences in receptivity to sexual offers. *Journal of Psychology and Human Sexuality* 2:39–55.

Clark, Sam, Elizabeth Colson, James Lee, and Thayer Scudder
1995 Ten thousand Tonga: A longitudinal anthropological study from Southern Zambia, 1956–1991. *Population Studies* 49:91–109.

Clarke-Stewart, K. Alison
1988 The "effects" of infant care reconsidered: Reconsidered. *Early Childhood Research Quarterly* 39:293–318.

Cloudsley, Anne
1983 *Women of Omdurman: Life, Love and the Cult of Virginity.* London: Ethnographica.

Clutton-Brock, T. H.
1991 *The Evolution of Parental Care.* Princeton, New Jersey: Princeton University Press.

Clutton-Brock, T. H., ed.
1988 *Reproductive Success: Studies of Individual Variation in Contrasting Breeding Systems.* Chicago: University of Chicago Press.

Clutton-Brock, T. H., and P. H. Harvey
1976 Evolutionary rules and primate societies. In *Growing Points in Ethology,* P.P.G. Bateson and R. A. Hinde, eds., 195–237. Cambridge: Cambridge University Press.

Clutton-Brock, T. H., and G. R. Iason
1986 Sex ratio variation in mammals. *Quarterly Review of Biology* 61:339–74.

Clutton-Brock, T. H., and G. A. Parker
1995a Punishment in animal societies. *Nature* 373:209–16.
1995b Sexual coercion in animal societies. *Animal Behavior* 49:1345–65.

Cohen, Joel E.
1995 *How Many People Can the Earth Support?* New York: Norton.

Cohen, Jon
1996 Does nature drive nurture? *Science* 273:577–78.

Coleman, Emily
1974 L'infanticide dans le Haut Moyen Age. *Annales: économies, societés, civilisations* 29:315–35.

Colp, Ralph, Jr.
1977 *To Be an Invalid: The Illness of Charles Darwin.* Chicago: University of Chicago Press.

Conniff, Ruth
1998 Democrats yield to profiteers. Further comment. *The Progressive* (September 12): 11–12.

Conquest, Robert
1986 *The Harvest of Sorrow: Soviet Collectivization and the Terror-Famine.* New York: Oxford University Press.

Conroy, Glenn C., and Kevin Kuykendall
1995 Paleopediatrics: or when did human infants really become human? *American Journal of Physical Anthropology* 98:121–31.

Cooper, Duff
    1986 *Talleyrand.* New York: Fromm International.

Corsini, Carol A., and Pier Paolo Viazzo, eds.
    1993 *The Decline of Infant Mortality in Europe, 1800–1950.* Florence: UNICEF and Instituto degli Innocenti.

Cosmides, Leda
    1989 The logic of social exchange: Has natural selection shaped how humans reason? *Cognition* 31:187–286.

Cosmides, Leda, and John Tooby
    1989 Evolutionary psychology and the generation of culture, part II: A computational theory of social exchange. *Ethology and Sociobiology* 10:51–97.

Coss, Richard G., and Ronald O. Goldthwaite
    1995 The persistence of old designs for perception. In *Behavioral Designs,* N. S. Thompson, ed., 83–148. Perspectives in Ethology, vol. 11. New York: Plenum Press.

Cowley, Geoffrey
    1996 The biology of beauty. *Newsweek* (June 3):60–66.

Cox, John L.
    1995 Postnatal depression in primate mothers: A human problem. In *Motherhood in Human and Nonhuman Primates,* C. R. Pryce, R. D. Martin, and D. Skuse, eds., 134–41. Basel: S. Karger.

Cranston, Maurice
    1997 *The Solitary Self: Jean Jacques Rousseau in Exile and Adversity.* Chicago: University of Chicago Press.

Creel, Scott, Steven L. Monfort, David E. Wildt, and Peter M. Waser
    1991 Spontaneous lactation is an adaptive result of pseudopregnancy. *Nature* 351:660–62.

Creel, Scott, Nancy Creel, David E. Wildt, and Steven Monfort
    1992 Behavioral and endocrine mechanisms of reproductive suppression in Serengeti dwarf mongooses. *Animal Behavior* 43:231–45.

Crocker, William, and Jean Crocker
    1994 *The Canela: Bonding Through Kinship, Ritual, and Sex.* New York: Harcourt Brace College.

Crockett, Carolyn M., and Ranka Sekulic
    1984 Infanticide in red howler monkeys (*Alouatta seniculus*). In *Infanticide: Comparative and Evolutionary Perspectives,* Glenn Hausfater and Sarah Blaffer Hrdy, eds., 173–91. Hawthorne, New York: Aldine.

Cronin, Carol
    1980 Dominance relations and females. In *Dominance Relations: An Ethological View of Human Conflict and Social Interaction,* Donald R. Omark, F. F. Strayer, and Daniel G. Freedman, eds. New York: Garland.

Cronin, Helena
    1991 *The Ant and the Peacock: Altruism and Sexual Selection from Darwin to Today.* Cambridge: Cambridge University Press.

Cronk, Lee

1993 Parental favoritism toward daughters. *American Scientist* 81:272–79.

1999 Female-biased investment and growth performance among the Mukogodo. In *Adaptation and Human Behavior: An Anthropological Perspective.* L. Cronk, N. Chagnon, and W. Irons, eds. Hawthorne, New York: Aldine de Gruyter. In press.

Cross, J. C., Z. Werb, and S. J. Fisher

1994 Implantation and the placenta: Key pieces of the development puzzle. *Science* 266:1508–18.

Cross, John W.

1885 *George Eliot's Life as Related in Her Letters and Journals,* 3 vols. Edinburgh and London: W. Blackwood and Sons.

Cucchiari, S.

1981 The gender revolution and the transition from bisexual horde to patrilocal bands: The origins of gender hierarchy. In *Sexual Meanings: The Cultural Construction of Gender and Sexuality,* Sherry B. Ortner and Harriet Whitehead, eds., 31–79. Cambridge: Cambridge University Press.

Cunningham, Allan S.

1995 Breastfeeding: Adaptive behavior for child health and longevity. In *Breastfeeding: Biocultural Perspectives,* Patricia Stuart-Macadam and Katherine A. Dettwyler, eds., 243–64. New York: Aldine de Gruyter.

Cunningham, Emma, and Tim Birkhead

1997 Female roles in perspective. *Trends in Ecology and Evolution (TREE)* 12(9):337–38.

Dagg, A. I.

1999 Infanticide by male lions: A fallacy influencing research into human behavior. *American Anthropologist* 100(4):940–50.

Dahl, J.

1985 The external genitalia of female pygmy chimpanzees. *Anatomical Record* 211: 24–28.

Dahl, J. R., R. Nadler, and D. G. Collins

1991 Monitoring the ovarian cycles of *Pan troglodytes* and *Pan paniscus:* A comparative approach. *American Journal of Primatology* 24:195–209.

Dahl, R. E.

1998 Pubertal timing, self-control, and adolescent psychopathology: An evolutionary view of potential maturational discrepancies. Paper presented at the Tenth Annual Meeting of the Human Behavior and Evolution Society, July 8–12, University of California, Davis.

Daley, Suzanne

1998 In Zambia, the abandoned generation. *New York Times* (September 18):A1, A12.

Daly, Martin

1979 Why don't male mammals lactate? *Journal of Theoretical Biology* 78:325–45.

Daly, Martin, and Margo Wilson

1978 *Sex, Evolution and Behavior: Adaptations for Reproduction.* North Scituate, Massachusetts: Duxbury.

1980 Discriminative parental solicitude: A biological perspective. *Journal of Marriage and the Family* 42:277–88.

1982 Whom are newborn babies said to resemble? *Ethology and Sociobiology* 3:69–78.

1988 *Homicide.* Hawthorne, New York: Aldine de Gruyter.

1995 Discriminative parental solicitude and the relevance of evolutionary models to the analysis of motivational systems. In *The Cognitive Neurosciences,* Michael Gazzaniga, ed., 1269–86. Cambridge: MIT Press.

Damasio, Antonio R.

1994 *Emotion, Reason and the Human Brain.* New York: Avon.

Danker-Hopfe, Heidi

1986 Menarcheal age in Europe. *Yearbook of Physical Anthropology* 29:81–112.

Darwin, Charles

1836–1844 *Charles Darwin's Notebooks,* Paul H. Barrett et al., eds. Ithaca: Cornell University Press, 1987.

1859 *On the Origin of Species.* London. (All references are to facsimile edition—New York: Atheneum, 1967.)

1871 *The Descent of Man, and Selection in Relation to Sex.* London. (All references are to reprint edition—Princeton: Princeton University Press, 1981.)

1872 *The Expression of the Emotions in Man and Animals.* Oxford: Oxford University Press (New 1998 edition with Commentaries by Paul Ekman).

1874 *The Descent of Man and Selection in Relation to Sex.* London. (All references are to reprint edition—Detroit: Gale Research, 1974.)

1876 Sexual selection in relation to monkeys. *Nature* 15:18–19.

1877 A biographical sketch of an infant. *Mind: Quarterly Review of Psychology and Philosophy* (July); reprinted in *The Portable Darwin,* Porter and Graham, eds., 1993:475–85.

1887 *The Autobiography of Charles Darwin, 1809–1882,* Nora Barlow, ed. New York: W. W. Norton, 1958.

Darwin, Francis, ed.

1887 *The Life and Letters of Charles Darwin,* 3 vols. London: John Murray.

Daschner, F.

1984 Infectious hazards in rooming-in systems. *Journal of Perinatal Medicine* 12:3–6.

Das Gupta, Monica

1987 Selective discrimination against female children in rural Punjab, India. *Population and Development Review* 13:77–100.

David, Henry P., Z. Dytrych, Z. Matejcek, and V. Schuller, eds.

1988 *Born Unwanted: Developmental Effects of Denied Abortion.* New York and Prague: Springer and Czechoslovakia Medical Press.

Davies, N. B.

1992 *Dunnock Behaviour and Social Evolution.* Oxford: Oxford University Press.

Davis, Elizabeth Gould

1971 *The First Sex.* New York: G. P. Putnam.

Dawkins, Richard

   1976 *The Selfish Gene.* New York: Oxford University Press.

   1982 *The Extended Phenotype.* Oxford: Oxford University Press.

Day, Corinne, and Bennett Galef

   1977 Pup cannibalism: One aspect of maternal behavior in golden hamsters. *Journal of Comparative and Physiological Psychology* 91:1179–89.

de Beauvoir, Simone

   1974 *The Second Sex,* H. H. Parshley, trans. New York: Vintage.

De Casper, A. J., and W. P. Fifer

   1980 Of human bonding: Newborns prefer their mothers' voices. *Science* 208:1174–76.

De Casper, A. J., and P. A. Prescott

   1984 Human newborns' perception of male voices: Preference, discrimination and reinforcing value. *Developmental Psychobiology* 17:481–91.

Delasselle, Claude

   1975 Les enfants abandonés à Paris au XVIIIe siècle. *Annales: économies, societés, civilisations* 30:187–218.

de Mause, Lloyd

   1974 The evolution of childhood. In *The History of Childhood,* L. de Mause, ed., 1–73. New York: Harper Torchbooks.

de Mause, Lloyd, ed.

   1974 *The History of Childhood.* New York: Harper Torchbooks.

de Meer, K., and H.S.A. Heymans

   1993 Child mortality and nutritional status of siblings. *Lancet* 342:313.

Dennett, Daniel

   1965 *Darwin's Dangerous Idea: Evolution and the Meaning of Life.* New York: Simon and Schuster.

De Parle, Jason

   1999 As welfare rolls shrink, load on relatives grows. *New York Times* (February 21).

Desmond, Adrian, and James Moore

   1991 *Darwin: The Life of a Tormented Evolutionist.* New York: W. W. Norton.

Dettwyler, Katherine A.

   1995 A time to wean: The hominid blueprint for the natural age of weaning in modern human populations. In *Breastfeeding: Biocultural Perspectives,* Patricia Stuart-Macadam and Katherine A. Dettwyler, eds., 39–73. New York: Aldine de Gruyter.

Devlin, B., Michael Daniels, and Katherine Roeder

   1997 The heritability of IQ. *Nature* 388:468–71.

DeVries, Marten W.

   1984 Temperament and infant mortality among the Masai of East Africa. *American Journal of Psychiatry* 141:1189–93.

   1987 Cry babies, culture and catastrophe: Infant temperament among the Masai. In *Anthropological Approaches to the Treatment and Maltreatment of Children,* Nancy Scheper-Hughes, ed., 165–86. Dordrecht: Reidel.

de Waal, Frans

1996 Survival of the kindest: A simian Samaritan shows nature's true heart. *New York Times* (August 22).

de Waal, Frans, with photographs by Frans Lanting

1997 *Bonobo: The Forgotten Ape.* Berkeley: University of California Press.

Dewey, Kathryn G.

1997 Energy and protein requirements during lactation. *Annual Review of Nutrition* 17:19–36.

Diamond, Jared

1997a *Why Is Sex Fun? The Evolution of Human Sexuality.* New York: Basic Books.

1997b *Guns, Germs, and Steel.* New York: W. W. Norton.

Dickemann, Mildred

1975 Demographic consequences of infanticide in man. *Annual Review of Ecology and Systematics* 6:109–37. (Author's name then spelled Dickeman.)

1979a Female infanticide and the reproductive strategies of stratified human societies. In *Evolutionary Societies and Human Social Behavior,* N. A. Chagnon and W. Irons, eds., 321–67. North Scituate, Massachusetts: Duxbury.

1979b The ecology of mating systems in hypergynous dowry societies. *Social Science Information* 18(2):163–95.

1984 Concepts and classification in the study of human infanticide: Sectional introduction and some cautionary notes. In *Infanticide: Comparative and Evolutionary Perspectives,* G. Hausfater and S. Blaffer Hrdy, eds., 427–37. Hawthorne, New York: Aldine.

Digby, Leslie

In press. Infanticide, infant care, and female reproductive strategies in a wild population of common marmosets. *American Journal of Physical Anthropology* (Supplement) 18:80–81.

Digby, Leslie L.

In press. Infanticide by female mammals: Implications for the evolution of social systems. In *Infanticide by Males and Its Implications,* C. van Schaik and C. Janson, eds. Cambridge: Cambridge University Press.

Diodorus of Sicily (Diodorus Siculus)

1935 *The Library of History,* vol. II, C. H. Oldfather, trans. Cambridge: Harvard University Press (reprinted 1953).

Dixson, Alan

1983 Observations on the evolution and behavioral significance of "sexual skin" in female primates. *Advances in the Study of Behavior* 13:63–106.

Dixson, A. F., and L. George

1982 Prolactin and parental behavior in a male New World primate. *Nature* 299:551–53.

Dolhinow, Phyllis

1977 Normal monkeys? *American Scientist* 65:266.

1980 An experimental study of mother loss in the Indian langur monkey (*Presbytis entellus*). *Folia Primatologica* 33:77–128.

Dolhinow, Phyllis, and Mark A. Taff

1993 Immature and adult langur monkey (*Presbytis entellus*) males: Infant-initiated adoption in a colony group. *International Journal of Primatology* 14:919–26.

Dong, Jinwen, David F. Albertini, Katsuhiko Nishimori, T. Rajendra Kumar, Naifang Lu, and Martin M. Matzuk

1996 Growth differentiation factor-9 is required during early ovarian folliculogenesis. *Nature* 383:531–35.

d'Orban, P. T.

1979 Women who kill their children. *British Journal of Psychiatry* 134:560–71.

Dorjahn, V.

1976 Rural-urban differences in infant and child mortality among the Temne of Kolifa. *Journal of Anthropological Research* 32(1):74–103.

dos Guimaraes Sa., Isabel

1992 The circulation of children in eighteenth century Southern Europe: The case of the foundling hospital of Porto. Ph.D. diss., Department of History of Civilization, European University Institute, Florence.

Downhower, F., and L. Brown

1980 Mate preferences of female mottled sculpins, *Cottus bairdi*. *Animal Behavior* 28:728–34.

Downhower, Jerry, L. Blumer, P. Lejeune, P. Gaudin, A. Marconats, and A. Bisazza

1990 Otolith asymmetry in *Cottus bairdi* and *C. gobio. Polski Archiwum Hydrobiologii* 37:209–20.

Draper, Patricia

1976 Social and economic constraints on child life among the !Kung. In *Kalahari Hunter Gatherers,* R. B. Lee and I. DeVore, eds., 199–217. Cambridge: Harvard University Press.

Draper, Patricia, and Pat Buchannon

1992 If you have a child you have a life: Demographic and cultural perspectives on fathering and old age in !Kung society. In *Father-Child Relations: Cultural and Biosocial Contexts,* B. Hewlett, ed., 131–52. Hawthorne, New York: Aldine de Gruyter.

Draper, Patricia, and Elizabeth Cashdan

1988 Technological change and child behavior among the !Kung. *Ethnology* 27:339–65.

Draper, Patricia, and R. Hames

1999 Birth order, sibling investment, and fertility among Ju/'hoans. (San) Paper presented at the Annual Meeting of the Human Behavior and Evolution Society, University of Utah, Salt Lake City, June 2–6, 1999.

Draper, Patricia, and Henry Harpending

1982 Father absence and reproductive strategy: An evolutionary perspective. *Journal of Anthropological Research* 38:255–73.

1987 Parent investment and the child's environment. In *Parenting across the Human Lifespan: Biosocial Dimensions,* Jane Lancaster, Jeanne Altmann, Alice Rossi, and Lonnie Sherrod, eds., 207–35. Hawthorne, New York: Aldine de Gruyter.

1988 A sociobiological perspective on the development of human reproductive strategies. In *Sociobiological Perspectives on Human Development,* Kevin MacDonald, ed., 340–72. New York: Springer-Verlag.

Drickamer, Lee

    1974 A ten-year summary of reproductive data of free-ranging *Macaca mulatta. Folia Primato-logica* 21:61–80.

Drotar, D.

    1991 The family context of nonorganic failure to thrive. *American Journal of Orthopsychiatry* 61:23–34.

Dullea, Georgia

    1987 In male-dominated Korea, an island of sexual equality. *NewYork Times* (July 9):C1, C10.

Dunbar, Robin

    1992 Neocortex size as a constraint on group size in primates. *Journal of Human Evolution* 20:469–93.

Duntley, D. M., and D. M. Buss

    1998 Evolved anti-infanticide modules. Paper presented at the Tenth Annual Meeting of the Human Behavior and Evolution Society, July 8–12, University of California, Davis.

Dupoux, A.

    1958 *Sur les pas de Monsieur Vincent: Trois cents ans d'histoire Parisienne de l'enfance abandonée.* Paris: Revue de l'Assistance Publique.

Durham, William H.

    1991 *Coevolution: Genes, Culture, and Human Diversity.* Stanford: Stanford University Press.

Duvernoy, Jean, trans.

    1965 *Le Register d'Inquisition de Jacque Fournier, eveque de Pamiers (1318–1325)*, 3 vols. Toulouse. (Latin Ms. 4030, Vatican Library)

Dwyer, Daisy, and Judith Bruce

    1988 *A House Divided: Women and Income in the Third World.* Stanford: Stanford University Press.

Dyson, Tim, and Mick Moore

    1983 On kinship structure, female autonomy and demographic behavior in India. *Population and Development Review* 9:35–60.

Dytrych, Zdenek, Zdenek Matejcek, and Vratislav Schuller

    1988 The Prague cohort: Adolescence and early adulthood. In *Born Unwanted: Developmental Effects of Denied Abortion,* H. P. David, Z. Dytrych, Z. Matejcek, and V. Schuller, eds., 87–102. New York and Prague: Springer and Czechoslovakia Medical Press.

Ealey, E.H.M.

    1963 The ecological significance of delayed implantation in a population of the hill kangaroo (*Macropus robustus*). In *Delayed Implantation,* A. C. Enders, ed. Chicago: University of Chicago Press.

    1967 Ecology of the *Macropus robustus* (Gould) in northwestern Australia. *CSIRO Wildlife Reserve* 12:27–51.

Early, J. D., and J. F. Peters

    1990 *The Population Dynamics of the Mucajai Yanomamö.* New York: Academic Press.

Eberhard, William

    1990 Inadvertent machismo. *Trends in Ecology and Evolution* 5(8):263.

    1996 *Female Control: Sexual Selection by Cryptic Female Choice.* Princeton: Princeton University Press.

Edwards, Carolyn Pope

1993 Behavioral sex differences in children of diverse cultures: The case of nurturance to infants. In *Juvenile Primates: Life History, Development and Behavior,* Michael Pereira and Lynn A. Fairbanks, eds., 327–38. Oxford: Oxford University Press.

Eibl-Eibesfeldt, Irenäus

1989 *Human Ethology.* New York: Aldine de Gruyter.

Eibl-Eibesfeldt, Irenäus, and Marie-Claude Mattei-Müller

1990 Yanomami wailing songs and the question of parental attachment in traditional kin-based societies. *Anthropos* 4–6:507–15.

Eisner, Thomas, Michael A. Goetz, David E. Hill, Scott R. Smedley, and Jerrold Meinwald

1997 Firefly "femme fatales" acquire defensive steroids (lucibufagins) from their firefly prey. *Proceedings of the National Academy of Sciences* 94:9723–28.

Eliot, George

1859 *Adam Bede.* London: Penguin Books, 1989.

1860 *The Mill on the Floss.* Boston: Houghton Mifflin, 1961.

1861 *Silas Marner.* Middlesex: Penguin Books, 1981.

1871–1872 *Middlemarch.* London: Penguin Books, 1965.

1876 *Daniel Deronda.* Middlesex: Penguin Books, 1979.

1990a Woman in France: Madame de Sablé (originally published in *Westminster Review,* October 1854). In *Selected Essays, Poems and Other Writings* (London: Penguin Classics), 8–37.

1990b The Natural History of German Life (originally published in *Westminster Review,* July 1856). In *Selected Essays, Poems and Other Writings,* 107–39.

1990c Silly Novels by Lady Novelists (originally published in *Westminster Review,* October 1856). In *Selected Essays, Poems and Other Writings,* 140–63.

1990d Margaret Fuller and Mary Wollstonecraft (originally published in *Leader,* October 13, 1855). In *Selected Essays, Poems and Other Writings,* 332–38.

Ellison, Peter T.

1995 Breastfeeding, fertility, and maternal condition. In *Breastfeeding: Biocultural Perspectives,* Patricia Stuart-Macadam and Katherine A. Dettwyler, eds., 305–45. Hawthorne, New York: Aldine de Gruyter.

In press. *On Fertile Ground.* Cambridge: Harvard University Press.

Ellsworth, Julie A., and Christopher Andersen.

1997 Adoption by captive parturient rhesus macaques: Biological vs. adopted infants and the cost of being a "twin" and rearing "twins." *American Journal of Primatology* 43: 259–64.

Elphick, M. C., and W. W. Wilkinson

1981 The effects of starvation and surgical injury on the plasma levels of glucose, free fatty acids, and neutral lipids in newborn babies suffering from various congenital anomalies. *Pediatric Research* 15:313–18.

Elwood, Robert W.

1994 Temporal-based kinship recognition: A switch in time saves mine. *Behavioural Processes* 33:15–24.

Elwood, Robert W., and Hazel Kennedy
  1990 The relationship between pregnancy block and the risk of infanticide from male mice. *Behavioral and Neural Biology* 53:277–83.
Ember, C.
  1975 Residential variation among hunter-gatherers. *Behavioral Science Research* 3:199–227.
  1978 Myths about hunter-gatherers. *Ethnology* 4:439–48.
Emlen, Stephen T.
  1995 An evolutionary theory of the family. *Proceedings of the National Academy of Sciences* 92:8092–99.
  1997 The evolutionary study of human family systems. *Social Science Information* 36:563–89.
Emlen, S. T., and L. W. Oring
  1977 Ecology, sexual selection, and the evolution of mating systems. *Science* 1297:215–23.
Emlen, Stephen T., N. J. Demong, and D. J. Emlen
  1989 Experimental induction of infanticide in female wattled jacanas. *Auk* 106:1–7.
Emlen, Stephen T., Peter H. Wrege, and Natalie J. Demong
  1995 Making decisions in the family: An evolutionary perspective. *American Scientist* 83:143–57.
Engels, F.
  1884 *The Origins of the Family, Private Property and the State.* New York: International Publishers, 1973.
Enquist, M., and A. Arak
  1994 Symmetry, beauty and evolution. *Nature* 372:169–72.
Essock-Vitale, Susan M.
  1984 The reproductive success of wealthy Americans. *Ethology and Sociobiology* 5:45–49.
Essock-Vitale, Susan M., and Michael T. McGuire
  1980 Predictions derived from the theories of kin selection and reciprocation assessed by anthropological data. *Ethology and Sociobiology* 1:233–43.
  1985a Women's lives viewed from an evolutionary perspective, I: Sexual histories, reproductive success and demographic characteristics of a random sample of American women. *Ethology and Sociobiology* 6:137–54.
  1985b Women's lives viewed from an evolutionary perspective, II: Patterns of helping. *Ethology and Sociobiology* 6:155–73.
Estrada, Alejandro
  1982 A case of adoption of a howler monkey infant (*Alouatta villosa*) by a female spider monkey (*Ateles geoffroyi*). *Primates* 23(1):135–37.
Etioko-Griffin, Agnes
  1986 Daughters of the forest. *Natural History* (May):36–42.
Evans, Roger M.
  1996 Hatching asynchrony and survival of insurance offspring in an obligate brood-reducing species, the American white pelican. *Behavioral Ecology and Sociobiology* 39:203–9.
Evans, Theodore A., Elycia J. Wallis, and Mark A. Elgar
  1995 Making a meal of mother. *Nature* 376:299.

Ewald, Paul W.

1996 Guarding against the most dangerous emerging pathogens: Insights from evolutionary biology. *Emerging Infectious Diseases* 2(4):245–57.

Eyer, Diane E.

1992a *Mother-Infant Bonding: A Scientific Fiction.* New Haven: Yale University Press.

1992b Infant bonding: A bogus notion. *Wall Street Journal* (November 24).

1993 The battle over bonding: How much must a baby bond with its mother? *USA Weekend* (May 7–9):4–5.

1996 *Motherguilt: How Our Culture Blames Mothers for What's Wrong with Society.* New York: Times Books.

Faerman, Marina, G. Kahila, P. Smith, C. Greenblatt, L. Stager, D. Filon, and A. Oppenheim

1997 DNA analysis reveals the sex of infanticide victims. *Nature* 385:212–13.

Fairbanks, Lynn A.

1988 Vervet monkey grandmothers: Interactions with infant grandoffspring. *International Journal of Primatology* 9:425–41.

1990 Reciprocal benefits of allomothering for female vervet monkeys. *Animal Behavior* 40:553–62.

1995 Maternal rejection is a U-shaped function of maternal condition in vervet monkeys (abstract). *American Journal of Primatology* 36(2):121.

Fairbanks, Lynn A., and M. T. McGuire

1986 Age, reproductive value, and dominance-related behaviour in vervet monkey females: Cross-generational influences on social relationships and reproduction. *Animal Behavior* 34:1710–21.

1995 Maternal condition and the quality of maternal care in vervet monkeys. *Behaviour* 132:733–54.

Faison, Seth

1995 Women as chattel: In China, slavery rises. *New York Times* (September 6).

Farnsworth, Clyde

1997 Facing pain of aborigines wrested from families, many Australians shrug. *New York Times* (June 8):10.

Fausto-Sterling, Ann

1985 *Myths of Gender: Theories about Women and Men.* New York: Basic Books.

Fedigan, Linda

1982 *Primate Paradigms.* Montreal: Eden Press.

Fedigan, Linda Marie, and Laurence Fedigan

1977 The social development of a handicapped infant in a free-living troop of Japanese monkeys. In *Primate Bio-social Development: Biological, Social and Ecological Determinants,* Suzanne Chevalier-Skolnikoff and Frank E. Poirier, eds., 205–22. New York: Garland.

1989 Gender and the study of primates. In *Gender and Anthropology: Critical Reviews for Teaching and Research,* Sandra Morgan, ed. Washington, D.C.: American Anthropological Association.

Fei, Hsiao-T'ung

1939 *Peasant Life in China: A field study of country life in the Yangtze Valley.* London: Routledge and Sons.

Fein, Esther B.

1998 For lost pregnancies, new rites of mourning. *New York Times* (January 25):1, 22.

Feldman, S. Shirley, and Sharon Churnin Nash

1986 Antecedents of early parenting. In *Origins of Nurturance: Developmental, Biological and Cultural Perspectives on Caregiving,* Alan Fogel and Gail F. Melson, eds., 209–32. Hillsdale, New Jersey: Lawrence Erlbaum Associates.

Felstiner, Mary Lowenthal

1994 *To Paint Her Life: Charlotte Salomon in the Nazi Era.* New York: HarperCollins.

Fernald, A.

1985 Four-month-old infants prefer to listen to motherese. *Infant Behavior and Development* 8:181–95.

Field, Tiffany, Robert Woodson, Reena Greenberg, and Debra Cohen

1982 Discrimination and imitation of facial expressions by neonates. *Science* 218: 179–81.

Fildes, Valerie

1986 *Breasts, Bottles and Babies.* Edinburgh: Edinburgh University Press.

1988 *Wet Nursing: A History from Antiquity to the Present.* Oxford: Basil Blackwell.

Firstman, Richard, and Jamie Talan

1997 *The Death of Innocents.* New York: Bantam Books.

Fisher, Helen

1992 *The Anatomy of Love.* New York: Norton.

Fisher, R. A.

1915 The evolution of sexual preference. *Eugenics Review* 7:184–92.

1930 *The Genetical Theory of Natural Selection.* Oxford: Clarendon Press.

Fleming, Alison S., Carl Corter, and Meir Steiner

1995 Sensory and hormonal control of maternal behavior in rat and human mothers. In *Motherhood in Human and Nonhuman Primates,* C. R. Pryce, R. D. Martin, and D. Skuse, eds., 106–14. Basel: S. Karger.

Fleming, Alison, Diane L. Ruble, Gordon L. Flett, and David L. Shaul

1988 Postpartum adjustment in first-time mothers: Relations between mood, maternal attitudes, and mother-infant interactions. *Developmental Psychology* 24(1):71–81.

Fleming, Alison, Diane N. Ruble, Gordon Flett, and Vicki van Wagner

1990 Adjustment in first-time mothers: Changes in mood and mood content during the early postpartum months. *Developmental Psychology* 26(1):137–43.

Flint, M.

1979 Is there a secular trend in age of menopause? *Maturitas* 1:133–39.

Foley, Robert

1996 The adaptive legacy of human evolution: A search for the Environment of Evolutionary Adaptedness. *Evolutionary Anthropology* 4(6):194–203.

Fonagy, Peter, Miriam Steele, George Moran, Howard Steele, and Anna Higgitt

1993 Measuring the ghost in the nursery: An empirical study of the relation between parents' mental representation of childhood experiences and their infants' security of attachment. *Journal of the American Psychoanalytic Association* 41(4):957–89.

Foote, John

1919 Ancient poems on infant hygiene. *Annals of Medical History* II(3):213–27.

Formby, David

1967 Maternal recognition of infant's cry. *Developmental Medicine and Child Neurology* 9:293–98.

Fossey, Dian

1984 Infanticide in mountain gorillas (*Gorilla gorilla beringei*) with comparative notes on chimpanzees. In *Infanticide: Comparative and Evolutionary Perspectives,* Glenn Hausfater and Sarah Blaffer Hrdy, eds., 217–36. NewYork: Aldine.

Fowke, Keith R., N. J. D. Nagelkerke, J. Kimani, J. N. Simonsen, A. O. Anzala, J. J. Bwayo, K. S. MacDonald, E. N. Ngugi, and F. A. Plummer

1996 Resistance to HIV-1 infection among persistently seronegative prostitutes in Nairobi, Kenya. *Lancet* 348:1347–51.

Fox, Charles W., Monica S. Thakar, and Timothy A. Mousseau

1997 Egg size plasticity in a seed beetle: An adaptive maternal effect. *American Naturalist* 149(1):150–63.

Fraiberg, S., E. Adelson, and V. Shapiro

1975 Ghosts in the nursery: A psychoanalytic approach to the problem of impaired mother-infant relationships. *Journal of the American Academy of Child Psychiatry* 14: 387–422.

Fraisse, Genevieve

1985 *Clémence Royer, philosophe et femme de science.* Paris: Editions La Decouverte.

Frame, L. H., J. R. Malcolm, G. W. Frame

1979 Social organization of African wild dogs (*Lycaon pictus*) in the Serengeti plains. *Zeitschrift Tierpsychologie* 50:225–49.

Francis, Charles M., Edythe Anthony, Jennifer A. Burnton, and Thomas H. Kunz

1994 Lactation in male fruit bats. *Nature* 367:691–92.

Frank, Laurence G.

1997 Evolution of genital masculinization: Why do female hyaenas have such a large "penis"? *Trends in Ecology and Evolution (TREE)* 12(2):58–62.

Frank, Laurence G., Mary L. Weldele, and Stephen E. Glickman

1995 Masculinization costs in hyaenas. *Nature* 377:584–85.

Frankel, Simon J.

1994 The eclipse of sexual selection theory. In *Sexual Knowledge, Sexual Science: The History of Attitudes Towards Sexuality,* Roy Porter and Mikulas Teich, eds., 158–83. Cambridge: Cambridge University Press.

Fredga, Karl, Alfred Gropp, Heinz Winking, and Fritz Frank

1977 A hypothesis explaining the exceptional sex ratio in the wood lemming (*Myopus schisticolor*). *Hereditas* 85:101–4.

Freedman, Daniel

1974 *Human Infancy: An Evolutionary Perspective.* New York: John Wiley.

1979 *Human Sociobiology: A Holistic Approach.* New York: Free Press.

French, Jeffrey A.

1997 Proximate regulation of singular breeding in callithrid primates. In *Cooperative Breeding in Mammals,* Nancy G. Solomon and Jeffrey A. French, eds., 34–75. Cambridge: Cambridge University Press.

Frisch, R. E.

1978 Populations, food intake, and fertility: Historical evidence for a direct effect of nutrition on reproductive ability. *Science* 199:22–29.

1988 Fatness and fertility. *Scientific American* 258:70–77.

Frith, H. J., and G. B. Sharman

1964 Breeding in wild populations of the red kangaroo *Megaleia rufa. CSIRO Wildlife Reserve* 9:86–114.

Frodi, A. M., M. E. Lamb, L. A. Leavitt, C. M. Donovan, C. Neff, and D. Sherry

1978 Fathers' and mothers' responses to the faces and cries of normal and premature infants. *Developmental Psychology* 14(5):40–49.

Fromm, Erich

1956 *The Art of Loving.* New York: Harper and Row.

Fuchs, Rachel Ginnis

1984 *Abandoned Children: Foundlings and Child Welfare in Nineteenth-Century France.* Albany: State University of New York Press.

1987 Legislation, poverty and child-abandonment in nineteenth-century Paris. *Journal of Interdisciplinary History* 18:55–80.

Fuchs, S.

1982 Optimality of parental investment: The influence of nursing on the reproductive success of mother and female young house mice. *Behavioral Ecology and Sociobiology* 10:39–51.

Furlow, F. Bryant

1996 The smell of love. *Psychology Today* (March–April):38–45.

Furlow, F. B., T. Armijo-Prewitt, S. W. Gangestad, and R. Thornhill

1997 Fluctuating asymmetry and psychometric intelligence. *Proceedings of the Royal Society of London,* Series B 264:823–29.

Gagneux, Pascal, David S. Woodruff, and Christophe Boesch

1997 Furtive mating in female chimpanzees. *Nature* 387:327–28.

1999 Female reproductive strategies, paternity and community structure in wild West African chimpanzees. *Animal Behavior* 57:19–32.

Galdikas, B.

1985a Adult male sociality and reproductive tactics among orangutans at Tanjung Puting. *Folia Primatologica* 45:9–24.

1985b Subadult male orangutan sociality and reproductive behavior at Tanjung Puting. *American Journal of Primatology* 9:101–19.

Galdikas, B., and J. Wood

1990 Birth spacing in humans and apes. *American Journal of Physical Anthropology* 83:185–91.

Galef, Bennett G., Jr.

1976 Social transmission of acquired behavior: A discussion of tradition and social learning in vertebrates. *Advances in the Study of Behavior* VI:77–100.

Gamble, Eliza Burt

1894 *The Evolution of Woman: An Inquiry into the Dogma of Her Inferiority to Man*. New York and London: G. P. Putnam.

Gandelman, R., and N. Simon

1978 Spontaneous pup-killing by mice in response to large litters. *Developments in Psychobiology* 11:235–41.

Gangestad, Steven, and Randy Thornhill

1997 Human sexual selection and developmental stability. In *Evolutionary Social Psychology*, Jeffrey A. Simpson and Douglas T. Kenrick, eds., 169–95. Mahwah, New Jersey: Lawrence Erlbaum.

Garden, Maurice

1970 La démographie de lyonnaise : l'analyse des compartements. In *Lyon et les Lyonnais au XVIIIe siècle*, 83–169. Bibiothèque de la Faculté des Lettres de Lyon. Paris: Edition "Les Belle Lettres."

Geddes, W. R.

1963 *Peasant Life in Communist China*. Monograph No. 6. Ithaca, New York: Society for Applied Anthropology.

Genevie, Louis, and Eva Margolies

1987 *The Motherhood Report: How Women Feel about Being Mothers*. New York: Macmillan.

Geronimus, Arline T.

1987 On teenage childbearing and neonatal mortality in the United States. *Population and Development Review* 13:245–79.

1996 What teen mothers know. *Human Nature* 7:323–52.

Gibber, Judith

1986 Infant-directed behavior of rhesus monkeys during their first pregnancy and parturition. *Folia Primatologica* 46:118–24.

Gibbons, Ann

1998a In mice, mom's genes favor brains over brawn. *Science* 280:1346.

1998b A blow to the "grandmother theory." *Science* 280:516.

Gilbert, Susan

1998a Raising grandchildren: Rising stress. *New York Times* (July 28):B-8.

1998b Infant homicide found to be rising in U.S. *New York Times* (October 27):F-10.

Gileva, Emily A., Isaac E. Benenson, Luidmila A. Konopistseva, V. F. Puchkov, and I. A. Makaranets

1982 XO females in the varying lemming, *Dicrostonyx torquatus:* Reproductive performance and its evolutionary significance. *Evolution* 36(3):601–9.

Gilibert, Jean Emmanuel

1770 Dissertation sur la depopulation causée par les vice, les prejuges et les erreurs des nourrices mercenaires. . . . In *Les chefs d'oeuvres de Monsieur de Sauvages*, Jean Emmanuel Gilibert, ed., vol. 2. Lyon.

Gillin, Frances D., David Reiner, and Chi-Sun Wang

1983 Human milk kills parasitic intestinal protozoa. *Science* 221:1290–92.

Gillogly, A. K.

1983 Changes in infant care and feeding practices in East Kwaio, Malita. Paper presented at the Symposium on Infant Care and Feeding in Oceania, Annual Meeting of the Association of Social Anthropology in Oceania, March 9–13, New Harmony, Indiana.

Gilman, Charlotte Perkins

1898 *Women and Economics.* Berkeley: University of California Press. (Reprinted 1998).

1901 *Concerning Children.* Boston: Small, Maynard and Co.

1979 *Herland.* New York: Pantheon (reprint of original 1915 publication).

Girard, J., and P. Ferre

1982 Metabolic and hormonal changes around birth. In *The Biochemical Development of the Fetus and Neonate,* C. T. Jones, ed., 517–51. Amsterdam: Elsevier.

Giray, Tugrul, and Gene Robinson

1997 Common endocrine and genetic mechanisms of behavioral development in male and worker honey bees and the evolution of division of labor. *Proceedings of the National Academy of Sciences* 93:11718–22.

Glander, Ken

1980 Reproduction and population growth in free-ranging mantled howler monkeys. *American Journal of Physical Anthropology* 53:25–36.

Glass, Nigel

1999 Infanticide in Hungary faces stiffer penalties. *Lancet* 353 (9152):570.

Glass, Roger I., Jan Holmgren, Charles E. Haley, M. R. Khan, et al.

1985 Predisposition for cholera of individuals with O blood group. *American Journal of Epidemiology* 121:791–96.

Goldschmidt, Walter

1996 Functionalism. In *Encyclopedia of Cultural Anthropology,* vol. 2, David Levinson and Melvin Ember, eds., 510–12. New York: Henry Holt.

Gomendio, Montserrat, Jorge Cassinello, Michael W. Smith, and Patrick Bateson

1995 Maternal state affects intestinal changes of rat pups at weaning. *Behavioral Ecology and Sociobiology* 37:71–80.

Gomendio, M., T. H. Clutton-Brock, S. D. Albon, F. E. Guinness, and M. J. Simpson

1990 Mammalian sex ratios and variation in costs of rearing sons and daughters. *Nature* 343:261–63.

Goodall, Jane

1977 Infant-killing and cannibalism in free-living chimpanzees. *Folia Primatologica* 28:259–82.

1986 *The Chimpanzees of Gombe.* Cambridge: Harvard University Press.

Goodall, Jane, Adriano Bandora, Emilie Bergmann, Curt Busse, Hilali Matama, Esilom Mpongo, Ann Pierce, and David Riss

1979 Intercommunity interactions in the chimpanzee population of the Gombe National Park. In *The Great Apes,* David A. Hamburg and Elizabeth R. McCown, eds., 13–54. Menlo Park, California: Benjamin-Cummings.

Goodman, Morris, Calvin A. Porter, John Czelusniak, H. Schneider, J. Shoshani, G. Gunnell, and C. P. Groves

1998 Toward a phylogenetic classification of primates based on DNA evidence complemented by fossil evidence. *Molecular Phylogenetics and Evolution* 9:585–598.

Goody, E.

   1984 Parental strategies: Calculation or sentiment? Fostering practices among West Africans. In *Interest and Emotions: Essays on the Study of Family and Kinship,* H. Medick and D. W. Sabean, eds., 266–77. Cambridge: Cambridge University Press.

Gordon Cumming, C. F.

   1900 *Wanderings in China.* Edinburgh: William Blackwood.

Gosden, Roger

   1996 The vocabulary of the egg. *Nature* 383:485–86.

Gosling, L. M.

   1986 Selective abortion of entire litters in the coypu: Adaptive control of offspring production in relation to quality and sex. *American Naturalist* 127(6):772–95.

Gould, Stephen J.

   1977 *Ontogeny and Phylogeny.* Cambridge: Belknap Press of Harvard University Press.

   1981 *The Mismeasures of Man.* New York: Norton.

Gowaty, Patricia Adair

   1985 Low probability of paternity or . . . something else? Commentary on "The human community as a primate society." *Behavioral and Brain Sciences* 8(4):675.

   1992 Evolutionary biology and feminism. *Human Nature* 3:217–49.

   1995 False criticisms of sociobiology and behavioral ecology: Genetic determinism, untestability, and inappropriate comparisons. *Politics and the Life Sciences* 14(2):174–80.

   1996 Battles of the sexes and origins of monogamy. In *Partnerships in Birds: The Study of Monogamy,* Jeffrey M. Black, ed., 21–52. Oxford: Oxford University Press.

   1997 Birds face sexual discrimination. *Nature* 385:486–87.

Gowaty, Patricia Adair, ed.

   1997 *Feminism and Evolutionary Biology: Boundaries, Intersections and Frontiers.* New York: Chapman and Hall.

Gowaty, Patricia Adair, and Michael R. Lennartz

   1985 Sex ratios of nestling and fledgling red-cockaded woodpeckers (*Picoides borealis*) favor males. *American Naturalist* 126:347–53.

Graglia, Carolyn

   1998 Feminism isn't antisex. It's only antifamily. *Wall Street Journal* (August 6).

Gragson, Theodore L.

   1989 Allocation of time to subsistence and settlement in a ciri khonome Pume village of the llanos of Apure, Venezuela. Ph.D. diss., Department of Anthropology, Pennsylvania State University.

Graham, C. E.

   1970 Reproductive physiology of the chimpanzees. In *The Chimpanzee,* vol. 3., G. Bourne, ed., 183–220. Basel: S. Karger.

   1986 Endocrinology of reproductive senescence. In *Comparative Primate Biology: Reproduction and Development,* vol. 3., W. R. Dukelow and J. Erwin, eds., 93–99. New York: Alan Liss.

Grammer, Karl

   1996 The human mating game: The battle of the sexes and the war of signals. Paper presented at the Annual Meeting of the Human Behavior and Evolution Society, June 26–30, Northwestern University, Evanston, Illinois.

Grammer, Karl, and Randy Thornhill

    1994 Human (*Homo sapiens*) facial attractiveness and sexual selection: The role of symmetry and averageness. *Journal of Comparative Psychology* 108(3):233–42.

Grant, Tom

    1989 *The Platypus: A Unique Mammal.* Kensington: New South Wales University Press.

Graves, Robert

    1955 *The Greek Myths,* vol. 2. Baltimore: Penguin Books.

Gray, Ronald H., Maria J. Wawer, D. Serwadda, N. Sewankambo, C. Li, F. Wabwire-Mangen, L. Paxton, N. Kiwanuka, G. Kogozi, J. Konde-Lule, T. C. Quinn, and C. A. Gaydos

    1998 Population-based study of fertility in women with HIV-1 infection in Uganda. *Lancet* 351:98–103.

Greaves, Rusty

    1996 Ethnoarchaeology of wild root collection among savanna foragers of Venezuela. Paper presented at the Fifty-fourth Annual Plains Anthropology Conference, Iowa City.

Greenberg, Martin, and Norman Morris

    1974 Engrossment: The newborn's impact upon the father. *American Journal of Orthopsychiatry* 44(4):520–31.

Greene, Erick

    1989 A diet-induced developmental polymorphism in a caterpillar. *Science* 243:643–46.

    1996 Effect of light quality and larval diet on morph induction in the polymorphic caterpillar *Nemoria arizonaria* (Lepidoptera: Geometridae). *Biological Journal of the Linnean Society* 58:277–85.

Gregor, Thomas

    1985 *Anxious Pleasures: The Sexual Lives of an Amazonian People.* Chicago: University of Chicago Press.

    1988 "Infants are not precious to us": The psychological impact of infanticide among the Mehinaku Indians. Paper presented by the Stirling Prize recipient, Annual Meeting of the American Anthropological Association, November 16–20, Phoenix.

Griffin, P. Bion, and Marcus B. Griffin

    1992 Fathers and childcare among the Cagayan Agta. In *Father-Child Relations: Cultural and Biosocial Contexts,* Barry S. Hewlett, ed. Hawthorne, New York: Aldine de Gruyter.

Grimes, David A.

    1998 The continuing need for late abortion. *Journal of the American Medical Association* 8:747–48.

Griminger, P.

    1983 Digestive system and nutrition. In *Physiology and Behaviour of the Pigeon,* Michael Abs, ed. New York: Academic Press.

Gross, M. R.

    1985 Disruptive selection for alternative life histories in salmon. *Science* 313:47–48.

Gross, Paul R., and Norman Levitt

1994 *Higher Superstition: The Academic Left and Its Quarrels with Science.* Baltimore: Johns Hopkins University Press.

Grossman, K. E., K. Grossman, F. Huber, and U. Wartner

1981 German children's behavior towards their mothers at 12 months and their fathers at 18 months in Ainsworth's Strange Situation. *International Journal of Behavioral Development* 4:157–81.

Gubernick, David J., and Randy Nelson

1989 Prolactin and paternal behavior in the biparental California mouse, *Peromyscus californicus. Hormones and Behavior* 23:203–10.

Gubernick, David J., Sandra Wright, and Richard E. Brown

1993 The significance of the father's presence for offspring survival in the monogamous California mouse, *Peromyscus californicus. Animal Behavior* 46:539–46.

Gusinde, Martin

1931 *Die Feuerland-Indianer,* vol. I. *Die Selk'nam.* Modling bei Wien: Anthropos Verlag.

Guttentag, M., and P. Secord

1983 *Too Many Women: The Sex Ratio Question.* Beverly Hills: Sage.

Guyer, Jane I.

1994 Lineal identities and lateral networks: The logic of polyandrous motherhood. In *Nuptiality in Sub-Saharan Africa: Current Changes and Impact on Fertility,* C. Bledsoe and G. Pison, eds., 231–52. Oxford: Clarendon Press.

Haffter, Carl

1968 The changeling: History and psychodynamics of attitudes to handicapped children in European folklore. *Journal of the History of Behavioral Sciences* 4(1):55–61.

Hagen, Edward H.

1996 Postpartum depression as an adaptation to paternal and kin exploitation. Paper presented at the Sixth Annual Meeting of the Human Behavior and Evolution Society, Northwestern University.

Hager, Barbara

1992 Get thee to a nunnery: Female religious claustration in Medieval Europe. *Ethology and Sociobiology* 13:385–407.

Hahn, D. Caldwell

1981 Asynchronous hatching in the laughing gull: Cutting losses and reducing rivalry. *Animal Behavior* 29:421–27.

Haig, David

1992 Genomic imprinting and the theory of parent-offspring conflict. *Seminars in Developmental Biology* 3:153–60.

1993 Genetic conflicts of human pregnancy. *Quarterly Review of Biology* 68:495–532.

1995 Prenatal power plays. *Natural History* 104:39.

1996a Alterations of generations: Genetic conflicts of pregnancy. *American Journal of Reproductive Immunology,* 35:226–32.

1996b Gestational drive and the green-bearded placenta. *Proceedings of the National Academy of Sciences* 93:6547–51.

1997 The social gene. In *Behavioural Ecology: An Evolutionary Approach,* 4th ed., J. R. Krebs and Nicholas B. Davies, eds., 284–304. Oxford: Blackwell Scientific.

Haig, David, and M. Westoby

1989 Parent-specific gene expression and the triploid endosperm. *American Naturalist* 134:147–55.

Haight, Gordon

1968 *George Eliot: A Biography.* New York: Oxford University Press.

Haight, Gordon, ed.

1954–78 *George Eliot Letters,* 9 vols. New Haven: Yale University Press.

Hakansson, T.

1988 *Bridewealth, Women and Land: Social Change Among the Gusii of Kenya.* Uppsala Studies in Cultural Anthropology 10.

Hames, Raymond B.

1988 The allocation of parental care among the Ye'kwana. In *Human Reproductive Behaviour: A Darwinian Perspective,* Laura Betzig, Monique Borgerhoff Mulder, and Paul Turke, eds., 237–52. Cambridge: Cambridge University Press.

Hamilton, W. D.

1963 The evolution of altruistic behavior. *The American Naturalist* 97:354–56

1964 The genetical evolution of social behavior. *Journal of Theoretical Biology* 7:1–16, 17–52.

1966 The moulding of senescence by natural selection. *Journal of Theoretical Biology* 12:12–45.

1967 Extraordinary sex ratios. *Science* 156:477–88.

1975 Gamblers since life began: Barnacles, aphids, elms. *Quarterly Review of Biology* 50:175–80.

1982 Pathogens as causes of genetic diversity in their host populations. In *Population Biology of Infectious Diseases,* R. M. Anderson and R. M. May, eds., 269–96. New York: Springer-Verlag.

1995 *The Narrow Roads of Gene Land.* Oxford: Spektrum/W. H. Freeman.

Hammel, Eugene A.

1996 Demographic constraints on population growth of early humans: Emphasis on the probable role of females in overcoming such constraints. *Human Nature* 7:217–55.

Hammer, M., and R. A. Foley

1996 Longevity, life history, and allometry: How long did humans live? *Human Evolution* 11:61–66.

Hampson, E., and D. Kimura

1988 Reciprocal effects of hormonal fluctuations on human motor and perceptual-spatial skills. *Behavioral Neuroscience* 102:456–59.

Haraway, Donna

1989 *Primate Visions: Gender, Race and Nature in the World of Modern Science.* New York: Routledge.

Harcourt, A. H., Kelly J. Stewart, and Dian Fossey

1981 Gorilla reproduction in the wild. In *Reproductive Biology of the Great Apes,* Charles E. Graham, ed., 265–79. New York: Academic Press.

Harcourt, A. H., P. H. Harvey, S. G. Larson, and R. V. Short

1981 Testis weight, body weight and breeding system in primates. *Nature* 293:55–57.

Harding, Sandra

1986 *The Science Question in Feminism.* Ithaca: Cornell University Press.

1992 After the neutrality ideal: Science, politics, and "strong objectivity." *Social Research* 59(3):567–87.

Hardy, A.

1960 Was man more aquatic in the past? *New Scientist* 17:642–45.

Hardy, Janet B., and E. David Mellits

1977 Relationship of low birth weight to maternal characteristics of age, parity, education and body size. In *The Epidemiology of Prematurity,* D. M. Reed and F. J. Stanley, eds., 131–55. Baltimore: Urban and Schwarzenberg.

Harlap, S.

1979 Gender of infants conceived on different days of the menstrual cycle. *New England Journal of Medicine* 300:1445–48.

Harley, Diane

1990 Aging and reproductive performance in langur monkeys (*Presbytis entellus*). *American Journal of Physical Anthropology* 83:253–61.

Harlow, Harry, Margaret K. Harlow, and Stephen J. Suomi

1971 From thought to therapy: Lessons from a private laboratory. *American Scientist* 659:538–49.

Harlow, H. K., M. K. Harlow, R. O. Dodsworth, and G. L. Arling

1966 Maternal behavior of rhesus monkeys deprived of mothering and peer association in infancy. *Proceedings of the American Philosophical Society* 110:58–66.

Harpending, Henry C., Stephen T. Sherry, Alan R. Rogers, and Mark Stoneking

1993 The genetic structure of ancient human populations. *Current Anthropology* 34:483–96.

Harpending, Henry C., Mark A. Batzer, Michael Gurven, Lynn B. Jorde, Alan R. Rogers, and Stephen T. Sherry

1998 Genetic traces of ancient demography. *Proceedings of the National Academy of Sciences* 95:1961–67.

Harris, Judith Rich

1998 *The Nurture Assumption.* New York: Free Press.

Hartmann, H.

1958 *Ego Psychology and the Problem of Adaptation.* London: Imago; New York: International Universities Press (originally published in German in 1939).

Hartung, John

1982 Polygyny and the inheritance of human wealth. *Current Anthropology* 23:1–12.

Harvey, J. R.

1970 *Victorian Novelists and Their Illustrators.* London: Sidgwick and Jackson.

Harvey, Joy

1987 Strangers to each other. In *Uneasy Careers and Intimate Lives: Women in Science, 1789–1979,* Pnina G. Abir-Am and Dorinda Outram, eds., 147–71. New Brunswick, New Jersey: Rutgers University Press.

1997 *Almost a Man of Genius: Clémence Royer, Feminism and Nineteenth-Century Science.* New Brunswick, New Jersey: Rutgers University Press.

Harvey, Paul H., R. D. Martin, and T. H. Clutton-Brock

1987 Life histories in comparative perspectives. In *Primate Societies,* Barbara Smuts et al., eds., 181–96. Chicago: University of Chicago Press.

Hashimoto, Chie, Takeshi Furuichi, and Osamu Takenaka

1996 Matrilineal kin relationship and social behavior of wild bonobos (*Pan paniscus*): Sequencing the D-loop region of mitochondrial DNA. *Primates* 37(3):305–18.

Hauser, Marc

1996 *The Evolution of Intelligence.* Cambridge: MIT Press.

In press *Wild Minds: What Animals Think.* New York: Henry Holt.

Hausfater, G., and S. B. Hrdy, eds.

1984 *Infanticide: Comparative and Evolutionary Perspectives.* New York: Aldine.

Hausfater, G., J. Altmann, and S. Altmann

1982 Long-term consistency of dominance rank among female baboons. *Science* 217:752–55.

Hawkes, Kristen

1991 Showing off: Tests of another hypothesis about men's foraging goals. *Ethology and Sociobiology* 11:29–54.

1993 Why hunter-gatherers work: An ancient version of the problem of public goods. *Current Anthropology* 34:341–61.

1997 What are men doing? Lecture presented in the Anthropology Colloquium, April 11, University of California, Davis.

n.d. Hunting and the evolution of egalitarian societies: Lesson from the Hadza. Ms. prepared for "Hierarchies in Action: Who Benefits?" Symposium organized by M. W. Dahl.

Hawkes, Kristen, J. F. O'Connell, and N. G. Blurton-Jones

1989 Hardworking Hadza grandmothers. In *Comparative Socioecology: The Behavioral Ecology of Humans and Other Mammals,* V. Standen and R. A. Foley, eds., 341–66. London: Basil Blackwell.

1995 Hadza children's foraging: Juvenile dependency, social arrangements, and mobility among hunter-gatherers. *Current Anthropology* 36:1–24.

1997 Hadza women's time allocation, offspring provisioning and the evolution of long post-menopausal life spans. *Current Anthropology* 38:551–77.

n.d. Hadza hunting and the evolution of the nuclear family. Ms. provided courtesy of K. Hawkes, Department of Anthropology, University of Utah. In preparation.

Hawkes, K., J. F. O'Connell, N. G. Blurton-Jones, H. Alvarez, and E. L. Charnov

1998 Grandmothering, menopause, and the evolution of human life histories. *Proceedings of the National Academy of Sciences* 95:1336–39.

1999 The grandmother hypothesis and human evolution. In *Evolutionary Anthropology and Human Social Behavior: Twenty Years Later,* L. Cronk, N. Chagnon, and W. Irons, eds. Hawthorne, New York: Aldine de Gruyter. In press.

Hays, Sharon

1996 *The Cultural Contradictions of Motherhood.* New Haven: Yale University Press.

Hayssen, Virginia

1993 Empirical and theoretical constraints on the evolution of lactation. *Journal of Dairy Science* 76:3213–33.

1995 Milk: It does a baby good. *Natural History* (December):36.

Hendrie, T. A., P. E. Peterson, J. Short, A. F. Tarantal, E. Rothgarn, M. I. Hendrie, and A. J. Hendrickx

1996 Frequency of prenatal loss in a macaque breeding colony. *American Journal of Primatology* 40:41–53.

Hepper, Peter, E. Alyson Shannon, and James C. Dornan

1997 Sex differences in fetal mouth movements. *Lancet* 350:1820.

Herbert, Bob

1994 China's missing girls. *New York Times* (October 30).

Herlihy, D., and C. Klapisch-Zuber

1985 *The Tuscans and Their Families: A Study of the Florentine Castrata of 1427*. New Haven: Yale University Press.

Herzog, Alfred, and Thomas Detre

1976 Psychotic reactions associated with childbirth. *Diseases of the Nervous System* 37:229–35.

Hewlett, Barry

1991 Demography and childcare in preindustrial societies. *Journal of Anthropological Research* 47:1–23.

1992 Husband-wife reciprocity and the father-infant relationship among Aka pygmies. In *Father-Child Relations: Cultural and Biosocial Contexts,* Barry S. Hewlett, ed., 153–76. Hawthorne, New York: Aldine de Gruyter.

Hewlett, Barry S., ed.

1992 *Father-Child Relations: Cultural and Biosocial Contexts.* Hawthorne, New York: Aldine de Gruyter.

Hildebrandt, K. A., and H. E. Fitzgerald

1979 Facial feature determinants of infant attractiveness. *Infant Behavior and Development* 2:329–39.

Hill, C. M., and H. L. Ball

1996 Abnormal births and other "ill omens": The adaptive case for infanticide. *Human Nature* 7:381–402.

Hill, Kim

1993 Life history theory and evolutionary anthropology. *Evolutionary Anthropology* 2(3):76–88.

Hill, Kim, and A. Magdalena Hurtado

1989 Hunter-gatherers of the New World. *American Scientist* 77(5):436–43.

1996 *Aché Life History: The Ecology and Demography of a Foraging People*. Hawthorne, New York: Aldine de Gruyter.

1997 The evolution of premature reproductive senescence and menopause in human females: An evaluation of the "grandmother hypothesis." In *Human Nature: A Critical Reader,* L. Betzig, ed., 118–39. Oxford: Oxford University Press.

Hill, Kim, and Hillard Kaplan

    1988 Tradeoffs in male and female reproductive strategies among the Aché. Parts 1 and 2. In *Human Reproductive Behaviour: A Darwinian Perspective,* Laura Betzig, Monique Borgerhoff Mulder, and Paul Turke, eds., I:277–89; II:291–305. Cambridge: Cambridge University Press.

Hilton, Charles E., and Rusty D. Greaves

    1995 Mobility patterns in modern human foragers. Paper presented at the Annual Meeting of the American Association of Physical Anthropologists, Oakland, California. (Abstract published in *American Journal of Physical Anthropology* supplement 20:11.)

Hilts, Philip J.

    1997 Misdiagnoses are said to mask lethal abuse. *New York Times* (September 11).

Hinde, Robert A.

    1969 Analyzing the roles of the partners in a behavioral interaction: Mother-infant relations in rhesus macaques. *Annals of New York Academy of Sciences* 159:651–67.

    1991 A biologist looks at anthropology. *Journal of the Royal Anthropological Institute* 26(4):583–608.

Hinde, Robert A., and Lynda McGinnis

    1977 Some factors influencing the effects of temporary mother-infant separation: some experiments with rhesus monkeys. *Psychological Medicine* 7:197–212.

Hiraiwa-Hasegawa, Mariko

    1987 Infanticide in primates and a possible case of male-biased infanticide in chimpanzees. In *Animal Societies: Theories and Facts,* Y. Ito, J. L. Brown, and J. Kikkawa, eds., 125–39. Tokyo: Scientific Societies Press.

Hiraiwa-Hasegawa, Mariko, and Toshikazu Hasegawa

    1994 Infanticide in nonhuman primates: sexual selection and local resource competition. In *Infanticide and Parental Care,* Stefano Parmigiani and F. vom Saal, eds., 137–84. Langhorne, Pennsylvania: Harwood Academic.

Ho, Ping-ti

    1959 *Studies on the Population of China, 1368–1953.* Cambridge: Harvard University Press.

Hoage, R. J.

    1978 Parental care in *Leontopithecus rosalia rosalia:* age and sex differences in carrying behavior and the role of prior experience. In *Biology and Conservation of the Callithrichidae,* D. Kleiman, ed. Washington, D.C.: Smithsonian Institution.

Hobcraft, J. M., J. W. O'Donald, and S. O. Rutstein

    1985 Demographic determinants of infant and early child mortality: A comparative analysis. *Population Studies* 39:363–85.

Hodder, H. F.

    1996 A few super women. *Harvard Magazine* (May–June): 13–14.

Holland, Brett, and William R. Rice

    1999 Experimental removal of sexual selection reverses intersexual antagonistic coevolution and removes a reproductive load. *Proceedings of the National Academy of Sciences* 96:5083–88.

Hölldobler, Bert, and Edward O. Wilson

1990 *The Ants.* Cambridge: Harvard University Press.

1994 *Journey to the Ants.* Cambridge: Harvard University Press.

Holmes, Donna, and Christine Hitchcock

1997 A feeling for the organism: An empirical look at gender and research choices of animal behaviorists. In *Feminism and Evolutionary Biology: Boundaries, Intersections and Frontiers,* P. Gowaty, ed., 184–202. New York: Chapman and Hall.

Holmes, Warren G., and Jill M. Mateo

1998 How mothers influence the development of litter-mate preferences in Belding's ground squirrels. *Animal Behavior* 55:1555–70.

Homer

1963 *The Odyssey,* Robert Fitzgerald, trans. Garden City, New York: Doubleday/Anchor.

1990 *The Iliad,* Robert Fagles, trans. New York: Penguin.

Honig, Barbara

1994 Components of lifetime reproductive success in communally and solitarily nursing house mice: A laboratory study. *Behavioral Ecology and Sociobiology* 34:275–83.

Hoogland, John L.

1994 Nepotism and infanticide among prairie dogs. In *Infanticide and Parental Care,* Stefano Parmigiani and F. vom Saal, eds., 321–37. Langhorne, Pennsylvania: Harwood Academic.

1995 *The Black-Tailed Prairie Dog: Social Life of a Burrowing Mammal.* Chicago: University of Chicago Press.

Hopkins, Nancy

1976 The high price of success in science. *Radcliffe Quarterly* (June):16–18.

Hopwood, J. S.

1927 Child murder and insanity. *Journal of Mental Science* 73:95–108.

Horgan, John

1995 The new social darwinists. *Scientific American* 273(4):174–81.

Horner, J. R., and B. Weishampel

1988 A comparative embryological study of two ornithischian dinosaurs. *Nature* 332:256–57.

Hornsveld, H. K., B. Garssen, M. J. C. Fiedeldij Dop, P. I. van Spiegel, and J. C. J. M. de Haas

1996 Double-blind placebo-controlled study of the hyperventilation provocation test and the validity of the hyperventilation syndrome. *Lancet* 348:154–58.

Hostetler, John

1974 *Hutterite Society.* Baltimore: Johns Hopkins University Press.

Howell, Nancy

1979 *Demography of the Dobe !Kung.* New York: Academic Press.

Hrdy, Sarah Blaffer

1974 Male-male competition and infanticide among the langurs (*Presbytis entellus*) of Abu, Rajasthan. *Folia Primatologica* 22:19–58.

1976 The care and exploitation of nonhuman primate infants by conspecifics other than the mother. *Advances in the Study of Behavior* VI:101–58.

1977 *The Langurs of Abu: Female and Male Strategies of Reproduction.* Cambridge: Harvard University Press.

1979 Infanticide among animals: A review, classification, and examination of the implications for the reproductive strategies of females. *Ethology and Sociobiology* 1:13–40.

1981a *The Woman That Never Evolved.* Cambridge: Harvard University Press.

1981b Nepotists and altruists: The behavior of senescent females in macaques and langur monkeys. In *Other Ways of Growing Old,* P. Amoss and S. Harrell, eds., 59–76. Stanford: Stanford University Press.

1984 "When the bough breaks": There may be method in the madness of infanticide. *The Sciences* 24(2):44–50.

1986a Empathy, polyandry and the myth of the "coy" female. In *Feminist Approaches to Science,* Ruth Bleier, ed., 119–46. New York: Pergamon.

1986b Sources of variance in the reproductive success of female primates. *Proceedings of the International Meeting on Variability and Behavioral Evolution,* 191–203. Problemi Attuali di Scienza e di Cultura, N. 259. Rome: Academia Nazionale dei Lincei.

1987 Sex-biased parental investment among primates and other mammals. In *Child Abuse and Neglect: A Biosocial Perspective,* Jane Lancaster and Richard Gelles, eds., 97–147. Hawthorne, New York: Aldine de Gruyter.

1990 Sex bias in nature and in history: A late 1980s examination of "the biological origins" argument. *Yearbook of Physical Anthropology* 33:25–37.

1992 Fitness tradeoffs in the history and evolution of delegated mothering. *Ethology and Sociobiology* 13:495–522.

1996 The contingent nature of maternal love and its implications for "adorable" babies. Paper presented at Wenner-Gren Foundation Symposium entitled "Is There a Neurobiology of Love?" organized by Kerstin Uvnas Moberg and Sue Carter, August 28–31, Stockholm.

1997 Raising Darwin's consciousness: Female sexuality and the prehominid origins of patriarchy. *Human Nature* 8:1–49.

Hrdy, Sarah Blaffer, and William Bennett

1979 The fig connection. *Harvard Magazine* (September–October):25–30.

Hrdy, Sarah Blaffer, and C. Sue Carter

1995 Hormonal cocktails for two. *Natural History* 104(12):34.

Hrdy, Sarah Blaffer, with Emily R. Coleman

1982 Why human secondary sex ratios are so conservative: A distant reply from ninth century France. Offprint no. 88. Wenner Gren Symposium on "Infanticide in Animals and Man," Ithaca, New York.

Hrdy, Sarah Blaffer, and Glenn Hausfater

1984 Comparative and evolutionary perspectives on infanticide. In *Infanticide: Comparative and Evolutionary Perspectives,* G. Hausfater and S. Hrdy, eds. Hawthorne, New York: Aldine de Gruyter, xiii–xxxv.

Hrdy, Sarah Blaffer, and Daniel B. Hrdy

1976 Hierarchical relations among female Hanuman langurs (Primates: Colobinae, *Presbytis entellus*). *Science* 193:913–15.

Hrdy, Sarah Blaffer, and Debra Judge

1993 Darwin and the puzzle of primogeniture: An essay on biases in parental investment after death. *Human Nature* 4:1–45.

Hrdy, Sarah Blaffer, and Patricia Whitten

1987 The patterning of sexual activity among primates. In *Primate Societies,* B. B. Smuts et al., eds., 370–84. Chicago: University of Chicago Press.

Hrdy, Sarah Blaffer, and George C. Williams

1983 Behavioral biology and the double standard. In *Social Behavior of Female Vertebrates,* S. K. Wasser, ed., 3–17. New York: Academic Press.

Hrdy, Sarah Blaffer, Charles Janson, and Carel van Schaik

1995 Infanticide: Let's not throw out the baby with the bath water. *Evolutionary Anthropology* 3:151–54.

Hubbard, Ruth

1979 Have only men evolved? In *Women Look at Biology Looking at Women,* Ruth Hubbard, Mary Sue Henifin, and Barbara Fried, eds., 7–35. Boston: G. K. Hall.

Hubert, Cynthia

1998 She sacrificed life for brief joy as a mom. *Sacramento Bee* (October 25):A1, A26.

Huck, U. William

1984 Infanticide and the evolution of pregnancy block in rodents. In *Infanticide: Comparative and Evolutionary Perspectives,* G. Hausfater and S. Blaffer Hrdy, eds., 349–65. Hawthorne, New York: Aldine de Gruyter.

Hufton, Olwen

1974 *The Poor in Eighteenth Century France, 1750–1789.* Oxford: Oxford University Press.

Hull, Terence H.

1990 Recent trends in sex ratios at birth in China. *Population and Development Review* 16:63–83.

Hinecke, V.

1985 Les enfants trouvés: Contexte Européen et cas Milanais (xviiie–xixe siècles). Revue d'histoire moderne et contemporaine. Tome xxxii. Janvier–Mars: 3–29.

Hurtado, A. M.

1985 Women's subsistence strategies among Aché hunter-gatherers of eastern Paraguay. Ph.D. diss., University of Utah, Salt Lake City.

Hurtado, A. M., K. Hawkes, K. Hill, and H. Kaplan

1985 Female subsistence strategies among Aché hunter-gatherers of eastern Paraguay. *Human Ecology* 13:1–28.

Hurtado, A. M., K. Hill, H. Kaplan, and I. Hurtado

1992 Tradeoffs between female food acquisition and childcare among Hiwi and Aché foragers. *Human Nature* 3:185–216.

Huss-Ashmore, Rebecca

1980 Fat and fertility: Demographic implications of differential fat storage. *Yearbook of Physical Anthropology* 23:65–91.

Huxley, Aldous

1992 *Ape and Essence.* Chicago: Elephant Paperbacks (originally published 1948).

Huxley, Julian

1914 The courtship habits of the great crested grebe (*Podiceps criatus*) with an addition to the theory of sexual selection. *Proceedings of the Zoological Society,* xxxv.

Insel, Thomas R.

1992 Oxytocin—a neuropeptide for affiliation: Evidence from behavioral, receptor auto-radiographic and comparative studies. *Psychoneuroendocrinology* 17:3–35.

Insel, Thomas R., and T. J. Hulihan

1995 A gender-specific mechanism for pair bonding: oxytocin and partner preference for-mation in monogamous voles. *Behavioral Neurosciences* 109:782–89.

Insel, Thomas R., and Lawrence E. Shapiro

1992 Oxytocin receptor distribution reflects social organization in monogamous and polyg-amous voles. *Proceedings of the National Academy of Sciences* 89:5981–85.

Irons, William

1979 Cultural and biological success. In *Evolutionary Biology and Human Social Behavior: An Anthropological Perspective,* Napoleon Chagnon and William Irons, eds., 257–72. North Scitu-ate, Massachusetts: Duxbury.

1997 Cultural and biological success. In *Human Nature: A Critical Reader,* Laura Betzig, ed., 36–49. Cambridge: Cambridge University Press (originally published 1979).

1998 Adaptively relevant environments versus the Environment of Evolutionary Adapted-ness. *Evolutionary Anthropology* 6:194–204.

Isaac, Barry

1980 Female fertility and marital form among the Mende of upper rural Bambara chief-dom, Sierra Leone. *Ethnology* 19(3):297–313.

Isiugo-Abanihe Uche, C.

1985 Child fosterage in West Africa. *Population and Development Review* 11:53–73.

James, William H.

1983 Timing of fertilization and the sex ratio of offspring. In *Sex Selection of Children,* N. Ben-nett, ed. New York: Academic Press.

Jay, Phyllis

1962 The social behavior of the langur monkey. Ph.D. diss., Department of Anthropology, University of Chicago.

1963 The female primate. In *The Potential of Woman,* S. Farber and R. Wilson, eds., 3–47. New York: McGraw Hill.

Jayaraman, K. S.

1994 India bans the use of sex screening tests. *Nature* 370:320.

Jefferey, R., and P. Jefferey

1984 Female infanticide and amniocentesis. *Social Science and Medicine* 11:1207–12.

Jesdanun, Anick

1997 Santorum: We chose not to abort. *Philadelphia Daily News* (May 16; published in Philadelphia Online).

Johansson, Sheila Ryan

1987 Status anxiety and demographic contraction of privileged populations. *Population and Development Review* 13:439–70.

Johansson, Sten, and Ola Nygren

1991 The missing girls of China: A new demographic account. *Population and Development Review* 17:35–51.

Johnson, Lorna D., A. J. Petto, and P. K. Sehgal

1991 Factors in the rejection and survival of captive cotton top tamarins (*Saquinus oedipus*). *American Journal of Primatology* 25:91–102.

Johnson, M. H., S. Dziurawiec, H. Ellis, and J. Morton

1991 Newborns' preferential tracking of face-like stimuli and its subsequent decline. *Cognition* 40:1–19.

Johnson, Orna R.

1981 The socioeconomic context of child abuse and neglect in native South America. In *Child Abuse and Neglect: Cross-Cultural Perspectives,* Jill Korbin, ed., 56–70. Berkeley: University of California Press.

Johnstone, Rufus A.

1994 Female preference for symmetrical males as a by-product of selection for mate recognition. *Nature* 372:172–75.

Jolly, Alison

1972a Hour of birth in primates and man. *Folia Primatologica* 18:108–21.

1972b *The Evolution of Primate Behavior.* New York: Macmillan.

Jones, Douglas

1996 *Physical Attractiveness and the Theory of Sexual Selection: Results from Five Populations.* Anthropological Papers 90. Ann Arbor: Museum of Anthropology, University of Michigan.

Jones, J. S.

1991 Farming is in the blood. *Nature* 351:97–98.

Jones, Steve

1992 Natural selection in humans. In *Cambridge Encyclopedia of Human Evolution,* Steve Jones, Robert Martin, and David Pilbeam, eds., 284–87. Cambridge: Cambridge University Press.

Jordan, Brigitte

1985 Biology and Culture: Some thoughts on universals in childbirth. Paper presented at the Eighty-fourth Annual Meeting of the American Anthropological Association, Washington, D.C.

1993 *Birth in Four Cultures: A Crosscultural Investigation of Childbirth in Yucatan, Holland, Sweden and the United States,* revised and expanded by Robbie Davis-Floyd. Prospect Heights, Illinois: Waveland Press.

Judge, Debra S.

1995 American legacies and the variable life histories of women and men. *Human Nature* 6:291–323.

Judge, Debra, and James R. Carey

In press Post-reproductive life predicted by primate patterns. *Journal of Gerontology: Biological Sciences.*

Judge, Debra, and S. Blaffer Hrdy

1992 Allocation of accumulated resources among close kin: Inheritance in Sacramento, California, 1890–1984. *Ethology and Sociobiology* 13:495–522.

Judson, George

    1995 Mother guilty in the killing of 5 babies: Infant death syndrome is at last discounted. *New York Times* (April 22):25, 28.

Kagan, Jerome

    1994 *Galen's Prophecy: Temperament in Human Nature.* New York: Basic Books.

Kagan, Jerome, Richard B. Kearsley, and Philip R. Zelago

    1978 *Infancy: Its Place in Human Development.* Cambridge: Harvard University Press.

Kalish, Susan

    1994 Rising costs of raising children. *Population Today* (July–August):4–5.

Kalscheur, V. M., E. C. Mariman, M. T. Schepens, H. Rehder, and H. H. Ropers

    1993 The insulin-like growth factor type-2 receptor gene is imprinted in the mouse but not in humans. *Nature Genetics* 5:74–78.

Kanbour-Shakir, A., Z. Zhang, A. Rouleau, D. T. Armstrong, H. W. Kunz, T. A. MacPherson, and T. J. Gill III

    1990 Gene imprinting and major histocompatibility complex class I antigen expression in the rat placenta. *Proceedings of the National Academy of Sciences* 87:444–48.

Kano, T.

    1982 *The Last Ape: Pygmy Chimpanzee Behavior and Ecology.* Stanford: Stanford University Press.

Kaplan, Hillard

    1994 Evolutionary and wealth flows theories of fertility: Empirical tests and new models. *Population and Development Review* 20(4):753–91.

    1996 A theory of fertility and parental investment in traditional and modern human societies. *Yearbook of Physical Anthropology* 39:91–135.

    1997 The evolution of the human life course. In *Between Zeus and the Salmon: The Biodemography of Longevity,* K. Wachter and C. Finch, eds., 175–211. Washington, D.C.: National Academy Press.

Kaplan, Hillard, Kim Hill, Kristen Hawkes, and Ana Hurtado

    1984 Food sharing among Aché hunter-gatherers of eastern Paraguay. *Current Anthropology* 25:113–15.

Kaplan, Hillard, Kim Hill, A. Magdalena Hurtado, and Jane B. Lancaster

    In prep. The theory of human life history. Ms., Department of Anthropology, University of New Mexico.

Kaplan, H. S., J. B. Lancaster, J. A. Bock, and S. E. Johnson

    1995 Fertility and fitness among Albuquerque men: A competitive labour market theory. In *Human Reproductive Decisions,* R.I.M. Dunbar, ed., 96–136. London: St. Martin's.

Karen, Robert

    1994 *Becoming Attached: Unfolding the Mystery of the Infant-Mother Bond and Its Impact on Later Life.* New York: Warner Books.

Karl, Frederick

    1995 *George Eliot: Voice of a Century.* New York: W. W. Norton.

Katz, M. M., and M. J. Konner

    1981 The role of the father: An anthropological perspective. In *The role of the father in child development,* M. E. Lamb, ed., 155–85. New York: Wiley.

Kaufman, I. C., and L. A. Rosenblum

1969 Effects of separation from mother on the emotional behavior of infant monkeys. *Annals of the New York Academy of Sciences* 159:681–95.

Kawai, M.

1958 On the system of social ranks in a natural troop of Japanese monkeys, parts 1 and 2. Translated into English and reprinted in *Japanese Monkeys,* S. Altmann, ed. Edmonton: University of Alberta.

Kawamura, S.

1958 The matriarchal social order in the Minoo-B troop: A study on the rank system of Japanese macaques. *Primates* 1:149–56.

Keeley, Lawrence H.

1996 *War Before Civilization.* New York: Oxford University Press.

Kempe, Henry C., F. N. Silverman, B. F. Steele, W. Droegmueller, and H. K. Silver

1962 The battered child syndrome. *Journal of the American Medical Association* 181:17–24.

Kempe, Ruth, and Henry Kempe

1978 *Child Abuse.* Cambridge: Harvard University Press.

Kenagy, G. J., and C. Tromulak

1986 Size and function of mammalian testes in relation to body size. *Journal of Mammalogy* 67:1–22.

Kendrick, K. M., F. Levy, and Eric B. Keverne

1992 Changes in the sensory processing of olfactory signals induced by birth in sheep. *Science* 256:833–36.

Kenrick, Douglas, Edward R. Sadalla, Gary Gorth, and Melanie R. Trost

1997 Where and when are women more selective than men? In *Human Nature: A Critical Reader,* Laura Betzig, ed., 223–24. Cambridge: Cambridge University Press.

Kertzer, David

1993 *Sacrificed for Honor: Italian Infant Abandonment and the Politics of Reproductive Control.* Boston: Beacon Press.

Keverne, Eric B.

1995 Neurochemical changes accompanying the reproductive process: Their significance for maternal care in primates and other mammals. In *Motherhood in Human and Nonhuman Primates,* C. R. Pryce, R. D. Martin, and D. Skuse, eds., 69–77. Basel: S. Karger.

Keverne, Eric B., Frances L. Martel, and Claire M. Nevison

1996 Primate brain evolution: Genetic and functional considerations. *Proceedings of the Royal Society of London,* Series B 263:689–96.

Keverne, Eric B., Claire M. Nevison, and Frances L. Martel

1997 Early learning and the social bond. In *The Integrative Neurobiology of Affiliation,* C. Sue Carter, Izja Lederhendler, and Brian Kirkpatrick, eds., 329–39. New York: New York Academy of Sciences.

Kilner, Rebecca

1997 Mouth colour is a reliable signal of need in begging canary nestlings. *Proceedings of the Royal Society of London,* Series B 264:963–68.

King, Mary-Claire
    1998 Human evolution and diversity. Public lecture, symposium on "Humankind's Evolutionary Roots: Our Place in Nature," October 9, Field Museum, Chicago.

King, Mary-Claire, and A. C. Wilson
    1975 Evolution at two levels in humans and chimpanzees. *Science* 188:107–16.

Kingsley, S.
    1982 Causes of non-breeding and the development of the secondary sexual characteristics in the male orangutan: A hormonal study. In *The Orangutan: Its Biology and Conservation,* L. de Boer, ed. The Hague: W. Junk.

Kirkpatrick, Mark, and Russell Lande
    1989 The evolution of maternal characters. *Evolution* 43(3):485–503.

Kitzinger, Sheila
    1980 *The Experience of Breastfeeding.* New York: Penguin.

Klapisch-Zuber, Christiane
    1986 Blood parents and milk parents: Wet-nursing in Florence, 1300–1530. In *Women, Family and Ritual in Renaissance Florence,* by Christiane Klapisch-Zuber, Lydia Cochrane, trans., 132–64. Chicago: University of Chicago Press.

Klaus, Marshall H., and John H. Kennell
    1976 *Maternal-Infant Bonding: The Impact of Early Separation and Loss on Family Development.* St. Louis: C. V. Mosby.

Klaus, M. H., R. Jerauld, N. C. Kreger, W. McAlpine, M. Steffa, and J. H. Kennell
    1972 Maternal attachment: The importance of the first post-partum days. *New England Journal of Medicine* 286:460–63.

Kleiman, D. G., and J. R. Malcolm
    1981 The evolution of male parental investment in mammals. In *Parental Care in Mammals,* D. J. Gubernick and P. H. Klopfer, eds., 347–87. New York: Plenum.

Klein, Richard G.
    1992 The archaeology of modern human origins. *Evolutionary Anthropology* 1(1):5–14.

Kleinman, Ronald, Linda Jacobson, Elizabeth Hormann, and W. A. Walker
    1980 Protein values of milk samples from mothers without biologic pregnancies. *The Journal of Pediatrics* 97:612–15.

Klopfer, Peter H.
    1971 Mother-love: What turns it on? *American Scientist* 59:404–7.
    1996 "Mother Love" revisited: On the use of animal models. *American Scientist* (July–August):319–21.

Knight, Chris
    1991 *Blood Relations: Menstruation and the Origins of Culture.* New Haven: Yale University Press.

Koenig, Walt
    1990 Opportunity of parentage and nest destruction in polygynandrous acorn woodpecker, *Melanerpes formicivorus. Behavioral Ecology* 1:55–61.

Kolata, Gina
    1982 New theory of hormones proposed. *Science* 215:1383–84.
    1988 Fetal sex test used as step to abortion. *New York Times* (December 25).

Komdeur, Jan, Serge Daan, Joost Tinbergen, and Christa Mateman
    1997 Extreme modification in sex ratio of the Seychelles warbler's eggs. *Nature* 385:522–25.

Konig, Barbara
    1994 Components of lifetime reproductive success in communally and solitarily nursing house mice—a laboratory study. *Behavioral Ecology and Sociobiology* 34:275–83.

Konig, B., J. Riester, and H. Markl
    1988 Maternal care in house mice (*Mus musculus*), II: The energy cost of lactation as a function of litter size. *Journal of Zoology* (London) 216:195–210.

Konner, Melvin J.
    1972 Aspects of the developmental ethology of a foraging people. In *Ethological Studies of Child Behavior,* N. Blurton-Jones, ed., 285–304. Cambridge: Cambridge University Press.
    1991 *Childhood.* Boston: Little, Brown.

Konner, Melvin, and Carol Worthman
    1980 Nursing frequency, gonadal function and birth spacing among !Kung hunter-gatherers. *Science* 207:788–91.

Konner, Melvin, and Marjorie Shostak
    1987 Timing and management of birth among the !Kung: Biocultural interaction in reproductive adaptation. *Cultural Anthropology* 2(1):11–28.

Korbin, Jill E., ed.
    1981 *Child Abuse and Neglect: Cross-cultural Perspectives.* Berkeley: University of California Press.

Kosikowski, F. V.
    1985 Cheese. *Scientific American* 252(5):88–99.

Kovaleski, Serge F., and Lorraine Adams
    1996 Kaczynski's mom still seeks cause of his anger and pain. *The Sacramento Bee* (June 17):A1, A7.

Koyama, Naoki
    1967 On dominance rank and kinship of a wild Japanese monkey in Arashiyama. *Primates* 8:189–216.

Kramer, Patricia Ann
    1998 The costs of human locomotion: Maternal investment in child transport. *American Journal of Physical Anthropology* 107:71–85.

Kristal, M. E.
    1991 Enhancement of opioid-mediated analgesia—A solution to the enigma of placentophagia. *Neuroscience and Behavioral Reviews* 15:425–35.

Kristof, Nicholas D.
    1991a A mystery of China's census: Where have the girls gone? *New York Times* (June 17):A1, A7.
    1991b Stark data on women: 100 million are missing. *New York Times* (November 5):B5, B9.

Kroeber, Theodora
    1989 *Ishi in Two Worlds: A Biography of the Last Wild Indian in North America.* Berkeley: University of California Press (reprint of 1961 edition).

Labov, Jay B., U. William Huck, Robert W. Elwood, and Ronald J. Brooks

1985 Current problems in the study of infanticidal behavior of rodents. *Quarterly Review of Biology* 60:1–20.

Lacey, Eileen A., and Paul W. Sherman

1997 Cooperative breeding in naked mole-rats: Implications for vertebrate and invertebrate sociality. In *Cooperative Breeding in Mammals,* Nancy Solomon and Jeffrey French, eds., 267–301. Cambridge: Cambridge University Press.

Lack, David

1941 *The Life of the Robin.* London: H. F. and G. Witherby, Ltd.

1947 The significance of clutch size. *Ibis* 89:302–52.

1966 Animal dispersion (Appendix 3). In *Population Studies of Birds,* by David Lack. Oxford: Clarendon Press.

1968 *Ecological Adaptations for Breeding in Birds.* London: Chapman and Hall.

Ladurie, Emmanuel Le Roy

1979 *Montaillou: The Promised Land of Error,* Barbara Bray, trans. New York: Vintage.

Laesthaeghe, Ron, Georgia Kaufmann, Dominique Meekers, and Johan Surkyn

1994 Postpartum abstinence, polygyny, and age at marriage: A macro-level analysis of sub-Saharan societies. In *Nuptiality in Sub-Saharan Africa: Contemporary Anthropological and Demographic Perspectives,* Caroline Bledsoe and Gilles Pison, eds., 25–54. Oxford: Oxford University Press.

Lamb, Michael E.

1982 The bonding phenomenon: Misinterpretations and their implications. *Journal of Pediatrics* 101:555–57.

Lamb, Michael E., and James A. Levine

1983 The Swedish parental insurance policy: An experiment in social engineering. In *Fatherhood and Family Policy,* M. E. Lamb and A. Sagi, eds. Hillsdale, New Jersey: Lawrence Erlbaum, 39–51.

Lamb, Michael E., Joseph H. Pleck, Eric L. Charnov, and James A. Levine

1987 A biosocial perspective on paternal behavior and involvement. In *Parenting Across the Life Span: Biosocial Dimensions,* Jane B. Lancaster, Jeanne Altmann, Alice S. Rossi, and Lonnie R. Sherrod, eds., 111–42. Hawthorne, New York: Aldine de Gruyter.

Lamb, Michael, R. Thompson, W. Gardner, and Eric Charnov, eds.

1985 *Infant-Mother Attachment: The Origins and Developmental Significance of Individual Differences in Strange Situation Behavior.* Hillsdale, New Jersey: Lawrence Erlbaum.

Lancaster, Jane

1971 Play-mothering: The relations between juvenile females and young infants among free-ranging vervet monkeys (*Cercopithecus aethiops*). *Folia Primatologica* 15:161–82.

1973 In praise of the achieving female primate. *Psychology Today* VII (September):30, 32, 34–36, 99.

1978 Caring and sharing in human evolution. *Human Nature* 1(2):82–89. Harcourt Brace, Jovanovich.

1986 School age pregnancy and parenthood. In *School Age Pregnancy and Parenthood: Biosocial Dimensions.* Jane B. Lancaster and Beatrix A. Hamburg, eds., 17–37. Hawthorne, New York: Aldine de Gruyter.

1997 The evolutionary history of human parental investment in relation to population growth and social stratification. In *Feminism and Evolutionary Biology,* P. Adair Gowaty, ed., 466–88. London: Chapman and Hall.

Lancaster, Jane B., and Beatrix A. Hamburg, eds.

1986 *School Age Pregnancy and Parenthood: Biosocial Dimensions.* Hawthorne, New York: Aldine de Gruyter.

Lancaster, Jane B., and B. King

1992 An evolutionary perspective on menopause. In *In Her Prime: A New View of Middle-Aged Women,* V. Kerns and J. Brown, eds., 7–15. Chicago: University of Illinois Press.

Lancaster, Jane, and Chet Lancaster

1983 Parental investment: The hominid adaptation. In *How Humans Adapt: A Biocultural Odyssey,* D. Ortner, ed., 33–65. Washington, D.C.: Smithsonian Institution Press.

1987 The watershed: Change in parental-investment and family formation strategies in the course of human evolution. In *Parenting Across the Human Lifespan: Biosocial Dimensions,* Jane B. Lancaster, Jeanne Altmann, Alice S. Rossi, and Lonnie R. Sherrod, eds., 187–205. Hawthorne, New York: Aldine de Gruyter.

Lang, O.

1946 *Chinese Family and Society.* New Haven: Yale University Press.

Langer, William L.

1972 Checks on population growth: 1750–1850. *Scientific American* 226:92–99.

1974 Further notes on the history of infanticide. *History of Childhood Quarterly,* 129–34. Supplement to 1(3):353–65.

Langlois, J., L. A. Roggma, R. J. Casey, J. M. Ritter, L. A. Rieser-Danner, and V. Y. Jenkins

1987 Infant preferences for attractive features: Rudiments of a stereotype? *Developmental Psychology* 23:363–69.

Lavely, William, William M. Mason, and Hiromi Ono

1992 Sex differentials in Chinese infant mortality. Lecture presented by William Lavely in the Program in East Asian Studies series on Family and Reproduction in Contemporary China, January 31, University of California, Davis.

Lawrence, Mark, Françoise Lawrence, W. A. Coward, Timothy J. Cole, and Roger G. Whitehead

1987 Energy requirements of pregnancy in The Gambia. *Lancet* 2:1072–75.

Leakey, L.S.B.

1977 *The Southern Kikuyu before 1903,* 3 vols. London: Academic Press.

Lee, P. C.

1987 Allomothering among African elephants. *Animal Behavior* 35:275–91.

Lee, P. C., and J. E. Bowman

1994 Influence of ecology and energetics on primate mothers and infants. In *Motherhood in*

Human and Nonhuman Primates: Biosocial Determinants, C. R. Pryce, R. D. Martin, and D. Skuse, eds., 47–58. Basel: S. Karger.

Lee, Richard Borshay

1979 *The !Kung San: Men, Women and Work in a Foraging Society.* Cambridge: Cambridge University Press.

Lefebvre, Louis, Stéphane Viville, Sheila C. Barton, Fumitoshi Ishino, Eric B. Keverne, and M. Azim Surani

1998 Abnormal maternal behaviour and growth retardation associated with loss of the imprinted gene *Mest. Nature Genetics* 20:163–69.

Leigh, Egbert

1971 *Adaptation and Diversity.* San Francisco: Freeman, Cooper.

Leighton, P. A., J. R. Seam, R. S. Ingram, and S. M. Tilghman

1996 Genomic imprinting in mice: Its function and mechanism. *Biology of Reproduction* 54(2):273–78.

LeNoir, Jean-Charles-Pierre

1780 Détails sur quelques établissements de la ville de Paris, demandés par sa Majesté Impériale, la reine de Hongrie. Pamphlet in the Bibliothèque Nationale de Paris, 68 pages.

Lerner, Gerder

1986 *The Creation of Patriarchy.* Oxford: Oxford University Press.

1993 *The Creation of Feminist Consciousness.* New York: Oxford University Press.

Lessells, Catherine M.

1991 The evolution of life histories. In *Behavioural Ecology: An Evolutionary Approach,* J. R. Krebs and N. B. Davies, eds., 32–65. Oxford: Blackwell Scientific.

Lévi-Strauss, Claude

1969 *The Raw and the Cooked,* trans. New York: Harper and Row (originally published in French, 1964).

Levine, Nancy E.

1987 Differential child care in three Tibetan communities: Beyond son preferences. *Population and Development Review* 13(2):281–304.

LeVine, Robert

1962 Witchcraft and co-wife proximity in southwestern Kenya. *Ethnology* 1:39–45.

LeVine, Robert, Suzanne Dixon, Sarah LeVine, Amy Richman, P. Herbert Leiderman, Constance H. Keefer, and T. Berry Brazelton

1996 *Child Care and Culture: Lessons from Africa.* New York: Cambridge University Press.

LeVine, Sarah, and Robert LeVine

1981 Child abuse and neglect in sub-Saharan Africa. In *Child Abuse and Neglect: Cross-cultural Perspectives,* J. Korbin, ed., 35–55. Berkeley: University of California Press.

LeVine, Sarah, in collaboration with Robert A. LeVine

1979 *Mothers and Wives: Gusii Women of East Africa.* Chicago: University of Chicago Press.

Lewes, George Henry

1877 *The Physical Basis of Mind.* London: Trubner.

1879 The study of psychology: Its object, scope, and method. London (n.p.).

Lewin, Tamar
　　1998 Birth rates for teenagers declined sharply in the 90s. *New York Times* (May 1):A-17.
Lewis, Judith Schneid
　　1986 *In the Family Way: Childbearing in the British Aristocracy, 1760–1860.* New Brunswick, New Jersey: Rutgers University Press.
Lewis, M.
　　1969 Infants' responses to facial stimuli during the first year of life. *Developmental Psychology* 2:75–86.
Lewis, N.
　　1985 *Life in Egypt Under Roman Rule.* Oxford: Oxford University Press.
Liesen, Laurette
　　1995 Feminism and the politics of reproductive strategies. *Politics and the Life Sciences* 14(2):145–97.
　　1998 The legacy of Woman the Gatherer: The emergence of evolutionary feminism. *Evolutionary Anthropology* 7(3):105–13.
Lindburg, D. G., and Lester Dessez Hazell
　　1972 Licking of the neonate and duration of labor in Great Apes and man. *American Anthropologist* 74:318–25.
Lipschitz, D. L.
　　1992 Profiles of oestradiol, progesterone, and luteinizing hormone during the oestrous cycle of female *Galago senegalensis Moholi.* Abstracts of the XIV Congress of the International Primatological Society, August 16–21, Strasbourg.
Lipson, S. F., and P. T. Ellison
　　1996 Comparison of salivary steroid profiles in naturally occurring conception and non-conception cycles. *Human Reproduction* 11(10):2090–96.
Litwack, Georgia
　　1979 Understanding sociobiology. *Boston Sunday Globe—New England Magazine* (April 8): 6ff.
Llewelyn-Davis, Melissa
　　1978 Two contexts of solidarity among pastoral Masai women. In *Women United, Women Divided: Cross-Cultural Perspectives on Female Solidarity,* Patricia Caplan and Janet M. Bujra, eds., 206–37. London: Tavistock.
Lloyd, James E.
　　1975 Aggressive mimicry in photuris fireflies: signal repertoires by femmes fatales. *Science* 187:452–53.
Lock, Stephen
　　1990 Right and wrong. *Nature* 345:397.
Lorence, Bogna W.
　　1974 Parents and children in eighteenth century Europe. *History of Childhood Quarterly* 2(1):1–30.
Lorenz, Konrad
　　1952 *King Solomon's Ring: A New Light on Animal Ways.* New York: Crowell.

Losos, Jonathan, Kenneth Warheit, and Thomas Schoener

1997 Adaptive differentiation following experimental island colonization in *Anolis* lizards. *Nature* 387:70–74.

Lovejoy, Owen

1981 The origin of man. *Science* 211:341–50.

Low, Bobbi S.

1978 Environmental uncertainty and the parental strategies of marsupials and placentals. *The American Naturalist* 112(983):197–213.

1979 Sexual selection and human ornamentation. In *Evolutionary Biology and Human Social Behavior: An Anthropological Perspective,* N. A. Chagnon and W. Irons, eds., 462–87. North Scituate, Massachusetts: Duxbury.

1991 Reproductive life in nineteenth-century Sweden: An evolutionary perspective on demographic phenomena. *Ethology and Sociobiology* 12:411–48.

Luce, Clare Boothe

1978 Only women have babies. *National Review* xxx (27, July 7): 824–27.

Lummas, V., E. Haukioja, R. Lemmetyinen, and M. Pikkola

1998 Natural selection on human twinning. *Nature* 394:533–34.

Luttbeg, Barney, Monique Borgerhoff Mulder, and Marc Mangel

1999 To marry again or not: A dynamic model for demographic transition. In *Human Behavior and Adaptation: An Anthropological Perspective,* Lee Cronk, Napoleon Chagnon, and William Irons, eds. Hawthorne, New York: Aldine de Gruyter. In press.

Lyon, Bruce E., John M. Eadie, and Linda D. Hamilton

1994 Parental choice selects for ornamental plumage in American coot chicks. *Nature* 371:240–43.

McBryde, Angus

1951 Compulsory rooming-in in the ward and private newborn service at Duke Hospital. *Journal of the American Medical Association* 145(9):625–28.

McCarthy, Dermod

1981 Effects of emotional disturbance and deprivation (maternal rejection) on somatic growth. In *Scientific Foundations of Pediatrics,* J. A. Davis and J. Dobbin, eds., 56–67. London: Heinemann.

McCarthy, M., and F. vom Saal

1985 The influence of reproductive state on infanticide by wild female house mice (*Mus musculus*). *Physiology and Behavior* 35:843–49.

Macdonald, David, ed.

1984 *Encyclopedia of Mammals.* New York: Facts on File.

Mace, Ruth

1996a Biased parental investment and reproductive success in Gabbra pastoralists. *Behavioral Ecology and Sociobiology* 38:75–81.

1996b When to have another baby: A dynamic model of reproductive decision-making and evidence from Gabbra pastoralists. *Ethology and Sociobiology* 17:263–73.

Mace, Ruth, and R. Sear

1999 Life history evolution in a rural Gambian population. (Abstract.) Paper presented at the Annual Meeting of the Human Behavior and Evolution Society, June 2–6, Salt Lake City.

McGrew, William C.

1979 Evolutionary implications of sex differences in chimpanzee predation and tool use. In *The Great Apes,* David A. Hamburg and Elizabeth R. McCown, eds., 441–64. Menlo Park, California: Benjamin-Cummings.

McGue, Matt

1977 The democracy of the genes. *Nature* 388:417–18.

McHenry, Henry

1992 How big were early hominids? *Evolutionary Anthropology* 1(1):15–20.

1996 Sexual dimorphism in fossil hominids and its socioecological implications. In *Power, Sex and Tradition,* James Steele and Stephan Shennan, eds., 91–109. The Archeology of Human Ancestry, vol. 24. London: Routledge.

Macilwain, Colin

1997a "Science Wars" blamed for loss of post. *Nature* 387:325.

1997b Campuses ring to a stormy clash over truth and reason. *Nature* 387:331–33.

McKee, Lauris

1984 Sex differentials in survivorship and the customary treatment of infants and children. *Medical Anthropology* 8(2):91–108.

McKenna, James J.

1979 The evolution of allomothering behavior among colobine monkeys: Function and opportunism in evolution. *American Anthropologist* 81:818–40.

McKinley, Catherine

1993 Infanticide and slave women. In *Black Women in America: An Historical Encyclopedia,* vol. 1, Darlene Clark Hine, ed., 607–9. Brooklyn, New York: Carlson.

Mackinnon, John

1979 Reproductive behavior in wild orangutan populations. In *The Great Apes,* D. Hamburg and E. McCown, eds., 257–73. Menlo Park, California: Benjamin-Cummings.

McLaughlin, Mary Martin

1974 Survivors and surrogates. In *The History of Childhood,* Lloyd de Mause, ed., 101–81. New York: Harper and Row.

1989 The suffering of little children. *New York Times Book Review* (March 19):16.

McNamara, John M., and Alasdair I. Houston

1996 State-dependent life histories. *Nature* 380:215–21.

Madan, T. N.

1965 *Family and Kinship.* Bombay: Asia Publishing House.

Maestripieri, Dario

1994 Infant abuse associated with psychosocial stress in a group-living pigtail macaque (*Macaca nemestrina*) mother. *American Journal of Primatology* 32:41–49.

1998 Parenting styles of abusive mothers in group-living rhesus macaques. *Animal Behavior* 55:1–11.

Maggioncalda, Anne N., Robert M. Sapolsky, and Nancy Czekala

1999 Reproductive hormone profiles in captive male orangutans: Implication for understanding developmental arrest. *American Journal of Physical Anthropology* 109:19–32.

Magrath, Robert D.

1991 Lack's solution? *Nature* 353:611.

Maier, Richard A., Deborah L. Holmes, Frank L. Slaymaker, and Jill Nagy Reich

1984 The perceived attractiveness of preterm infants. *Infant Behavior and Development* 7:403–14.

Main, Mary

1981 Avoidance in the service of attachment: A working paper. In *Behavioral Development: The Bielefeld Interdisciplinary Project,* Klaus Immelmann, George W. Barlow, Lewis Petrinovich, and Mary Main, eds., 651–93. Cambridge: Cambridge University Press.

1991 Metacognitive knowledge, metacognitive monitoring, and singular (coherent) vs. multiple (incoherent) models of attachment: findings and directions for further research. In *Attachment across the Life Cycle,* Colin Murray Parkes, Joan Stevenson-Hinde, and Peter Marris, eds., 127–59. New York: Routledge.

1995 Recent studies in attachment: Overview, with selected implications for clinical work. In *Attachment Theory: Social, Development, and Clinical Perspectives,* S. Goldberg, John Kerr, and R. Muir, eds., 407–74. Hillsdale, New Jersey: Analytic Press.

Main, Mary, and Erik Hesse

1990 Parents' unresolved traumatic experiences are related to infant disorganized attachment status: Is frightened and/or frightening parental behavior the linking mechanism? In *Attachment in the Preschool Years: Theory, Research and Intervention,* M. T. Greenberg, D. Cicchetti, and E. M. Cummings, eds., 161–82. Chicago: University of Chicago Press.

Main, Mary, and Judith Solomon

1986 Discovery of a new, insecure-disorganized/disoriented attachment pattern. In *Affective Development in Infancy,* B. Brazelton and M. Yogman, eds., 95–124. Norwood, New Jersey: Ablex.

1990 Procedures for identifying infants as disorganized/disoriented during the Ainsworth Strange Situation. In *Attachment in the Preschool Years: Theory, Research and Intervention,* Mark T. Greenberg, Dante Cicchetti, and E. Mark Cummings, eds., 121–60. Chicago: University of Chicago Press.

Main, Mary, and D. Weston

1982 Avoidance of the attachment figure in infancy: Descriptions and interpretations. In *The Place of Attachment in Human Behavior,* Colin Murray Parkes and Joan Stevenson-Hinde, eds., 31–59. New York: Basic Books.

Mairson, Alain

1993 America's beekeepers: Hives for hire. *National Geographic* 183(5):73–93.

Malcolm, J., and K. Marten

    1982 Natural selection and the pups in wild dogs (*Lycaon pictus*). *Behavioral Ecology and Sociobiology* 10:1–13.

Mallardi, A., A. C. Mallardi, and D. G. Freedman

    1961 Studio su primo manifestarso dell paura dell'estraneo n el bambino: Osservazioni compartive tra sogetti allevati in famiglia e soggentti allevati in communita chiusa. *Atti de VI Congresso Nazionale della S.I.A.M.E., Bari,* 254–56.

Mann, Janet

    1992 Nurturance or negligence: Maternal psychology and behavioral preference among preterm twins. In *The Adapted Mind,* J. Barkow, L. Cosmides, and J. Tooby, eds., 367–90. New York: Oxford University Press.

Mann, Susan

    1997 *Precious Records: Women in China's Long Eighteenth Century.* Stanford: Stanford University Press.

Manning, C. J., D. A. Dewsbury, E. K. Wakeland, and W. K. Potts

    1995 Communal nesting and communal nursing in house mice, *Mus musculus domesticus. Animal Behavior* 50:741–51.

Manning, J. T., D. Scutt, G. H. Whitehouse, and S. J. Leinster

    1997 Breast asymmetry and phenotypic quality in women. *Evolution and Human Behavior* 18:223–36.

Manzoni, Alessandro

    1961 *The Betrothed (I promessi sposi).* New York: E. P. Dutton (translation originally published in 1825–27).

Marlowe, Frank

    1998 Showoffs or providers? The parenting effort of Hadza men. Paper presented at the Annual Meeting of the Human Behavior and Evolution Society, July 8–12, University of California, Davis.

Marsh, H., and T. Kasuya

    1986 Evidence for reproductive senescence in female cetaceans. *Report of the International Whaling Commission,* special issue 8:57–74.

Marshall, Lorna

    1976 Sharing, talking and giving: relief of social tensions among !Kung Bushmen. In *Kalahari Hunter-Gatherers: Studies of the !Kung San and Their Neighbors,* R. B. Lee and I. DeVore, eds., 349–71. Cambridge: Harvard University Press.

Martin, R. M.

    1847 *China: Political, Commercial and Social, in an Official Report to Her Majesty's Government,* vol. I. London: James Madden.

Martin, Robert D.

    1995 Phylogenetic aspects of primate reproduction: The context of advanced maternal care. In *Motherhood in Human and Nonhuman Primates: Biosocial Determinants,* C. R. Pryce, R. D. Martin, and D. Skuse, eds., 16–26. Basel: S. Karger.

Marvick, Elizabeth Wirth

   1974 Nature versus nurture: Patterns and trends in seventeenth century French child-rearing. In *The History of Childhood*, Lloyd de Mause, ed., 259–301. New York: Harper Torchbooks.

Mascia-Lees, Frances E., John Relethford, and Tom Sorger

   1986 Evolutionary perspectives on permanent breast enlargement in human females. *American Anthropologist* 88:423–29.

Mason, M. A.

   1966 Social organization of the South American monkey, *Callcebus moloch:* A preliminary report. *Tulane Studies in Zoology* 13:23–28.

Masters, William H., and Virginia E. Johnson

   1966 *Human Sexual Response.* Boston: Little, Brown.

Mastrodiacomo, I., M. Fava, G. Fava, A. Kellner, R. Grismondi, and C. Cetera

   1982–83 Postpartum hostility and prolactin. *International Journal of Psychiatry in Medicine* 12:289–94.

Matejcek, Zdenek, Zdenek Dytrych, and Vratislav Schuller

   1988 The Prague cohort through age nine. In *Born Unwanted: Developmental Effects of Denied Abortion,* H. P. David, Z. Dytrych, Z. Matejcek, and V. Schuller, eds., 53–86. New York and Prague: Springer and Czechoslovakia Medical Press.

Matteo, Sherri, and Emilie F. Rissman

   1984 Increased sexual activity during the midcycle portion of the human menstrual cycle. *Hormones and Behavior* 18:249–55.

Matthews Grieco, Sara F.

   1991 *Breastfeeding, Wet Nursing and Infant Mortality in Europe (1400–1800).* Florence: UNICEF and Insituto degli Innocenti.

Mayer, Alexander

   1865 De la création d'une société protectrice de l'enfance pour l'amélioration de l'espèce humaine par l'éducation du premier âge. Paris: Librairie des science sociales.

Mead, M.

   1962 A cultural anthropologist's approach to maternal deprivation. In *Deprivation of Maternal Care: A Reassessment of Its Effects.* Public Health Papers no. 14. Geneva: World Health Organization.

Meadow, Roy

   1990 Suffocation, recurrent apnea, and sudden infant death. *Journal of Pediatrics* 117(3):351–57.

Mealey, Linda

   1995 The sociobiology of sociopathy: An integrated evolutionary model. *Behavioral and Brain Sciences* 18:523–99.

Meikle, D. B., B. L. Tilford, and S. H. Vessey

   1984 Dominance rank, secondary sex ratio and reproduction of offspring in polygynous primates. *American Naturalist* 124:173–88.

Meltzer, David, ed.

   1981 *Birth: An anthology of ancient texts, songs, prayers and stories.* San Francisco: North Point Press.

Meltzoff, A. N.

1993 The centrality of motor coordination and proprioception in social and cognitive development: From shared action to shared minds. In *The Development of Coordination in Infancy,* G. J.P. Savelsbergh, ed., 463–96. Amsterdam: North Holland.

Meltzoff, A. N., and M. K. Moore

1977 Imitation of facial and manual gestures by human neonates. *Science* 198:75–78.

1997 Explaining facial imitation: A theoretical model. *Early Development and Parenting* 6:179–92.

Mendoza, Sally P., and William A. Mason

1986 Parental division of labour and differentiation of attachments in a monogamous primate (*Callicebus moloch*). *Animal Behavior* 34:1336–47.

Merino, Santiago, Jaime Potti, and Juan Moreno

1996 Maternal effort mediates the prevalence of trypanosomes in the offspring of a passerine bird. *Proceedings of the National Academy of Sciences* 93:5726–30.

Mestel, Rosie

1995 Monkey "murderers" may be falsely accused. *New Scientist* (July 15):17.

Meyburg, B.-U.

1974 Sibling aggression and mortality among eagles. *Ibis* 116:224–28.

Middleton, John F.

1973 The Lugbara of north-western Uganda. In *Beliefs and Practices. Cultural Source Materials for Population Planning in East Africa,* vol. 3, Angela Molnos, ed., 289–98. Nairobi: East African Publishing.

Miller, Andrew T.

1998 Child fosterage in the United States: Signs of an African heritage. *The History of the Family* 3:35–62.

Miller, Barbara

1981 *The Endangered Sex: Neglect of Female Children in Rural North India.* Ithaca: Cornell University Press.

Miller, K. F., J. M. Goldberg, and T. Falcone

1996 Follicle size and implantation of embryos from in vitro fertilization. *Obstetrics and Gynecology* 88:583–86.

Mineka, S.

1987 A primate model of phobic fears. In *Theoretical Foundations of Behavior Therapy,* H. Eysenck and I. Martin, eds. New York: Plenum.

Minturn, Leigh, and Jerry Stashak

1982 Infanticide as a terminal abortion procedure. *Behavior Science Research* 17(1 and 2):70–90.

Mitani, John C., and David Watts

1997 The evolution of non-maternal caretaking among anthropoid primates: Do helpers help? *Behavioral Ecology and Sociobiology* 40:213–20.

Mock, Douglas W.

1984 Infanticide, siblicide, and avian nestling mortality. In *Infanticide: Comparative and Evolutionary Perspectives,* Glenn Hausfater and Sarah Blaffer Hrdy, eds., 2–30. New York: Aldine.

Mock, D. W., and L. S. Forbes
  1995 The evolution of parental optimism. *Trends in Ecology and Evolution* 10:130–34.
Mock, Douglas W., and Geoffrey Parker
  1997 *The Evolution of Sibling Rivalry.* Oxford: Oxford University Press.
Moffitt, Terrie E., Avshalom Caspi, Jay Belsky, and Phil A. Silva
  1992 Childhood experience and the onset of menarche: A test of a sociobiological model. *Child Development* 63:47–58.
Mohnot, S. M.
  1971 Some aspects of social changes and infant-killing in the Hanuman langur, *Presbytis entellus,* in Western India. *Mammalia* 35:175–98.
Møller, Anders P.
  1992a Female swallow preference for symmetrical male sexual ornaments? *Nature* 357: 238–40.
  1992b Parasites differentially increase the degree of fluctuating asymmetry in secondary sexual characters. *Journal of Evolutionary Biology* 5:691–700.
Møller, Anders P., and J. Hoglund
  1991 Patterns of fluctuating asymmetry in avian feather ornaments: Implications for models of sexual selection. *Proceedings of the Royal Society of London,* Series B 245:1–5.
Møller, A. P., M. Soler, and R. Thornhill
  1995 Breast asymmetry, sexual selection and human reproductive success. *Ethology and Sociobiology* 16:207–19.
Molleson, Theya
  1994 The eloquent bones of Abu Hureyra. *Scientific American* 271(2):70–75.
Montgomery, H. E., R. Marshall, H. Hemingway, S. Myerson et al.
  1998 Human gene for physical performance. *Nature* 393:221.
Moore, Celia L., H. Dou, and J. Juraska
  1992 Maternal stimulation affects the number of motor neurons in a sexually dimorphic nucleus of the lumbar spinal cord. *Brain Research* 572:52–56.
Moore, James
  1984 The evolution of reciprocal sharing. *Ethology and Sociobiology* 5:5–14.
  1985 Demography and Sociality in Primates. Ph.D. diss., Department of Anthropology, Harvard University.
Morbeck, Mary Ellen, A. Galloway, and A. L. Zihlman, eds.
  1997 *The Evolving Female: A Life History Perspective.* Princeton: Princeton University Press.
Morell, Virginia
  1993a Seeing nature through the lens of gender. *Science* 260:428–29.
  1993b Anthropology: Nature-culture battleground. *Science* 261:1798–1802.
  1995 A 24-hour circadian clock is found in the mammalian retina. *Science* 272:349.
Morgan, Elaine
  1982 *The Aquatic Ape: A Theory of Human Evolution.* London: Souvenir Press.
  1995 *The Descent of the Child: Human Evolution from a New Perspective.* New York: Oxford University Press.

Mori, U., and Robin Dunbar

  1985 Changes in the reproductive condition of female gelada baboons following the takeover of one-male units. *Zeitschrift für Tierpsychologie* 67:215–24.

Morris, Desmond

  1967 *The Naked Ape.* New York: Dell.

Morris, John D.

  1996 Will infanticide follow abortion as "acceptable behavior"? *Acts and Facts* (November).

Morris, N. M., and J. R. Udry

  1970 Variations in pedometer activity during the menstrual cycle. *Obstetric Gynecology* 35:199–201.

Morton, N. E., C. S. Chung, and M. P. Mi

  1967 *Genetics of Interracial Crosses in Hawaii.* Monographs in Human Genetics 3. New York: S. Karger.

Mousseau, Timothy, and Charles W. Fox

  1998 The adaptive significance of maternal effects. *Trends in Ecology and Evolution* 13: 403–6.

Muhuri, Pradip, and Jane Menken

  1993 Child survival in rural Bangladesh. Paper prepared for the Stanford-Berkeley Colloquium in Demography, June 10, Berkeley.

Muhuri, Pradip, and Samuel Preston

  1991 Effects of family composition on mortality differentials by sex among children in Matlab, Bangladesh. *Population and Development Review* 17:415–34.

Mull, Dorothy S.

  1991 Traditional perceptions of marasmus in Pakistan. *Social Science Medicine* 32:175–91.

  1992 Mother's milk and pseudoscientific breastmilk testing in Pakistan. *Social Science Medicine* 34(11):277–90.

Murdock, G. P.

  1967 *Ethnograpic Atlas.* Pittsburgh: University of Pittsburgh Press.

Murdock, R.

  1934 *Our Primitive Contemporaries.* New York: Macmillan.

Murphy, Yolanda, and Robert F. Murphy

  1974 *Women of the Forest.* New York: Columbia University Press.

Myers, B.

  1984 Mother-infant bonding: The status of the critical period hypothesis. *Developmental Review* 4 (September):240–74.

Nathan, David

  1995 *Genes, Blood, and Courage: A Boy Called Immortal Sword.* Cambridge, Massachusetts: Harvard University Press, Belknap Press.

Nathanielsz, Peter W.

  1996 *Life before Birth: The Challenges of Fetal Development.* New York: W. H. Freeman.

Neal, Robert

  1998 Female suicides in China point to burden for U.S. men. *Focus* (Summer):1, 6.

Neel, J.

1970 Lessons from a "primitive" people. *Science* 170:815–22.

Nelson, Edward Wilson

1901 *The Eskimo About the Bering Strait. Annual Report of the Bureau of American Ethnology, 1896–1897.* Washington, D.C.: Government Printing Office.

Nelson, Randy J.

1995 *An Introduction to Behavioral Endocrinology.* Sunderland, Massachusetts: Sinauer Associates.

Nesse, Randolph, and George C. Williams

1994 *Why We Get Sick: The New Science of Darwinian Medicine.* New York: Times Books.

Newburg, D. S., J. A. Peterson, G. M. Ruiz-Palacios et al.

1998 Role of human-milk lactadherin in protection against symptomatic rotavirus infection. *Lancet* 351:1160–64.

Newton, Niles

1955 *Maternal Emotions: A Study of Women's Feelings toward Menstruation, Pregnancy, Childbirth, Breast Feeding, Infant Care, and Other Aspects of Their Femininity.* New York: Paul Hoeber.

1977 Interrelationships between sexual responsiveness, birth and breastfeeding. In *Contemporary Sexual Behavior: Critical Issues in the 1970s,* J. Zubin and John Money, eds., 77–98. Baltimore: Johns Hopkins University Press.

Newton, Niles, and M. Newton

1962 Mothers' reactions to their newborn babies. *Journal of the American Medical Association* 181:206–11.

New York Times

1993 What's behind odd idea of war in the womb? Editorials and Letters Section (August 7).

1996a In tests of mice, a gene seems to hold clues to the nature of nurturing (July 26).

1996b Congress plays doctor (April 1).

1997 Forbes puts on anti-abortion mantle. National Report (September 14).

Nicoll, C. S.

1974 Physiological actions of prolactin. In *Handbook of Physiology,* vol. IV, R. O. Greep and E. B. Astwood, eds., 253–92. Washington, D.C.: American Physiological Society.

Nicolson, Nancy

1987 Infants, mothers and other females. In *Primate Societies,* B. Smuts et al., eds., 330–42. Chicago: University of Chicago Press.

Normile, Dennis

1997 Yangtze seen as earliest rice site. *Science* 275:309.

Numan, Michael

1988 Maternal behavior. In *The Physiology of Reproduction,* E. Knobil et al., eds. New York: Raven Press.

Oberlander, T. F., R. G. Barr, S. N. Young, and J. A. Brian

1992 The short-term effects of feed composition on sleeping and crying in newborn infants. *Pediatrics* 90(5):733–40.

O'Brien, Tim

1988 Parasitic nursing behavior in the wedge-capped capuchin monkey (*Cebus olivaceus*). *American Journal of Primatology* 16:341–44.

O'Brien, T. G., and J. G. Robinson

1991 Allomaternal care by female wedge-shaped capuchin monkeys: Effects of age, rank and relatedness. *Behaviour* 119:30–50.

Odling-Smee, F. J.

1994 Niche construction, evolution and culture. In *Companion Encyclopedia of Anthropology,* T. Ingold, ed., 162–96. London: Routledge.

Oftedal, O.

1980 Milk and mammalian evolution. In *Comparative Physiology of Primitive Mammals,* K. Schmidt-Nielsen et al., eds., 31–42. Cambridge: Cambridge University Press.

Okimura, Judy T., and Scott A. Norton

1998 Jealousy and mutilation: Nose biting as retribution for adultery. *Lancet* 352: 2010–11.

Origo, Iris

1986 *The Merchant of Prato.* Boston: Godine (originally published 1957).

Overpeck, Mary D., Ruth A. Brenner, Ann C. Trumble, Lara B. Trifiletti, and Heinz W. Berendes

1998 Risk factors for infant homicide in the United States. *New England Journal of Medicine* 339:1211–16.

Packer, Craig, and Anne Pusey

1984 Infanticide in carnivores. In *Infanticide: Comparative and Evolutionary Perspectives,* Glenn Hausfater and Sarah Blaffer Hrdy, eds., 31–42. New York: Aldine.

Packer, Craig, Susan Lewis, and Anne Pusey

1992 A comparative analysis of non-offspring nursing. *Animal Behavior* 43:265–81.

Packer, Craig, Mark Tatar, and Anthony Collins

1998 Reproductive cessation in female mammals. *Nature* 392:807–11.

Page, E. W.

1939 The relation between hydatid moles, relative ischemia of the gravid uterus and the placental origin of eclampsia. *American Journal of Obstetrics and Gynecology* 37:291–93.

Page, H. J.

1989 Childrearing vs. childbearing: Coresidence of mother and child in sub-Saharan Africa. In *Reproduction and Social Organization in sub-Saharan Africa,* R. Laesthaeghe, ed., 401–41. Berkeley: University of California Press.

Pagel, Mark

1994 Parents prefer pretty plumage. *Nature* 371:200–201.

1997 Desperately concealing father: A theory of parent-infant resemblance. *Animal Behavior* 53:973–81.

Palombit, Ryne A., Robert M. Seyfarth, and Dorothy Cheney

1997 The adaptive value of "friendships" to female baboons: Experimental and observational evidence. *Animal Behavior* 54:599–614.

Panigrahi, Lalita

1976 *British Social Policy and Female Infanticide in India.* New Delhi: Munshiram Manoharlal.

Panter-Brick, Catherine
1989 Motherhood and subsistence work: The Tamang of rural Nepal. *Journal of Biosocial Science* 23:137–54.

Panter-Brick, C., and M. Smith, eds.
In press Nobody's Children: A reconsideration of child abandonment. Cambridge: Cambridge University Press.

Paradis, James, and George C. Williams
1989 *T. H. Huxley's Evolution and Ethics, with New Essays on Its Victorian and Sociobiological Context.* Princeton: Princeton University Press.

Parish, Amy
1994 Sex and food control in the "uncommon chimpanzee": How bonobo females overcome a phylogenetic legacy of male dominance. *Ethology and Sociobiology* 15:157–79.
1996 Female relationships in bonobos (*Pan paniscus*): Evidence for bonding, cooperation, and female dominance in a male philopatric species. *Human Nature* 7:61–96.

Parish, Amy, and E. Voland
1998 Reciprocal altruism in bonobos (*Pan paniscus*): Evidence from food sharing and affiliative interactions. Paper presented at the Annual Meeting of the Human Behavior and Evolution Society, July 8–12, University of California, Davis.

Parish, Amy, and Frans de Waal
1992 Bonobos fish for sweets: The female sex-for-food connection. Paper presented at the XIVth Congress of the International Primatological Society, August 16–21, Strassbourg, France.

Park, Chai Bin
1983 Preference for sons, family size, and sex ratio: An empirical study in Korea. *Demography* 209(3):333–52.

Parker, Rozsika
1995 *Mother Love/Mother Hate: The Power of Maternal Ambivalence.* New York: Basic Books.

Parks, Fanny
1975 *Wanderings of a Pilgrim in Search of the Picturesque Karachi,* reprint ed. Oxford: Oxford University Press (originally published 1850).

Parmigiani, Stefano, and Frederick S. vom Saal, eds.
1994 *Infanticide and Parental Care.* Langhorne, Pennsylvania: Harwood Academic.

Parmigiani, Stefano, Paola Palanza, Danilo Mainardi, and Paul F. Brain
1994 Infanticide and protection of young in house mice (*Mus domesticus*): Female and male strategies. In *Infanticide and Parental Care,* Stefano Parmigiani and F. vom Saal, eds., 341–63. Langhorne, Pennsylvania: Harwood Academic.

Parry, J.
1979 *Caste and Kinship in Kangra.* Cambridge: Cambridge University Press.

Patino, Exequiel M., and Juan T. Borda
1997 The composition of primates' milk and its importance in selecting formulas for hand-rearing. *Laboratory Primate Newsletter* 36(2):8–9.

Paul, A., and J. Kuester

1987 Dominance, kinship and reproductive value in female barbary macaques (*Macaca sylvanus*) at Affenberg Salem. *Behavioral Ecology and Sociobiology* 21:323–31.

1988 Life-history patterns of barbary macaques (*Macaca sylvanus*) at Affenberg Salem. In *Ecology and Behavior of Food-Enhanced Primate Groups,* J. Fa, ed., 199–228. New York: Alan Liss.

Pavelka, Mary S. M., and Linda Marie Fedigan

1991 Menopause: A comparative life history perspective. *Yearbook of Physical Anthropology* 34:13–38.

Pawlowski, Boguslaw

1998 Why are human newborns so big and fat? *Human Evolution* 13:65–72.

Paxton, Nancy L.

1991 *George Eliot and Herbert Spencer: Feminism, Evolutionism, and the Reconstruction of Gender.* Princeton: Princeton University Press.

Peláez-Nogueras, Martha, Tiffany M. Field, Ziarat Hossain, and Jeffrey Pickens

1996 Depressed mothers' touching increases infants' positive affect and attention in stiff-face interactions. *Child Development* 67:1780–92.

Pennington, René

1991 Child fostering as a reproductive strategy among southern African pastoralists. *Ethology and Sociobiology* 12:83–104.

1996 Causes of early human population growth. *American Journal of Physical Anthropology* 99:259–74.

Pennisi, Elizabeth

1996 Research News: A look at maternal guidance. *Science* 273:1334–36.

Pereira, Joseph

1994 Oh, Boy! In Toyland, you get more if you're male. *Wall Street Journal* (September 23):B-1.

Pereira, M. E.

1983 Abortion following immigration of an adult male baboon (*Papio cynocephalus*). *American Journal of Primatology* 4:93–98.

Perera, Judith

1987 Sex seals the fate of fetuses in Britain. *New Scientist* (January 22).

Perrigo, Glenn, and Frederick S. vom Saal

1994 Behavioral cycles and the neural timing of infanticide and parental behavior in male house mice. In *Infanticide and Parental Care,* Stefano Parmigiani and F. vom Saal, eds., 365–96. Langhorne, Pennsylvania: Harwood Academic.

Perry, Susan

1996 Female-female social relationships in wild white-faced capuchin monkeys, *Cebus capucinus. American Journal of Primatology* 40:167–82.

Perusse, D.

1993 Cultural and reproductive success in industrial societies: testing the phenomenon at the proximate and ultimate levels. *Behavioral and Brain Sciences* 16:267–322.

Petrie, Marion
    1994 Improved growth and survival of offspring of peacocks with more elaborate trains. *Nature* 371:598–99.
Petrie, Marion, and A. Williams
    1993 Peahens lay more eggs for peacocks with larger trains. *Proceedings of the Royal Society of London,* Series B 251:127–31.
Petrie, Marion, Claude Dooms, and Anders Pape Møller
    1998 The degree of extra-pair paternity increases with genetic variability. *Proceedings of the National Academy of Sciences* 95:9390–95.
Pettigrew, Joyce
    1986 Child neglect in rural Punjabi families. *Journal of Comparative Family Studies* 17:63–85.
Phillips, Deborah, Kathleen McCartney, Sandra Scarr, and Carollee Howes
    1987 Selective review of infant day care research: A cause for concern! *Zero to Three* (February):18–25.
Piercy, Marge
    1986 Magic Mama. In *Mother's Body: Poems by Marge Piercy,* 78. New York: Alfred A. Knopf.
Piers, Maria
    1978 *Infanticide: Past and Present.* New York: W. W. Norton.
Pinker, Steven
    1997 *How the Mind Works.* New York: W. W. Norton.
Pinneau, S. R.
    1955 The infantile disorders of hospitalism and anaclitic depressions. *Psychological Bulletin* 52:429–52.
Pittard, W. B., III
    1979 Breast milk immunology. *American Journal of Diseases of Children* 133:83–87.
Plooij, Frans X.
    1984 *The Behavioral Development of Free-Living Chimpanzee Babies and Infants.* Norwood, New Jersey: Ablex.
Pollitt, Ernesto, and Rudolph Leibel
    1980 Biological and social correlates of failure to thrive. In *Social and Biological Predictors of Nutritional Status, Physical Growth and Neurological Development,* Lawrence S. Greene and Francis Johnston, eds., 173–200. New York: Academic Press.
Pollock, Linda
    1983 *Forgotten Children: Parent-Child Relations from 1500 to 1900.* Cambridge: Cambridge University Press.
Pomeroy, Sarah
    1984 *Women in Hellenistic Egypt.* New York: Schocken.
Pond, Caroline
    1977 The significance of lactation in the evolution of mammals. *Evolution* 31:177–99.
    1978 Morphological aspects and the ecological and mechanical consequences of fat deposition in wild vertebrates. *Annual Review of Ecology and Systematics* 9:519–70.

Porter, D. M., and P. W. Graham, eds.

1993 *The Portable Darwin*. New York: Penguin.

Porter, R. H.

1991 Mutual mother-infant recognition in infants. In *Kin Recognition,* P. G. Hepper, ed., 413–32. Cambridge: Cambridge University Press.

Posner, Richard A.

1992 *Sex and Reason*. Cambridge: Harvard University Press.

Prechtl, Heinz F. R.

1950 Das Verhalten von Kleinkindern gegenuber Schlangen. *Wiener Zeitschrift für Philosophie, Psychologie, Pedagogik* 2:68–70.

1965 Problems of behavioral studies in the newborn infant. *Advances in the Study of Behavior* 1:75–98.

Prentice, Andrew, et al.

1996 Energy requirements of pregnant and lactating women. *European Journal of Clinical Nutrition,* Supplement 1, no. 501.

Prevost, J., and V. Vilter

1963 Histologie de la secretion oesophagienne du manchot empereur. *Proceedings of the XIIIth International Ornithology Congress,* 1085–94.

Pringle, Heather

1998 North America's wars: new analyses suggest that prehistoric North America, once considered peaceful, was instead a bitter battlefield where tribes fought over land and water. *Science* 279:2038–40.

Pritchard, James B., ed.

1955 *Ancient Near Eastern Texts Relating to the Old Testament,* 2nd ed. Princeton: Princeton University Press.

Pryce, C. R., R. D. Martin, and D. Skuse, eds.

1995 *Motherhood in Human and Nonhuman Primates: Biosocial Determinants*. Basel: S. Karger.

Pryce, Christopher R.

1995 Determinants of motherhood in human and nonhuman primates. In *Motherhood in Human and Nonhuman Primates: Biosocial Determinants,* Christopher R. Pryce, Robert D. Martin, and D. Skuse, eds., 1–15.

Pugh, George E.

1977 *The Biological Origins of Human Values*. New York: Basic Books.

Pusey, Anne

1979 Intercommunity transfer of chimpanzees in Gombe National Park. In *The Great Apes,* David A. Hamburg and Elizabeth R. McCown, eds., 465–79. Menlo Park, California: Benjamin-Cummings.

Pusey, Anne, and Craig Packer

1994 Infanticide in lions: Consequences and counter-strategies. In *Infanticide and Parental Care,* Stefano Parmigiani and F. vom Saal, eds., 277–300. Langhorne, Pennsylvania: Harwood Academic.

Pusey, Anne, Jennifer Williams, and Jane Goodall

    1997 The influence of dominance rank on the reproductive success of female chimpanzees. *Science* 277:828–31.

Queller, David C.

    1994 Extended parental care and the origin of eusociality. *Proceedings of the Royal Society of London,* Series B 256:105–11.

    1996 The origin and maintenance of eusociality: The advantage of extended parental care. In *Natural History and Evolution of Paper Wasps,* S. Turellazzi and M. J. West-Eberhard, eds., 218–34. Oxford: Oxford University Press.

Quinn, Naomi

    1977 Anthropological studies on women's status. *Annual Review of Anthropology* 6:181–225.

Ralls, Katherine

    1976 Mammals in which females are larger than males. *Quarterly Review of Biology* 51:245–76.

Ramanamma, A., and Usha Bambawale

    1980 The mania for sons: An analysis of social values in South Asia. *Social Science and Medicine* 14B:107–10.

Ramsay, M. A., and R. L. Dunbrack

    1986 Physiological constraints on life history phenomena: The example of small bear cubs at birth. *American Naturalist* 127:735–43.

Ransel, David

    1988 *Mothers of Misery: Child Abandonment in Russia.* Princeton: Princeton University Press.

Ransom, T. W., and T. E. Rowell

    1972 Adult male-infant relations among baboons (*Papio anubis*). *Folia Primatologica* 16:179–95.

Rao, R.

    1986 Move to stop sex-test abortion. *Nature* 324:202.

Rasa, Anne E.

    1994 Altruistic infant care or infanticide: The dwarf mongoose's dilemma. In *Infanticide and Parental Care,* Stefano Parmigiani and Frederick vom Saal, eds., 301–20. Langhorne, Pennsylvania: Harwood Academic.

Rasekh, Zorah, Heidi M. Bauer, M. Michele Manos, and Vincent Iacopino

    1998 Women's health and human rights in Afghanistan. *Journal of the American Medical Association* 280:449–55.

Raverat, Gwen

    1952 *Period Piece: A Cambridge Childhood.* London: Faber and Faber.

Reagan, Leslie J.

    1997 *When Abortion Was a Crime: Women, Medicine and Law in the United States, 1867–1973.* Berkeley: University of California Press.

Reeder, Ellen D.

    1995a Representing women. In *Pandora: Women in Classical Greece,* Ellen D. Reeder, ed., 123–94. Baltimore: Walters Art Gallery.

    1995b Women as the metaphor of wild animals. In *Pandora: Women in Classical Greece,* Ellen D. Reeder, ed., 299–372. Baltimore: Walters Art Gallery.

Reeves, P. D., ed.

1971 *Sleeman in Oudh: An Abridgement of W. H. Sleeman's "A Journey Through the Kingdom of Oudh in 1849–50."* London: Cambridge University Press.

Reite, Martin, and R. Short

1994 Nocturnal sleep in separated monkey infants. *Archives of General Psychiatry* 35: 1247–53.

Reite, Martin, R. Short, C. Seiler, and J. D. Pauley

1981 Attachment, loss and depression. *Journal of Child Psychology and Psychiatry* 22:141–69.

Relethford, John H.

1998 Mitochondrial DNA and ancient population growth. *American Journal of Physical Anthropology* 105:1–7.

Resnick, Philip J.

1970 Murder of the newborn: A psychiatric review of filicide. *American Journal of Psychiatry* 126:325–34.

Resnick, N. N., F. H. Shaw, F. Helen Rodd, and Ruth G. Shaw

1997 Evaluation of the rate of evolution in natural populations of guppies (*Poecilia reticulata*). *Science* 275:1934–37.

Rheingold, Harriet, ed.

1963 *Maternal Behavior in Mammals.* New York: Wiley.

Rheingold, Harriet, and Carol O. Eckerman

1970 The infant separates himself from his mother. *Science* 168:78–83.

Rhine, Ramon, S. K. Wasser, and G. W. Norton

1988 Eight-year study of social and ecological correlates of mortality among immature baboons of Mikumii National Park, Tanzania. *American Journal of Primatology* 16:199–212.

Rice, William

1996 Sexually antagonistic male adaptation triggered by experimental arrest of female evolution. *Nature* 381:232–34.

Rich, Adrienne

1986 *Motherhood as Experience and Institution.* New York: W. W. Norton.

Richards, Audrey

1939 *Land, Labour and Diet in Northern Rhodesia: An Economic Study of the Bemba Tribe.* London: International Institute of African Languages and Cultures (through Oxford University Press).

Riddle, Oscar, and Pela Fay Braucher

1931 Studies on the physiology of reproduction in birds: Control of the special secretion of the crop-gland in pigeons by an anterior pituitary hormone. *American Journal of Physiology* 97:617–25.

Ridley, M.

1993 *The Red Queen: Sex and the Evolution of Human Nature.* New York: Macmillan.

Riesman, Paul

1992 *First Find Your Child a Good Mother: The Construction of Self in Two African Communities.* New Brunswick: Rutgers University Press.

Roberts, C., and C. Lowe

1975 Where have all the conceptions gone? *Lancet* 1:498–99.

Robertson, J.

1953 *A Two-Year-Old Goes to Hospital* (film). Tavistock Child Development Research Unit, London. Available through the Penn State Audiovisual Services, University Park, Pennsylvania.

Robertson, J., and J. Bowlby

1952 Some responses of young children to loss of maternal care. *Nursing Care* 49:382–86.

Robine, Jean-Marie, and Michel Allard

1998 The oldest French woman. *Science* 279:1834–35.

Robinson, Paul

1981 Does Mom not care? *New York Times Book Review* (October 4):11.

Robinson, Roger J.

1992 Introduction. In *Fetal and Infant Origins of Adult Disease,* D.J.P. Barker and Roger J. Robinson, eds. London: British Medical Journal Publishing.

Robson, K. M., and R. Kumar

1980 Delayed onset of maternal affection after childbirth. *British Journal of Psychiatry* 136:347–53.

Rockhill, William W.

1895 Notes on the ethnology of Tibet. *U.S. National Museum Report for the Year 1893,* 665–747.

Rode, S. S., P.-N. Chang, P. O. Fisch, and L. A. Sroufe

1981 Attachment patterns of infants separated at birth. *Developmental Psychology* 17:188–91.

Rodman, Peter, and Henry McHenry

1980 Bioenergetics and the origin of hominid bipedalism. *American Journal of Physical Anthropology* 52(1):103–6.

Rogers, Alan

1998 The molecular record of human population history. Plenary lecture presented at the Tenth Annual Meeting of the Human Behavior and Evolution Society, July 8–12, University of California, Davis.

Roiphe, Anne

1996 *Fruitful: A Real Mother in the Modern World.* Boston: Houghton Mifflin.

Rosenberg, Karen, and Wenda Trevathan

1996 Bipedalism and human birth: The obstetrical dilemma revisited. *Evolutionary Anthropology* 4(5):161–68.

Rosenqvist, Gunilla, and Anders Berglund

1992 Is female sexual behaviour a neglected topic? *Trends in Ecology and Evolution* 7(6):174–76.

Rosenthal, A. M.

1997 Killing Iraqi children. *New York Times,* Op-ed section (December 9):A-21.

Rosenthal, Elizabeth

1998 In North Korean hunger, legacy is stunted children. *New York Times* (December 10) A1, A12.

Ross, C., and A. Maclarnon

1995 Ecological and social correlates of maternal expenditure on infant growth in hap-
lorhine primates. In *Motherhood in Human and Nonhuman Primates,* C. R. Pryce, R. D. Martin,
and D. Skuse, eds., 37–46. Basel: S. Karger.

Ross, James Bruce

1974 The middle-class child in urban Italy, fourteenth to early sixteenth century. In *The History of Childhood,* Lloyd de Mause, ed., 183–228. New York: Harper and Row.

Rosser, Sue V.

1997 Possible implications of feminist theories for the study of evolution. In *Feminism and
Evolutionary Biology: Boundaries, Intersections and Frontiers,* Patricia Adair Gowaty, ed., 21–41.
New York: Chapman and Hall.

Rossi, Alice

1973 Maternalism, sexuality and the new feminism. In *Contemporary Sexual Behavior: Critical
Issues in the 1970s,* Joseph Zubin and John Money, eds., 145–73. Baltimore: Johns Hopkins
University Press.

1977 A biosocial perspective on parenting. *Daedalus* (Spring):1–31.

1997 The impact of family structure and social change on adolescent sexual behavior. *Child
and Youth Services Review* 19:369–400.

Rossi, Alice, ed.

1978 *The Feminist Papers.* New York: Bantam.

Rossiter, Mary Carol

1994 Maternal effects hypothesis of herbivore outbreak. *Bioscience* 44:752–63.

1996 Incidence and consequences of inherited environmental effects. *Annual Review of Ecology and Systematics* 27:451–76.

Roth, H. L.

1896 *Natives of Sarawak and British North Borneo.* London.

Rousseau, J.-J.

1762 *Émile.* Barbara Foxley, trans. New York: Dutton, 1977.

Royer, Clémence

1870 *Origine de l'homme et des sociétés.* Paris: Guillaumin.

Rudran, R.

1973 Adult male replacement in one-male troops of purple-faced langurs and its effect on
population structure. *Folia Primatologica* 19:166–92.

Russett, Cynthia Eagle

1989 *Sexual Science: The Victorian Construction of Womanhood.* Cambridge: Harvard University
Press.

Sade, Donald

1967 Determinants of dominance in a group of free-ranging rhesus monkeys. In *Social Communication Among Primates,* S. Altmann, ed. Chicago: University of Chicago Press.

Sagi, Abraham, Marinus H. van IJzendoorn, Ora Aviezer, Frank Donnell, and Ofra Mayseless

1994 Sleeping out of home in a kibbutz communal arrangement: It makes a difference for
infant-mother attachment. *Child Development* 65:992–1004.

Salzman, Freda

1977 Are sex roles biologically determined? *Science for the People* (July–August):27–32, 43.

Sanders, D., and J. Bancroft

1982 Hormones and the sexuality of women—the menstrual cycle. *Clinics in Endocrinology and Metabolism* 11(3):639–59.

Sargent, Carolyn F.

1988 Born to die: Witchcraft and infanticide in Bariba culture. *International Journal of Cultural and Social Anthropology* 27(1):79–95.

Sayers, Janet

1982 *Biological Politics: Feminist and Anti-feminist Perspectives,* 1982. London: Tavistock.

Saylor, A., and M. Salmon

1971 Communal nursing in mice: influence of multiple mothers on the growth of the young. *Science* 164:1309–10.

Scarr, Sandra

1984 *Mother Care/Other Care.* New York: Basic Books.

Schaal, B., H. Montagner, E. Hertling, D. Bolzoni, A. Moyse, and A. Quichon

1980 Les stimulations olfactives dans les relations entre l'enfant et la mere. *Reproduction, Nutrition, Development* 20:843–58.

Schapera, Isaac

1933 Premarital pregnancy and native opinion: A note on social change. *Africa* 6:59–89.

Scheper-Hughes, Nancy

1992 *Death Without Weeping: The Violence of Everyday Life in Brazil.* Berkeley: University of California Press.

Schiebinger, Londa

1994 Mammals, primatology and sexology. In *Sexual Knowledge, Sexual Science,* Roy Porter and Mikulas Teich, eds., 184–209. Cambridge: Cambridge University Press.

1995 *Nature's Body: Gender in the Making of Modern Science.* Boston: Beacon Press.

1999 *Has Feminism Changed Science?* Cambridge: Harvard University Press.

Schiefenhövel, G., and W. Schiefenhövel

1978 Eipo, Iran Java (West-Neuguinea): Vorgange bei der Geburt eines Madchens und Anderung der Infantizid-Absicht. *Homo* 29:121–38.

Schiefenhövel, W.

1989 Reproduction and sex ratio manipulation through preferential female infanticide among the Eipo, in the highlands of western New Guinea. In *The Sociobiology of Sexual and Reproductive Strategies,* A. Rasa, C. Vogel, and E. Voland, eds., 170–93. London: Chapman and Hall.

Schlegel, Alice

1972 *Male Dominance and Female Autonomy.* New Haven: Human Relations Area Files.

Schmitt, Jean-Claude

1983 *The Holy Greyhound: Guinefort, Healer of Children* (trans. of *Le saint levrier: Guinefort, guerisseur d'enfants depuis le XIIIe siecle*), Martin Thom, trans. Cambridge: Cambridge University Press (originally published in French, 1979).

Schneider, David M., and Kathleen Gough, eds.

1961 *Matrilineal Kinship.* Berkeley: University of California.

Schneider, J., and P. Schneider

1984 Demographic transitions in a Sicilian rural town. *Journal of Family History* 9: 245–72.

Schoenmakers, R., I. H. Shah, R. Lesthaeghe, and O. Tambashe

1981 The child-spacing tradition and the postpartum taboo in tropical Africa: Anthropological evidence. In *Child-Spacing in Tropical Africa,* H. J. Page and R. Lesthaeghe, eds. New York: Academic Press.

Schoesch, Stephan J.

1998 Physiology of helping in Florida scrub jays. *American Scientist* 86:70–77.

Schraudolph, Ellen

1996 Sculpture and architectural fragments (cat. nos. 1–35). In *Pergamon: The Telephos Frieze from the Great Altar,* vol. 1, Renee Dreyfus and Ellen Schraudolph, eds., 53–112. San Francisco: Fine Arts Museum.

Schubert, Glendon

1982 Infanticide by usurper hanuman langur males: A sociobiological myth. *Social Science Information* 21(2):199–244.

Schultz, Adolph H.

1969 *The Life of Primates.* New York: Universe Books.

Schuster, Ilsa M. Glazer

1979 *New Women of Lusaka.* Palo Alto, California: Mayfield.

Schwabl, Hubert

1994 Yolk is a source of maternal testosterone for developing birds. *Proceedings of the National Academy of Sciences* 90:11446–50.

Schwabl, Hubert, Douglas Mock, and Jennifer A. Gieg

1996 A hormonal mechanism for parental favoritism. *Nature* 386:231.

1995 Adaptive hatching. *Science* 268:371.

Science

1995 Adaptive hatching. *Science* 268:371.

Scrimshaw, Susan C. M.

1984 Infanticide in human populations: Societal and individual concerns. In *Infanticide: Comparative and Evolutionary Perspectives,* G. Hausfater and S. Blaffer Hrdy, eds., 439–62. Hawthorne, New York: Aldine de Gruyter.

Scutt, D., J. T. Manning, G. H. Whitehouse, S. J. Leinster, and C. P. Massey

1997 The relationship between breast asymmetry, breast size, and the occurrence of breast cancer. *British Journal of Radiology* 70:1017–21.

Seelye, Katharine

1997a Medical group supports ban on a type of late abortion. *New York Times* (May 20).

1997b A partial-victory abortion vote. *New York Times* (May 25).

1997c Senators reject Democrats' bills to limit abortion: Loopholes cited by foes. *New York Times* (May 16).

Seger, Jon

1977 Model of gene action and the problem of behavior. Unpublished ms. in the author's possession, Department of Biology, University of Utah, Salt Lake City.

Segerstrale, Ullica

1997 Science by worst cases. *Science* 263:837–38.

Sellen, Daniel

1995 The socioecology of young child growth among the Datoga pastoralists of northern Kenya. Ph.D. diss., Department of Anthropology, University of California, Davis.

Sharp, Henry S.

1981 Old age among the Chipewyan. In *Other Ways of Growing Old,* Pamela T. Amoss and Stevan Harrell, eds., 99–110. Stanford: Stanford University Press.

Shenon, Philip

1994 China's mania for baby boys creates surplus of bachelors. *New York Times* (August 16):A1, A4.

Sherfey, Mary Jane

1966 *The Nature and Evolution of Female Sexuality.* 1972 reprint edition, New York: Vintage/Random House.

Sherman, Paul

1981 Reproductive competition and infanticide in Belding's ground squirrels and other animals. In *Natural Selection and Social Behavior: Recent Research and New Theory,* R. D. Alexander and D. W. Tinkle, eds., 311–31. New York: Chiron Press.

1998 The evolution of menopause. *Nature* 392:759–61.

Sherwood, John

1988 *Poverty in Eighteenth Century Spain: The Women and Children of the Inclusa.* Toronto: University of Toronto Press.

Shields, Stephanie

1982 The "Variability Hypothesis": The history of a biological model of sex differences in intelligence. *Signs: Journal of Women in Culture and Society* 7(4):769–97.

1984 To pet, coddle, and "do for": caretaking and the concept of maternal instinct. In *In the Shadow of the Past: Psychology Portrays the Sexes,* Miriam Lewin, ed., 256–73. New York: Columbia University Press.

Short, Roger V.

1977a Sexual selection and the descent of man. In *Reproduction and Evolution,* J. H. Calaby and C. H. Tyndale-Biscoe, eds., 3–19. Canberra: Australian Academy of Science.

1977b The discovery of the ovaries. In *The Ovary,* Solly Zuckerman and Barbara J. Weir, eds. New York: Academic Press.

1984 Breast feeding. *Scientific American* 250:35–41.

1997 The testis: The witness of the mating system, the site of mutation and the engine of desire. *Acta Paediatrica* Supplement 422:3–7.

Shorter, Edward

1975 *The Making of the Modern Family.* New York: Basic Books.

1982 *A History of Women's Bodies.* New York: Basic Books.

Shostak, Marjorie

 1981 *Nisa.* Cambridge: Harvard University Press.

Sieff, Daniela

 1990 Explaining biased sex ratios in human populations. *Current Anthropology* 31:25–48.

Sieratzki, J. S., and B. Wolf

 1996 Why do mothers cradle babies on their left? *Lancet* 347:1746–48.

Sih, A., and R. D. Moore

 1993 Delayed hatching of salamander eggs in response to enhanced larval predation risk. *American Naturalist* 142:947–60.

Silk, Joan B.

 1980 Kidnapping and female competition in captive bonnet macaques. *Primates* 21: 100–110.

 1983 Local resource competition and facultative adjustment of sex ratios in relation to competitive abilities. *American Naturalist* 121:56–66.

 1987a Activities and diet of free-ranging pregnant baboons, *Papio cynocephalus. International Journal of Primatology* 8:593–613.

 1987b Adoption and fosterage in human societies: Adaptations or enigmas? *Cultural Anthropology* 2:39–49.

 1987c Social behavior in evolutionary perspective. In *Primate Societies,* B. Smuts et al., eds., 318–29. Chicago: University of Chicago Press.

 1988 Maternal investment in captive bonnet macaques, *Macaca radiata. American Naturalist* 132(1):1–19.

 1990 Human adoption in evolutionary perspective. *Human Nature* 1:25–52.

Silk, Joan B., C. B. Clark-Wheatley, P. S. Rodman, and A. Samuels

 1981 Differentiated reproductive success and facultative adjustment of sex ratios among captive female bonnet macaques (*Macaca radiata*). *Animal Behavior* 29:1106–20.

Simoons, F. J.

 1978 The geographic hypothesis and lactose malabsorption. *American Journal of Digestive Diseases* 15:695–710.

Singer, Lynn T., Ann Salvator, Shenyang Guo, Marc Collin, Lawrence Lilien, and Jill Baley

 1999 Maternal psychological distress and parenting stress after the birth of a very low-birth-weight infant. *Journal of the American Medical Association* 281:799–805.

Singh, Devendra, and Suwardi Luis

 1995 Ethnic and gender consensus for the effect of waist-to-hip ratio on judgment of women's attractiveness. *Human Nature* 6:51–65.

Skinner, William

 1993 Conjugal power in Tokugawa Japanese families: A matter of life or death. In *Sex and Gender Hierarchies,* Barbara D. Miller, ed., 236–70. New York: Cambridge University Press.

 1997 Family systems and demographic processes. In *Anthropological Demography: Towards a New Synthesis,* David Kertzer and Tom Fricke, eds., 53–114. Chicago: University of Chicago Press.

Skinner, G. William, and Yuan Jianhua

 1998 Reproductive strategizing in the face of China's birth-planning policies: The lower

Yangzi macroregion, 1966–1990. Paper prepared for the Center for Chinese Studies Seminar, April 14, University of Michigan, Ann Arbor.

Skuse, D. H., R. S. James, D. V. M. Bishop, B. Coppin, P. Dalton, G. Aamodt-Leeper, M. Bacarese-Hamilton, C. Creswell, R. McGurk, and P. A. Jacobs

1997 Evidence from Turner's syndrome of an imprinted X-linked locus affecting cognitive function. *Nature* 287:705–8.

Slob, A. K., M. Ernste, and J. J. van der Werff ten Bosch

1991 Menstrual cycle phase and sexual arousability in women. *Archives of Sexual Behavior* 20(6):567–76.

Slob, A. K., W. H. Groeneveld, and J. J. van der Werff ten Bosch

1986 Physiological changes during copulation in male and female stumptail macaques (*Macaca arctoides*). *Physiology and Behavior* 38:891–95.

Slob, A. Koos, Cindy M. Bax, Wim C. J. Hop, David L. Rowland, and Jacob J. van der Werff ten Bosch

1996 Sexual arousability and the menstrual cycle. *Psychoendocrinology* 21:545–58.

Slob, A. K., and J. J. van der Werff ten Bosch

1991 Orgasm in nonhuman species. In *Proceedings of the First International Conference on Orgasm,* P. Kothari and R. Patel, eds. Bombay: VRP.

Slocum, Sally (later Sally Linton)

1971 Woman the gatherer. In *Women in Perspective: A Guide for Cross-cultural Studies,* Sue-Ellen Jacobs, ed., 9–21. Urbana: University of Illinois Press. (Reprinted in 1975 in *Towards an Anthropology of Women,* R. Rapp Reiter, ed., 36–50. New York: Monthly Review Press.)

Small, Meredith F.

1990 Promiscuity in barbary macaques (*Macaca sylvana*). *American Journal of Primatology* 20:267–82.

1992 What's love got to do with it? Sex among our closest relatives is a rather open affair. *Discovery* (June):46–51. (Reprinted in *Physical Anthropology: 1997–98 Annual Editions,* 96–99. Guilford, Connecticut: Benchmark.)

1994 *Female Choices.* Ithaca: Cornell University Press.

1998 *Our Babies, Ourselves.* New York: Anchor Books.

Small, Meredith F., ed.

1984 *Female Primates: Studies by Women Primatologists.* New York: A. R. Liss.

Smith, A. H.

1899 *Village Life in China: A Study in Sociology.* New York: F. H. Revell.

Smith, Bruce D.

1997 The initial domestication of *Cucurbita pepo* in the Americas 10,000 years ago. *Science* 276:932–34.

Smith, Eric Alden

1996 Human life history comes of age. *Evolutionary Anthropology* 5(5):181–85.

Smith, Eric Alden, and Bruce Winterhalder, eds.

1992 *Evolutionary Ecology and Human Behavior.* Hawthorne, New York: Aldine de Gruyter.

Smith, Fabiene

1990 Charles Darwin's health problems. *Journal of Historical Biology* 23:443–49.

Smith, G., M. Smith, M. McNay, and J. Fleming

1998 First-trimester growth and the risk of low birth weight. *New England Journal of Medicine* 339(25):1817–22.

Smith, H. F.

1984 Notes on the history of childhood. *Harvard Magazine* (July–August), *Discovery Supplement,* 64c.

Smith, Holly

1993 The physiological age of KNM-WT 1500. In *The Nariokotome* Homo erectus *Skeleton,* Alan Walker and Richard Leakey, eds., 195–220. Cambridge: Harvard University Press.

1994 Patterns of dental development in *Homo, Australopithecus, Pan,* and *Gorilla. American Journal of Physical Anthropology* 94:307–25.

Smith, T. C.

1977 *Nakahara: Family Farming and Population in a Japanese Village.* Stanford: Stanford University Press.

Smuts, Barbara

1985 *Sex and Friendship in Baboons.* New York: Aldine.

1995 The evolutionary origins of patriarchy. *Human Nature* 6:1–32.

Smuts, Barbara B., and Robert T. Smuts

1993 Male aggression and sexual coercion of females in nonhuman primates and other mammals: evidence and theoretical implications. *Advances in the Study of Behavior* 22:1–63.

Soffer, O., J. M. Adovasio, D. C. Hyland, B. Klima, and J. Svoboda

1998 Perishable industries from Dolni Vestonice I: New insights into the nature and origin of the Gravettian. Paper presented at the Sixty-third Annual Meeting of the Society for American Archaeology, March 25–29, Seattle.

Solomon, Nancy, and Jeffrey A. French, eds.

1997 *Cooperative Breeding in Mammals.* Cambridge: Cambridge University Press.

Sommer, Volker

1994 Infanticide among the langurs of Jodhpur: Testing the sexual selection hypothesis with a long-term record. In *Infanticide and Parental Care,* Stefano Parmigiani and F. vom Saal, eds., 155–98. Langhorne, Pennsylvania: Harwood Academic.

1996 Infanticide in Hanuman langurs: Female counter-strategies. Paper presented at the XVth Congress of the International Primatological Society, August 11–16, Ann Arbor.

Soranus

1956 *Gynaecology.* Oswei Temkin, trans. Baltimore: Johns Hopkins University Press.

Soroker, V., and J. Terkel

1988 Changes in the incidence of infanticidal and parental responses during the reproductive cycle of male and female house mice, *Mus musculus. Animal Behavior* 36:1275–81.

Spangler, G., and K. E. Grossmann

1993 Biobehavioral organization in securely and insecurely attached infants. *Child Develop-ment* 64:152–62.

Spencer, Herbert

1859 Physical training. *British Quarterly Review* (April):362–97.

1864–67 *Principles of Biology,* 2 vols. London: William and Norgate.

1865 Personal beauty. In *Essays: Moral, Political and Aesthetic.* New York: D. Appleton.

1868 The development hypothesis. (First published in 1852.) Reprinted in *Essays: Scientific, Political and Speculative,* vol. I, 377–83. London: William and Norgate.

1873 Psychology of the sexes. *Popular Science Monthly* 4:30–38.

1904 *An Autobiography,* 2 vols. New York: D. Appleton.

Spencer-Booth, Yvette, and R. A. Hinde

1971a Effects of brief separation from mother on rhesus monkeys. *Science* 173:111–18.

1971b Effects of 6 days' separation from mother in 18- to 32-week-old rhesus monkeys. *Animal Behavior* 19:174–91.

Speth, John

1990 Seasonality, stress, and food-sharing in so-called "egalitarian" foraging societies. *Journal of Anthropological Archaeology* 9:148–88.

Spitz, R.

1945 Hospitalism: An inquiry into the genesis of psychiatric conditions in early childhood. *Psychoanalytic Study of the Child* 1:53–74.

1946 Anaclitic depression. *Psychoanalytic Study of the Child* 2:313–42.

Srinivasan, K.

1998 Demography and reproductive health. *Lancet* 351:1274.

Srivastave, J. N., and D. M. Saksena

1981 Infant mortality differentials in an Indian setting: Follow-up of hospital deliveries. *Journal of Biosocial Science* 13:467–78.

Sroufe, L. A., and J. Fleeson

1986 Attachment and the construction of relationships. In *Relationships and Development,* W. Hartup and Z. Rubin, eds., 51–72. Hillsdale, New Jersey: Erlbaum.

Stacey, P. B.

1979 Kinship, promiscuity, and communal breeding in the acorn woodpecker. *Behavioral Ecology and Sociobiology* 6:53–66.

Stacey, Peter, and Walter D. Koenig

1990 *Cooperative Breeding in Birds.* Cambridge: Cambridge University Press.

Stack, Carol

1974 *All Our Kin.* New York: Harper and Row.

Stallings, J. C., C. Panter-Brick, and C. M. Worthman

1994 Prolactin levels in nursing Tamang and Kami women: Effects of nursing practices on lactational amenorrhea (abstract). *American Journal of Physical Anthropology* (supplement) 18:185–86.

Stallings, J. F., A. S. Fleming, C. M. Worthman, M. Steiner, C. Corter, and M. Coote

1997 Mother/Father differences in response to infant crying (abstract). *American Journal of Physical Anthropology* (supplement) 24:217.

Stanford, Craig B.

1992 Costs and benefits of allomothering in wild capped langurs (*Presbytis pileata*). *Behavioral Ecology and Sociobiology* 30:29–34.

Stanislaw, H., and F. J. Rice

1988 Correlation between sexual desire and menstrual cycle characteristics. *Archives of Sexual Research* 17(6):499–508.

Stearns, Stephen

1977 The evolution of life history traits: A critique of the theory and a review of the data. *Annual Review of Ecology and Systematics* 8:145–71.

1982 The role of development in the evolution of life histories. In *Evolution and Development*, John Bonner, ed., 237–58. New York: Springer-Verlag.

1992 *The Evolution of Life Histories*. New York: Oxford University Press.

Stenseth, Nils Christian

1978 Is the female biased sex ratio in wood lemming *Myopus schisticolor* maintained by cyclic inbreeding? *Oikos* 30:83–89.

Stern, Daniel

1990 *Diary of a Baby*. New York: Basic Books.

Stern, Judith, Mel Konner, Talia N. Herman, and S. Reichlin

1986 Nursing behaviour, prolactin and postpartum amenorrhoea during prolonged lacation in American and !Kung mothers. *Clinical Endocrinology* 25:247–58.

Stern, Kathleen, and Martha McClintock

1998 Regulation of ovulation by human pheromones. *Nature* 393:177–79.

Stewart, Andrew

1995 Rape? In *Pandora: Women in Classical Greece,* Ellen D. Reeder, ed., 74–90. Baltimore: Walters Art Gallery.

Stewart, Kelly

1981 Social development of wild mountain gorillas. Ph.D. diss., Cambridge University.

1984 Parturition in wild gorillas: Behavior of mothers, neonates, and others. *Folia Primatologica* 42d:62–69.

Stolberg, Sheryl Gay

1997a Definition of fetal viability is focus of debate in Senate. *New York Times* (May 16):A13.

1997b Women's issue and wariness in Congress: Men's motivations are being questioned on health legislation. *New York Times* (May 26).

Stone, Lawrence

1977 *The Family, Sex and Marriage in England, 1500–1800*. London: Weidenfeld and Nicolson.

Storey, Anne E.

1990 Pregnancy disruption by unfamiliar males in meadow voles: A comparison of chemical and behavioural cues. *Physiology and Behavior* 5:199–208.

Strassmann, Beverly I.

1993 Menstrual hut visits by Dogon women: A hormonal test distinguishes deceit from honest signaling. *Behavioral Ecology* 7(3):304–15.

1997 Polygyny as a risk factor for child mortality among the Dogon. *Current Anthropology* 38:688–95.

Strassmann, Beverly, and Keith Hunley

1996 Polygyny, sorcery and child mortality among the Dogon of Mali. Paper presented at the Ninety-fifth Annual Meeting of American Anthropological Association, November 20–24, San Francisco.

Strassmann, Beverly I., and John Warner

1998 Predictors of fecundability and conception waits among the Dogon of Mali. *American Journal of Physical Anthropology* 105:167–84.

Strindberg, August (Elizabeth Sprigge, trans.)

1955 *Six Plays of Strindberg.* New York: Doubleday.

Strum, Shirley

1987 *Almost Human: A Journey into the World of Baboons.* New York: Random House.

Strum, Shirley, and Linda Fedigan

1996 Theory, method and gender: What changed our view of primate society? Target paper prepared for Wenner-Gren Foundation Symposium No. 120, Changing Images of Primate Societies, June 15–23, Teresopolis, Brazil.

Stuart-Macadam, Patricia, and Katherine A. Dettwyler, eds.

1995 *Breastfeeding: Biocultural Perspectives.* Hawthorne, New York: Aldine de Gruyter.

Studer-Thiersch, A.

1975 Basle Zoo. In *Flamingos,* J. Kear and N. Duplaiz-Hall, eds. Berkhamsted: T. and A. D. Poyser.

Sugiyama, Y.

1965 On the social change of Hanuman langurs (*Presbytis entellus*) under natural conditions. *Primates* 6:381–417.

1967 Social organization of hanuman langurs. In *Social Communication Among Primates,* S. Altmann, ed., 221–36. Chicago: University of Chicago Press.

Sulloway, Frank

1991 Darwinian psychobiography. *New York Review of Books* (October 10):30.

1996 *Born to Rebel.* New York: Pantheon.

Suomi, S. J., and H. J. Harlow

1972 Social rehabilitation of isolated-reared monkeys. *Developmental Psychobiology* 6:487–96.

Suomi, Stephen J.

1999 Attachment in rhesus monkeys. In *Handbook of Attachment: Theory, Research, and Clinical Applications,* J. Cassidy and P. R. Shaver, eds., 845–87. New York: Guilford Press.

Surani, M. Azim

1993 Silence of the genes. *Nature* 236:302–3.

Suransky, Valerie Polakow

1982 *The Erosion of Childhood.* Chicago: University of Chicago Press.

Surbey, Michele K.

1990 Family composition, stress and the timing of human menarche. In *Socioendocrinology of Primate Reproduction,* Toni E. Ziegler and Fred B. Bercovitch, eds., 11–32. New York: Wiley-Liss.

1998 Parent and offspring strategies in the transition to adolescence. *Human Nature* 9:67–94.

Sussman, George

1982 *Selling Mother's Milk: The Wet-Nursing Business in France, 1715–1914.* Urbana: University of Illinois Press.

Sussman, Robert W., James M. Cheverud, and Thad Q. Bartlett

1995 Infant killing as an evolutionary strategy: reality or myth? *Evolutionary Anthropology* 3:149–51.

Svare, Bruce, and Ronald Gandelman

1976 Suckling stimulation induces aggression in virgin female mice. *Nature* 260:606–8.

Svensson, Erik

1997 The speed of life-history evolution. *Trends in Ecology and Evolution* 12:380–81.

Swarns, Rachel

1998 Mothers poised for workfare face acute lack of day care. *New York Times* (April 14).

Sykora, P.

1998 Is there male-selective infanticide in Slovakia after the fall of communism? Paper presented at the Annual Meeting of the Human Behavior and Evolution Society, July 8–12, University of California, Davis.

Symington, Meg McFarland

1987 Sex ratio and maternal rank in wild spider monkeys: When daughters disperse. *Behavioral Ecology and Sociobiology* 20:421–25.

Symons, Don

1979 *The Evolution of Human Sexuality.* New York: Oxford University Press.

Tait, D. E. N.

1980 Abandonment as a tactic in grizzly bears. *American Naturalist* 115:800–808.

Tanner, Nancy, and Adrienne Zihlman

1976 Women in evolution (Part 1). *Signs* 1:585–608.

Tansillo, Luigi

1798 The Nurse. Translated from the Italian by William Roscoe. Liverpool.

Tardieu, Christine

1998 Short adolescence in early hominids: Infantile and adolescent growth of the human femur. *American Journal of Physical Anthropology* 107:163–78.

Tardif, Suzette

1997 The bioenergetics of parental behavior and the evolution of alloparental care in marmosets and tamarins. In *Cooperative Breeding in Mammals,* Nancy G. Solomon and Jeffrey A. French, eds., 11–33. Cambridge: Cambridge University Press.

Taub, David

1980 Female choice and mating strategies among wild barbary macaques (*Macaca sylvanus*

L.). In *The Macaques: Studies in Ecology, Behavior and Evolution,* D. Lindburg, ed., 287–344. NewYork: Van Nostrand Reinhold.

Taub, David, ed.

1984 *Primate Paternalism.* NewYork: Von Nostrand Reinhold.

Tavris, Carol

1992 *The Mismeasures of Woman.* NewYork: Touchstone.

Temple, C. L., ed.

1965 *Notes on the Tribes, Provinces, Emirates and States of the Northern Provinces of Nigeria Compiled from Official Reports by O. Temple.* London: Frank Cass (originally published 1919).

Terkel, Joseph, and Jay Rosenblatt

1968 Maternal behavior induced by maternal blood plasma injected into virgin rats. *Journal of Comparative and Physiological Psychology* 65:479–82.

Thierry, B., and J. R. Anderson

1986 Adoption in anthropoid primates. *International Journal of Primatology* 7:191–216.

Thoman, E. B., A. M. Turner, C. R. Barnett, and P. H. Leiderman

1971 Neonate-mother interaction: Effects of parity on feeding behavior. *Child Development* 42:1471–83.

Thomas, Elizabeth Marshall

1958 *The Harmless People.* NewYork: Random House/Vintage.

Thomas, Evan

1996 Blood brothers—The Unabomber saga: A family's history of loneliness, fear and betrayal. *Newsweek* (April 22):28–38.

Thompson, N. S.

1967 Primate infanticide: A note and request for information. *Laboratory Primate Newsletter* 6:18–19.

Thornhill, Randy

1979 Male and female sexual selection and the evolution of mating systems in insects. In *Sexual Selection and Reproductive Competition in Insects,* M. S. Blum and N. A. Blum, eds., 81–121. NewYork: Academic Press.

1992a Fluctuating asymmetry and the mating system of the Japanese scorpionfly, *Panorpa japonica. Animal Behavior* 44:867–79.

1992b Female preference for the pheromone of males with low fluctuating asymmetry in the Japanese scorpionfly (*Panorpa japonica: Mecoptera*). *Behavioral Ecology* 3:277–83.

Thornhill, Randy, and John Alcock

1983 *The Evolution of Insect Mating Systems.* Cambridge: Harvard University Press.

Thornhill, Randy, and StevenW. Gangestad

1994 Human fluctuating asymmetry and sexual behavior. *Psychological Science* 5(5):297–302.

Thornhill, Randy, and NancyWilmsenThornhill

1983 Human rape: An evolutionary analysis. *Ethology and Sociobiology* 4:137–73.

Thornhill, R., S. Gangestad, and R. Comer

1995 Human female orgasm and male fluctuating asymmetry. *Animal Behavior* 50:1601–15.

Thurer, Shari

1994 *The Myths of Motherhood: How Culture Reinvents the Good Mother.* Boston: Houghton Mifflin.

Thurston, Anne

1996 In a Chinese orphanage. *Atlantic Monthly* 227(4):28–41.

Tilley, Louise A., Rachel G. Fuchs, David I. Kertzer, and David L. Ransel

1992 Child abandonment in European history: A symposium. *Journal of Family History* 17:1–23.

Tilly, Louise

1988 Terrible tales: Coping with fear of infanticide. *Time* (November 4):140.

1996 The gorilla of America's dreams. *Time* (September 2).

Tokuda, M.

1935 An eighteenth-century Japanese guide-book on mouse-breeding. *Journal of Heredity* 26:481–84.

Tooby, John, and Leda Cosmides

1990 The past explains the present. Emotional adaptations and the structure of ancestral environments. *Ethology and Sociobiology* 11:375–424.

1992 Psychological foundations of human culture. In *The Adapted Mind,* Jerome Barkow, Leda Cosmides, and John Tooby, eds., 19–136. New York: Oxford University Press.

Treloar, A. E., Kim-Anh Do, and Nicholas G. Martin

1998 Genetic influences on the age of menarche. *Lancet* 352:1084–85.

Trevathan, Wenda

1987 *Human Birth: An Evolutionary Perspective.* New York: Aldine de Gruyter.

Trexler, Richard C.

1973a Infanticide in Florence: New sources and first results. *History of Childhood Quarterly* 1:98–116.

1973b The foundlings of Florence, 1395–1455. *History of Childhood Quarterly* 1:259–84.

Trinkaus, E.

1987 Neandertal pubic morphology and gestation length. *Current Anthropology* 25:509–13.

Trivers, Robert L.

1971 The evolution of reciprocal altruism. *Quarterly Review of Biology* 46(4):35–57.

1972 Parental investment and sexual selection. In *Sexual Selection and the Descent of Man, 1871–1971,* B. Campbell, ed., 136–79. Chicago: Aldine.

1974 Parent-offspring conflict. *American Zoologist* 14:249–64.

1985 *Social Evolution.* Menlo Park, California: Benjamin Cummins.

Trivers, Robert L., and D. E. Willard

1973 Natural selection of parental ability to vary the sex ratio of offspring. *Science* 179:90–92.

Troisi, Alfonso, F. R. D'Amato, R. Fuccillo, and S. Scucchi

1982 Infant abuse by a wild-born group-living Japanese macaque mother. *Journal of Abnormal Psychology* 91:451–56.

Trollope, A.

   1986 *The Last Chronicle of Barset.* London: Penguin (originally published 1867).

Tronick, E. Z., G. Morelli, and S. Winn

   1987 Multiple caretaking of Efé (Pygmy) infants. *American Anthropologist* 89:96–106.

Turke, Paul W.

   1988 Helpers at the nest: Childcare networks on Ifaluk. In *Human Reproductive Behaviour: A Darwinian Perspective,* L. Betzig, M. Borgerhoff Mulder, and P. Turke, eds., 173–89. Cambridge: Cambridge University Press.

   1989 Evolution and the demand for children. *Population and Development Review* 15:61–89.

   1991 Theory and evidence on wealth flows and old-age security: A reply to Fricke. *Population and Development Review* 17(4):687–702.

Turke, Paul, and Laura Betzig

   1985 Those who can, do: Wealth, status and reproductive success on Ifaluk. *Ethology and Sociobiology* 6:79–87.

Tutin, Caroline

   1975 Sexual behavior and mating patterns in a community of wild chimpanzees (*Pan troglodytes schweinfurthii*). Ph.D. diss., University of Edinburgh.

Uglow, Jenny

   1987 *George Eliot.* London: Virago.

U.S. Dept. of Health and Human Services

   1995 *A Nation's Shame: Fatal Abuse and Neglect in the United States.* Executive Summary, Report of the U.S. Advisory Board on Child Abuse and Neglect (21).

USA Today

   1998 Switched at birth. *USA Today* (April 17):2A.

Uvnas-Moberg, Kerstin

   1997 Physiological and endocrine effects of social contact. In *The Integrative Neurophysiology of Affiliation,* C. Sue Carter, I. Izja, and Brian Kirkpatrick, eds., 146–63. New York: New York Academy of Sciences.

Valenzuela, Marta

   1990 Attachment in chronically underweight young children. *Child Development* 61: 1984–96.

van Hoof, J.A.R.M.

   1962 Facial expressions in higher primates. *Symposia of the Zoological Society of London* 8:97–125.

van IJzendoorn, Marinus H.

   1995 Adult attachment representations, parental responsiveness, and infant attachment: A meta-analysis on the predictive validity of the adult attachment in interview. *Psychological Bulletin* 117(3):387–403.

van IJzendoorn, M. H., and P. M. Kroonenberg

   1988 Cross-cultural patterns of attachment. A meta-analysis of the Strange Situation. *Child Development* 59:147–56.

van Lawick, Hugo

   1973 *Solo: The Story of an African Wild Dog Puppy and His Pack.* London: Collins.

van Schaik, Carel, and Robin Dunbar

1990 The evolution of monogamy in large primates: A new hypothesis and some crucial tests. *Behaviour* 115:30–62.

van Schaik, Carel, and Sarah Blaffer Hrdy

1991 Intensity of local resource competition shapes the relationship between maternal rank and sex ratios at birth in Certopithecine primates. *American Naturalist* 138:1555–62.

van Schaik, Carel, Maria A. van Noordwijk, and Charles L. Nunn

1999 Sex and social evolution in primates. In *Comparative Primate Socioecology*, Phyllis C. Lee, ed. Cambridge: Cambridge University Press.

Vaux, The Reverend J.

1894 *Church Folklore: A Record of Some Post-Reformation Usages in the English Church, Now Mostly Obsolete.* London: Griffith and Farran.

Vila, Bryan

1997 Human nature and crime control: Improving the feasibility of nurturant strategies. *Politics and Life Sciences* 16:3–21.

Visaria, Pravin

1967 Sex ratio at birth in territories with a relatively complete registration. *Eugenics Quarterly* 14:132–42.

Vitzthum, Virginia J.

1989 Nursing behavior and its relation to duration of post-partum amenorrhea in an Andean community. *Journal of Biosocial Science* 21:145–60.

1997 Flexibility and paradox: The nature of adaptation in human reproduction. In *The Evolving Female: A Life-History Perspective,* Mary Ellen Norbeck, Alison Galloway, and Adrienne L. Zihlman, eds., 242–58. Princeton: Princeton University Press.

Voegelin, Ermine

1942 *Culture Elements Distribution XX: Northeast California.* Anthropological Records 7(2). Berkeley: University of California Press.

Vogel, Christian

1979 Der Hanuman-Langur (*Presbytis entellus*), ein Parade-Exempel für die theoretischen Konzepte der "Soziobiologie"? In *Verhandlungen der Deutschen Zoologischen Gesellschaft* 1979 in Regensburg (W. Rathmayer, ed.), 73–89. Stuttgart: Gustav Fischer Verlag.

Vogel, Gretchen

1998 Fly development genes lead to immune find. *Science* 281:1942–44.

Voland, Ekart

1988 Differential infant and child mortality in evolutionary perspective: Data from the late 17th to 19th century Ostfriesland (Germany). In *Human Reproductive Behaviour: A Darwinian Perspective,* L. Betzig, M. Borgerhoff Mulder, and P. Turke, eds., 253–62. Cambridge: Cambridge University Press.

1990 Differential reproductive success within the Krummhorn population (Germany, 18th and 19th centuries). *Behavioral Ecology and Sociobiology* 26:54–72.

vom Saal, Frederick S., Patricia Franks, Michael Boechler, Paola Palanza, and Stefano Parmigiani

1995 Nest defense and survival of offspring in highly aggressive wild Canadian female house mice. *Physiology and Behavior* 58(4):669–78.

Wade, Nicholas
    1998 Human or chimp? Fifty genes are the key. *New York Times* (October 20):D1–D4.
Wadley, S.
    1988 More children, fewer daughters: Family building strategies of the urban poor in North India. Paper presented at the Eighty-seventh Annual Meeting of the American Anthropological Association, November, Phoenix.
Walker, Alan, and Richard Leakey, eds.
    1993 The Nariokotome *Homo erectus* Skeleton. Cambridge: Harvard University Press.
Walker, Alan, and Pat Shipman
    1996 *The Wisdom of the Bones: In Search of Human Origins.* New York: Alfred A. Knopf.
Walker, Margaret
    1995 Menopause in female rhesus monkeys. *American Journal of Primatology* 35:59–71.
Wallen, Kim
    1980 Desire and ability: Hormones and the regulation of female sexual behavior. *Neuroscience and Biobehavioral Reviews* 14:233–41.
    1995 The evolution of female desire. In *Sexual Nature/Sexual Culture,* P. R. Abramson and S. D. Pinkerton, eds., 57–79. Chicago: University of Chicago Press.
Wallis, J. A.
    1997 Survey of reproductive parameters in the free-ranging chimpanzees of Gombe National Park. *Journal of Reproduction and Fertility* 109:297–307.
Wallis, J., and Y. Almasi
    1995 A survey of reproductive parameters in free-ranging chimpanzees (*Pan troglodytes*). Paper presented at the Eighteenth Annual Meeting of the American Society of Primatologists, June 21–24.
Walsh, Anthony
    1998 Human reproductive strategies and life history theory. Paper presented at the conference "The Role of Theory in Sex Research," organized by the Kinsey Institute, May 14–17, Bloomington, Indiana.
Walton, Susan
    1986 How to watch monkeys. *Science 86* (June):22–27.
Warkentin, Karen
    1995 Adaptive plasticity in hatching age: A response to predation risk trade-offs. *Proceedings of the National Academy of Sciences* 92:3507–10.
Warren, J. M.
    1967 Discussion of social dynamics. In *Social Communication Among Primates,* S. Altmann, ed., 255–57. Chicago: University of Chicago Press.
Wartner, Ulrike G., Karin Grossmann, Elisabeth Fremmer-Bombik, and Gerhard Suess
    1994 Attachment patterns at age six in south Germany: Predictability from infancy and implications for pre-school behavior. *Child Development* 65:1014–27.
Wasser, Samuel K., ed.
    1983 *Social Behavior of Female Vertebrates.* New York: Academic Press.

Wasser, Samuel K., and David Y. Isenberg

1986 Reproductive failure among women: Pathology or adaptation? *Journal of Psychosomatic Obstetrics and Gynecology* 5:153–75.

Watson, P. W., and R. Thornhill

1994 Fluctuating asymmetry and sexual selection. *Trends in Ecology and Evolution* 9:21–25.

Watts, David P.

1989 Infanticide in mountain gorillas: New cases and a reconsideration of the evidence. *Ethology* 81:1–18.

n.d. Karisoke orphans. Annotated data from unpublished fieldnotes.

Watts, David P., and Jorg Hess

1988 Twin births in wild mountain gorillas. *Oryx* 22:5–6.

Weaver, D. R., and S. M. Reppert

1986 Maternal melatonin communicates day length to the fetus in the Djungerian hamster. *Endocrinology* 119:2861.

Weber, Gerhard W., Hermann Prossinger, and Horst Seidler

1998 Height depends on month of birth. *Nature* 39:754–55.

Weinberg, M. K., and E. Z. Tronick

1996 Infant affective reactions to the resumption of maternal interaction after the still face. *Child Development* 67(3):905–14.

Weiner, Jonathan

1994 *The Beak of the Finch: A Story of Evolution in Our Time*. New York: Vintage.

Weinrich, James

1977 Human sociobiology: Pair-bonding and resource predictability (effects of social class and race). *Behavioral Ecology and Sociobiology* 2:91–118.

Weiss, Kenneth M.

1981 Evolutionary perspectives on aging. In *Other Ways of Growing Old: Anthropological Perspectives,* Pamela T. Amoss and Stevan Harrell, eds., 25–58. Stanford: Stanford University Press.

Werren, John

1988 Manipulating mothers. *Natural History* 97(4):68–69.

West-Eberhard, Mary Jane

1967 Foundress associations in polistine wasps: Dominance hierarchies and the evolution of social behavior. *Science* 157:1584–85.

1969 The social biology of polistine wasps. *Miscellaneous Publications* 140:1–101. Museum of Zoology, University of Michigan, Ann Arbor.

1978 Temporary queens in *Metapolybia* wasps: Nonreproductive helpers without altruism. *Science* 200:441–43.

1983 Sexual selection, social competition and speciation. *Quarterly Review of Biology* 58(2): 155–83.

1986 Dominance relations in *Polistes canadensis* (L.), a tropical social wasp. *Monitore zoologico italiano* (n.s.) 20:263–81.

1989 Phenotypic plasticity and the origins of diversity. *Annual Review of Ecology and Systematics* 20:249–78.

In prep. *Developmental Plasticity and Evolutionary Change.* Oxford: Oxford University Press.

Westmoreland, David, Louis B. Best, and David E. Blockstein

1986 Multiple brooding as a reproductive strategy: Time-conserving adaptations in mourning doves. *The Auk* 103:196–203.

Whelan, Christine B.

1998 No honeymoon for covenant marriage. *Wall Street Journal* (August 17):A-14.

White, Frances

1988 Party composition and dynamics in *Pan paniscus. International Journal of Primatology* 9:179–93.

Whitten, Patricia L.

1982 Female reproductive strategies among vervet monkeys. Ph.D. diss., Harvard University.

1983 Diet and dominance among female vervet monkeys (*Cercophithecus aethiops*). *American Journal of Primatology* 5:139–59.

1987 Infants and adult males. In *Primate Societies,* B. B. Smuts et al., eds., 343–47. Chicago: University of Chicago Press.

Widdowson, E. M.

1950 Chemical composition of newly born mammals. *Nature* 166:769–74.

Wieschhoff, H. A.

1940 Artificial stimulation of lactation in primitive cultures. *Bulletin of the History of Medicine* VIII(10):1403–15.

Wiessner, Pauline W.

1977 *Hxaro: A regional system of reciprocity for reducing risk among the !Kung San.* Ph.D. dissertation, University of Michigan, UMI, Ann Arbor.

Wiley, Andrea S.

1994 Neonatal size and infant mortality at high altitude in the Western Himalaya. *American Journal of Physical Anthropology* 94:289–305.

Wilkinson, G. S.

1992 Communal nursing in the evening bat, *Nycticeius humeralis. Behavioral Ecology and Sociobiology* 31:225–35.

Wille, R., and K. M. Beier

1994 Denial of pregnancy and infanticide. *Sexologie* 1:75–100.

Williams, George C.

1957 Pleiotrophy, natural selection and the evolution of senescence. *Evolution* 11:398–411.

1966a Natural selection, the costs of reproduction and a refinement of Lack's principle. *American Naturalist* 100:687–90.

1966b *Adaptation and Natural Selection.* Princeton: Princeton University Press.

1979 The question of adaptive sex ratios in outcrossed vertebrates. *Proceedings of the Royal Society of London,* Series B 205:567–80.

Wilson, A., R. Cann, S. Carr, M. George, U. Gyllensten, K. Helm-Bychowski, R. Higuchi, S. Palumbi, E. Prager, R. Sage, and M. Stoneking

    1985 Mitochondrial DNA and two perspectives on evolutionary genetics. *Biological Journal of the Linnean Society* 26:375–400.

Wilson, Edward O.

    1971a *The Insect Societies.* Cambridge: Harvard University Press.

    1971b Competitive and aggressive behavior. In *Man and Beast: Competitive Social Behavior,* J. F. Eisenberg and W. Dillon, eds., 183–217. Washington, D.C.: Smithsonian Institution Press.

    1975 *Sociobiology.* Cambridge: Harvard University Press.

    1978 *On Human Nature.* Cambridge: Harvard University Press.

    1994 *Naturalist.* Washington, D.C.: Shearwater.

Winterhalder, Bruce

    1986 Diet choice, risk and food sharing in a stochastic environment. *Journal of Anthropological Archaeology* 5:369–92.

Wissow, Lawrence S.

    1998 Infanticide (editorial). *New England Journal of Medicine* 339:1241–42.

Witkowsi, Stanley R., and William T. Divale

    1996 Kin groups, residence and descent. In *Encyclopedia of Cultural Anthropology,* vol. 2, David Levinson and Melvin Ember, eds., 673–80. New York: Henry Holt.

Wollstonecraft, Mary

    1978 *Vindication of the Rights of Woman.* London: Hammondsworth (originally published 1792).

Wood, James

    1994 *Dynamics of Human Reproduction.* Hawthorne, New York: Aldine de Gruyter.

Woolf, Virginia

    1938 *The Three Guineas.* New York: Harcourt Brace and World.

World Health Organization

    1976 *Family Formation Patterns and Health.* Geneva: World Health Organization.

Worthman, C. M.

    1978 Psychoendocrine study of human behavior: Some interactions of steroid hormones with affect and behavior in the !Kung San. Ph.D. diss., Harvard University, Cambridge.

    1988 Concealed ovulation and the eye of the beholder. Paper presented at the annual meeting of the Human Behavior and Evolution Society, April 8–10, University of Michigan, Ann Arbor.

Wrangham, Richard W.

    1993 The evolution of sexuality in chimpanzees and bonobos. *Human Nature* 4:47–79.

Wrangham, Richard, and Dale Peterson

    1996 *Demonic Males: Apes and the Origins of Human Violence.* Boston: Houghton Mifflin.

Wrangham, Richard, James Holland Jones, G. Laden, David Pilbeam, and N. Conklin-Brittain

    In press. The theft hypothesis: Cooking and the evolution of sexual alliances. *Current Anthropology.*

Wray, Herbert

  1982 The evolution of child abuse. *Science News* 122:24–26.

Wright, Robert

  1994a Feminists, meet Mr. Darwin. *New Republic* (November 28):34–46.

  1994b *The Moral Animal: Why We Are the Way We Are. The New Science of Evolutionary Psychology.* New York: Pantheon.

WuDunn, Sheryl

  1997 Korean women still feel demands to bear a son. *New York Times* (January 14).

Wynne-Edwards, V. C.

  1959 The control of population-density through social behaviour: A hypothesis. *Ibis* 101:436–41.

  1962 *Animal Dispersion in Relation to Social Behaviour.* Edinburgh: Oliver and Boyd.

Wyon, J. B., and J. E. Gordon

  1971 *The Khanna Study.* Cambridge: Harvard University Press.

Xiao, Shuhai, Yun Zhang, and Andrew W. Knoll

  1998 Three-dimensional preservation of algae and animal embryos in a Neoproterozoic phosphorite. *Nature* 391:553–58.

Yalom, Margaret

  1997 *The History of the Breast.* New York: Alfred A. Knopf.

Yoshiba, K.

  1968 Local and intertroop variability in ecology and social behavior of common Indian langurs. In *Primates,* P. Jay, ed., 217–42. New York: Holt, Rinehart and Winston.

Zhao, Zhongwei

  1997 Deliberate birth control under a high-fertility regime: Reproductive behavior in China before 1970. *Population and Development Review* 23:729–67.

Ziegler, Toni E., and Charles T. Snowdon

  1997 Role of prolactin in paternal care in a monogamous New World primate, *Saguinus oedipus.* In *The Integrative Neurobiology of Affiliation,* C. Sue Carter, I. Izja Lederhendler, and Brian Kirkpatrick, eds., 599–601. New York: New York Academy of Sciences.

Ziegler, Toni E., A. M. O'Donnell, S. E. Nelson, and S. J. Fomon

  1976 Body composition of the reference fetus. *Growth* 40:329–41.

Zihlman, Adrienne

  1978 Women and evolution (Part 2): Subsistence and social organization among early hominids. *Signs* 4:4–20.

Zuckerman, Sir Solly

  1932 *The Social Life of Monkeys and Apes.* London: Routledge and Kegan Paul.

Zuk, Marlene

  1993 Feminism and the study of animal behavior. *Bio Science* 43(11):774–78.

Index

Page numbers in italics refer to illustrations

A

abandonment, 297–308, *300*, 310, 315,
    371–72, 375, 376, 451, 452, 454,
    459, 533, 534, 536
    attachment and, 315, 316
    data on, 299, 300, 301, 302–4
    foundling homes, 299–308, *301*, 312, *313*,
        352, 369, 372
      depositories in, 304, *305, 306, 317*
      and names of foundlings, 305–7
      unintended consequences of, 302
    infanticide and, 297, 299
    lactation and, 472
    mortality rates from, 297, 298, 299, 300,
        301, 302, 303, 304, 305, 307–8, 369
    multiple births and, 450, 469
    and physical appearance of infant, 449
      see also infant survival, and infant's physi-
        cal traits
    self-delusion of parents in, 307–8
    sociopaths and, 513, 527
    Spitz's documentary on, 395–96, 404
    survivors of, in prehistoric times, 517–20
    in Third World, 312
abortion, 21, 355, 372
    implantation and, 433–34
    infanticide and, 391, 470
    politics and, 4–6, 470
    sex-selective, 319, 320, 322, 324, 349
abortion, spontaneous, 124, 420, 423, 437
    early-pregnancy reabsorption in rodents, 89
    miscarriage, 186, 276
    in monkeys, 90
    sex-selective, in animals, 328, 329, 330–31
Aché, 197, 199, 202, 236–37, *248,* 268, 270,
    270
    attitudes toward old people in, 282
    deaths in, 414–15, 416–18
    infanticide in, 416–18, 453–54
    multiple fathers in, 246–47, *248,* 249
    orphaned infants in, 408–9
Achilles, *240*
Adam Bede (Eliot), 17, 288–89
adolescence
    subfertility in, 185–86, 187
    see also puberty; teenagers
adoption, 57, 114, 116, 157, 158–59, *160,*
    302, 509, 538
    in monkeys and apes, 156, 478
      by males, 207–8
      maternal commitment and, 180
    of needy infants, 460
Aesop's fables, 441
Afghanistan, 6
Africa, 7, 196, 249, 254, 480, 511
    attitudes toward children in, 372–76
    proverb in, 269
Agamemnon, *240, 526, 528*
aggression, lactational, 152–53
    postpartum depression and, 172–73
aggression-increasing hormones, in birds,
    443–44

aging, 274–76
    and cultural attitudes toward old people,
        282, 283
    increased maternal altruism and, 276–78,
        281–82, 285
    see also lifespan
agriculture, 202, 250, 252
Agta, 498–99
AIDS, 107–8, 373, 376
Ainsworth, Mary, 401–2, *403,* 404, 405, 461,
    508
Aka, 226, 227, 409, 495
Alexander, Richard, 55, 57, 275
aliens, in films, 472
Allegorie de la charité (Daret), *11*
alligators, 328–29
alloparents, *see* childcare, allomothers, and
        cooperative breeding
alphalactalbumin, 135
Altmann, Jeanne, 44–46, *45,* 81, *216,* 334,
        437–38, *439,* 440
altruism
    Hamilton's Rule on, 62–65, 69, 244, 276,
        363–64, 371, 378, 504, 505, 519,
        527, 536, 539
    mother's encouragement of, among off-
        spring, 428–29
    in postmenopausal females, 276–78,
        281–82, 285
Amazonia, 246
ambition, motherhood and, 110–13
American Anthropological Society, 293
amphidromia, 468
Anaxagoras, 348
Andaman Islanders, 270, 357
androgens, 444
anisogamy, 421
Annie Hall, 223
antelopes, 217, 228
antibiotic hypothesis of origin of lactation,
        135–36
antibodies, in milk, 137
ants, 57, 60, 64–65

apes and monkeys
    abandonment of infants in, 181–82
    abortion in, 90
    adoption in, 156, 478
        by males, 207–8
        maternal commitment and, 180
    alloparenting in, 161–63, *163,* 447–48,
        478, 503–4
    attachment in, 98, 178, 398, 399–401
    baboons, *see* baboons
    bonobos, 85, *87,* 185, 220, 259, 477
        human genes and, 392
        lack of infanticide in, 243
    cercopithecine, 192, 333, 446
    chimpanzees, *see* chimpanzees
    Colobine monkeys, 447–48
    dusky leaf monkeys, 446, 448
    fixed action patterns in, 96
    flexibility of female sexuality in, 217–18
    gorillas, 35, *128,* 180, 258, 284, 478
        Binti Jua, 157, *157,* 207
        birth weight in, 476
        infanticide in, 243
        Jambo, 207
        maternal abandonment in, 181–82
        pregnancy, labor, and delivery in, *165,*
            167, 175
        testicle size in, *219*
    howler, 35, 156
    inexperienced mother, 155–56, 190–91
    infant death in, 178–79
    infant fat in, 476
    infant sharing in, 161–63, *163,* 447–48,
        495, 498
    labor and delivery in, vs. human mothers,
        165, 478
    langurs, 36, 88, 100, 179, 258, 477–78
        female sexuality in, 35, *218*
        infanticide in, xv–xvi, 32–34, *34, 35,* 36,
            40–41, 86–87, 181–82, 237, 293, 504
        infant-mother separation in, 399
        infant-sharing in, 162–63, *163, 164, 238,*
            447

male caretaking in, 209
old females, 277
physical appearance of infants, 446
Sol, 277, *278*
macaques, 82, 85, 111, 124–25, 181, 186, 192, 224, 259, 274
infant-sharing in, 161
rank and sex ratios in, 333–36
rhesus, in separation experiments, 399–400, 430–31
in surrogate mother experiments, 398, *399*
male caretaking in, 207–9, 213–17
marmosets, 130, 180
maternal investment in, 176, 450
unconditional, 177–79, 180–81, 182, 447, 451, 482
menarche in, 185
mother-infant relationships in, 100
multiple births in, 180
orangutans, 175, 209, *219*
"Peter Pan," 75–76, *76*, 511, 514
oxytocin in, 154
placenta eaten by, 167
proboscis monkeys, 446
snakes feared by, 416
strangers feared by, 416
tamarins, *93*, 131, 180, 203, 316
testicle size in, 218–19, *219*
titis, 213–14, *214*, 226, 227, 498
vervets, 190–91, 192, 278–79, 316, 446
see also primates
apnea, 291, 292
Apollo, *239*
Aquatic Ape, The (Morgan), 476
aquatic origins hypothesis, 476–77
Arab cultures, kinship in, 359
arcutio, 291, *291*
Aristotle, 262, 348, 464
Asia, family configurations in, 344
infanticide and, 321, 322
in China, 318–20, *320*, 321, 323, 344, 349–50

atresia, 421
attachment, xiii, xiv, 24, 26, 98–99, 116, 213, 377–78, 394–407, 487–93, 501, 531, 534–39
abandonment and, 315, 316
ambivalent, 403
avoidant, 403, 406, 522–23, *523, 526, 527*
backlash against, 488–89, 491–95, 508, 509
bonding vs., 487–88
Bowlby's theory of, 387–88, 393, 394–96, 397, 399, 401, 404–5, 407, 408, 410, 411, 412, 487, 488–89, 493, 495, 509, 518
see also Bowlby, John
and daycare improvements, 509–10
defined, 24
disorganized/disoriented, 404, *405*
emotional development and, xiii, 116
feminists and, 407, 489, 535, 539
food and, *397, 398*
to frightening figure, 403–4
in goslings, *397*
growth and, 461
and guilt of mothers, 490–93, 494, 539
infanticide and, *455*
insecure, 403–4, *405*, 406, 516–17, 533
to male caretakers, 214
maternal, before birth, 458
post-Bowlby understanding of, 522
as predator protection, 411–13
primary, factors in, 501
in primates, 98, 178, 398, 399–401
sociopaths and, 512
strange situation test and, 401–3, *403, 404, 405*, 461, 522
and subsequent performance of individual, 404–6, 506
in surrogate mother experiment, 398, *399*
see also maternal love; parental investment
Attachment (Bowlby), 394, *405*
attraction of adults to infants, 156, 157, 158, 161, *164*, 390, 444–48, 452, 481, 483–84, 536, 539–40

attraction of adults to infants (*continued*)
  see also infant survival, and infant's physical
      traits
Austen, Jane, 357
australopithecines, 48–49, 267, *269*
avoidance, in attachment, 403, 406, 522–23,
      523, 526, 527
Awakening of an Abandoned Infant, The, 300
Ayoreo Indians, 314, 469

B
babies, *see* infants
Baboon Mothers and Infants (Altmann), 46
baboons, 35, 41, 85, 100, 111, 190, 192, 258,
      259, 273
  Altmann's study of, 44–46, *45,* 81, *216,*
      334, 438, 440
  female orgasm in, 222
  grandmother, 278
  infant-sharing and, 161
  male caretaking in, 208–9, 215
  menopause in, 275
  mortality rates in, 438
  physical appearance of infants, 446
  rank and reproductive success in, 81–82,
      192, 333–36, 514
  relationships between mothers and males
      they have mated with, *216*
  sex ratios in, 333–36
Bachofen, Johann, 257
back-load hypothesis, 197
Badinter, Elisabeth, 309, 352, 363
"Balia, La" (Tansillo), 351
Bangladesh, 323, 325, 344
Bantu, 374–75
baptism, 473–74
Bari, 246, 247–48, 249
bats, 485
bears, 124, 480
beauty, motherhood and, 17–18, 23–24
Beauvoir, Simone de, 17, 24, 490, 491

bees, 55–56, 57, 59, 60–63, *61,* 64, 65, 92,
      273, 274
Belloc, Hilaire, 408, 411, 412
Beloved (Morrison), *290*
Belsky, Jay, 507–8, 522, 525
Betskoi, Ivan, 300
bias, in science, 53, 496–97, 535
Bible, *241*
big mothers, 47–50, 285–86, 329
Binti Jua (gorilla), 157, *157,* 207
Biographical Sketch of an Infant (Darwin), 523
"biological," use of term, 57
biology, 23, 24, 26, 308
  vs. culture, parent-offspring conflict as, 427
  and motherhood as social construction,
      309, 310
  sociobiology, 57, *58,* 67, 69, 82, 231, 308,
      496
    child development and, 522
birds, 203, 272
  blackbirds, 253
  broodiness in, 133
  brood manipulation in, 29–32
  canaries, 443, 444
  coots, 441–42, 443, 444, 448, *482*
  cowbirds, 132
  crop milk in, 133, 134
  dunnocks, 88
  egrets, 442, 443–44
  peacocks, 36, 38, *38*
  predator-distraction tactics in, 131
  prolactin in, 130, 131, 133
  scrub jays, 131
  sex ratios in Seychelles warblers, 332–33
  sibling rivalry in, 441–45, 448
  staggered hatching in, 442, 443–44
  swallows, *40*
birth
  announcements of, 481
  emotions and, 165, 166
  infant's weight at, 436, 457, 459, 464, 476,
      479, 480, 481–82

as indicator of survival, 479, 481–82
labor and delivery
    brain changes caused by, 94–95
    and size of pelvic canal, *165,* 478, 479,
        480
maternal attachment to baby before, 458
maternal depression following, 170–73
maternal responses to, 166–70
in mice, 146–47
mortality connected to, 471
mother's relationship with infant immedi-
        ately following, 486, 487–88
    see also attachment
multiple
    parental commitment and, 179–80, 450,
        459, 469
    twins, 180, 202–3, 450, 459, 480
naturalization of, in U.S., 486, 487
order of, personality and, 523–25
oxytocin and, 138, 139
pain in, cultural expectations about, 164–65
rituals following, as transition to person-
        hood, 468–70
    baptism, 473–74
and size of pelvic canal, *165,* 478, 479, 480
spacing of, 175, 184, 202–4, 258, 284, 373,
        438
    aging mothers and, 276
    breast-feeding and, 104–5, 194–95, 196
    evolutionary changes in, maternal dis-
        crimination and, 449–51
    feedback loops in, 196, 197, 239
    in hunter-gatherers, 193–94, 200–201,
        453
    infanticide and, 345, 453
    maternal commitment and, 179–80, 188
    wet-nursing and, 360–62, 364–66, 368
    time of, 434–35
birth control, 283, 356, 366, 370, 373
    infanticide and, 296, 349–50
    pills, and sense of smell, 193
    teenage pregnancy and, 188, 191, 283

birth defects, 276, 469
birth rates, 9–10, 20, 202
    sex ratios and, 335–36
blackbirds, 253
Blackwell, Antoinette Brown, 20, 22, 53
Bloch, Dorothy, 533, 534
Blurton-Jones, Nick, 197, 284
body size, of males vs. females, 47–50, *49,*
        269, 325
Bolivia, 314
bonding, 114, 486
    attachment vs., 487–88
    backlash against, 488–89
    see also attachment
bone development, 380
"bonny babes," 481–82, 536
bonobos, 85, *87,* 185, 220, 259, 477
    human genes and, 392
    lack of infanticide in, 243
Boone, James, 341, 342
Born to Rebel (Sulloway), 524
Boswell, John, 297–99, 373, 505, 517
Bowlby, John, xiii, xiv, 25, 96, 104, 383–88,
        391, 397, 403, 430, 486, *510, 512,*
        513–14, 533, 534, 535, 536
    attachment theory of, 387–88, 393,
        394–96, 397, 399, 401, 404–5, 407,
        408, 410, 411, 412, 487, 488–89,
        493, 495, 509, 518
    backlash against, 488–89, 492, 495
    see also attachment
    on Darwin, 383–87
    Environment of Evolutionary Adaptedness
        concept of, 97–99, *99,* 101, 409, 410,
        411, 412, 500, 517, 520, 521, 522
    expansion of, 99–102
    sociopaths as viewed by, 512, 516
    on working mothers, 495–96, 509
brain, 141
    and length of childhood, 267, 284
    neocortex of, 141, 155
    neonatal fat and, 478–79, 480, 481–82, 483

brain (*continued*)
   neurons in, 131–32
   nurturing behaviors and, 149
   older part of, 141
   size of, birth process and, 479
   social behavior and, 141, 143
brain stem, 141
breast development, 126–27, *128*
breast-feeding, 409, 501
   and bias against daughters, 322, *323,* 324
   infant survival and, 297, 298, 301–2,
      310–12, 322, *323,* 324, 362–63
   intimacy promoted by, 137, 140–41, 144
   mothers' bodily responses in, 95, 536–38
   nighttime, 104–5, 195
   nipple preference in, 105
   ovulation suppression and, 104–5, 194–95,
      196, 239, 409–10, 438, 449
   oxytocin and, 139, 536, 537, 538
   sensual responses in, 139, 536–38
   weaning from, 127, 129, 176, 177, 197,
      362–63, 409, 438, 449
      conflicts in, 429
      infants' preparation for, 439
      time of, 439, 460–61
   wet-nursing, *see* wet-nursing
   see also lactation; milk
breeding
   cooperative, *see* childcare, allomothers, and
      cooperative breeding
   systems of, *see* mating systems
   see also reproduction
broodiness, in birds, 133
brood manipulation, in birds, 29–32
brothels, maternal abandonment and, 297, 299
Brown, Jennifer, 149, 150, *150*
Bruce effect, 89

            C
calcium, 175
California, 7–8
Calment, Jeanne, 280

Cama Sotz, 259
camouflage, of babies, 446
canaries, 443, 444
Canela, 246, 247, 248–49, 250, 252, 262, 346
cannibalism, 52, 294, 540
   in chimpanzees, 161, 502–3
   in spiders, 43–44, 133
caretaking, *see* childcare, allomothers, and
      cooperative breeding
Carnauba, Madalena, *84*
carrying babies, 197, *198, 199, 200, 203, 204,*
      496, 499, 502
Carter, Sue, 139
Cassatt, Mary, *381*
caterpillars, 73, *74,* 76, 511, 524
Catherine II, Empress of Russia, 300
cats, big, as predators, 411, 415, 416
cercopithecine monkeys, 192, 333
Chagnon, Napoleon, 242, 296
changelings, 465–68, *466, 469*
Charles Darwin: A New Life (Bowlby), 383
Charnov, Eric, 284, 522
chastity, 263
Cheju Do, 343–44
child abuse and neglect, 188, 236, 237, 349
   of infants, 25, 180, 190, 349, 370, 376,
      452, 454, 518, 536
      in Third World, 312
   see also abandonment; infanticide
childbirth, *see* birth
childcare, allomothers, and cooperative breed-
      ing, 64–65, 90–93, 101, 109, 121,
      158, 180, 197, 201, 203, 204,
      265–77, 370–72, 409, 497–510, 518,
      519, 521, 534
   in Agta, 498–99
   attachment and, *see* attachment
   brain and, 149
   circumstances facilitating multiple caregiv-
      ing, 499
   daycare (paid communal), 490–96, 504,
      505–10, *508*
      improvements in, 509–10

as modified wet-nursing, 369–70
definition of allomother or alloparent, 91
discrimination in, 181
division of labor in, 209–13
   and sensitivity to infant needs, 212–13,
     227, 500, 501
in Efé, 271, 357, 409, 495, 499, 500, 502,
   503
effect of social conditions on maternal
   investment, 88–89
fathers and other male caretakers, 91,
   205–17, 226–30, 498, 499–501, 509
   division of labor between mothers and,
     209–13
   intermittent care by, 214–15
   sensitivity to infant needs in, 212–13,
     227
   show-off hypothesis and, 228–30,
     268–69
   as sole caretakers, 214
grandmothers and, 275, 278–79, 283
"grannies," in Africa, 373–74, 375, 375, 376
in Hadza, 198–99, 503
infanticide and, 237
in insects, see insects, social
kin and, 271, 278–79, 372, 509
lactation and, 358, 500
and manipulation of family configurations,
   345
milk sharing in, 356–57, 499
   see also wet-nursing
in monkeys, 161–63, 163, 447–48, 478
mothers as likeliest primary caretakers,
   500–501
mothers' reasons for not using allomothers,
   502–3
nannies, 371
natal coats and, 447, 448, 449
postmenopausal females as caretakers, 267,
   273–74, 275
problems in, 495–96, 497–98, 502–3
prolactin and, 130–31, 132, 134
and sex ratios in birds, 333

and shortage of alloparents, 494, 495–96,
   497, 499, 503, 504–6
in tamarins, 93, 131, 203
teenage caretakers, 271–72
unrelated alloparents, 302
see also mothers, motherhood
childhood
   concept of, 310, 312, 520
   fears in, 533–34
   length of, 267–68, 284–87, 483
childless women, 7–8
child mortality, 416, 418
   in Dogon, 255
   see also infant mortality
children
   abandonment of, see abandonment
   last-born, 276
   learning biases in, 521–22
   murder of, 236, 416, 418
     see also infanticide
   regional differences in rearing of, 406
   socialization of, importance of peers vs.
     parents in, 514–16
   in sub-Saharan Africa, attitude toward,
     372–76
   see also infants
chilling, of newborns, 464, 465, 473–74
chimpanzees, 43, 175, 183, 202, 258, 259,
   260, 264–65, 284, 478, 482, 495
   birth weight in, 476, 477
   division of labor in, 199
   female rank in, 52, 110
   female sexuality in, 222, 224
   Flo, 27–28, 29, 30, 50–52, 54, 86, 185
     last baby of, 275
     status of, 110
   human genes and, 392
   infanticide in, 52, 86, 161, 181, 243, 502–3
   infant-sharing and, 161, 502–3
   manipulation in, 528–29
   mating in, 85–86, 185
   menopause in, 186, 274
   orphaned, 408

chimpanzees (*continued*)
  sexual maturation in, 185–86, 190
  size of genitalia in, *219, 220*
  strangers feared by, 416
China, 263–64
  infanticide in, 318–20, *320,* 321, 323, 344,
    349–50
  scarcity of women in, 321–22
Chipewyan, 282
cholera, 106
chorionic gonadotropin, 433
Christianity, and attitudes toward infants, 464
  baptism, 473–74
chromosomes, *422*
  and sex bias in lemmings, 329
  of social insects, 62
  X, 142–43
  XO girls, 142, 143
Churchill, Winston, 497
clitoridectomy, 255, 258
clitoris, *220, 222*
cold water, newborns subjected to, 464, *465,*
    473–74
Colobine monkeys, 447–48
colostrum, 136, 137, 357, 475
Colp, Ralph, 384
compassion, 145, 393
  lack of, 530–31
competition, 81
  local mate, 65–66
  male-male, 82, 140, 325
  sexual selection, 36–37, 81, 83, 445, 496
    female mate choice in, 36–42
    large females and, 47–50
    parental investment and, 37
conception, 421, 423
  fertilization in, 70, 71, 207, 222, 421, 423,
    538
  sex selection and, 348–49
Concerning Children (Gilman), *177*
contingent commitment, 92–93, 176, 182–83,
    315–16, 388, 389–90, 519, 535

contraception, 283, 356, 366, 370, 373
  infanticide and, 296, 349–50
  oral, and sense of smell, 193
  teenage pregnancy and, 188, 191, 283
convents, 341–42
cooperative breeding, *see* childcare, allomoth-
    ers, and cooperative breeding
coots, 441–42, 443, 444, 448, 482
Coram, Thomas, 300, *301*
cortisol, 212, 400–401, 536
cosmology, 471
cowbirds, 132
coypu, 330–31, 338
cradleboards, *170*
cradling ceremonies, 468
Cronos Devouring His Children (Goya), *417*
crop milk, 133, 134
culturally transmitted information (memes),
    77–78
culture
  vs. biology, parent-offspring conflict as, 427
  as factor in evolutionary development,
    520–22
  mothering and, 309–12
cystic fibrosis, 107

D

Daly, Martin, 236, 314, 458, 463
Daniel Deronda (Eliot), 491
Daret, Pierre, *11*
Darwin, Catherine, *385*
Darwin, Charles, xviii, 3, 13–14, 15, 17,
    18–20, 23, 36–37, *87,* 184, 224, *240,*
    267, 268, 383, *385,* 388, 394, 401,
    406, 419, 496
  Bowlby's study of, 383–87
  childhood of, 383, 384–87
  *The Descent of Man and Selection in Relation to
    Sex,* 20, 36
  on his son William, 523
  illness of, 384

lactation hypothesis of, 134–35
on male instincts for caretaking, 205, 206,
   207, 209
The Origin of Species, xvii, 13, 17, 20, 55, 386
Royer and, 20
Darwin, Emma, 365
Darwin, William, *523*
Darwinism, 13, 17, 21–22, 80, 308
   bias in, 53, 496, 535
   changes in, 388
   social, 13, 15, *19,* 24
Datoga, 460
daughters
   abortion of, 319, 320, 322, 324, 349
   conception methods for producing, 348–49
   infanticide of, 318–27, 338–39, 344,
      349–50
      in China, 318–20, *320, 321, 323,* 344,
         349–50
      in Eipo, 455
      in India, 326–27, 339
   maternal compliance with, 322–24
   Mukogodo, 342–43
   parental investment in, in U.S., 338
   preference for, and social status, 339–40,
      341, 342, 343–44
   reasons for preferring sons over, 321,
      324–25
   reproductive potential, 325–27
Daumier, Honoré, *16*
daycare
   paid communal, 490–96, 504, 505–10, 508
      improvements in, 509–10
      as modified wet-nursing, 369–70
   see also childcare, allomothers, and coopera-
      tive breeding
death and mortality
   childbirth connected to, 471
   infant, *see* infant mortality
   murder, 179, 238, 243
      of children, 236, 416, 418
      of infants, *see* infanticide

senescence, 274–76
   and postreproductive longevity,
      279–87
   siblicide, 442, 443–44
Death Without Weeping (Scheper-Hughes), 312
deer, 331
delayed maturation, 184–86
   adolescent subfertility, 185–86, *187*
   length of childhood, 267–68, 284–87,
      483
   stress and, 189, 190
delivery and labor
   brain changes caused by, 94–95
   and size of pelvic canal, *165, 478, 479,* 480
   see also birth
Dennett, Daniel, 406–7
depression, maternal, 170–73, 411
Descent of Man and Selection in Relation to Sex, The
   (Darwin), 20, 36
development, 55–78
   alternative outcomes in, 74–76, 511–31
   bees, 55–56, 59, 60–63, *61, 64, 65*
   embryonic, 70, 71–72
   factors in, 56–57, 59
      social environment, 520–22
   flexibility in, 514, 520–22
   genes in, *see* genes
   genotypes in, 55–56, 59
      defined, 55–56
   maternal effects and, *see* maternal effects
   phenotypes in, 56, 59, 71, 72, 514, 515
      defined, 56
   polyphenism in, 72–74
   role of peers vs. parents in socialization,
      514–16
   "wretched" courses of, 525–27
   see also gestation
DeVore, Irven, xv, 253
DeVries, Marten, 461, 470
diabetes, 436
diaries, of parents, 312
Diary of a Baby (Stern), 410

Dickemann, Mildred, 338–39, 343
diet and nutrition
  fertility and, *187,* 188, 191, 196, 288
  natural selection and, 107
dinosaurs, 123
  Maiasaura, 123
Diodorus Siculus, 511, 517, *528*
diploid reproduction, 62
diseases, 106, 107–8, 485
division of labor, 199
  in caretaking, 209–13
    and sensitivity to infant needs, 212–13,
      227, 500, 501
Dogon, 254–57, *256,* 258, 376
dogs, 167, 177, 358, 504
  pseudopregnancy in, *134*
dolls, 272, 338
donative intent, 277, 278
dopamine, 194
Doré, Gustave, *xvi*
dowries, 324, 325–26, 339
Draper, Patricia, 189, 502, 520–22
drosophila (fruit flies), 41–42, 231–32
dry-nursing, 359
dunnocks, 88
dusky leaf monkeys, 446, 448

        E

Efé, 271, 357, 409, 495, 499, 500, 502, 503
eggs
  behavior of, 420–21
  developmental differences in, 420
  fertilization of, 70, 71, 207, 222, 421, 423,
    538
  follicle surrounding, 420–21
  human, 70, 419–23
  implantation of, 124, 153, 420, 423
    pregnancy maintenance and, 433–34
  ingredients added to, 56, 70–71
  insect, 59–60, 62, 64, 65, 68, 69
  number of, 419–20, 421–23
  oogenesis, *422*

predator-influenced hatching of, 424–25,
    425
  staggered hatching of, 442, 443–44
  survival of, 419–20, 421
  see also ovulation
egrets, 442, 443–44
Egypt, 359, 366, 540
Ehrhardt, Anke, 407
Eipo, 454–56, *456*
elephants, 273, 279
Eliot, George, vii, xvii, 14–15, 17–18, 22, 23,
    27, 53, 55, 79, *115,* 121, 131, 205–6,
    212, 264, 346, 392, 394, 460,
    490–91, 493
  Adam Bede, 17, 288–89
  Daniel Deronda, 491
  Middlemarch, 19, 235, 318
  portrait of, *18*
  Silas Marner, 205–6, 214
  Spencer and, 14–15, *15,* 17–18, *19*
embryo(s), 388, 389, *422,* 452
  as active agents, 425–26
  beginning of life in, 391–92, 457
  "decisions" made by, 423–25, *425*
  development of, 70, 71–72
  fossil, *421*
  implantation of, 124, 153, 420, 423
    pregnancy maintenance and, 433–34
  placenta and, 434
  reabsorption of, 89, 90
  see also fetus
emotions, 95, 141
  in childbirth, 165
  cultural attitudes and, 310
  evolutionary origins of, 114, 116
  see also love
empathy, 141, 392–93, 527–29, 531
endearments, 539–40
endorphins, 103
Engels, Friedrich, *240,* 249, 250
Environment of Evolutionary Adaptedness
    (EEA), 97–99, *99,* 101, 409, 410,
    411, 412, 500, 517, 520, 521, 522

expansion of, 99–102

Épinay, Madame d', 312, 313

Erice, 294–96

Ericsson, Ronald, 348–49

Eskimos, 249, 282, 452

Essay de Dioptrique (Hartsoeker), 71

essentialism, 17, 27, 308, 495

estradiol, 434

Estrées, Gabrielle d', 362

estrogen, 153, 154, 173, 274, 420, 434
  synthetic, 125

ethics, parental investment and, 460

Euripides, 239

eusocial societies
  defined, 62
  see also insects, social

evolution, evolutionary theory, 183, 406–7
  alternate morphologies in, 73, 74, 75, 76,
    511, 514, 524
  of altruism, Hamilton's Rule on, 62–65, 69,
    244, 276, 363–64, 371, 378, 504,
    505, 519, 527, 536, 539
  defined, 13
  emotions and, 114, 116
  Environment of Evolutionary Adaptedness,
    97–99, 99, 101, 409, 410, 411, 412,
    500, 517, 520, 521, 522
  expansion of, 99–102
  expanded paradigms of, 52–54
  feminists and, 15, 17, 18, 20–23, 24, 490,
    535
  of humans, 257, 264–65
    infanticide and, 413–14, 416
  of lactation, 121, 122–24, 129, 130,
    133–36, 140
  memes and, 77–78
  mothers' role in, xi–xii, xvii, 12–25,
    80–83
  pace of, 105–6
  phenotypes and, 56, 450
  post-Pleistocene, 105–9
  research bias in, 53, 496, 535
  reuse of elements in, 103

of the sexes in response to each other,
    41–42
  see also natural selection

Evolution of Desire, The (Buss), 24

"Extraordinary Sex Ratios" (Hamilton), 328

Eyer, Diane, 491–92, 508

F

faces, newborns' preference for, 410

facial expressions, infants' imitating of, 412

Fairbanks, Lynn, 190, 316

fairy tales, 539

family planning, 175–76, 193–204
  birth control, 283, 356, 366, 370, 373
    infanticide and, 296, 349–50
    pills, 193
    teenage pregnancy and, 188, 191, 283
  birth spacing, 175, 184, 202–4, 258, 284,
    373, 438
    aging mothers and, 276
    breast-feeding and, 104–5, 194–95, 196
    evolutionary changes in, maternal dis-
      crimination and, 449–51
    feedback loops in, 196, 197, 239
    in hunter-gatherers, 193–94, 200–201,
      453
    infanticide and, 345, 453
    maternal commitment and, 179–80, 188
    wet-nursing and, 360–62, 364–66, 368
  China's one-child policy, 319, 321
  saliva changes as aid in, 195–96
  wet-nursing and, 364–66

family structures, 327
  gender configurations, 318, 344–45
    see also sex bias
  nuclear, 216
    breakdown of, 250–51, 379–80
  see also mating systems

famine, population and, 463

fat, body, 124–27, 125, 129
  on breasts, 126–27
  on buttocks, 126

fat, body (*continued*)
  in fetuses, 476, 477, 479, 480–81, 482
  in infants, 457, 458, 474, 475–83, 477, 536
    brain development and, 478–79, 480, 481–82, 483
    hypotheses explaining, 476–79, 481
  reproduction and, 125, 125, 126, 188
fat, in milk, 127, 129, 409, 476, 480, 482
Father, The (Strindberg), 257
fatherlessness, 235–36
fathers, 205–17, 206, 226–30, 235–36, 409
  and age of menarche, 189–90
  Aka, 226, 227
  as caretakers, 214, 498, 499–501, 509
    division of labor between maternal caretakers and, 209–13
    sensitivity to infant needs in, 212–13, 227
  desirable traits in, 244–45
  diaries of, 312
  infants' resemblance to, 458–59, 459
  monogamy and, 231–34
  multiple, 245–49, 250, 251
  parental investment by, 37, 129, 224, 252, 255, 258, 379, 458–59
  paternity of, 250
    certainty of, 207, 217–19, 226–28, 246, 248, 252, 255, 264, 458–59
    multiple, 246–49, 252
    sexuality and, 538
  reproductive tradeoffs by, 207, 233
  temporary-hero type, 215
  see also male caretakers
fears, childhood, 533–34
"Female Primate, The" (Jay), 27
feminism
  attachment theory and, 407, 489, 535, 539
  evolutionary theory and, 15, 17, 18, 20–23, 24, 490, 535
  family and, 250, 379–80
  maternal instinct and, 25–26, 308–9, 352, 535

fertility
  adolescent subfertility, 185–86, 187
  fat and, 125, 125, 185, 188
  increase in, in human evolution, 449–50
  nutrition and, 187, 188, 191, 196, 288
  pattern of, 186, 275–76
  wet-nursing and, 360–62, 364–66, 368
fertilization, 70, 71, 207, 222, 421, 423, 538
fetus, 389
  body fat of, 476, 477, 479, 480–81, 482
  death of, 423
    see also abortion, spontaneous
  debate over when life begins in, 391–92, 457
  growth of, and parent-specific genes, 432–33
  malnourished, 436
  maternal interests and contracts with, 431–33, 434–40
  melatonin and, 104
  and mother's immune response, 94
  placenta and, 433–34, 435, 436, 481
    see also placenta
  and reproductive rights of women, 391–92, 457
  strangeness and symbolism of, 471, 472–73, 473
  voices and, 411
  see also embryo
Finland, 202–3
fireflies, 71, 72
fish, 48, 130, 274, 328, 329, 539–40
  cichlid, 134
  salmon, 43
Fisher, Ronald, 336
Fisher's principle of the sex ratio, 336–37, 346
fitness (reproductive success), 9, 14, 42, 79, 80–83, 99, 114, 286, 318
  and bias against females, 497
  inclusive, 63, 427
  of infants, assessment of, 463, 481
    viability tests in, 464, 473–74

sex bias and, 330–36, 337
  see also sex bias
status and, 81–82, 110–12, 366
variation in, 83, 90
Fitz, Grancel, *426*
fixed action patterns, 96, 167, 410
Flo (chimpanzee), 27–28, 29, 30, *50*–52, 54, 86, 185
  last baby of, 275
  status of, 110
Florence, 291, *291*, 299, 304, 307
fluctuating asymmetries, 39
follicles, egg, 420–21
follicle-stimulating hormone, 185
fontanel, *169*, 479
food, attachment and, *397*, 398
"food for thought" hypothesis, 478–79
foragers, *see* hunter-gatherers
Formation of Eve, The (Doré), *xvi*
"Forty-four Juvenile Thieves: Their Characters and Home-life" (Bowlby), 395
fos genes, 149–51, 173
foster homes, 374, 505–6
  in Africa, 373–74, 375, *375, 376*
  see also adoption
foundling homes, 299–308, *301, 312*, 352, 369, 372
  depositories in, 304, *305, 306, 317*
  and names of foundlings, 305–7
  unintended consequences of, 302
Fraiberg, Selma, 405, 490
France
  birth rates in, 11, 20, 366
  daycare in, *508*
  Holy Greyhound cult in, 466
  wet-nursing in, 11, 351–54, 355, 366, 367–69
fraternal interest groups, 251
Freud, Anna, 396
Freud, Sigmund, 15, 396, 398, 405, 430, 512
frogs, 71–72, 424, *425, 462*
Fromm, Erich, 146, 152, 174, 180, 457

fruit flies (drosophila), 41–42, 231–32
Fuchs, Rachel, 315, 487–88

G

galagos, 46, 217, 434
Garden, Maurice, 369
Garner, Margaret, *290*
genes, 55, 56–57, 58, 60, 80, 108, 204, 222, 280, 431–33, 439, 463, 511, 514
  and age of menarche, 189
  of bees, 55
  context and, 56–57
  disease protection and, 106, 107
  embryonic development and, 70, 71–72
  female mate choice and, 36–42
  fos, 149–51, 173
  human vs. chimpanzee, 392
  kin selection and, 63–64
  for lactose tolerance, 108
  mothering and, 148–51, *150*
  mouse breeding and, *148*
  parent-offspring conflict and, 431, 436, 437–38
  parent-specific expression of (genetic imprinting), 141–42, 432–33, 435, 436, 459
  phenotypes and, 56, 59, 72, 127
  and phenotypic flexibility, 74, 76, 514
  polyphenism and, 72–74
  selfish, 431
  social, 431
  social relationships and, 142–43
  see also natural selection
"genetic," "biological" compared with, 57
genetic determinism, 57, *58*
genetic imprinting, 141–42, 432–33, 435, 436, 459
genitals
  modesty and, 259
  of rodent pups, 147
  testicle size of Great Apes, 218–19, *219*

genome, 107

genotypes, 55–56, 59
    defined, 55–56
    polyphenism and, 72–74

Germany, 406

gestation, 389, 419, 422, 425, 434
    beginning of life in, 391–92, 457
    birth weight and, 476, 481
    parent-specific genes and, 432–33
    see also embryo; fetus; pregnancy

Gilibert, Jean-Emmanuel, 1, 10–12, 43, 94

Gilman, Charlotte Perkins, 177

glucose, 435

goats and sheep, imprinting in, 114, 157–59,
    159, 487, 488

godparents, 246, 247, 248

Golden, Claudia, 227, 228

gonadotropin-releasing hormone, 185

Goodall, Jane, 27, 30, 50, 51–52, 110, 275

gorillas, 35, 128, 180, 258, 284, 478
    Binti Jua, 157, 157, 207
    birth weight in, 476
    infanticide in, 243
    Jambo, 207
    maternal abandonment in, 181–82
    pregnancy, labor, and delivery in, 165, 167,
    175
    testicle size in, 219

goslings, attachment in, 397

Gowaty, Patricia Adair, 27, 41

Goya, Francisco de, 417

grandmothers, 275, 278–79, 282, 283
    lactation in, 358
    see also postmenopausal females

grandmother's clock hypothesis, 284–86

"grannies," in Africa, 373–74, 375, 375, 376

Greece, ancient, 261–62, 468, 511, 517

Greenberg, Michael, 149, 150

Greene, Erick, 73, 74, 524

Greuze, Jean-Baptiste, 352, 353, 526

grief, over infant death, 470

"Grief: A Peril in Infancy," 395–96

Grimaldi Venus, 125

Grimm brothers, 465

growth
    maternal love and, 460–61
    parent-specific genes and, 432–33

growth hormone inadequacy, 338

guilt, attachment theory and, 490–93, 494,
    539

Guinefort, Saint, 466, 467

gulls, 32

Gynaecology (Soranus), 463

H

Hadza, 198–99, 228–29, 250, 268, 281, 282,
    286, 287, 495, 503

Hahn, Caldwell, 31–32

Haig, David, 389, 391, 431–33, 434, 435,
    437, 439, 459, 481, 539

Hamilton, William, 57, 60, 62–69, 68, 96,
    271, 274, 328, 388, 427, 431, 445
    Rule of, 62–65, 69, 244, 276, 363–64, 371,
    378, 504, 505, 519, 527, 536, 539

hamsters, 46, 177

Hansel and Gretel, 534

haplodiploid reproduction, 62, 64

haploid reproduction, 62

Harlow, Harry, 398, 399

Harpending, Henry, 189, 462–63, 521–22

Harris, Judith Rich, 507, 514–15

Hawkes, Kristen, 229, 284, 286

Henri IV, King of France, 362

Heracles, 208, 517–18

Herod, King, 241

Hesse, Erik, 404, 485

high blood pressure, 436

h'ikal, 259

Hill, Kim, 193, 246, 248, 282, 414, 416, 453

Himalayas, 106

Hinde, Robert, 396–98, 397, 430, 486

History of Women's Bodies (Shorter), 365

HIV, 107–8, 376, 452

Hogarth, William, 301

Holland, 472

Holy Greyhound, 466
Homer, *240*, 262, 358
homicide, 179, 238, 243
   of children, 236, 416, 418
   of infants, *see* infanticide
Homo
   H. erectus, 48–49, *49*, 268, 285–86, 478–79
   H. ergaster, 48–49, *49*
   H. sapiens, 77, 183, 264–65, 268, 390, 392, 478
   length of childhood in, 267–68, 284–87, 483
homunculus, 70, *71*
honeybees, 55–56, 57, 59, 60–63, *61*, 64, 65, 92, 273, 274
hormones, 131–32
   behavior and, 131–32
   and division of labor in caretaking, 212
   estrogen, 153, 154, 173, 274, 420, 434
      synthetic, 125
   leptin, 125, 185
   melatonin, 104, 105
   in menarche, 185, 188–89
   oxytocin, 137–40, *138*, 153–54, 158, 357, 536, 537, 538
      birth and, 434–35
      pair-bonding and, 139–40
   pheromones, 59, 220–21, 357
   in pregnancy, 151, 153–54, 178, 377, 433–34
   progesterone, 153, 154, 173
   prolactin, 95, 105, 128, 133, 135, 153, 154, 194–95, 212, 357, 537
      in birds, 130, 131, 133
      caretaking and, 130–31, 132, *134*
      lactational aggression and, 173
      ovulation suppression and, 195, 409–10, 449
      pseudopregnancy and, *134*
   testosterone, 212, 443
hospital nurseries, 485–86
Howell, Nancy, 184, 374
Hoyt, Waneta, 291–92, 297

humans, qualities unique to, 392, 527–29
hunger
   attachment and, *397*, 398
   early weaning and, 460–61
hunter-gatherers, 98, 99, 100–102, 106, 198, 498, 518, 520
   acknowledgment of human identity in infants in, 468
   attitudes toward old people in, 282
   bias in studies of, 496
   birth spacing in, 193–94, 200–201, 440, 453
   children's death and, 312–13, 440
   constraints on using allomothers in, 503, 505
   desirable traits in husbands, 245
   division of labor in, 199
   length of childhood in, 267–68, 285, 286
   lifespan in, 279–81, 285, 286
   marriage and kin in, 192–93
   Neolithic revolution and, 199–201
   population size and, 183, 184
   show-off hypothesis and, 228–29, 268–69
   see also !Kung San
Hurtado, Magdalena, 193, 270, 282, 414, 416, 453
Huxley, Aldous, 328
Huxley, Thomas Henry, *59*
hyenas, 33–34, 48, *51*
hymenopteran insects, *66*, 67
   see also insects, social
hypergamy, 326, 339, 340, 343
hypertension, 436
hypothalamo-hypophyseal-gonadal axis, 196
hypothalamus, 141, 149, 150–51, 185, 188, 194, 196, 272

I

Ifaluk Atoll, 344
immune system, 103
   maternity wards and, 485
   and women's sense of smell, 193

immunological protection provided by milk,
      136–37
implantation, 124, 153, 420
   pregnancy maintenance and, 433–34
imprinting, 486
   in goslings, 397
   in sheep, 114, 157–59, 159, 487, 488
   see also attachment
imprinting, genetic, 141–42, 432–33, 435,
      436, 459
inclusive fitness, 63, 427
India, 511
   bias against daughters in, 321, 322, 324,
      326–27, 339, 347
   birth rituals in, 468
   caste system in, 340
   deaths from wolves in, 411–12
infant care, see childcare, allomothers, and
      cooperative breeding
infanticide, 25, 36, 37, 216
   cooperative breeding and, 92
   in humans, 179, 180, 184, 188, 190,
      236–44, 288–97, 290, 310, 370, 372,
      374, 413, 417, 451, 519
      abandonment and, 297, 299
      abortion and, 391, 470
      and accusing other peoples of depraved
         behavior, 294–97, 295
      in Aché, 416–18, 453–54
      as act of insanity, 289
      in Adam Bede, 288–89
      attachment and, 455
      in Ayoreo Indians, 314
      birth control and, 296, 349–50
      birth spacing and, 345, 453
      changelings and, 465–68, 469
      children's fears of, 533, 534
      in China, 318–20, 320, 321, 323, 344,
         349–50
      Christianity and, 464
      of daughters, 318–27, 338–39, 344,
         349–50, 455

      effect of changing circumstances on,
         314–15
      in Eipo, 454–56, 456
      evolution and, 413–14, 416
      in India, 326–27, 339
      and infant's identity as human, 468–70,
         472
      and infant's physical traits, 453–56
      overlaying and, 290–91
      and physical appearance of newborn, 449
      postpartum depression and, 172
      resistance in acknowledging, 289–97
      sex-selective, 318–28, 338–39, 340–41,
         344, 349–50
      SIDS and, 291–93
      slavery and, 290, 299
      of sons, 341, 344
      by stepfathers or mothers' boyfriends,
         236–37, 413
      in warfare, 237–42, 241, 243–44, 413
   lactational aggression and, 152, 153
   in mice, 89–90
   in prairie dogs, 93
   in primates, 34–35, 179, 215, 236, 237,
      238, 244, 293–94
      chimpanzees, 52, 86, 161, 181, 243,
         502–3
      langurs, xv–xvi, 32–34, 34, 35, 36,
         40–41, 86–87, 181–82, 237, 293, 504
      maternal abandonment and, 181–82
Infanticide (Piers), 356
infant mortality, 7, 107, 202, 462, 479, 498
   from abandonment, 297, 298, 299, 300,
      301, 302, 303, 304, 305, 307–8, 369
   in Dogon, 255
   grief over, 470
   inexperienced mothers and, 155
   in monkeys and apes, mother's attachment
      and, 178–79
   monogamy and, 233
   wet-nursing and, 11, 360–61, 365, 368, 369
   see also infanticide; infant survival

infants
  abandonment of, *see* abandonment
  abuse and neglect of, 25, 180, 190, 349,
      370, 376, 452, 454, 518, 536
    in Third World, 312
    see also abandonment; infanticide
  as active agents, 426
  adoption of, *see* adoption
  ape and monkey
    attachment and, 98, 178
    death of, 178–79
    fixed action patterns in, 96
    maternal abandonment of, 181–82
    sharing of, 161–63, *163,* 447–48, 495,
      498
  appeal of, 156, 157, 158, 161, *164,* 390,
      444–48, 452, 481, 483–84, 536,
      539–40
    see also infant survival, and infant's physi-
      cal traits
  attachment and, *see* attachment
  birth of, *see* birth
  camouflage of, 446
  carrying of, 197, *198,* 199, *200, 203, 204,*
      496, 499, 502
  as connoisseurs of mothering, 388, 389, 536
  in continuous contact with mother, 98, *99,*
      100–101, 409, 485, 498
  faces preferred by, 410
  facial expressions imitated by, *412*
  fat in, 457, 458, 474, 475–83, *477,* 536
    brain development and, 478–79, 480,
      481–82, 483
    hypotheses explaining, 476–79, 481
  fixed action patterns in, 96, 410
  hospital mix-up of, 158–59
  laughter of, 483
  learning systems in, 410
  low-birth-weight, 436, 457, 459, 464, 482
  male aggression against, 412–18
  maternal indifference to, 166–67
  Moro reflex in, 97, 410

  mother's relationship with, immediately fol-
      lowing birth, 486, 487–88
    see also attachment
  natal coats in, 156, 446–49, 482
  needs of, 535
    responsibility for, 490–95, 503
    sensitivity to, 212–13, 227, 500, 501
  parental investment in, *see* parental invest-
      ment
  parentally manufactured foods for, 133–34
    crop milk, 133, 134
    milk, *see* milk
  and parent-offspring conflict, 388–91,
      426–31, *426,* 483, 493, 539
  patterns preferred by, 310
  physical appearance as newborns, 448–49
    resemblance to father, 458–59, *459*
  in Pleistocene epoch, 7, 98, 99, 101, 103,
      105, 109, *204,* 518, 519, 520–21
  predators and, 411–13
  premature, 458, 460, 464, 482
  separation of, from caretakers, 97, 109–10,
      399–401, 408, 430–31, 521
    strange situation test and, 401–3, 404,
      405, 522
  sharing of, 161–63, *163,* 447–48, 495, 498
    see also childcare, allomothers, and coop-
      erative breeding
  singleton, 122, 178
    maternal commitment to, 179–80
  spacing of, *see* birth, spacing of
  strangers feared by, 414, *415,* 416–18
  swaddling of, *169, 170*
  tantrums in, 429–30, 438
  twin, 180, 202–3, 450, 459, 480
  vulnerability of, 389, 390, 493
  see also childcare, allomothers, and coopera-
      tive breeding; mothers, motherhood
infant survival, 390, 408–9, 519
  birth weight as indicator of, 479, 482
  breast-feeding and, 297, 298, 301–2,
      310–12, 322, *323,* 324, 362–63

infant survival (*continued*)
   and infant's physical traits, 452–74, 481–83
     in Aché, 453–54
     changelings and, 465–68, *466, 469*
     in Eipo, 454–56, *456*
     factors affecting maternal assessment,
       457–60, 463, 481–82
     infanticide and, 453–56
     infant's distress signals and, 460–62
     and infant's identity as human, 468–70,
       472
     natural selection and, 387, 390, 454,
       462, 463, 479, 481–82, 483, 484,
       516, 533, 536
     viability tests and, 464, 473–74
   see also infanticide; infant mortality
insects
   fireflies, 71, 72
   fruit flies, 41–42, 231–32
   hymenopteran, *66, 67*
     see also insects, social
   polyphenism in, 72–74
   spiders, 43–44, 133
insects, social, 57, 60, 62, 64, 65
   as allomothers, 64–65
   ants, 57, 60, 64–65
   bees, 55–56, 57, 59, 60–63, *61,* 64, 65, 92,
     273, 274
   eusocial, *61,* 62, 64, 273
   reproductive attributes of, 55–56, 59–63,
     64, 65–69
   termites, 64, 65
   wasps, 57, 60, 62, 63, 64, 65–69
     fig, 65–67, *66,* 68, *68,* 328, 329
     parasitic jewel, 68–69
     sex bias in, 65–69, 328, 337, 338, 345
instinct, maternal, 3, 10, 12, 26, 146, 173–74,
     377–78, 483
   lack of, 25–26, 183
   questioning of, 308–9, 351, 352, 535
   Royer's views on, 20, 21
   wet-nursing and, 351, 352, 356
insulation hypothesis, 477–78

insulin, 435–36
insurance policy hypothesis, 478
intelligence, 480
intimacy, breast-feeding and, 137, 140–41,
     144
investment, parental, *see* parental investment
Ishi in Two Worlds (Kroeber), 529–30
Islamic cultures, kinship in, 359
Israeli kibbutzim, 406
iteroparity, 43, 44, 90, 127

J

Jambo (gorilla), 207
James, William, 233
Japan, 283, 345, 406
jays, scrub, 131
Journal of Obstetrics and Gynaecology of the British
     Empire, 435

K

Kaczynski, David, 512
Kaczynski, Theodore, 512, 513
Kaczynski, Wanda, 512–13
Kalahari, 197–98
kangaroos, 128–29
Kashmir, 106
Kertzer, David, 304, *306,* 371–72
Keverne, Eric, 143, *144*
kibbutzim, 406
kin, 63–64, 144, 515, 531
   alloparenting and, 271, 278–79, 372, 509
     see also childcare, allomothers, and coop-
       erative breeding
   and marriage and pregnancy, 192–93
Kindness of Strangers, The: The Abandonment of
     Children in Western Europe from Late
     Antiquity to the Renaissance (Boswell),
     297–99
Kipsigis, 196, 253
Klapisch-Zuber, Christiane, 363
Klein, Melanie, 396

Konner, Mel, 194, 221

Korea, 322, 324, 343

Kroeber, Theodora, 529–30

!Kung San, 108, 136, *168,* 192, 202, *211,* 448, 520

    acknowledgment of human identity in infants in, 468

    adolescent subfertility in, *187*

    birth spacing in, 197–98, *198,* 199–201, 226, 440

    characterization of, 296

    child mortality in, 312, 440

    constraints on use of allomothers in, 502, 503, 505

    infanticide in, 172, 184, 374–75

    mother-infant contact in, *99,* 100, 101, 109, 194, 495

    Neolithic revolution and, 199–201

    Nisa, 7, 111, *168,* 230–31, 235, 245, 312, 429

    old age in, 280, *281,* 282

    pair-bonds in, 230–31

    show-off hypothesis and, 229

    work done by teenagers in, 272

L

labor (work)

    division of, 199

        in caretaking, 209–13

    motherhood and, xiv–xv, 109–10, 112, 113–14, *270,* 370, 490–91

        childcare and, 490–96, 504, 505–10

    young people and, 271–72

labor and delivery

    brain changes caused by, 94–95

    and size of pelvic canal, *165,* 478, 479, 480

    see also birth

Lack, David, 29–32, 441–42

lactadherin, 136

lactation, 121–45, 425, 500, 534, 536

    aggression and, 152–53

        postpartum depression and, 172–73

    benefits of, 123–25, 135–36

    breast development and, 126–27, *128*

    calories needed for, *126, 270*

    colostrum and, 136, 137, 357, 475

    disinfectant hypothesis of origin of, 135–36

    duration of, 175

    evolution of, 121, 122–24, 129, 130, 133–36, 140

    fat storage and, 124–27, *126,* 129

    as female specialty, 121–22, 129, 134, 140

    flexibility of, 357–58

    induction of, 357–58

    and infant's incorporation into human world, 471–72

    lifestyle and, 124–25, 127–29

    male competition and, 140

    in monotremes, 134–35

    postbirth delay in, 471–72, 475

    prolactin and, 95, 105, 128, 537

        caretaking and, 130–31, 132, *134*

    in pseudopregnancy, *134*

    see also breast-feeding; milk; wet-nursing

lactogen, 435

lactose tolerance, 108

Lamarck, Jean Baptiste, 80

Lancaster, Jane, 79, 162, 199

language, 392

langurs, 36, 88, 100, 179, 258, 477–78

    female sexuality in, 35, *218*

    infanticide in, xv–xvi, 32–34, *34,* 35, 36, 40–41, 86–87, 181–82, 237, 293, 504

    infant-mother separation in, 399

    infant-sharing in, 162–63, *163, 164, 238,* 447

    male caretaking in, 209

    old females, 277

    physical appearance of infants, 446

    Sol, 277, *278*

lanugo, 448

last-born children, 276

learning biases, in children, 521–22

learning to mother, 156, 162

    sensitization in, 151, 152

Lee, Richard, 200–201, 520

Leigh, Egbert, 431

lemmings, wood, 329

lemurs, ruffed, 176, 203, 399

LeNoir, Charles-Pierre, 351, 360, 369

leptin, 125, 185

let-down reflex, *138, 139*

Levine, Nancy, 524

Lewes, George Henry, 264, 392

life, human, beginning of, 391–92, 457

lifespan, 274

    and length of childhood, 267, 284–87, 483

    postreproductive longevity, 279–87

    of primates, 280–81

lifestyle, lactation and, 124–25, 127–29

limbic system, 141

Linnaeus, Carolus, 12

lions, 262, 273

litters, 203

    pruning of, 177–78, 294

local mate competition, 65–66

local resource enhancement, 332

Lodge, David, 119

Lorenz, Konrad, 396–97, *397,* 486

Loss (Bowlby), 394

love, 225

    bonding and, 488

        see also attachment

    of infant for mother, 398

    of mother for infant, 24–25, 116, 310, 312,
        315, 388, 451, 457, 535

        infant growth and, 460–61

        legislation of, 116–17

        as social construction, 309–12

        and unequal treatment of children,
            363–64

        see also attachment; maternal instinct;
            parental investment

low-birth-weight babies, 436, 457, 459, 464,
    482

Luce, Clare Boothe, 321

Lucy (fossil), 49, 267, *269*

lullabies, 532–33, 539, 540–41

luteinizing hormone, 185

lysozymes, 135

M

macaques, 82, 85, 111, 124–25, 181, 186,
    192, 224, 259, 274

    infant-sharing in, 161

    rank and sex ratios in, 333–36

    rhesus, in separation experiments,
        399–400, 430–31

    in surrogate mother experiments, 398, *399*

Machiguenga, 166

McLaughlin, Mary Martin, 299

"Magic Mama" (Piercy), 28

Main, Mary, 403–4, 405–6, *405,* 508, 522–23

malaria, 107

male caretakers, 91, 205–17, 226–30

    division of labor between mothers and,
        209–13

    intermittent care by, 214–15

    sensitivity to infant needs in, 212–13, 227

    show-off hypothesis and, 228–30, 268–69

    as sole caretakers, 214

    see also fathers

mammals, 12, 538

    alternative outcomes of development in,
        74–76

    body fat in, 476–77

    marsupial, 127–28

    milk production in, 121, 123–24

        see also lactation

    placental, 127

    see also primates; rodents

mammary glands

    evolutionary development of, 121–22, 129,
        134

    nippleless, 134, *135*

    oxytocin and, *138, 139*

Manzoni, Alessandro, 341–42

marmosets, 130, 180

marriage

    patriarchal, 249–50, 251

see also patriarchal societies

wealth and status in, 245, 326, 339, 340, 343

see also mating systems

marsupials, 127–28

Masai, 342–43, 461, 470

masturbation, 259

maternal effects, 69–72, 76–78, 80, 81, 190, 480

  immunological protection, 136–37

  sexuality and, 86–88, 263

maternal instinct, 3, 10, 12, 26, 146, 173–74, 377–78, 483

  lack of, 25–26, 183

  questioning of, 308–9, 351, 352, 535

  Royer's views on, 20, 21

  wet-nursing and, 351, 352, 356

maternal love, 24–25, 116, 310, 312, 315, 388, 451, 457, 535

  infant growth and, 460–61

  legislation of, 116–17

  as social construction, 309–12

  and unequal treatment of children, 363–64

  see also attachment; maternal instinct; parental investment

maternity, *see* mothers, motherhood

maternity wards, 485–86, 488

mating systems, 232–33, 264

  conflicting maternal and paternal interests in, 249–50, 257–58, 264

  convincing women to accept poor terms in, 252, 258–59

  hypergamy, 326, 339, 340, 343

  matrilineal, 248, 250, 252, 257, 343

  matrilocal, 248, 252

  monogamy, 215, 216, 226–27, 230–34, 250, 258

  patrilineal, 250, 251, 252, 254, 257, 264

  patrilocal, 251–52, 254, 325

  polyandry, 249, 250

  polygyny, 251–57, 269

    as beneficial to women, 252–54

    in blackbirds, 253

    in Dogon, 254–57

    female defense and resource defense, 253

matriarchal societies, 252, *261*

  and statues of mother goddesses, *256*

matrilineal descent, 248, 250, 252, 257, 343

matrilocal residence, 248, 252

maturation, delayed, 184–86

  adolescent subfertility, 185–86, *187*

  length of childhood, 267–68, 284–87, 483

  stress and, 189, 190

Maya-speaking people, 166–67, 259, *260, 472*

Mayer, Alexander, 354, 356

Mead, Margaret, 495

Mehinaku, 246, 247

meiosis, *422*

melatonin, 104, 105

Meltzoff, Andrew, 412

memes, 77–78

menarche, 185, 186, 272, 275, 288

  age of, *187,* 189–190

  body fat and, 125–26, 185, 188

  hormones in, 185, 188–89

  nutrition and, *187,* 188, 288

Mendel, Gregor, 80

menopause, 186, 273, 274–76, 279

  see also postmenopausal females

menstruation, 220, 472

  Dogon monitoring of, 254–55

  see also menarche; menopause

mest gene, 433, 436

mice, 46, 130, 146, 193

  birth in, 146–47

  breeding of, *148*

  communal caretaking in, 158

  effects of males on mothers' investment in offspring, 89

  fetal growth in, 432–33

  infanticide in, 89–90

  "knockout," 149–51, *150*

  lactational aggression in, 152–53

  maternal hormones in, 153–54

  parent-specific gene expression in, 143, *144*

Middlemarch (Eliot), *19,* 235, 318

milk, 121–45, 388
 colostrum and, 136, 137, 357, 475
 composition of, 127, 129, 409, 476
  fat, 127, 129, 409, 476, 480, 482
 immunological protection provided by,
  136–37
 lactose tolerance and, 108
 let-down of, *138, 139*
 see also breast-feeding; lactation; wet-nursing
milk, crop, 133, 134
miscarriage, 186, 276
 see also abortion, spontaneous
mitosis, *422*
modesty, womanly, 259–62, *260*
mongooses, dwarf, 92
monkeys, *see* apes and monkeys
monogamy (pair-bonds), 215, 216, 226–27,
 230–34, 250, 258
monotremes, 134–35
morality, 392–93
Morelli, Giovanni, 312
Moro reflex, 97, 410
morphologies, 73, 74, 75, 76, 511, 514, 524
Morrison, Toni, *290*
mortality, *see* death and mortality
mother goddesses, statues of, *256*
*Motherguilt: How Our Culture Blames Mothers for
 What's Wrong with Society* (Eyer), 491,
 493
Mother Love: Myth and Reality (Badinter), 352
Mother Nursing Her Baby (Cassatt), *381*
mothers, motherhood, 3–26, 27–54
 aggression in, 152–53
  postpartum depression and, 172–73
 aging and increased altruism in, 276–78,
  281–82, 285
 allomothers and, *see* childcare, allomothers,
  and cooperative breeding
 ambition and, 110–13
 ambivalence in, xiv–xv, 178, 290, 296, 377,
  390–91, 533, 539
 attachment and, *see* attachment

 autonomy of, 102, 370
 balancing of subsistence and childrearing by,
  44–46, *47*, 109–10, 113–14, 367,
  370, 494
 beauty and, 17–18, 23–24
 behaviors that benefit individuals over
  groups in, 29–34
 bias in ideas about, 496–97, 535
 big mothers, 47–50, 285–86, 329
 "biological," use of term, 57
 blaming of, for aberrations in development,
  511–13, 517, 518
 conclusions about, from early animal studies, 28–29
 constraints on reproductive choices of,
  40–42, 114, 116–17
 division of labor between fathers and,
  209–13, 227
 evolutionary emotions and, 114, 116
 evolutionary role of, xi–xii, xvii, 12–25,
  80–83
 expectations of, 164–65
 fetal interests and contracts with, 431–33,
  434–40
 Flo (chimpanzee mother), 27–28, 29, *30,*
  50–52, *54*, 86, 185
  last baby of, 275
  status of, 110
 former mothers as allomothers, 267,
  273–74, 275
 genes and, 148–51, *150*
 hormonal changes and, 151, 153–54
 indifference of, following birth, 166–67
 inexperienced, 155–56, 190–91, 452
  social support for, 191–93
 infants as connoisseurs of, 388, 389, 536
 infants carried by, 197, *198, 199, 200, 203,
  204,* 496, 499, 502
 infants in continuous contact with, 98, *99,*
  100–101, 409, 485, 498
 infant's interests vs., 388–91
 infant's relationship with, immediately following birth, 486, 487–88

see also attachment

instinct in, 3, 10, 12, 26, 146, 173–74, 377–78, 483
    lack of, 25–26, 183
    questioning of, 308–9, 351, 352, 535
    Royer's views on, 20, 21
    wet-nursing and, 351, 352, 356
intellectual abilities of, as viewed by evolutionists, 14, 15, *15,* 16, 17, 20, 27
learning in, 156, 162
as likeliest primary caretakers, 500–501
love in, 24–25, 116, 310, 312, 315, 388, 451, 457, 535
    infant growth and, 460–61
    legislation of, 116–17
    as social construction, 309–12
    and unequal treatment of children, 363–64
    see also attachment; maternal instinct; parental investment
maternal effects, 69–72, 76–78, 80, 81, 190, 480
    immunological protection, 136–37
    sexuality and, 86–88, 263
memes and, 77–78
neurochemical changes in, 94–95
parental investment by, *see* parental investment
and parent-offspring conflict, 388–91, 426–31, *426,* 483, 493, 539
    maternal-fetal, 431–33, 434–40, 483
in Pleistocene epoch, 7, 24, 98, 99, 101, 103, 105, 109, *204,* 266, 518, 519, 520–21
politics and, 4–6
polygyny and, 253–54
postpartum depression in, 170–73
pregnancy and, *see* pregnancy
quality vs. quantity in reproductive choices of, 8–10, 361, 365
responses to birth, 166–70
responsibility of, for infant needs, 490–95, 503

self-sacrifice in, 10–12, 42–44, 94, 110
sensitivity of, to infant needs, 212–13, 227, 500, 501
sensitization in, 151, 152
separation of infants from, 97, 109–10, 399–401, 408, 430–31, 521
    strange situation test and, 401–3, 404, 405, 522
sex ratios of offspring controlled by, *see* sex bias
sexual history of, as maternal effect, 86–88
sexuality and, xiii, xvii, 537–38
as social construction, 309–12
tradeoffs made by, 31–32, 46–47, *47,* 109, 309, 351–80, 390, 437
    childcare as, 369–72
    options and decisions in, 376–80
    wet-nursing, *see* wet-nursing
unnatural, 28, 288–317, 355, 451, 533
    see also abandonment; infanticide
variations in, 79, 83
work and, xiv–xv, 109–10, 112, 113–14, *270,* 370, 490–91
    childcare and, 490–96, 504, 505–10
see also infants; reproduction
moths, *Nemoria arizonaria, 73,* 74
Moulay Ismail the Bloodthirsty, *84*
Moynihan, Daniel Patrick, 470
Mukogodo, 342–43
Mulder, Monique Borgerhoff, 196, 253
murder, 179, 238, 243
    of children, 236, 416, 418
    of infants, *see* infanticide
Mutterrecht, Das (*The Mother Right*) (Bachofen), 257
mythology, 517

N

naming rituals, 468
nannies, *371*
Napoleon I, Emperor of France, *306,* 364

Nasonia vitripennis (parasitic jewel wasp),
      68–69
natal coats, 446–49, 482
National Lampoon, 160
National Science Foundation, 294
naturalistic fallacy, 23
natural selection, 106, 401, 419, 431–32
      childhood length and, 284–87
      directional, 462
      diseases and, 107–8
      embryos and, 423
      females' role in, 14, 42, 46, 52–54, 82–83
      group vs. individual, 29–34, 293
      as impersonal process, xviii, 13
      infant survival and, 387, 390, 454, 462,
            463, 479, 481–82, 483, 484, 516,
            533, 536
      and parental favoring of offspring, 444, 445
      phenotypes and, 56
      postmenopausal females and, 277–79, 285
      post-Neolithic, 106–7
      recycling in, 104, 132–33, 450
      rejection of, 535
      runaway, 445
      sex ratios and, 336–38
      variation and, 83
      see also evolution, evolutionary theory
nature-nurture dichotomy, 57, 147–48, 174
Nemoria arizonaria moths, 73, 74
neocortex, 141, 155
Neolithic period, 106–7, 199–201, 204, 409
neonaticide, *see* infanticide
Nepal, 480
neurochemistry, maternal, 94–95
neurons, 131–32
newborns
      swaddling of, 169, 170
      see also birth; infants
New England Journal of Medicine, 496
New Guinea, 321, 454
Newsweek, 23, 512
New York Times, 283, 369
Nicolas of Jawor, 466–67

night, 104–5, 195, 435
Nisa (!Kung woman), 7, 111, 168, 230–31,
      235, 245, 312, 429
nuclear family, 216
      breakdown of, 250–51, 379–80
nurseries, hospital, 485–86
nursing, *see* breast-feeding
Nurture Assumption, The (Harris), 507, 514–15
nurture-nature dichotomy, 57, 147–48, 174
nutria (coypu), 330–31, 338
nutrition
      fertility and, 187, 188, 191, 196, 288
      natural selection and, 107

O

ontogeny, *see* development
oocytes, *see* eggs
oogenesis, 422
orangutans, 175, 209, 219
      "Peter Pan," 75–76, 76, 511, 514
orgasm, female, 139, 221, 222–23, 225
Origine de l'homme et des sociétés (Royer), 21
Origin of Species, The (Darwin), xvii, 13, 17, 20,
      55, 386
Outside Over There (Sendak), 303, 465
ovaries, 185, 196, 272
      menopause and, 274–76, 279
overlaying, 290–91
ovulation, 258, 419, 420, 421
      Dogon society and, 255
      sexual desire and, 220–22
      sexual swellings as indicators of, 87, 185,
            217, 218
      suppression of
            breast-feeding and, 104–5, 194–95, 196,
                  239, 409–10, 438, 449
            in insects, 59–60, 64
            in mongooses, 92
      synchrony of, 357
      see also eggs
oxytocin, 137–40, 138, 153–54, 158, 357,
      536, 537, 538

birth and, 434–35
pair-bonding and, 139–40

P

pacifiers, 388
pair-bonds (monogamy), 215, 216, 226–27,
    230–34, 250, 258
Paleolithic period, 183
Palombit, Ryne, 215, 416
pangenesis, 20
pap, 359
parental effects, 514, 516
    maternal, 69–72, 76–78, 80, 81, 190, 480
        immunological protection, 136–37
        sexuality and, 86–88, 263
parental instinct, 533
    see also maternal instinct
parental investment, 8, 37, 42, 232, 315, 516,
    519
    in apes and monkeys, 176
        unconditional, 177–79, 180–81, 182,
            447, 451, 482
    birth spacing and, 203
    contingent, 92–93, 176, 182–83, 315–16,
        388, 389–90, 519, 535
    discrimination in, 363–64, 444, 445, 447,
        450–51, 470–71, 483
        birth order and, 524
        see also parental investment, and infant's
            physical traits
    duration, extent, and forms of, 176–77,
        346, 389
    ethical behavior and, 460
    family configurations and, 344–45
    by fathers, 37, 129, 224, 252, 255, 258,
        379, 458–59
        see also fathers; male caretakers
    and inequalities in offspring, 443–44
    and infant's physical traits, 452–74,
        481–83
        in Aché, 453–54
        changelings and, 465–68, 466, 469

in Eipo, 454–56, 456
    factors affecting maternal assessment,
        457–60, 463, 481–82
    infanticide and, 453–56
    infant's distress signals and, 460–62
    and infant's identity as human, 468–70,
        472
    viability tests and, 464, 473–74
lactation and, 129, 152, 500
maternal ambivalence about, 290, 296, 377,
    390–91
    abandonment and, 297
in multiple young, 179–80, 450, 459, 469
and parent-offspring conflict, 388–91,
    426–31, 426, 483, 493, 539
postpartum depression and, 171–72
sex bias in, see sex bias
social conditions and, 88–89
tradeoffs and, see tradeoffs
see also attachment; maternal love
"Parental Investment and Sexual Selection"
    (Trivers), 37
parents
    interests and family goals of, 318
        see also sex bias
    parent-offspring conflicts and negotiations,
        388–91, 426–31, 426, 483, 493, 539
        maternal-fetal, 431–33, 434–40, 483
    parent-specific gene expression (genetic
        imprinting), 141–42, 432–33, 435,
        436, 459
    peers vs., in socialization, 514–16
Parker, Rozsika, 378, 409
Parks, Fanny, 326
Parmigiani, Stefano, 152, 153, 296
partible paternity, 246–49, 252
paternal care, see fathers; male caretakers
paternal sex ratio element, 69
paternity, 250
    certainty of, 207, 217–19, 226–28, 246,
        248, 252, 255, 264, 458–59
    multiple, 246–49, 252
    sexuality and, 538

patriarchal societies, 251, 252, 254, 255, 257, 258, 262, 263, 264, 325
  and bias in ideas about females, 496–97, 535
  marriage in, 249–50, 251
  mother-centered societies' contact with, 249–50
  polygynous structures in, as beneficial to women, 252–54
  sex bias in, 321, 345, 346
patrilineal descent, 250, 251, 252, 254, 257, 264
patrilocal residence, 251–52, 254, 325
Pawlowski, Boguslaw, 477
peacocks, 36, 38, *38*
peers, socialization role of, 514–16
pelvic canal, size of, *165,* 478, 479, 480
penis, *220*
Period Piece (Raverat), 386
personality, birth order and, 523–25
Petrie, Marion, 38, 40
phenotypes, 56, 59, 71, 72, 127, 450, 514, 515
  and alternative outcomes of development, 74–76, 514
  defined, 56
  polyphenism and, 72–74
pheromones, 59, 220–21, 357
Philippines, 498
philopatry, 50–51
Piercy, Marge, 28
Piers, Maria, 356
pineal gland, 104
pituitary gland, 130, 173, 185, 188, 194, 196, 433
placenta, 153, 154, 167, 389, 433–34, 435–36
  eating of, 167, 433, 436
  evolution of, 435
  fetus and, 433–34, 435, 436, 481
  pregnancy maintenance and, 433–34
placental mammals, 127
plague, 106, 107
Planned Parenthood, 370–71
Plato Comicus, 262

platypuses, 134, *135*
pleiotrophic effects, 274
Pleistocene epoch, 7, 24, 98, 99, 101, 103, 105, 109, 197–98, 199, 201, *204,* 226, 266, 288, 518, 519, 520–21
  evolution since, 105–9
  population size in, 183
  wet-nursing and, 357
  see also hunter-gatherers
Pliocene epoch, 98
politics, motherhood and, 4–6
polyandry, 249, 250
polygyny, 251–57, *269*
  as beneficial to women, 252–54
  in blackbirds, 253
  in Dogon, 254–57
  female defense and resource defense, 253
polyphenism, 72–74
population, 183–84, 462–63, 517
  agriculture and, 202
population genetics, 463
  and Fisher's principle of the sex ratio, 336–37
  in primates, 183
Popul Vuh, 259
Portugal, 341, 342
postmenopausal females, 274–75, 279–87
  as alloparents, 267, 273–74, 275
  altruism in, 276–78, 281–82, 285
  grandmothers, 275, 278–79, 282, 283
    lactation in, 358
  natural selection and, 277–79, 285
postpartum depression, 170–73
Potential of Women, The, 27
prairie dogs, 93, 177, 178
predators
  birds' distraction tactics and, 131
  hatching and, 424–25, *425*
  infants and, 411–13
preeclampsia, 436
pregnancy
  calories needed for, *126*
  costs of, to mother, *175*

fetal fat and, 477, 480–81, 482
gestation, 389, 419, *422,* 425, 434
  beginning of life in, 391–92, 457
  birth weight and, 476, 481
  parent-specific genes and, 432–33
  see also embryo; fetus
hormones in, 151, 153–54, 178, 377,
    433–34
labor and delivery
  brain changes caused by, 94–95
  and size of pelvic canal, *165,* 478, 479,
    480
  see also birth
length of, in primates, 175
maternal attachment to baby in, 458
physical changes caused by, 94–95
preeclampsia in, 436
pseudopregnancy, *134*
teenage, 188, 190–91, 283, 288, 423
  infanticide and, 289
premature infants, 458, 460, 464, 482
primates, 12
adoption in, 157, 158–59, 180
  by males, 207–8
alternative outcomes of development in,
    74–76
attachment in, 98, 178, 398, 399–401
  see also attachment
body fat in, 476, 477, 478
delayed puberty in, 189, 190
durations of pregnancy in, 175
family planning in, 194
female orgasm in, 222, 225
fixed action patterns in, 96
infant-caretaker separation in, 97,
    399–401, 408
infanticide in, 34–35, 179, 215, 236, 237,
    238, 244, 293–94
  in chimpanzees, 52, 86, 161, 181, 243,
    502–3
  in langurs, xv–xvi, 32–34, *34,* 35, 36,
    40–41, 86–87, 181–82, 237, 293,
    504

maternal abandonment and, 181–82
labor and delivery in, *165,* 478
lifespans in, 280–81
male caretaking in, 207–10
  see also male caretakers
motherhood and status in, 110–11, 192
natal coats in, 446–49
neocortices in, 141
night and, 104, 435
population genetics in, 183
pregnancy hormones in, 154–55
"promiscuity" in, 85, 86, 185
sexual maturation in, 184–86
sexual swellings in, *87,* 185, 217, *218*
singleton offspring of, 122, 178
  maternal commitment to, 179–80
snakes feared by, 416
uncertainty of paternity in, 207, 217–19
  see also apes and monkeys
primatology, stereotypes in, 27
Privation sensible, La (The Painful Deprivation)
    (Greuze), *353*
proboscis monkeys, 446
progesterone, 153, 154, 173
prolactin, 95, 105, 128, 133, 135, 153, 154,
    194–95, 212, 357, 537
  in birds, 130, 131, 133
  caretaking and, 130–31, 132, *134*
  lactational aggression and, 173
  ovulation suppression and, 195, 409–10,
    449
  pseudopregnancy and, *134*
promiscuity, 83–86, 87–88, *87,* 185, 250
prosimians, 217, 399, 434
prostaglandins, 434
prostitution, maternal abandonment and, 297,
    299
puberty
  breast development in, 126–27
  delayed, stress and, 189, 190
  menarche, 185, 186, 272, 275, 288
    age of, *187,* 189–90
    body fat and, 125–26, 185, 188

puberty, menarche (*continued*)
  hormones in, 185, 188–89
  nutrition and, 187, *187,* 188, 288
Pumé, 271, 272, *273*

Q

quality vs. quantity, in reproduction, 8–10,
      361, 365
Quindlen, Anna, 461

R

raiders, infanticide by, 237–42, 243–44, 413
Ralls, Katherine, 47–48, 50
rats, 146, 439
Raverat, Gwen Darwin, 386–87
reproduction, 127
  abortion and, *see* abortion
  adolescent workers and, 272
  aging and, 276
  big mothers and, 47–50, 285–86, 329
  birth rates, 9–10, 20, 202
    sex ratios and, 335–36
  childhood length and, 284–87
  conception in, 421, 423
    sex selection and, 348–49
  constraints on female choice in, 40–42,
      114, 116–17
  cooperative breeding, *see* childcare, allo-
      mothers, and cooperative breeding
  delayed maturation and, 184–86
    stress and, 189, 190
  diploid, 62
  eggs in, *see* eggs
  and evolution of behaviors that benefit indi-
      viduals over groups, 29–34
  fat and, 125, *125, 126,* 185, 188
  female mate choice in, 36–42
  fertility
    adolescent subfertility, 185–86, *187*
    fat and, 125, *125,* 185, 188

  increase in, in human evolution,
      449–50
  nutrition and, *187,* 188, 191, 196, 288
  pattern of, 186
  wet-nursing and, 360–62, 364–66, 368
fertilization in, 70, 71, 207, 222, 421, 423,
      538
genes and, *see* genes
grandmother's clock hypothesis and,
      284–86
haplodiploid, 62, 64
haploid, 62
infanticide and, 293
iteroparous, 43, 44, 90, 127
learning biases and, 522
longevity and, 279–87, 284–87
male attempts to control, 6, 87
male-imposed constraints on female choice
      in, 40–42
mammalian, three types of relationship in,
      538
menarche and, *see* menarche
menopause and, *see* menopause
menstruation and, 220, 472
  Dogon monitoring of, 254–55
mental maturation and, 188
parental investment and, *see* parental invest-
      ment
postmenopausal females and, 274–75
  natural selection and, 277–79, 285
pregnancy in, *see* pregnancy
promiscuity and, 83–86, 87–88, *87,* 185
quality vs. quantity in, 8–10, 361, 365
semelparous, 43–44
sex bias in, *see* sex bias
sexual selection and, 36–37, 81, 83, 445,
      496
  female mate choice in, 36–42
  large females and, 47–50
  parental investment and, 37
show-off hypothesis and, 229, 268–69
in social insects, 55–56, 59–63, 64, 65–69

sperm in, 70, 421–23, *422*
> homunculus in, 70, *71*
> insect, 62–63, 69
> production of, 224
> testicle size and, 218–19, *219*

tradeoffs in, *see* tradeoffs
viviparous, 122–23
and women's intellect, evolutionists' views
> on, 14, 15, *15, 16*, 17, 20, 27

see also mothers, motherhood
reproductive rights, 4–6, 102, 370, 393, 470
> and debate over when life begins, 391–92,
>> 457

see also abortion
reproductive strategies, male vs. female, 83
reproductive success (fitness), 9, 14, 42, 79,
> 80–83, 99, 114, 286, 318
> and bias against females, 497
> inclusive, 63, 427
> of infants, assessment of, 463, 481
>> viability tests in, 464, 473–74
> sex bias and, 330–36, 337
>> *see also* sex bias
> status and, 81–82, 110–12, 366
> variation in, 83, 90

reproductive value, 276
reptiles, *123,* 124
Retour de Nourrice (Greuze), 526
Rice, William, 41, 42, 232
Rich, Adrienne, 25, 26
rodents
> effects of males on mothers' investment in
>> offspring, 89
> hamsters, 46, 177
> mice, *see* mice
> prairie dogs, 93, 177, 178
> rats, 146, 439
> sensitization in, 151

Roland, Jeanne-Marie Phlipon de, 352
Rousseau, Jean-Jacques, 133, 300, 310, *313,*
> 352, 366

Royer, Clémence, 20–21, 22, *22,* 53

S

salamanders, 424–25
salmon, 43
Sardinia, 371–72
Scarr, Sandra, 507
Scheper-Hughes, Nancy, 310–12, 314–15,
> 467

Schiefenhövel, Wulf and Grete, 454, 455
schizophrenia, 513
Schultz, Adolph, 476
science, bias in, 496–97, 535
scleroderma, 94
scorpionflies, 39
scrub jays, 131
sea horses, 122, *122,* 131
seals, 129, 476
Second Sex, The (Beauvoir), 17
"self-advertising" hypothesis, 481
self-sacrifice
> in mothers, 10–12, 42–44, 94, 110
> in postmenopausal females, 276–78,
>> 281–82, 285

Selk'nam, *261*
semelparity, 43–44
Sendak, Maurice, 303, 465
senescence, 274–76
> aging, 274–76
>> and cultural attitudes toward old people,
>>> 282, 283
>> increased maternal altruism and,
>>> 276–78, 281–82, 285
>> *see also* lifespan
> and postreproductive longevity, 279–87
> *see also* death and mortality

sensitization, 151, 152
separation, 97, 109–10, 399, 408, 521
> effects of, in primates, 399–401, 430–31
> strange situation test and, 401–3, 404, 405,
>> 522

Separation: Anxiety and Anger (Bowlby), 394
sex bias, 67, 318–50
> abortion and, 319, 320, 322, 324, 349

sex bias (continued)
    in animals, 328–36, 346
    in birds, 332–33
    conception methods and, 348–49
    and equality of human sex ratios, 336–38
    family configurations and, 318, 344–45
    in humans, as changeable behavior, 346–47
    in humans, as postbirth decision, 345–46
    infanticide and, 318–28, 338–39, 340–41,
        344, 349–50
        in China, 318–20, 320, 321, 323, 344,
            349–50
        of daughters, 318–27, 338–39, 344,
            349–50, 455
        earliest evidence for, 340–41
        in India, 326–27, 339
        of sons, 341
    reasons for preferring sons over daughters,
        321, 324–25
        reproductive potential, 325–27
    sex ratio theory and, 67, 328–39
    Trivers-Willard hypothesis and, 331, 332,
        335, 338, 339
    in wasps, 65–69, 328, 337, 338, 345
Sexes Throughout Nature, The (Blackwell), 20
sex ratios, 67, 328–39
    Fisher's principle of, 336–37, 346
    in humans, equality of, 336–38
sex roles, 259, 260
sex taboos, wet-nursing and, 365
sexual dimorphism, 47–50, 49, 269
sexuality
    female, 262
        clitoris in, 220, 222
        cyclicity of, 219–22, 225–26
        desire in, 220–22, 223–26
        Dogon monitoring of, 254–55, 256, 258
        flexibility in, in monkeys and apes,
            217–18
        infant protection and, 217
        in langurs, 35, 218
        as maternal effect, 86–88, 263
        maternity and, xiii, xvii, 537–38

        modesty and, 259–62, 260
        orgasm in, 139, 221, 222–23, 225
        and uncertainty of paternity, 207,
            217–19
    male
        desire in, compared with women,
            223–25
        paternity and, 538
sexual selection, 36–37, 81, 83, 445, 496
    female mate choice in, 36–42
    large females and, 47–50
    parental investment and, 37
sexual swellings, in primates, 87, 185, 217,
    218
Seychelles warblers, 332–33
Shakespeare, William, 458, 540
Shaw, George Bernard, 252
sheep, imprinting in, 114, 157–59, 159, 487,
    488
Sherman, Paul, 52, 275
Shettles, Landrum, 348
shopping habits, of men vs. women, 229–30
Shorter, Edward, 310, 365
show-off hypothesis, 228–30, 268–69
shrews, tree, 129, 480
siblicide, 442, 443–44
sibling rivalry, 498
    in birds, 441–45, 448
sickle-cell gene, 107
SIDS (sudden infant death syndrome),
    291–93
Silas Marner (Eliot), 205–6, 214
size difference, between males and females,
    47–50, 49, 269
slavery
    infanticide and, 290, 299
    wet-nursing and, 359
sleep apnea, 291, 292
slime grafting, 159
smell, sense of, 193, 221
snakes, as predators, 414–16
    and hatching of frogs, 424, 425, 462
social construction, mother love as, 309–12

social Darwinism, 13, 15, *19,* 24

social environment, as factor in evolutionary development, 520–22

social insects, *see* insects, social

socialization, role of peers vs. parents in, 514–16

social rank, *see* status

social relationships, 141–45
    genes and, 142–43

sociobiology, 57, *58,* 67, 69, 82, 231, 308, 496
    child development and, 522

Sociobiology (Wilson), 496

sociopaths, 512–13, 516, 517–18, 527

Solomon Islanders, 270–71

sons
    conception methods for producing, 348–49
    female infanticide and, 318–27, 338–39
    infanticide of, 341, 344
    Mukogodo, 342–43
    parental investment in, in U.S., 338
    preference for daughters over, social status and, 339–40, 341, 342, 343–44
    reasons for preference for, over daughters, 321, 324–25
    reproductive potential, 325–27

Sophocles, 262

Soranus, 452, 463, 464, 474, 482

So the Witch Won't Eat Me (Bloch), 534

South America, 166, 321

sparrows, 88

Spencer, Herbert, 13, 14–15, *15,* 16, 17, 18, 19, 20, 23, 31, 82, 83, 310
    Eliot and, 14–15, *15,* 17–18, *19*

sperm, 70, 421–23, *422*
    homunculus in, 70, *71*
    insect, 62–63, 69
    production of, 224
    testicle size and, 218–19, *219*

spermatheca, 62

spiders, matriphagous, 43–44, 133

Spitz, Rene, 395–96, 404

spontaneous abortion, *see* abortion, spontaneous

Stapleton, Lorna, *428*

statues of mother goddesses, 256

status (social rank), 110–11, 112
    in baboons, 81–82, 192, 333–36, 514
    in cercopithecine monkeys, 192, 333
    in chimpanzees, 52, 110
    reproductive success and, 81–82, 110–12, 366

steatopygia, *126*

stepfathers, 236–37, 413

Stephen, Saint, *469*

Stern, Daniel, *410*

strangers, fear of, 414, *415,* 416–18

strange situation test, 401–3, *403,* 404, 405, 461, 522

Strassmann, Beverly, 255, 256, 257

stress, 130
    delayed puberty and, 189, 190

Strindberg, August, 257

sturgeon, 274

sucrase, 439

sudden infant death syndrome (SIDS), 291–93

Sugiyama, Yukimaru, 32

Sulloway, Frank, 524

"Sur la natalité" ("On Birth") (Royer), 21

survival of the fittest, 13–14
    see also natural selection

Sussman, George, 368

swaddling, *169, 170*

swallows, *40*

symmetry, of physical traits, 38–40, *40*

T

tadpoles, 424

Talleyrand-Périgord, Charles-Maurice de, 364, *365*

tamarins, *93,* 131, 180, 203, 316

Tansillo, Luigi, 351

tantrums, 429–30, 438

Tanzania, 198, 228

teenagers
    as alloparents, 271–72

teenagers (*continued*)
  pregnancy and motherhood in, 188,
      190–91, 283, 288, 423
    infanticide and, 289
teeth, childbearing and, 175
Telephos, *208, 525–26, 528*
television, 77
Tereus (Sophocles), 262
termites, 64, 65
testicles, size of, 218–19, *219*
testosterone, 212, 443
Thornhill, Randy, 39, 40
Three Men and a Baby, 209, *210*
Thurer, Shari, 110, 512
Time, 58, 157
Tinbergen, Niko, 396–97
titi monkeys, 213–14, *214,* 226, 227, 498
Titus Andronicus (Shakespeare), 458
Tonga, 343
torgovki, 302
tortillas, Mayan symbolism surrounding, 472
toys, 338
tradeoffs, 31–32, 46–47, *47,* 109, 309,
      351–80, 390, 437
  childcare, 369–72
    see also wet-nursing
  by males, 207, *233*
  options and decisions in, 376–80
Treatise of Superstitions (Nicolas of Jawor),
      466–67
Treatise on the Family (Becker), 252
Trevathan, Wenda, 166, 167
Trexler, Richard, 359
tribal raiders, infanticide by, 237–42, 243–44,
      413
Trivers, Robert, 32, 37, *68,* 224, 253, 266,
      419, 426–31, 436, 437, 438, 539
  comment of, on Hrdy's work, 490, 492
  in Jamaica, 427, *428*
  parent-offspring conflict theory of, 388,
      389, 426–31, *426*
  Trivers-Willard hypothesis, 331, *332,* 335,
      338, 339

trophy wives, *240*
Turner's syndrome, 142
Tut, King, 359
twins, 180, 202–3, 450, 459
  aging in, 280
  IQ scores of, 480
typhoid fever, 107

U

urination, in rodent pups, 147

V

Valero, Elena, 239–42, 244, 469
vervets, 190–91, 192, 278–79, 316, 446
viability tests, 464, 473–74
viviparity, 122–23
vom Saal, Frederick, 89, 296

W

Wall Street Journal, 250, 379, 488
war
  infanticide in, 237–42, *241,* 243–44, 413
  trophy wives and, *240*
War Between the Tates, The (Lurie), 216
warblers, 332–33
Washington Post, 513
wasps, 57, 60, 62, 63, 64, 65–69
  fig, 65–67, *66,* 68, *68,* 328, 329
  parasitic jewel, 68–69
  sex bias in, 65–69, 328, 337, 338, 345
weaning, 127, 129, 176, 177, 197, 362–63,
      409, 438, 449
  conflicts in, 429
  infants' preparation for, 439
  time of, 439, 460–61
Werren, John, 68–69
West-Eberhard, Mary Jane, 56, 57, 60, 63,
      71–72, 75–76, 113, *115*
wet-nursing, 11, 12, 91, 194, 197, 271, 302,
      309, 351–69, *355, 362,* 372, 498, *526*

allomaternal sharing of milk, 91, 356–57,
499
coerced, 358–59
commercialization of, 356, 366–67
costs to wives from, 364–66
daycare compared with, 369–70
decisions about, 363–64
distancing of mother from infant in, 352–54
and flexibility of lactation, 357–58
in France, 11, 351–54, 355, 366, 367–69
infanticide and, 354–55, 356
infant mortality and, 11, 360–61, 365, 368,
369
in-house, 363, 368, 369
maternal instinct and, 351, 352, 356
sex taboos and, 365
shorter birth intervals produced by,
360–62, 364–66, 368
and unequal treatment of children, 363–64
in U.S. hospitals, 361
wet nurses in, 359–60
whales, 48, 273
Willard, Dan, 331
Trivers-Willard hypothesis, 331, 332, 335,
338, 339
Williams, George, 32, 77, 274, 275, 338
Wilson, Edward O., 63, 91, 211, 496
Wilson, Margo, 236, 314, 458, 463

Winnicott, D. W., 392
wives, trophy, 240
Wollstonecraft, Mary, 175, 194
wolves, 134, 167, 411–12
Woman That Never Evolved, The (Hrdy), 21
Woolf, Virginia, xvii
work
division of labor in, 199
in caretaking, 209–13
motherhood and, xiv–xv, 109–10, 112,
113–14, 270, 370, 490–91
childcare and, 490–96, 504, 505–10
young people and, 271–72
Worthman, Carol, 194, 212, 221

X

XO girls, 142, 143

Y

Yanamamo, 239–42, 243, 246, 250, 346, 469
characterization of, 296
Yucatan, 166

Z

Zaire, 343

About the Author

Sarah Blaffer Hrdy, emeritus professor of anthropology at the University of California at Davis, has been elected to the National Academy of Sciences and is a fellow of the American Academy of Arts and Sciences. Her previous book, *The Woman That Never Evolved,* was chosen by the *New York Times* as a "Notable Book of the Year in Science and Social Science." She lives with her family in northern California where they combine habitat restoration with farming.

Courtesy of J. A. Wallis